# ADVANCED SPECTROSCOPIC METHODS TO STUDY BIOMOLECULAR STRUCTURE AND DYNAMICS

# ADVANCED SPECTROSCOPIC METHODS TO STUDY BIOMOLECULAR STRUCTURE AND DYNAMICS


Edited by

**PRAKASH SAUDAGAR**

Assistant Professor, Department of Biotechnology, National Institute of Technology Warangal, Warangal, India

**TIMIR TRIPATHI**

Senior Assistant Professor, Molecular and Structural Biophysics Laboratory, Department of Biochemistry, North-Eastern Hill University, Shillong, India

Regional Director, Regional Director's Office, Indira Gandhi National Open University (IGNOU), Regional Center Kohima, Kohima, India



ELSEVIER

ACADEMIC PRESS

An imprint of Elsevier

Academic Press is an imprint of Elsevier
125 London Wall, London EC2Y 5AS, United Kingdom
525 B Street, Suite 1650, San Diego, CA 92101, United States
50 Hampshire Street, 5th Floor, Cambridge, MA 02139, United States
The Boulevard, Langford Lane, Kidlington, Oxford OX5 1GB, United Kingdom



**Notices**

Knowledge and best practice in this field are constantly changing. As new research and experience broaden our understanding, changes in research methods, professional practices, or medical treatment may become necessary.

Practitioners and researchers must always rely on their own experience and knowledge in evaluating and using any information, methods, compounds, or experiments described herein. In using such information or methods they should be mindful of their own safety and the safety of others, including parties for whom they have a professional responsibility.

To the fullest extent of the law, neither the Publisher nor the authors, contributors, or editors, assume any liability for any injury and/or damage to persons or property as a matter of products liability, negligence or otherwise, or from any use or operation of any methods, products, instructions, or ideas contained in the material herein.

ISBN 978-0-323-99127-8

For information on all Academic Press publications
visit our website at https://www.elsevier.com/books-and-journals

*Publisher:* Jonathan Simpson
*Acquisitions Editor:* Glyn Jones
*Editorial Project Manager:* Franchezca A. Cabural
*Production Project Manager:* Niranjan Bhaskaran
*Cover Designer:* Mark Rogers

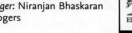

Typeset by STRAIVE, India

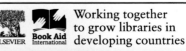

To our parents and family for their love and enduring support.

# Contents

# Contributors

**Swati Aggarwal**
BioMAX, MAXIV Laboratory, Lund, Sweden

**Eva Rose M. Balog**
School of Mathematical and Physical Sciences, University of New England, Biddeford, ME, United States

**Bharti**
Department of Biosciences and Bioengineering, Indian Institute of Technology Roorkee, Roorkee, Uttarakhand, India

**Sourya Bhattacharya**
Department of Biotechnology, Indian Institute of Technology Roorkee, Roorkee, India

**Björn M. Burmann**
Wallenberg Centre for Molecular and Translational Medicine; Department of Chemistry and Molecular Biology, University of Gothenburg, Gothenburg, Sweden

**Guillermo Javier Copello**
Universidad de Buenos Aires, Facultad de Farmacia y Bioquímica, Departamento de Ciencias Químicas; CONICET—Universidad de Buenos Aires, Instituto de Química y Metabolismo del Fármaco (IQUIMEFA), Buenos Aires, Argentina

**Xiaorong Fan**
State Key Laboratory of Crop Genetics and Germplasm Enhancement, Nanjing Agricultural University, Nanjing, Jiangsu, China

**Alireza Ghanbarpour**
Department of Biology, Massachusetts Institute of Technology, Cambridge, MA, United States

**Manjul Gondwal**
Department of Chemistry, Laxman Singh Mahar Government Post Graduate College (Soban Singh Jeena University, Almora), Pithoragarh, India

**Michael L. Gross**
Department of Chemistry, Washington University in Saint Louis, St. Louis, MO, United States

**Joanna Grzyb**
Department of Biophysics, Faculty of Biotechnology, University of Wrocław, Wroclaw, Poland

**Tingting Gu**
State Key Laboratory of Crop Genetics and Germplasm Enhancement, Nanjing Agricultural University, Nanjing, Jiangsu, China

**Sebanti Gupta**
Yenepoya Research Centre, Yenepoya (Deemed to be University), Mangalore, India

**Saugata Hazra**
Department of Biotechnology; Centre of Nanotechnology, Indian Institute of Technology Roorkee, Roorkee, India

**Prashant N. Jethva**
Department of Chemistry, Washington University in Saint Louis, St. Louis, MO, United States

**Ravi Kant**
Department of Chemistry, Washington University in Saint Louis, St. Louis, MO, United States

**Ashish A. Kawale**
Wallenberg Centre for Molecular and Translational Medicine; Department of Chemistry and Molecular Biology, University of Gothenburg, Gothenburg, Sweden

**Ali A. Kermani**
Department of Molecular, Cellular, and Developmental Biology, University of Michigan, Ann Arbor, MI, United States

**Khosrow Khalifeh**
Department of Biology, Faculty of Sciences; Department of Biotechnology, Research Institute of Modern Biological Techniques, University of Zanjan, Zanjan, Iran

**Varsha Kumari**
Department of Animal Biology, School of Life Sciences, University of Hyderabad, Telangana, India

**Qiao Li**
College of Veterinary Medicine, Nanjing Agricultural University, Nanjing, Jiangsu, China

**Zhen Li**
College of Resources and Environmental Sciences, Nanjing Agricultural University, Nanjing, Jiangsu; Jiangsu Provincial Key Lab for Organic Solid Waste Utilization, Nanjing Agricultural University, Nanjing, China

**Andrea Mathilde Mebert**
Universidad de Buenos Aires, Facultad de Farmacia y Bioquímica, Departamento de Ciencias Químicas, Buenos Aires; CONICET—Centro de Investigación y Desarrollo en Criotecnología de Alimentos (CIDCA), La Plata, Argentina

**Jiajie Mei**
College of Resources and Environmental Sciences, Nanjing Agricultural University, Nanjing, Jiangsu, China

**Parul Mishra**
Department of Animal Biology, School of Life Sciences, University of Hyderabad, Telangana, India

**Austin B. Moyle**
Department of Chemistry, Washington University in Saint Louis, St. Louis, MO, United States

**Niharika Nag**
Molecular and Structural Biophysics Laboratory, Department of Biochemistry, North-Eastern Hill University, Shillong, India

**Maya S. Nair**
Department of Biosciences and Bioengineering, Indian Institute of Technology Roorkee, Roorkee, Uttarakhand, India

**Amrita Arpita Padhy**
Department of Animal Biology, School of Life Sciences, University of Hyderabad, Telangana, India

**Shang Pan**
College of Resources and Environmental Sciences; College of Agro-grassland Sciences, Nanjing Agricultural University, Nanjing, Jiangsu, China

**Himanshu Pandey**
Department of Chemistry, Ben-Gurion University of the Negev, Beer-Sheva, Israel

**Seema R. Pathak**
Department of Chemistry, Biochemistry and Forensic Science, Amity School of Applied Sciences, Amity University Haryana, Gurugram, India

**Jonás José Perez Bravo**
CONICET—Universidad de Buenos Aires, Instituto de Química y Metabolismo del Fármaco (IQUIMEFA); Grupo de Aplicaciones de Materiales Biocompatibles, Departamento de Química, Facultad de Ingeniería, Universidad de Buenos Aires (UBA), CABA; CONICET-Universidad de Buenos Aires, Instituto de Tecnología de Polímeros y Nanotecnología (ITPN-UBA-CONICET), Buenos Aires, Argentina

**Kavya Prince**
Department of Biotechnology, National Institute of Technology Warangal, Warangal, India

**Subhashree Sahoo**
Department of Animal Biology, School of Life Sciences, University of Hyderabad, Telangana, India

**Riya Sahu**
Department of Biotechnology, National Institute of Technology Warangal, Warangal, India

**Anand Salvi**
Department of Chemistry, Biochemistry and Forensic Science, Amity School of Applied Sciences, Amity University Haryana, Gurugram, India

**Shreya Sarkar**
Department of Chemistry, Biochemistry and Forensic Science, Amity School of Applied Sciences, Amity University Haryana, Gurugram, India

**Santanu Sasidharan**
Department of Biotechnology, National Institute of Technology Warangal, Warangal, India

**Prakash Saudagar**
Department of Biotechnology, National Institute of Technology Warangal, Warangal, India

**Manish Shandilya**
Department of Chemistry, Biochemistry and Forensic Science, Amity School of Applied Sciences, Amity University Haryana, Gurugram, India

**Nidhi Sharma**
Department of Chemistry, School of Applied and Life Sciences, Uttaranchal University, Dehradun, India

**Akram Shirdel**
Department of Biochemistry, Faculty of Biological Sciences, Tarbiat Modares University, Tehran, Iran

**Kummari Shivani**
Department of Animal Biology, School of Life Sciences, University of Hyderabad, Telangana, India

**Seema Singh**
Department of Chemistry, School of Applied and Life Sciences, Uttaranchal University, Dehradun, India

**Jakub Sławski**
Department of Biophysics, Faculty of Biotechnology, University of Wrocław, Wroclaw, Poland

**Amit Kumar Sonkar**
Department of Biochemistry, All India Institute of Medical Sciences, Guwahati, India

**Banesh Sooram**
Department of Biotechnology, National Institute of Technology Warangal, Warangal, India

**Mu Su**
College of Resources and Environmental Sciences, Nanjing Agricultural University, Nanjing, Jiangsu, China

**Gabriel Ibrahin Tovar**
Universidad de Buenos Aires, Facultad de Farmacia y Bioquímica, Departamento de Ciencias Químicas; CONICET—Universidad de Buenos Aires, Instituto de Química y Metabolismo del Fármaco (IQUIMEFA), Buenos Aires, Argentina

**Timir Tripathi**
Molecular and Structural Biophysics Laboratory, Department of Biochemistry, North-Eastern Hill University, Shillong; Regional Director's Office, Indira Gandhi National Open University (IGNOU), Regional Center Kohima, Kohima, India

**María Emilia Villanueva**
CONICET—Universidad de Buenos Aires, Instituto de Química y Metabolismo del Fármaco (IQUIMEFA); Departamento de Ciencias Básicas, Universidad Nacional de Luján (UNLu), Buenos Aires, Argentina

**Yuxuan Wang**
Key Laboratory of Meat Processing, MOA, Key Laboratory of Meat Processing and Quality Control, MOE, Jiang Synergetic Innovation Center of Meat Production, Processing and Quality Control, Nanjing Agricultural University, Nanjing, PR China

**Junjie Xu**
College of Resources and Environmental Sciences, Nanjing Agricultural University, Nanjing, Jiangsu, China

**Di Zhao**
Key Laboratory of Meat Processing, MOA, Key Laboratory of Meat Processing and Quality Control, MOE, Jiang Synergetic Innovation Center of Meat Production, Processing and Quality Control, Nanjing Agricultural University, Nanjing, PR China

**Xue Zhao**
Key Laboratory of Meat Processing, MOA, Key Laboratory of Meat Processing and Quality Control, MOE, Jiang Synergetic Innovation Center of Meat Production, Processing and Quality Control, Nanjing Agricultural University, Nanjing, PR China

# Editors biography

**Dr. Prakash Saudagar** is an active researcher and a sterling classroom teacher, currently working as Assistant Professor at the Department of Biotechnology, National Institute of Technology Warangal, Warangal, India. He obtained his PhD degree from the Indian Institute of Technology, Guwahati, India, in 2013. His research has immensely contributed to exploring potential drug target proteins and inhibitors. His research interests include molecular and biochemical parasitology, infectious disease, and protein biochemistry. He has a strong command over computational techniques and in vitro techniques used to study proteins. He has published more than 40 research articles in reputed journals such as the *International Journal of Biological Molecules*, *The FEBS Journal*, *FEBS OpenBio*, *PLOS One*, *Scientific Reports*, *The Journal of Biological Chemistry*, *Parasitology International*, *Molecular Simulation*, and so on and book chapters in Elsevier and Springer to his name. He has been PI/Co-PI in research grants from SERB and DST. He has been the recipient of the B.S. Narasinga Rao Award SBC, India (2011); Best Presentation Award, ICIDN, Nepal (2015); Young Faculty Award, VIF India (2016); and Young Scientist Award, Telangana Academy of Science (2018). He is an associate fellow of the Telangana Academy of Science (2018) and a life member of the Indian Science Congress and Society of Biological Chemists. He has guided several PhD and M. Tech students, postdoctoral fellows, project fellows, and trainees. He has many interdisciplinary collaborating partners in prestigious institutions in India and abroad.

**Dr. Timir Tripathi** is the Regional Director of Indira Gandhi National Open University (IGNOU), Regional Center Kohima, Nagaland, India. Earlier, he served as Senior Assistant Professor and Principal Investigator at the Department of Biochemistry, North-Eastern Hill University, Shillong, India. He holds a PhD from the Central Drug Research Institute, Lucknow, India. He was a visiting faculty at ICGEB, New Delhi, India (2011), and Khon Kaen University, Thailand (2015). He is known for his research in the fields of protein biophysics, biochemistry, structural biology, and drug discovery. He has more than 16 years of experience in teaching and research on protein structure, function, and dynamics at the postgraduate and doctoral levels. He has developed and improved methods to investigate and analyze proteins. His research areas include protein interaction dynamics and understanding the roles of noncatalytic domains in regulating the catalytic activity of proteins. The common theme in his research is an interest in understanding biological phenomena involving proteins at the molecular, structural, and mechanistic levels. He has handled several research grants as a principal investigator from various national and international funding agencies, including DST-Russian Foundation for Basic Research, UGC-Israel Science Foundation, DBT, SERB, DHR, and ICMR. He has received several awards, including Prof. B.K. Bachhawat Memorial Young Scientist Lecture Award (2020) by the National Academy of Sciences, India; ISCB-Young Scientist Award (2019); ICMR-Shakuntala Amir Chand Prize (2018); BRSI-Malviya Memorial Award (2017); DST Fasttrack Young Scientist Award (2012); DBT Overseas Associateship Award (2012); Dr. D.M. Bose Award (2008); and so on. He is an elected member of the National Academy of Sciences, India, and the Royal Society of Biology, United Kingdom. He has published more than 100 research papers, reviews, commentaries, viewpoints, and editorial articles in international journals; has edited three books; and has published several book chapters. He currently serves on the editorial boards of several international journals, including the *International Journal of Biological Macromolecules*, *Acta Tropica*, *Scientific Reports*, *PLoS One*, etc.

# Foreword

The functional activity of a protein is governed by its three-dimensional structure, whereas its conformation and dynamics determine its interactions with other molecules. Understanding the structure, function, and dynamics of proteins is thus crucial for determining the molecular basis of cellular functions. Spectroscopic methods are the foundation techniques used for the analysis of biomolecular structure and dynamics. During the past few decades, significant methodological advancements have occurred in biomolecular spectroscopy to study the structure, function, folding, and dynamics. The advances in technology, computational methods, and hardware have allowed the determination of molecular and structural properties of proteins within a short period that was not possible to achieve before. In this book, Prakash Saudagar and Timir Tripathi cover both traditional and high-throughput spectroscopic approaches to comprehensively characterize proteins.

The book covers various spectroscopic techniques used to study biomolecular structure and dynamics. The book contains 17 chapters, each discussing specific spectroscopic methods for analyzing biomolecular structure and dynamics. Chapter 1 introduces the fundamentals of spectroscopy for biomolecular structure and dynamics. Chapters 2 and 3 focus on the fluorescence- and circular dichroism-based spectroscopic techniques. Chapters 4 and 5 discuss the basic and advanced NMR spectroscopy methods to study protein structure, folding, and dynamics. Chapters 6 and 7 discuss applications of infrared and Raman spectroscopy to study biomolecules. Chapter 8 focuses on the spectroscopic investigation of biomolecular dynamics using light scattering methods. Chapter 9 provides detailed information on the principles and applications of protein footprinting by contemporary mass spectrometry methods. Chapters 10 and 11 discuss the small-angle scattering and X-ray crystallography techniques to study the structural dynamics of biomolecules. Chapters 12 and 13 discuss the spectroscopic methods to study the kinetics and thermodynamics of biomolecular interactions. Chapter 14 focuses on the spectroscopic techniques to detect and analyze protein oligomerization, aggregation, and fibrillation. Chapter 15 discusses the spectroscopic methods for the analysis of carbohydrates. Chapter 16 elaborates how the spectroscopic and computational data can be integrated to analyze protein structure, function, folding, and dynamics.

Chapter 17 discusses advanced data handling tools for easy, fast, and accurate interpretation of spectroscopic data.

The chapters are prepared by leading scientists across the globe. The authors present a broad picture of the current, emerging, and evolving spectroscopic methods for analyzing biomolecules. Each chapter discusses the principle and application to perform a particular technique. I am certain that the book will serve as a valuable manual for researchers and students of biophysics, structural biology, computational biology, biochemistry, and molecular biology. It will help the basic and applied scientists in conducting biomolecular research.

In the past 40 years, I have trained a generation of young, energetic researchers in modern branches of structural biology. I consider this book a fantastic, comprehensive, and handy guide on contemporary spectroscopic methods. This book will benefit scientists, researchers, and students as they progress in their fields. I highly appreciate and thank both the editors for bringing out this exceptional piece of compendium for the benefit of a wider scientific community.

**Prof. T. P. Singh**
New Delhi, India

# Preface

Understanding the biochemistry of biomolecules, such as proteins, carbohydrates, and nucleic acids, is crucial as their structure plays a critical role in the functioning of a cell. The biological function of biomolecules depends on their intrinsic structural dynamics; therefore, it is imperative to understand their structure and dynamics comprehensively. Following the completion of the human genome project, it is now established that proteins are the primary executors of biological functions. As a result, the structure, folding, dynamics, and function of proteins are being expeditiously analyzed to understand how cellular biochemistry works in health and disease. Among the range of methods used, spectroscopic techniques have been at the heart of the research interests of scientists for several decades to help them study biomolecular structure and dynamics. These techniques can be used individually or in tandem to derive useful information about the precise biomolecular characteristics. Several advancements in technology and algorithms have allowed researchers to characterize proteins at the atomic level rapidly and accurately. This book presents a compilation of various well-established spectroscopic techniques used to understand and analyze the structure and dynamics of biomolecules, particularly proteins.

The fundamental principles underlying spectroscopic techniques are introduced in Chapter 1. Once the basic concepts are made perceptible, Chapters 2 and 3 discuss the analysis of the structure and dynamics of macromolecules with the help of fluorescence and circular dichroism spectroscopy, respectively. Chapters 4 and 5 discuss the basic and advanced nuclear magnetic resonance spectroscopy methods, respectively, for studying protein structure, folding, and dynamics. Chapters 6 and 7 focus on the applications of infrared and Raman spectroscopy to understand the biomolecular structure and dynamics. Chapter 8 exclusively concentrates on the investigation of biomolecular dynamics using dynamic light scattering techniques. Chapter 9 discusses high-throughput mass spectroscopic techniques, including HDX, FPOP, and amino acid footprinting. Chapters 10 and 11 discuss the techniques of small-angle scattering and X-ray crystallography, respectively, for understanding the structural dynamics of proteins. Chapters 12 and 13 focus on how spectroscopic methods can be used to study the kinetics of protein folding and the thermodynamics of biomolecular interactions, respectively. Chapter 14 describes the spectroscopic techniques used to detect, analyze, and quantify the oligomerization, aggregation, and fibrillation of proteins. Chapter 15 presents different spectroscopic methods to analyze carbohydrates. Chapter 16 focuses on integrating spectroscopic data with computational tools to better understand the protein structure and dynamics at the macro and atomic levels. Finally, Chapter 17 focuses on the advanced data handling software tools for quick, accurate, and easy interpretation of spectroscopic data.

This book is a result of the dedicated and sincere contributions of eminent researchers in the field of spectroscopy across the globe. The chapters have been meticulously crafted such that the reader gets the best out of each chapter. We strongly believe that the book will serve as a guiding light for faculties, students, and researchers equally, who are engaged in the field of biochemistry, structural biology, computational biology, biophysics, spectroscopy, and analytical and medicinal chemistry. The book will help the readers understand the current and emerging trends in the fields of biomolecular spectroscopy.

We sincerely thank all the authors who took their time and effort to contribute to this book on spectroscopic techniques. Finally, we also thank our families for their constant support and unconditional love, without which this book would not have been possible.

**Prakash Saudagar**
Warangal, India

**Timir Tripathi**
Kohima, India

# CHAPTER 1

# Fundamentals of spectroscopy for biomolecular structure and dynamics

**Niharika Nag[a], Santanu Sasidharan[b], Prakash Saudagar[b], and Timir Tripathi[a,c]**

[a]Molecular and Structural Biophysics Laboratory, Department of Biochemistry, North-Eastern Hill University, Shillong, India
[b]Department of Biotechnology, National Institute of Technology Warangal, Warangal, India
[c]Regional Director's Office, Indira Gandhi National Open University (IGNOU), Regional Center Kohima, Kohima, India

## 1. Introduction to spectroscopy

Spectroscopy is the study of the interaction between electromagnetic radiation (light) and matter as a function of wavelength or frequency. A spectroscopy experiment uses light of a specific wavelength that passes through the sample, leading to either absorption or emission. A range of techniques employ visible and ultraviolet (UV) light for the experimental use of spectroscopy; however, there are more spectroscopic techniques that utilize other wavelengths of the electromagnetic spectrum having higher and lower energies. The former methods are used mainly for qualitative and quantitative measurements and identification of compounds, while the latter are aimed more toward the study of the structure of molecules. Spectroscopy experiments are carried out in an instrument called the spectrophotometer, which is used to view and analyze the "spectrum" representing the characteristics of the sample. The spectrum typically depicts the wavelength of the light absorbed, transmitted, reflected, or emitted by the sample. The very basic build of a spectrometer consists of a source of radiation, an apparatus for filtering a specific wavelength, and a sensor for detecting the resulting radiation. These components do vary over a wide range depending on the application of the instruments and also on the resolution expected. The result from a spectrometer is obtained in the form of a spectrum. A spectrum shows the intensity of radiation at varying wavelengths or the response of a system to different wavelengths of radiation

*Advanced Spectroscopic Methods to Study Biomolecular Structure and Dynamics*
https://doi.org/10.1016/B978-0-323-99127-8.00002-7

in the form of a graph. The measurement of the resulting spectrum is called spectrometry, which is thus the application of spectroscopy.

Spectroscopic methods have a wide range of applications in various fields, including biology, medicine, chemistry, and physics. Moreover, they are used extensively in analytical chemistry to qualitatively identify unknown compounds, even in trace amounts, quantify the concentration of compounds, and distinguish compounds from one another. In astronomy, it is used to determine the composition, density, motion, and temperature of objects. The techniques can extensively reveal the structure of macromolecules, including the information on the associated molecular properties of the molecules like the bonds and energies that stabilize the structure [1,2]. Besides being a vital component in any research laboratory, spectroscopy is an indispensable tool in many industries, like medical, chemical, health, food, paint, petroleum, etc.

## 1.1 Electromagnetic spectrum

Electromagnetic radiation or light may be described either as packets of energy, called photons (particle theory) or as waves of energy (wave theory). Both the theories are related through Plank's law, which is given by the following equation:

$$E = h\nu$$

where $E$ is the photon energy (Joules, J), $\nu$ is the radiation frequency (Hz or $s^{-1}$), and $h$ stands for the Plank's constant ($6.63 \times 10^{-34}$ J s).

As the name suggests, the electromagnetic wave comprises both electric and magnetic fields, each of which oscillates perpendicular to the direction of propagation of the wave and each other. The wave of electromagnetic radiation has a sinusoidal waveform with a wavelength $\lambda$ and a frequency $\nu$. The wavelength is the spatial distance between two consecutive peaks of the wave, and in the case of light, it is in the order of nanometers. The frequency is the number of oscillations made by the wave per second and thus has the unit of Hz ($s^{-1}$). The wavelength and the frequency are related by the following equation:

$$\nu = c\lambda^{-1}$$

where $c$ is the speed of light, i.e., $2.998 \times 10^{8}$ m $s^{-1}$.

The maximum field strength of the electric and magnetic fields is called the amplitude of the wave, which is used to calculate the wave energy. The wavenumber is the number of wave cycles per unit distance and is denoted by $\bar{\nu}$.

The range of frequencies or wavelengths of electromagnetic radiation forms the electromagnetic spectrum. The spectrum comprises electromagnetic waves having frequencies from <1 to >$10^{25}$ Hz. The spectrum is divided into separate bands or portions, beginning from low frequencies followed by higher frequencies. Since the wavelength and frequency of a wave have an inverse proportional relation, the radiation having the lowest frequency has the longest wavelength. Almost all wavelengths of the spectrum can be used for spectroscopy. Broadly, the electromagnetic spectrum is divided into gamma rays, X-rays, UV radiation, visible light, infrared radiation (IR), microwaves, and radio waves.

**(i)** **Radio waves** are low-energy waves and, as the name suggests, are a type of radiofrequency radiation. The frequency of these waves ranges from 3 kHz to 300 MHz corresponding to the wavelength of 100 km to 1 m. Radio waves are generated by accelerating charged particles. Natural radio waves are generated by lightning and astronomical objects like planets, stars, gas clouds, etc. However, radio waves are also artificially generated using transmitters, which are used for radio communication, broadcasting, navigation, wireless computer networks, etc.

**(ii)** **Microwaves** are the second type of radiofrequency radiation, which have a higher frequency than radio waves (300 MHz to 300 GHz). Natural sources of these waves include the sun and cosmic microwave background. These waves are extensively used in modern-day life for terrestrial and satellite communication, wireless networks, microwave ovens, etc. Both radio waves and microwaves are nonionizing, which means that they do not carry enough energy to ionize atoms or molecules.

**(iii)** **IR** is the band in the spectrum that comes after microwaves in the electromagnetic spectrum. It has frequencies ranging from 300 GHz to 400 THz (1 mm to 750 nm). Based on their wavelengths, IR is broadly divided into the near-, mid-, and far-infrared regions. The near-IR band consists of wavelengths between 750 and 1300 nm, and this region is known to be responsible for the heat generated by IR due to its shorter wavelength. The mid- and far-IR bands cover the wavelengths of 1300–3000 nm and 3000 nm to 1 mm, respectively. Most of the radiation emitted by a moderately heated surface is in the form of IR. IR absorption occurs in compounds with minor energy differences in the possible rotational and vibrational states. Interestingly, when the frequency of the radiation and the vibrational frequency of the molecule match, a change in the amplitude of the vibration is observed. In addition, IR can cause molecular rotations and molecular vibrations like stretching and bending. Apart from these, IR has applications in astronomy, thermal imaging, medicine, surveillance, etc.

**(iv)** **Visible light**, as the name suggests, is the portion of the electromagnetic spectrum that is observable by the human eye. This region ranges from the wavelengths of 400–750 nm. The visible region of the spectrum is used extensively in spectroscopy.

**(v)** **UV light** includes another range of wavelengths used commonly in spectroscopy. UV radiation lies in the region between 10 and 400 nm, and based on their energies, they are divided into three types: UVA, UVB, and UVC rays. UVA rays have the longest wavelengths (400–315 nm) and thus have the lowest energy. UVB and UVC rays have shorter wavelengths of 315–280 and 280–100 nm. There are other classifications that account for the even shorter wavelengths of UV radiation. The most common source of natural UV radiation is the sun. Most of the UV rays that reach the Earth are usually in the form of UVA rays, followed by UVB rays. UVC rays have the highest energy, due to which they react with the ozone layer and do not reach the Earth. Though long-wavelength UV is nonionizing, shorter wavelengths of UV radiation are capable of ionizing atoms.

**(vi)** **X-rays** are another type of radiation in the electromagnetic spectrum. X-rays have a wavelength range of 10 pm to 10 nm. X-rays have very high energies, thus capable of ionizing atoms and disrupting molecular bonds. They are produced by accelerating or decelerating charged particles, such as in an X-ray tube or a synchrotron.

**(vii)** **Gamma rays** constitute the end of the electromagnetic spectrum and have the shortest wavelengths, i.e., <20 pm. These rays originate from the radioactive decay of atomic nuclei and are ionizing. Due to their high energies, they can penetrate through most everyday materials, including the human body, thus making them extremely hazardous.

## 1.2 Electron energy level and Jablonski diagram

Electrons in atoms and molecules are distributed among energy levels, though they are principally found in the ground state, the lowest energy level. When energy is absorbed by an atom/molecule, an electron can jump to a higher energy level, called the excited state. When this energy is obtained from electromagnetic radiation, the phenomenon is called absorption, as the electron jumps from the electronic ground state ($S_0$) to the first electronic excited state ($S_1$). Each electronic state of the molecule also possesses its own set of vibrational and rotational levels with distinct energies. Thus the molecule is also excited at the vibrational and rotational state. Eventually, the molecule relaxes back to the ground state, which is accompanied by the emission of light and/or heat.

The Jablonski diagram depicts the various electronic states of a molecule and the possible transitions that can occur after interaction with light (Fig. 1). It is named after Aleksander Jabłoński, a Polish physicist. In the Jablonski diagram, the molecular energy levels are shown as horizontal lines, grouped by spin multiplicity, as the energy increases vertically. The electronic ground states are represented by thicker lines and the higher vibrational levels with thinner lines. The vibrational levels are spaced more and more closely as the energy increases, eventually forming a continuum. The singlet states, having a total spin angular momentum of zero, are denoted by $S$, whereas the triplet states having a total spin angular momentum of 1 are denoted by $T$. Therefore $S_0$ denotes the singlet ground stat, and $S_n$ is the $n$th excited singlet state. Similarly, $T_0$ denotes the triplet ground state and $T_n$ is the nth excited triplet state. The transitions that can transfer energy between the states are divided into the radiative and nonradiative transitions.

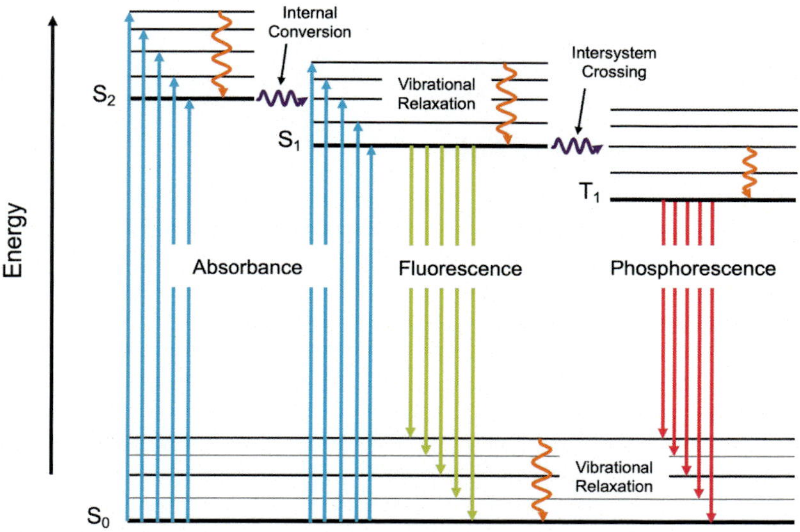


**Fig. 1** The Jablonski diagram. The diagram depicts various possible radiative and nonradiative transitions.

The radiative transitions involve transitions wherein the difference in energies is either emitted (higher to lower) or absorbed (lower to higher) in the form of a photon. This transition is represented by straight arrows. Nonradiative transitions are transitions that do not involve the emission or absorption of photons. These are represented by squiggly arrows. Radiative transitions include the following types:

1. **Absorption:** It involves a transition phenomenon from a lower to higher electronic state due to the absorption of the energy of a photon. The energy helps the molecule excite from $S_0$ to one of the higher singlet vibrational levels.
2. **Fluorescence:** It is a transition phenomenon between two electronic states having the same spin multiplicity. The light emitted in the process has a longer wavelength than the light absorbed.
3. **Phosphorescence:** It is a transition phenomenon between electronic states having different spin multiplicities, e.g., $T_1 \rightarrow S_0$.

The nonradiative transitions include the following types:

1. **Vibrational relaxation:** It involves a nonradiative transition to a lower vibrational level within the same electronic state.
2. **Internal conversion:** It is a nonradiative transition between electronic states having the same spin multiplicity.
3. **Intersystem crossing:** It is a nonradiative transition between isoenergetic vibrational levels that belong to electronic states having different spin multiplicities.

A timescale of these events is shown in Table 1.

## 1.3 Basic properties of light

In addition to causing the phenomenon mentioned previously, light waves themselves undergo processes like reflection, refraction, diffraction, scattering, and polarization.

**Table 1** Timescale of radiative and nonradiative transitions.

| Transition | Timescale (s) | Type |
|---|---|---|
| Absorption | $10^{-15}$ | Radiative |
| Vibrational relaxation | $10^{-14}$–$10^{-11}$ | Nonradiative |
| Internal conversion | $10^{-14}$–$10^{-11}$ | Nonradiative |
| Fluorescence | $10^{-9}$–$10^{-7}$ | Radiative |
| Intersystem crossing | $10^{-8}$–$10^{-3}$ | Nonradiative |
| Phosphorescence | $10^{-4}$–$10^{-1}$ | Radiative |

### 1.3.1 Reflection

Reflection is a process that occurs when a beam of light bounces off of a smooth polished surface. In reflection, the ray of light that hits the surface is called the incident ray, and the ray that bounces off is called the reflected ray. A perpendicular drawn between the two rays is called the normal. According to the laws of reflection, the incident ray, the reflected ray, and the normal, all lie on the same plane and the angle between the incident ray and the normal (angle of incidence) and the reflected ray and the normal (angle of reflection) are equal. Reflection is of two main types, specular and diffused. In specular reflection, the incident light is reflected in only one outgoing direction, whereas in diffused reflection, the incident rays are reflected in a range of directions. Reflection occurs when the ray of light originates from one medium and reflects into the same medium.

### 1.3.2 Refraction

In cases where the light travels from one medium to another, the process is called refraction. Refraction is the change in the direction of light that occurs when it passes from one medium to another or a gradual change in the medium. There are two laws of refraction which state that the incident ray, the refracted ray, and the normal all lie on the same plane and that the ratio of the sine of the angle of incidence to the sine of the angle of refraction is constant for the light of a given color and a given pair of media (refer to Snell's law). The constant value is the refractive index of the second medium with respect to the first medium. The refractive index of a medium is the extent to which the refractive medium changes the speed of light.

### 1.3.3 Diffraction

Diffraction of light is a phenomenon in which light waves bend around the corners of an obstacle or through an aperture. Diffraction occurs due to interference, a phenomenon in which two or more superimposing waves give rise to a resultant wave that may have higher or lower amplitude. Diffraction is most noticeable when the linear dimensions of the obstacle are comparable to the wavelength of light.

### 1.3.4 Scattering

Another important phenomenon of light is called the scattering of light. Scattering of light is a process in which nonuniformities suspended in a medium deviate light from its straight trajectory. There are two main types of scattering, elastic scattering and inelastic scattering. In elastic scattering, there is a conservation of energy while its direction is changed, whereas, in inelastic scattering, both the energy and direction of light are changed. Elastic scattering is of two types: Rayleigh scattering and Mie scattering. Rayleigh scattering is the process of scattering in which radiation is scattered by particles with a radius less than approximately one-tenth of the wavelength of radiation. Mie scattering is scattering where particle sizes are larger than the wavelength of light. Inelastic scattering is classified as Raman scattering and Crompton scattering. Raman scattering is scattering where the molecule which causes the scattering gains vibrational energy. In Crompton scattering, light is scattered by a free-charged particle, which causes a decrease in the energy of the light photon. The probability of scattering, $P$, is given by

$$P \propto \frac{1}{\lambda^4}$$

where $\lambda$ is the wavelength of radiation. Therefore the scattering probability of light is higher for light having a shorter wavelength.

### *1.3.5 Polarization*

Polarized light is made up of waves that vibrate in a single plane (Fig. 2). There are three types of polarized lights, linearly, circularly, and elliptically polarized lights. In linearly polarized light, the electric vector of light moves in a single plane along the direction of propagation. In circularly polarized light, the electric vector of light is made of two linear components perpendicular to each other, having equal amplitudes but with a phase difference of $\pi/2$. In elliptically polarized light, the electric field propagates elliptically, with unequal amplitudes and phase differences. Unpolarized light can be converted to polarized light using methods like transmission, reflection, scattering, and refraction.

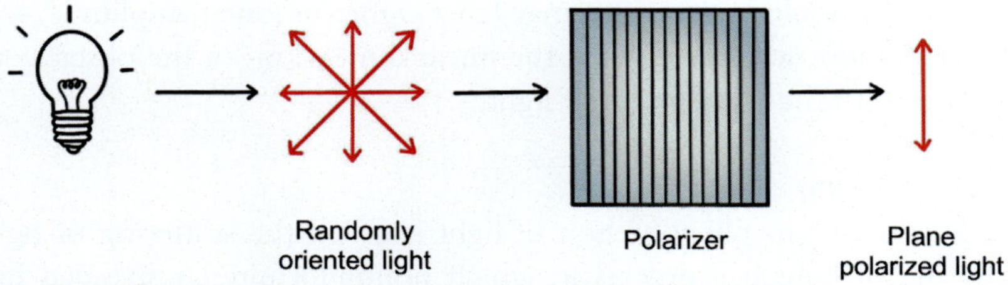


**Fig. 2** Generation of plane-polarized light. The plane-polarized light vibrates in a single plane.

## 1.4 Spectroscopy: Types and parameters

Spectroscopy can be broadly classified into absorption and emission spectroscopy. *Absorption spectroscopy* involves the use and passage of radiation through a sample and measuring the attenuation of the radiation, which is called absorption. Attenuation is described by two terms called absorbance and transmittance. Transmittance is defined as the ratio of the source radiation intensity exiting the sample to that incident on the sample. The transmission multiplied by 100 gives the percentage of transmission, which varies between 0% and 100%. The absorption of radiation is dependent on the wavelength of radiation and does not occur at all wavelengths. Absorption spectroscopy is used primarily in analytical chemistry to determine the presence of a substance in a sample and the amount present in the sample. It is also used in molecular, atomic, and astronomical studies. The most common forms of absorption spectroscopy used are ultraviolet-visible spectroscopy and infrared spectroscopy. *Emission spectroscopy* is a technique where the radiation emitted by atoms or molecules during their decay from higher excited states to lower energy states is examined. The emitted radiation can be analyzed to give rise to an emission spectrum which is used to determine a number of factors, including the presence or absence of a substance in the sample.

The analysis of the resulting spectrum obtained through spectroscopy is called spectral or spectrum analysis. The radiant flux is the radiant energy per unit time, also known as radiant power (W, mW, μW). Radiant intensity is the radiant flux emitted, transmitted, reflected, or received per unit solid angle. The SI unit of radiant intensity is W/sr (watt per steradian). Radiance is the radiant intensity per unit of the emitting area. The SI unit of radiance is watt per steradian per square meter ($\frac{W}{sr \cdot m^2}$). Spectral intensity is the radiant intensity per unit of wavelength or frequency of the radiation. The SI unit is watt per steradian per hertz ($\frac{W}{sr \cdot Hz}$) or watt per steradian per meter ($\frac{W}{sr \cdot m}$). Another term is irradiance, which is the power per unit area of radiation incident on a surface. Its SI unit is watts per square meter ($\frac{W}{m^2}$) or milliwatts per square millimeter ($\frac{mW}{mm^2}$). Other terms also associated are spectral radiant flux, which is the radiation power per unit wavelength ($\frac{W}{m}$), spectral irradiance, which is used for radiation incident on a surface ($\frac{W}{m^3}$), and spectral radiance, used for radiation power within a unit solid angle from a unit emitting area and unit wavelength ($\frac{W}{m^2\ nm\ sr}$). A few other important parameters in the spectral analysis include the peak, which is the highest point in the curve, the peak height, and width. The peak width is the range of wavelengths between the points where the intensity falls to half of the maximum intensity.

The spectral resolution dictates the maximum number of spectral peaks that a spectrometer can resolve. It is the capability to differentiate between two wavelengths separated by $\Delta\lambda$. It is given by the dimensionless quantity, $R$, as

$$R = \frac{\lambda}{\Delta\lambda}$$

Another critical parameter to keep in consideration is the signal-to-noise ratio (SNR). The SNR can be defined as the ratio of the average over time of the peak signal to the root-mean-square noise of the peak signal over the same time.

## 1.5 Biomolecular structure and dynamics

Biomolecules are not static structures but are dynamic in nature, influenced by the environment where they are present. The four major groups of biomolecules are carbohydrates, lipids, nucleic acids, and proteins. Among them, proteins are a class of highly complex and versatile biomolecules. The sequence of amino acids determines the structure of proteins, which in turn dictates their functionality. Proteins interact with each other and a number of other biomolecules to perform functions. The structure of proteins has four levels. The primary structure is formed by the linear polymerization of amino acid residues, which make up the protein. The secondary structure is derived by the local folding of the polypeptide chain into ordered structures like α-helices and β-sheets, stabilized by hydrogen bonds. The tertiary structure refers to the three-dimensional (3D) arrangement of the polypeptide chain. These structures are stabilized by hydrophobic interactions, electrostatic interactions, hydrogen bonds, etc. Finally, many proteins are made up of more than one polypeptide chain in their functional state, wherein each polypeptide chain is called a subunit of the protein. The arrangement of the subunits through intermolecular interactions constitutes the quaternary structure of the protein.

The transition from an unfolded polypeptide chain to a structured protein does not occur randomly. Proteins follow a semidefined folding pathway that consists of intermediates between the fully unstructured to the native structure. A folding funnel or an energy landscape is used to understand the process of protein folding (Fig. 3). The breadth of the funnel represents all the possible conformations of the unstructured protein, with each point on the surface representing a 3D structure and its energy. The diffusion-collision model of protein folding suggests that local interactions take place in an unfolded protein, leading to the formation of the secondary structures,

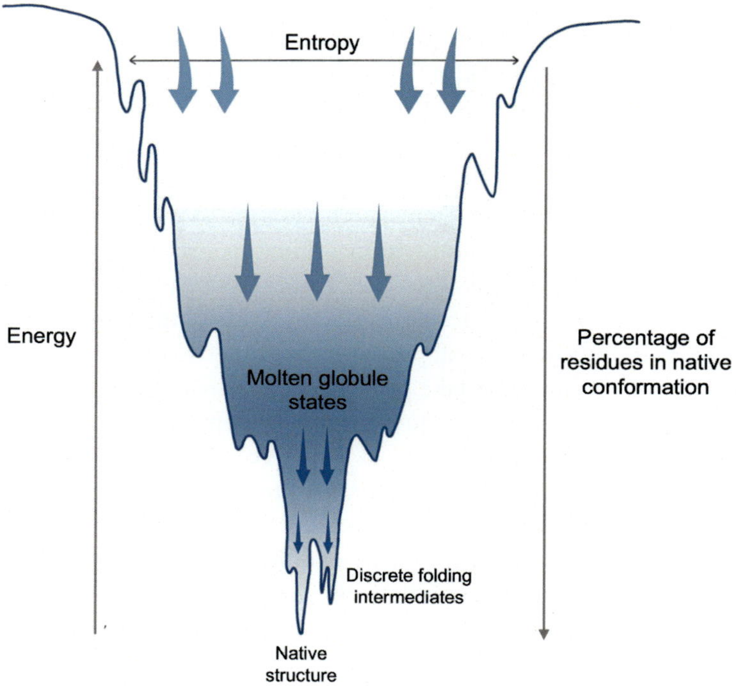


**Fig. 3** A typical protein folding funnel.

followed by the tertiary structure [3]. The hydrophobic collapse model suggests that the hydrophobic interactions among the amino acid residues bring them together to collapse into the interior of the protein and thus form a globular structure called a molten globule [4]. The nucleation-condensation model suggests that the interactions proposed by the two previous models take place together to form the native structure of a protein [5–7]. The nucleation-condensation model can settle the Levinthal paradox [8], which states the contradiction between the time required for random sampling through the astronomical number of protein conformations and the folding rate constants (in microseconds) that are observed in experiments [9]. The structure of a protein is critical for its functionality since the structure determines the ability of the protein to interact with other proteins and molecules [10,11].

It should be noted that despite the essential relationship between the sequence of a protein and its folding, some proteins do not need to form structured 3D shapes in order to be functional. These proteins are unstructured in their native state and are called intrinsically disordered proteins

(IDPs) [12,13]. The intrinsic disorder of IDPs is closely related to their functionality, which usually involves interaction with more than one other protein and/or molecule [14,15].

Though structural proteins have a 3D structure, it does not essentially mean that the structures are rigid and fixed. In fact, proteins are conformationally flexible and dynamic, and this flexibility and dynamics are crucial to the functionality of structured proteins, such as in the case of enzymes [16,17]. The allosteric nature and conformational alterations of proteins can be influenced by conformational substates. The conformational changes are defined as atomic displacements that result in different local configurations and interconversions in the same overall protein structure [18–20]. The timescale of various dynamic events that take place in proteins is shown in Fig. 4.

Though the amino acid sequence of a protein plays a crucial role in determining the structural features of a protein, the solution environment also affects the structure and dynamics of proteins. The properties of the solution change the propensity of amino acid residues to form a certain type of secondary structure [21,22]. The use of different solvents has shown that a properly chosen solvent can influence a peptide to form an α-helix, a β-strand, or a random coil. Since proteins are stabilized by many forces like hydrogen bonds, electrostatic interactions, hydrophobic interactions, etc., a change in the temperature, pH, salt concentration, etc., can alter the conformation and dynamics of a protein [23–28]. Protein-protein interaction, which utilizes the property of dynamics of proteins, is also influenced by changes in the solution, such as molecular crowding [29]. Macromolecular crowding has two components that have physical consequences: hard-core repulsions and soft chemical interactions [30]. Hard-core repulsions are steric effects that reduce the volume of the protein available for the reaction.

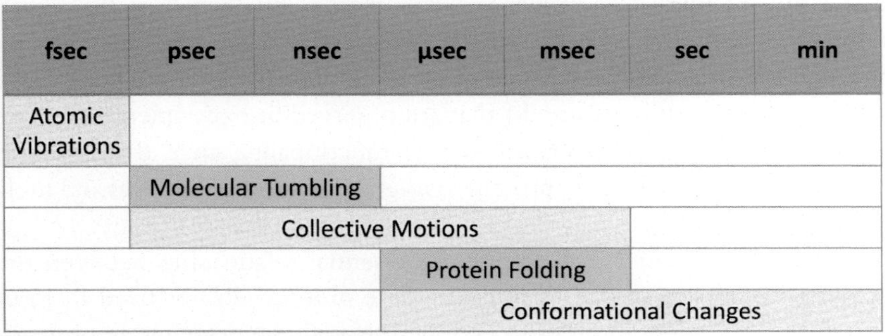

fsec
psec
nsec
µsec
msec
sec
min
Atomic Vibrations
Molecular Tumbling
Collective Motions
Protein Folding
Conformational Changes


**Fig. 4** Timescale of dynamic events that take place in proteins.

Soft chemical interactions include hydrogen bonds, electrostatic, hydrophobic, water-protein, and solute-protein interactions.

Along with proteins, carbohydrates and lipids also constitute major classes of biomolecules. The primary role of carbohydrates and lipids is to provide the energy for all the biochemical reactions that take place in the cell. Carbohydrates are carbon-rich molecules with the empirical formula $(CH_2O)_n$. The simplest forms of carbohydrates are called monosaccharides. Polymers of monosaccharides are called polysaccharides. The smallest monosaccharides are dihydroxyacetone and glyceraldehyde, each made up of three carbon atoms. Disaccharides are carbohydrates made up of the covalent linkage of two monosaccharides. The monomers are linked by an *O*-glycosidic bond. The most well-known disaccharides are sucrose, lactose, and maltose. Larger polysaccharides include glycogen and starch. Glycogen and starch are the storage forms of glucose in animals and plants, respectively. Starch is composed of two forms, amylose, which is the unbranched starch, and amylopectin, which is the branched starch and is similar to glycogen but with a lower degree of branching. Another prominent polysaccharide of glucose in plants is cellulose, which is an important structural component found in the plant cell wall. Besides carbohydrates existing independently, they can also covalently bind to proteins to give rise to proteins known as glycoproteins.

Lipids are molecules that are water insoluble but highly soluble in organic solvents. They store energy and serve various other roles in membranes and signal transduction pathways. Lipids do not form polymers and consist of five classes: free fatty acids, triacylglycerides, phospholipids, glycolipids, and steroids. Fatty acids act as the primary source of energy and are made up of chains of hydrocarbons that end with a carboxylic group. They also serve as the building blocks of lipid membranes. The storage form of fatty acids is called triacylglycerides, in which the three fatty acids are attached to a glycerol molecule. During fasting, the fatty acids are cleaved from the glycerol and used as fuel. Fatty acids are richer in energy than carbohydrates due to their hydrophobicity. A gram of anhydrous fat stores more than six times the energy stored in a gram of hydrated glycogen. Phospholipids are membrane lipids that consist of fatty acids attached to a scaffold and contain a phosphoryl group at its head. Glycolipids are lipids bound to carbohydrates and are a constituent of biomembranes. Lastly, steroids are polycyclic hydrocarbons that function as hormones and regulate several physiological functions. Cholesterol, which is an essential component of membranes, is a steroid.

## 2. General design of a spectrometer instrumentation

The instrumental design of a spectrometer varies with the type of radiation being utilized for a certain kind of spectroscopy. For absorption spectroscopy, instruments available are usually of two types: single-beam and double-beam spectrophotometers. A single-beam instrument (Fig. 5A) consists of a source of optical radiation (usually made of tungsten, halogen, deuterium, LED, xenon lasers, etc.), a monochromator (such as a diffraction grating to select the wavelength), a sample chamber (to place the sample in), and a detector (to detect the exiting light), which can be photomultiplier tubes (PMTs), photodiode arrays, photodiode, charge-coupled devices (CCDs), multichannel analyzers (MCAs), etc. The transmittance data collected by the detector is finally digitized and plotted onto a graph by software. Two types of diffraction gratings are available, i.e., the ruled grating and the holographic grating. While the former is produced by the physical etching of grooves onto a reflective surface using a diamond-form tool, the latter is produced by the interference lithography process, which involves the construction of an interference pattern using two UV beams. In a double-beam spectrophotometer, the path of light alternates between a sample and a reference sample (Fig. 5B). The spectrophotometer for emission spectroscopy consists of a source of radiation, an excitation monochromator, a sample chamber, and an emission monochromator, followed by a detector (Fig. 6).

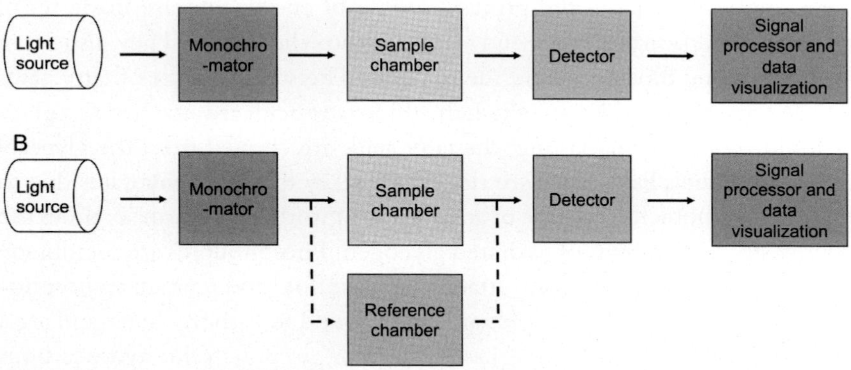


**Fig. 5** Schematic representation of the basic components of an absorption spectrophotometer. (A) Single-beam spectrophotometer. (B) Double-beam spectrophotometer.

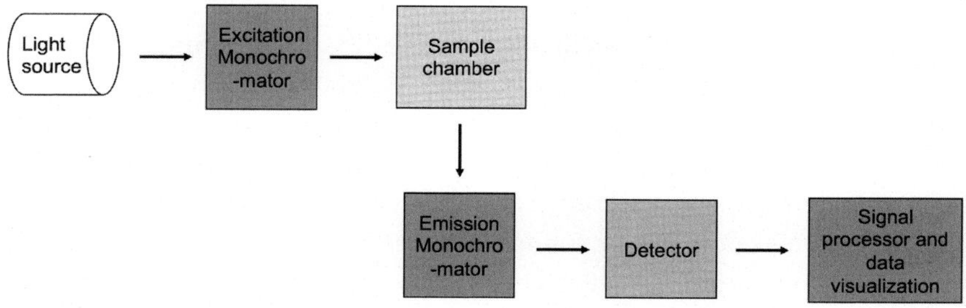


**Fig. 6** Schematic representation of the basic components of an emission spectrophotometer.

## 3. Various spectroscopic methods

Spectroscopic methods have been used extensively to determine and study the structure and dynamics of biomolecules, particularly proteins. Different methods utilize light of various wavelengths in different ways to get a range of information about the structure and dynamics of proteins. In the subsections later, we briefly introduce the spectroscopic methods used to study the structure and dynamics of biomolecules.

### 3.1 Fluorescence spectroscopy-based methods

Among the simplest and commonly used methods to study the structure and dynamics of proteins are fluorescence-based methods that include fluorescence spectroscopy, fluorescence resonance energy transfer (FRET), fluorescence correlation spectroscopy (FCS), fluorescence anisotropy, and fast relaxation imaging (FReI). In general, in fluorescence-based spectroscopy, the fluorescence of a molecule is analyzed, which provides information on the structural, conformational, and geometrical properties of a biomolecule. Here, a sample is illuminated using a beam of light of a particular wavelength to excite a molecule to a higher energy state, and the light of the longer wavelength emitted is recorded and analyzed. In traditional fluorescence spectroscopy experiments, the intrinsic tryptophan fluorescence of a protein can be utilized to perform structural studies of the protein as tryptophan emits at different wavelengths depending on its location in the protein structure. Tryptophan emits maximally between 330 and 335 nm in a hydrophobic environment, between 340 and 345 nm when partially exposed, and between 350 and 355 nm when completely exposed. Similarly, extrinsic fluorophores like 1-anilino-8-naphthalene sulfonate (ANS), 4,4′-dianilino-1,1′-binaphthyl-5,5′-sisulfonic acid (Bis-ANS), Thioflavin

T (ThT), etc., can be attached to proteins to track unfolding and refolding, detect molten globule intermediates, etc. [31–33]. FRET is used to measure the dynamics of biomolecules. It is based on the transfer of energy between a donor fluorescent probe and its acceptor fluorescent probe; the efficiency of which determines the distance between the fluorescent molecules. FRET is an excellent method to detect the distance between the probes, which can be between protein domains or subunits or other proteins, etc., thereby providing information about the conformational dynamics of the protein(s) and their interactions. More recently, FRET has been used to analyze the conformational dynamics and interactions of proteins under steady-state conditions [34–37]. The principle behind fluorescence anisotropy is the photoselective excitation of fluorophores by polarized light, which often results in polarized emission. The emitted radiation is influenced by rotational diffusion, which is the motions that take place within the lifetime of the excited fluorophore. The data is used to study protein-protein interactions, binding dynamics, conformational dynamics, etc. FCS analyses the fluctuations in the fluorescence intensity in an open volume of a solution containing fluorophores. It provides kinetic information by the temporal relaxation of the fluctuations in fluorescence and information on the thermodynamic properties by the fluctuation amplitudes [38]. The FReI technique records the biomolecular stability, dynamics, and kinetics in living cells [39], by the use of combined time-resolved fluorescence and fast temperature jump-induced kinetics [40].

Fluorescence spectroscopy was used to study protein interaction and dynamics in a study involving bovine serum albumin (BSA) and a synthesized compound 2-amino-6-hydroxy-4-(4-*N*,*N*-dimethylaminophenyl)-pyrimidine-5-carbonitrile (AHDMAPPC). Fig. 7 represents the fluorescence quenching of BSA when the AHDMAPPC compound is added in increasing concentrations while keeping the protein concentration constant. The "a" is the lowest concentration of AHDMAPPC, and the "i" is the highest concentration. The decrease in fluorescence can be observed as a result of the increasing concentration of the compound. This results from the binding of the compound to the protein BSA and the subsequent decrease in the fluorescence. The experiment was further exploited to derive the binding constant ($K$) and the number of binding sites per BSA molecule ($n$). Interestingly, it is possible to determine these values at both dynamic and static quenching interactions [41]. For a detailed discussion on the use of fluorescence spectroscopy-based methods to study biomolecular structure and dynamics, please refer to Chapter 2.

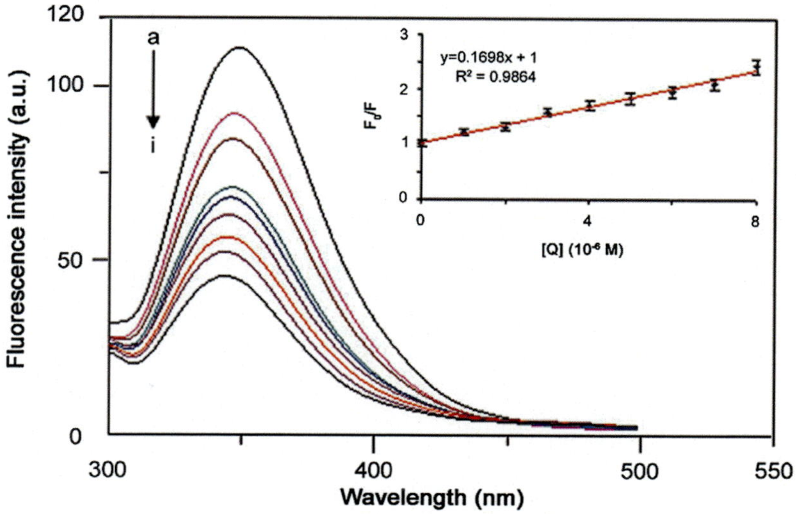


**Fig. 7** The spectra from the fluorescence quenching experiment involving BSA and AHDMAPPC. The concentration of BSA was kept constant at $2.0 \times 10^{-5}$ M, while the concentration of AHDMAPPC was varied from 0 (a) to $8 \times 10^{-6}$ M (i). The inset shows the $F/F_0$, which is calculated from the fluorescence intensity at the $\lambda_{\text{emission max}}$ in the presence and absence of the quencher. *(Reproduced with permission from Suryawanshi VD, Walekar LS, Gore AH, Anbhule PV, Kolekar GB. Spectroscopic analysis on the binding interaction of biologically active pyrimidine derivative with bovine serum albumin. J Pharm Anal 2016;6(1):56–63.)*

## 3.2 Circular dichroism spectroscopy

Circular dichroism (CD) spectroscopy is a technique based on the differential absorption of right and left polarized light and provides information about the structure and conformation of biomolecules. Due to their intrinsic chirality, biomolecules absorb right and left polarized light differently. CD is commonly used to study the secondary and tertiary protein structures and the conformational changes of proteins. In the far-UV region, protein absorption gives rise to a spectrum depending on the protein's secondary structures. The α-helices display a positive band at ~190 nm and two negative bands at ~208 and ~222 nm; β-sheets show a positive band at ~198 nm and a negative band at ~216 nm, while random coils or disordered proteins show a negative band below 200 nm (Fig. 8). The spectrum in the near-UV region provides information about the tertiary structure of a protein due to the CD bands obtained by the aromatic amino acid residues phenylalanine (255–270 nm), tyrosine (275–285 nm), and tryptophan (290–305 nm).

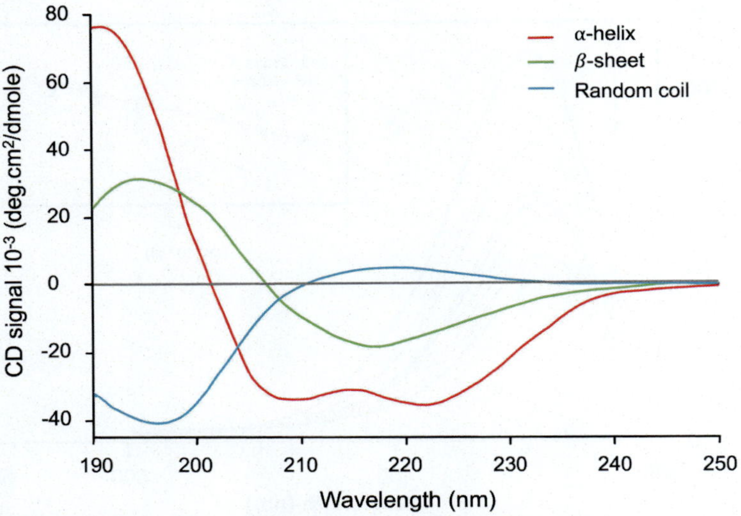


**Fig. 8** A characteristic far-UV CD spectrum for a protein with all α-helix, all β-sheet, and a random coil.

In the same study mentioned in the earlier example of fluorescence spectroscopy, the authors used CD spectroscopy to assess the alterations in the secondary structure of the protein (Fig. 9). The data showed that the binding of AHDMAPPC to BSA induced structural changes in the protein. An increase in ellipticity indicated the stability of BSA upon binding to AHDMAPPC. It can be observed that when the molar ratio of BSA:AHDMAPPC was 1:1, the α-helical content of BSA increased to 58.6%, while the molar ratio of 1:4 increased the α-helical content of BSA to 66.3% [41]. For a detailed discussion on the use of CD spectroscopy to study biomolecular structure and dynamics, please refer to Chapter 3.

## 3.3 Nuclear magnetic resonance spectroscopy

Nuclear magnetic resonance (NMR) spectroscopy is a technique based on the change in the nuclear spin energy in the presence of an external magnetic field. In NMR, the absorption of radiation takes place in the radio wave region. Nuclei with an odd mass number, atomic number, or both have a spin value. The number of allowed spin states is determined by its nuclear spin quantum number (I), which in turn is determined by the number of unpaired protons and neutrons in the nucleus. The nuclei of some isotopes like $^{1}H$ and $^{13}C$ behave like tiny magnets due to their spin. Their nuclei have

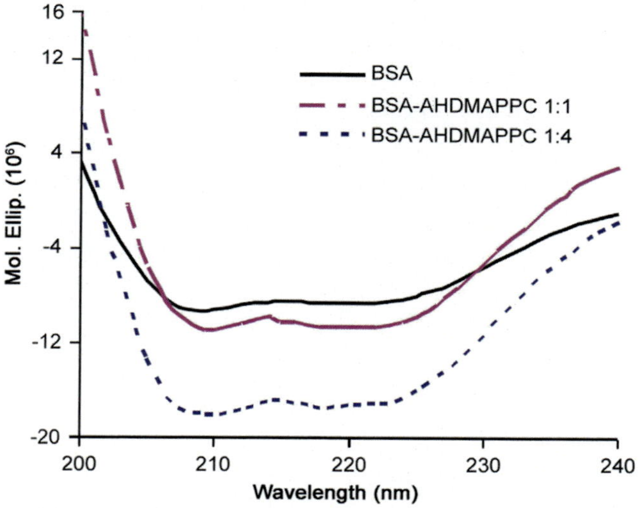


**Fig. 9** CD spectra of free BSA and BSA-AHDMAPPC complex at 1:1 and 1:4 M ratio. The concentration of BSA and AHDMAPPC was kept constant at $2.0 \times 10^{-5}$ M and $2.0 \times 10^{-6}$ M, respectively. The mean residue ellipticity at 222 nm was calculated to understand the α-helical content, and the increase can be observed in the figure. *(Reproduced with permission from Suryawanshi VD, Walekar LS, Gore AH, Anbhule PV, Kolekar GB. Spectroscopic analysis on the binding interaction of biologically active pyrimidine derivative with bovine serum albumin. J Pharm Anal 2016;6(1):56–63.)*

different levels of energy when they are placed in a magnetic field. $^{1}$H and $^{13}$C nuclei have two different energy levels, and when placed in a magnetic field, they can either align themselves against the field or along with it, which would be the highest energy state or the lowest energy state, respectively. In NMR, radio wave energy is absorbed by nuclei that are in alignment with the magnetic field, and the spin orientation changes with respect to the applied magnetic field. Since the nucleus is now in a higher energy level, it returns to the lower energy state, and the energy released is measured in NMR.

The use of NMR for the structural and conformational studies of biomolecules is based on nuclear shielding, which is basically the difference in the magnetic field experienced by a nucleus due to the local surrounding electronic environment. This, in turn, causes a slight variation in the energy levels and thus needs different frequencies to drive the flip in the spin, which creates a new peak in the NMR spectrum. Therefore different peaks correspond to a different chemical environment, and the variation in frequencies is called a chemical shift.

NMR as a tool is extensively used in the determination of protein structure, dynamics, interactions, folding, etc. [42–45]. Using NMR, Tyler et al. reported the 3D structure of L75F-TrpR, which is a temperature-sensitive mutant of the tryptophan repressor protein in *E. coli* [46]. The 2D $^{1}H$-$^{15}N$ HSQC (heteronuclear single quantum coherence) correlation spectra of apo-L75F-TrpR and apo-WT-TrpR showed that under identical temperature, pH, and buffer conditions, the two proteins were found to have comparable overall structures, with only a few differences (Fig. 10). The backbone and side-chain resonances of the mutant protein were also assigned using a combination of 2D and 3D heteronuclear NMR. For a detailed discussion on the use of NMR spectroscopy to study biomolecular structure and dynamics, please refer to Chapters 4 and 5.

## 3.4 Infrared spectroscopy

Infrared spectroscopy is a type of absorption spectroscopy in which the molecules are irradiated with IR. When molecules absorb IR radiation, electronic transitions are not induced as IR does not have enough energy to cause transitions. However, it can induce vibrational motion in molecules that have a dipole moment (IR-active molecules). Vibrational motions are of two main types: stretching and bending. In stretching, the bond length is altered, whereas in bending, the angle between the bonds is altered. Since the energy involved in vibrational motions are dependent on factors like the bond length and the mass of the atoms, different bonds vibrate in different ways, and different energies are required to vibrate depending on the bonds. IR spectroscopy is used to identify certain functional groups since they have characteristic absorbances, for e.g., carbonyl (C=O) stretching vibrations absorb between frequencies of 1800 and 1600 $cm^{-1}$. The resulting infrared spectrum from IR spectroscopy provides information on various structural aspects like the mass of atoms (position of bands), change in dipole moment (strength of bands), and hydrogen bonding (width of bands). Most modern IR spectroscopy is carried out in the form of Fourier transform infrared (FTIR) spectroscopy. In FTIR, detector signals are related by a Fourier transformation to the measured spectrum. FTIR spectroscopy is a fast technique wherein the spectrum is collected in ~10 ms. IR spectroscopy is a powerful technique used for the dynamics and structural analysis of proteins [47–52]. It is also used in the study of the molecular mechanism of protein reactions [53–56], protein folding and unfolding [52,54,57,58], intermediate states of proteins [55,59,60], and their dynamics [61–63].

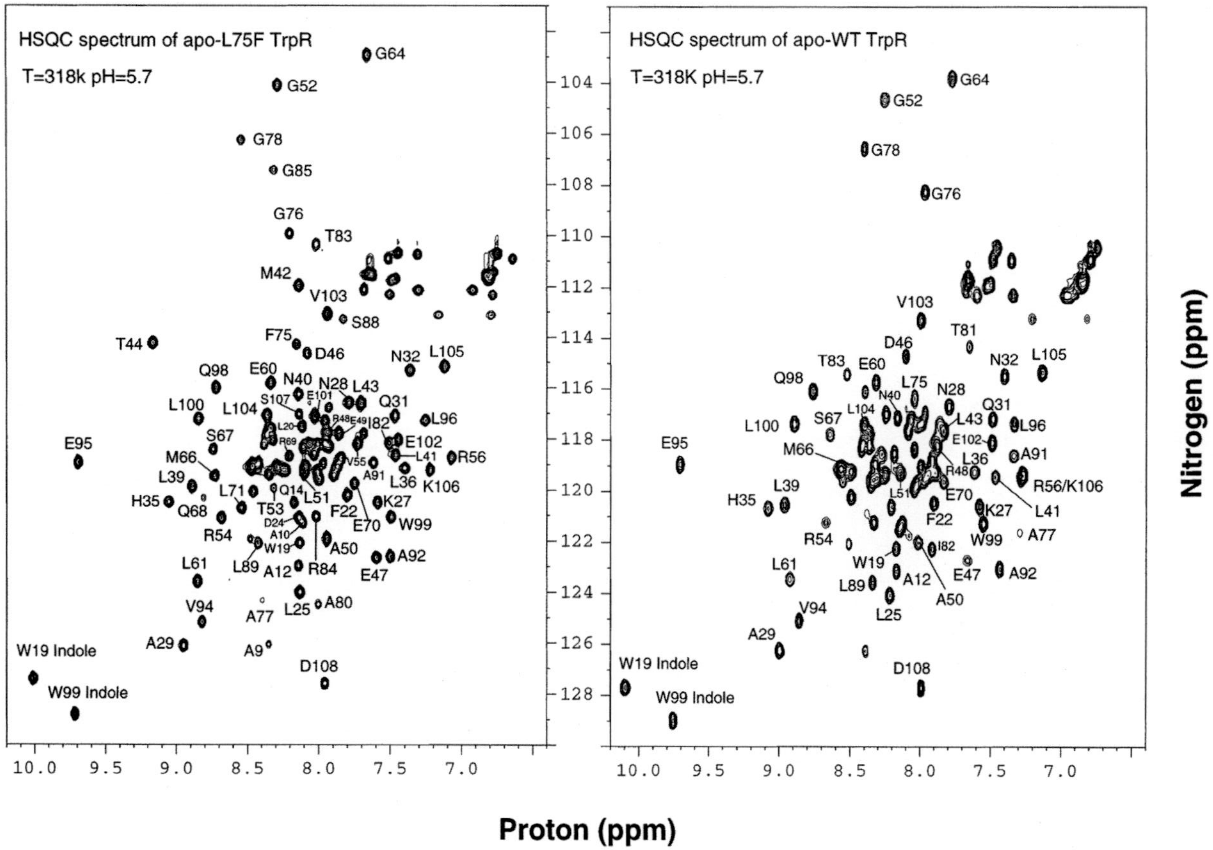


**Fig. 10** 2D 1H-15N HSQC spectra. Spectra of uniformly [15]N-labeled apo-L75F-TrpR and apo-WT-TrpR were recorded at 45°C in 500 mM NaCl, 50 mM $Na_2HPO_4$, and 90% $H_2O$/10% $D_2O$ at pH 5.7. *(Reproduced with permission from Tyler R, Pelczer I, Carey J, Copié V. Three-dimensional solution NMR structure of Apo-L75F-TrpR, a temperature-sensitive mutant of the tryptophan repressor protein. Biochemistry 2002;41 (40):11954–62.)*

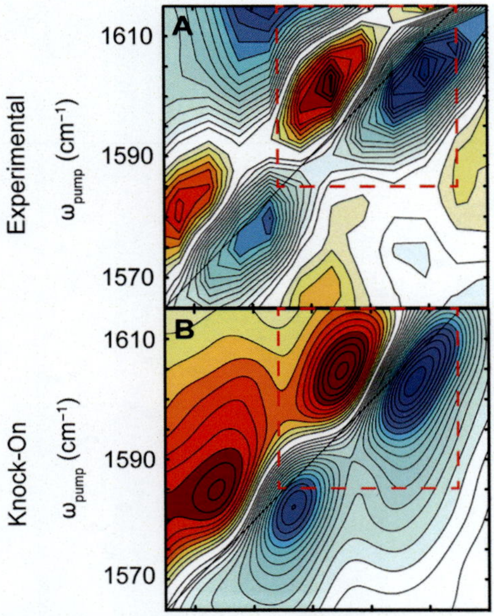


**Fig. 11** 2D IR spectrum. (A) The experimental 2D IR spectrum of KcsA. (B) The simulated 2D spectrum of the knock-on model. *(Reproduced with permission from Ghosh A, Ostrander JS, Zanni MT. Watching proteins wiggle: mapping structures with two-dimensional infrared spectroscopy. Chem Rev 2017;117(16):10726–59.)*

Recently, 2D IR spectroscopy was used to reveal ion configurations in the selectivity filter of KcsA, a potassium channel [64]. Kratochvil et al. used the technique to capture an instantaneous picture of the multiion configurations and structural distributions that are found at the selectivity filter. They found that the experimental IR spectrum pointed toward a "knock-on" model, in which the filter is occupied by two potassium ions simultaneously (Fig. 11A). The spectrum was also in agreement with the molecular dynamics (MD) simulations of the knock-on model (Fig. 11B). For a detailed discussion on the use of IR spectroscopy to study biomolecular structure and dynamics, please refer to Chapter 6.

## 3.5 Raman spectroscopy

Raman spectroscopy is also a form of vibrational spectroscopy, which is based on the inelastic scattering of monochromatic light. Inelastic scattering is a type of scattering in which there is a change in energy after light interacts with a sample. Such inelastic scattering is called Raman scattering. Upon interaction with matter, the scattered wavelength of light can be longer (Stokes Raman scattering), shorter (anti-Stokes Raman scattering), or remain the same (Rayleigh scattering). The difference in the energies of the incident photon of light and the scattered photon called the Raman shift corresponds to the energy difference between the vibrational energy levels. The Raman shifts or bands are visualized in the spectrum from which the vibrational modes of the sample molecule can be identified. The number of peaks, intensity, and wavelength provide information about the molecular bond vibrations, including individual and group bonds like C—C and the benzene ring.

Raman spectroscopy is also used in the study of the structure and dynamics of proteins. The difference between IR spectroscopy and Raman spectroscopy is that in IR spectroscopy, infrared energy over a range of frequencies is incident on the sample and the frequency which matches that of the vibration is absorbed and detected. In Raman spectroscopy, a single frequency of radiation is used, and the scattered radiation is detected. Raman spectroscopy is used to study protein secondary structures, determine side-chain conformations, and detect intermolecular interactions [65]. It is also used to study protein dynamics and conformational changes, such as in protein folding studies [66–72].

Due to different secondary structures and amino acids found in BSA and lysozyme, Raman spectroscopy was used to differentiate them [73]. The Raman spectrum of lysozyme displayed an amide I peak maximum at $1665\,cm^{-1}$, depicting the β-sheets found in lysozymes, whereas the BSA spectrum had a sharp amide III band at $1300\,cm^{-1}$, which occurred due to the high helix content of BSA (Fig. 12). The high tryptophan content of lysozymes was also depicted by the multiple sharp peaks for tryptophan. Similarly, the higher tyrosine content of BSA was visible through the intense tyrosine peaks at 830 and $850\,cm^{-1}$. For a detailed discussion on the use of Raman spectroscopy to study biomolecular structure and dynamics, please refer to Chapter 7.

## 3.6 Light scattering methods

Scattering of light takes place when light is deviated from its straight trajectory due to the presence of nonuniform particles in the medium. There are two main types of light scattering techniques, called dynamic light scattering (DLS) and static light scattering (SLS), which are used to determine parameters like molecular size, mass, and intermolecular interactions. SLS is used

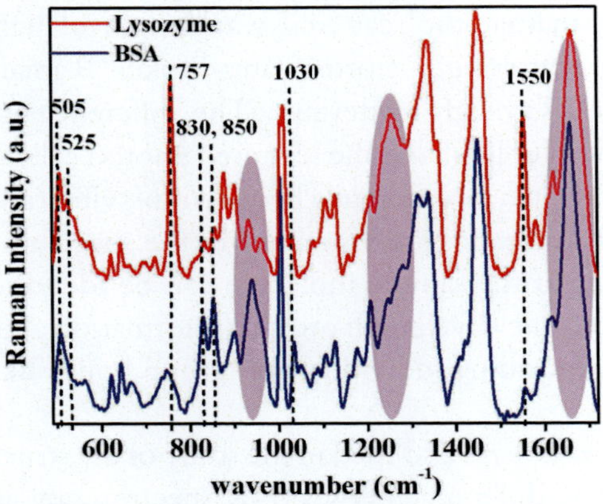


**Fig. 12** Raman spectra of BSA and lysozyme dissolved in water. *(Reproduced with permission from Kuhar N, Sil S, Umapathy S. Potential of Raman spectroscopic techniques to study proteins. Spectrochim Acta A Mol Biomol Spectrosc 2021;258:119712.)*

to determine the molecular weight of a compound using the Rayleigh theory, which states that larger molecules scatter more light than smaller molecules and that the intensity of the light which is scattered is directly proportional to the molecule's molecular weight. In DLS, the temporal fluctuations of the scattered light are analyzed at a known scattering angle $\theta$. These fluctuations are caused by the hydrodynamic motions of the sample. The Stokes radius, $R_S$, or the hydrodynamic radius is determined using this technique which is used to detect the expansion or compactness of proteins. SLS and DLS are usually carried out sequentially.

The applications of SLS and DLS include the determination of the mass of biomolecules [74,75], the study of unfolding and aggregation of proteins [76–78], intermolecular interactions [79–82], etc. Light scattering was used to determine the intrinsic disorder of late embryogenesis abundant (LEA) proteins in *Arabidopsis thaliana* [83]. The authors extended their previous studies on the cold-regulated proteins COR15A and COR15B to include proteins LEA11, LEA25, and the truncated version of LEA25, 2H, and 4H [83]. SLS and DLS experiments were used to determine the hydrodynamic radius and single-particle mass of the proteins in the solution. The apparent molecular mass ($M_{app}$) of the proteins was determined, and it was found that all native proteins were monomeric (Fig. 13). DLS determined the Stokes radii of all proteins and indicated that they were disordered since the value observed was relatively larger when compared to globular proteins (Fig. 14). For a detailed discussion on the use of light scattering techniques to study biomolecular structure and dynamics, please refer to Chapter 8.

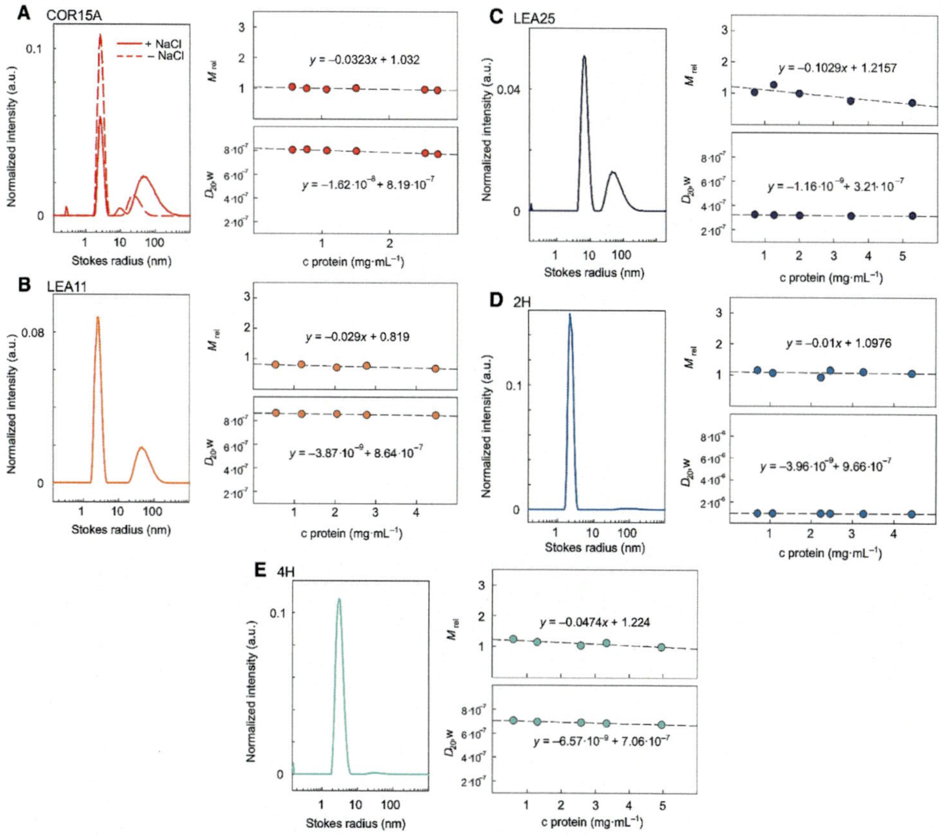


**Fig. 13** Size distribution of LEA proteins ($1\,g\,L^{-1}$) (left panels A–E). Concentration dependence is shown for the diffusion coefficients ($D_{20}$, *W*) for each protein (lower right panels A–E) and for relative particle masses (upper right panels A–E). *(Reproduced with permission from Bremer A, Wolff M, Thalhammer A, Hincha DK. Folding of intrinsically disordered plant LEA proteins is driven by glycerol-induced crowding and the presence of membranes. FEBS J 2017;284(6):919–36.)*

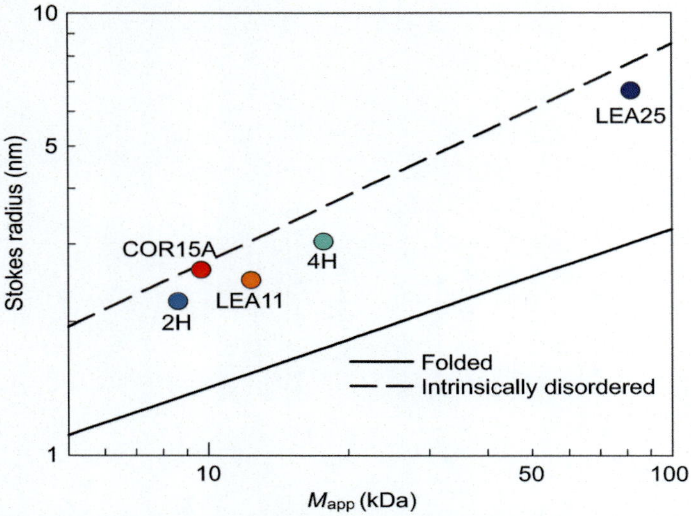


**Fig. 14** Compactness of the LEA proteins. The Stokes radius was calculated from diffusion coefficients extrapolated to $0\,g\,L^{-1}$ protein concentration determined by DLS measurements and was plotted against the $M_{app}$ determined from SLS measurements. For comparison, reference datasets including folded *(solid line)* and natively unfolded *(dashed line)* proteins have been used. *(Reproduced with permission from Bremer A, Wolff M, Thalhammer A, Hincha DK. Folding of intrinsically disordered plant LEA proteins is driven by glycerol-induced crowding and the presence of membranes. FEBS J 2017;284(6):919–36.)*

## 3.7 Mass spectrometry

Mass spectrometry is used to determine the molecular mass by measuring the mass-to-charge ratio ($m/z$) of ions in the gas phase. In mass spectrometry, firstly, the sample is converted to an ionized form in the gas phase. This is followed by the acceleration of these ions in an electric field where each ion emerges with a velocity that is proportional to its mass-to-charge ratio ($m/z$). The ions then pass into a field-free zone, and the time of arrival of the ions, i.e., the time-of-flight (TOF) is detected. The mass analyzers, where the ions are separated based on their $m/z$ ratio, are of different types, with the most common one being the TOF. This analyzer is based on the principle that ions having the same kinetic energy will vary in their velocity depending on the mass of the ions. The smaller ions reach the detector first, followed by the larger ions.

Mass spectrometry is generally used for protein identification [84,85], determination of the molecular mass, and analysis of posttranslational modification of proteins [86]. Mann et al. used mass spectrometry to detect SARS-CoV-2 variants [87]. A high-resolution MALDI spectrum was obtained for the doubly digested (trypsin + endoproteinase GluC) Spike protein (Fig. 15). MALDI or matrix-assisted laser desorption/ionization is a technique used for ionization before mass spectrometry. The technique uses a matrix mixed with a sample, which can absorb laser energy to create ions [88]. The 17 proteolytic peptides that were obtained in the spectrum constituted 20% of the S-protein. Most of the peptides represented completely cleaved products, with the six peptides highlighted in bold containing regions through which the five different variants of the virus (alpha, beta, delta, delta plus, and gamma) can be distinguished [87]. For a detailed discussion on the use of mass spectrometry to study biomolecular structure and dynamics, please refer to Chapter 9.

## 3.8 X-ray spectroscopy

X-ray spectroscopy is a term used for spectroscopic techniques that use X-ray excitation to study molecules. The principle of this technique is based on the fact that when an atom is irradiated with X-rays, electrons in the atom get excited and reach higher energy levels. When these electrons return to the lower energy levels, characteristic X-rays are emitted, which can be measured and used to study the spatial positions of atoms. Two main types of X-ray spectroscopic techniques are wavelength-dispersive X-ray spectroscopy (WDXS) and energy-dispersive X-ray spectroscopy (EDXS). In WDXS, the X-rays of a single wavelength are measured after being diffracted by a crystal, whereas in EDXS, the X-ray radiation emitted by electrons after being excited is measured. Both techniques are used to elucidate the atomic structure of macromolecules.

The most well-known X-ray spectroscopy method is X-ray crystallography, which is based on the principle of X-ray diffraction. As the name suggests, X-ray diffraction is the diffraction of X-ray beams on striking a crystal. The analysis of the angles and intensities of these diffracted beams allows for the determination of the 3D structure of biomolecules such as proteins. Bragg's law is the underlying principle of X-ray crystallography and is given by the equation:

$$n\lambda = 2d \sin \theta$$

where $n$ is an integer, $\lambda$ is the wavelength of X-ray, $d$ is the spacing between planes of atoms, and $\theta$ is the angle of reflection between the incident ray and the scattered planes.

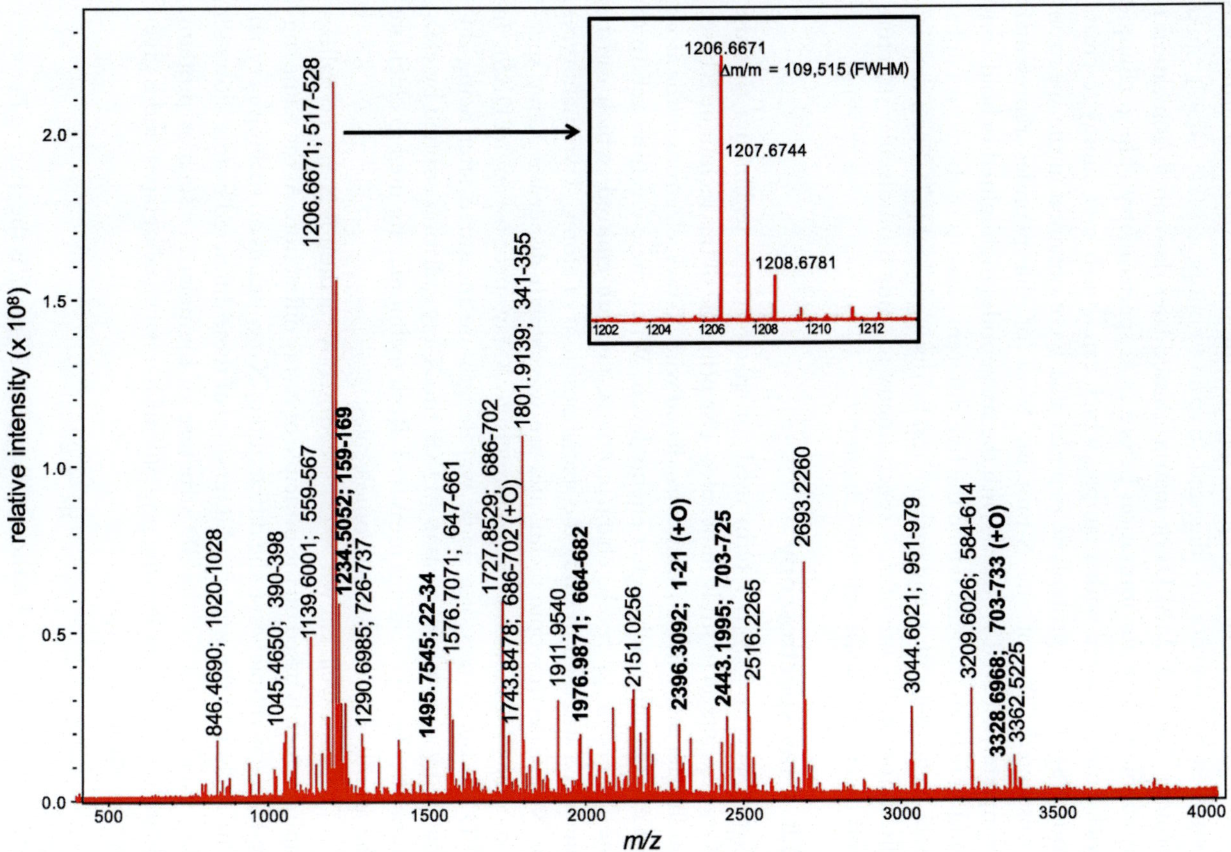


**Fig. 15** High-resolution MALDI mass spectra for the doubly digested (trypsin + GluC) S-protein obtained from a laboratory-grown virus. Peaks highlighted in *bold* represent regions containing mutations in the five variants of the virus. *(Reproduced with permission from Mann C, Griffin JH, Downard KM. Detection and evolution of SARS-CoV-2 coronavirus variants of concern with mass spectrometry. Anal Bioanal Chem 2021;413 (29):7241–9.)*

When a beam of monochromatic X-rays strikes a crystal, constructive interference takes place, which diffracts the rays at specific angles from each lattice plane of the crystal. The diffracted X-rays, which are detected by the detector, form peaks that provide information about various structural parameters of the sample. X-ray crystallography is one of the best ways to determine the 3D structures of biomolecules, and it is now also increasingly used for drug discovery and drug design [89–92]. The knowledge of the structures of protein targets can significantly enhance the speed of drug discovery.

Zhang et al. [93] used X-ray crystallography in combination with NMR to show that GNF-2, a selective allosteric inhibitor of oncogenic Bcr-Abl protein, binds to the myristate-binding site of Abl (Fig. 16). The authors found that GNF-2 was bound to the myristate pocket in an extended conformation, with the $CF_3$ group buried at the same depth as the last two carbon atoms of myristate are. They also found weak favorable interactions of GNF-2 with the protein, with most interactions being hydrophobic. Three

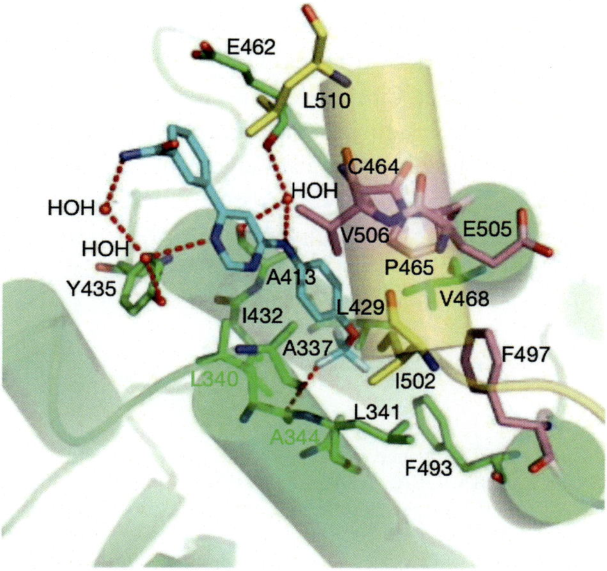


**Fig. 16** Crystal structure of GNF-2 bound to the Abl myristoyl pocket. The Abl kinase is shown in *green*, the GNF-2 resistance mutations in *pink*, and GNF-2 carbons in *cyan*. The *dotted lines* represent H-bonding and other polar interactions. *(Reproduced with permission from Zhang J, Adrián FJ, Jahnke W, Cowan-Jacob SW, Li AG, Iacob RE, et al. Targeting Bcr–Abl by combining allosteric with ATP-binding-site inhibitors. Nature 2010;463(7280):501-6.)*

mutations created near the myristate binding site were found to cause GNF-2 binding resistance. For a detailed discussion on the use of X-ray spectroscopy to study biomolecular structure and dynamics, please refer to Chapters 10 and 11.

## 4. Conclusions

Spectroscopy involves the use of electromagnetic radiation and its properties to elucidate information about any sample. Spectroscopic methods present a wide range of techniques that exploit different properties of electromagnetic radiation to provide specific details about biomolecules. This introductory chapter delved into the fundamentals of spectroscopy and how it is utilized to study the structure and dynamics of biomolecular structures, particularly proteins. Proteins are one of the most complex and crucial biomolecules found in the living system, having a variety of structures and many varied functions. The study and understanding of the structural and conformational dynamics and functions of proteins are areas of research that still have much to offer. With the recent spike in the knowledge about intrinsically disordered proteins, which defy the previous understanding of protein structure-function, the need for research into this area has increased. As discussed, there are many techniques that fall under the broad term of spectroscopy. Most of these techniques are relatively simple to use, and they can be used cooperatively to obtain different information about the same sample. Thus a basic understanding of the fundamentals of spectroscopy, the principle, instrumentation, and its applications is an essential prerequisite for conducting biomolecular research.

## References

[1] Tripathi T, Dubey VK. Advances in protein molecular and structural biology methods. 1st ed. Cambridge, MA, USA: Academic Press; 2022. p. 1–714.
[2] Singh DB, Tripathi T. Frontiers in protein structure, function, and dynamics. Singapore Springer Nature; 2020. p. 1–458.
[3] Karplus M, Weaver DL. Protein folding dynamics: the diffusion-collision model and experimental data. Protein Sci 1994;3(4):650–68.
[4] Dill KA, Fiebig KM, Chan HS. Cooperativity in protein-folding kinetics. Proc Natl Acad Sci U S A 1993;90(5):1942–6.
[5] Abkevich VI, Gutin AM, Shakhnovich EI. Specific nucleus as the transition state for protein folding: evidence from the lattice model. Biochemistry 1994;33(33):10026–36.
[6] Fersht AR. Optimization of rates of protein folding: the nucleation-condensation mechanism and its implications. Proc Natl Acad Sci U S A 1995;92(24):10869–73.

[7] Nölting B, Golbik R, Neira JL, Soler-Gonzalez AS, Schreiber G, Fersht AR. The folding pathway of a protein at high resolution from microseconds to seconds. Proc Natl Acad Sci U S A 1997;94(3):826–30.
[8] Levinthal C. Are there pathways for protein folding? J Chim Phys 1968;65:44–5.
[9] Finkelstein AV, Badretdinov A. Rate of protein folding near the point of thermodynamic equilibrium between the coil and the most stable chain fold. Fold Des 1997; 2(2):115–21.
[10] Koshland DE. Application of a theory of enzyme specificity to protein synthesis. Proc Natl Acad Sci U S A 1958;44(2):98–104.
[11] Tripathi T. Calculation of thermodynamic parameters of protein unfolding using far-ultraviolet circular dichroism. J Proteins Proteomics 2013;4(2):85–91.
[12] Dunker AK, Lawson JD, Brown CJ, Williams RM, Romero P, Oh JS, et al. Intrinsically disordered protein. J Mol Graph Model 2001;19(1):26–59.
[13] Tompa P. Intrinsically unstructured proteins. Trends Biochem Sci 2002; 27(10):527–33.
[14] Ward JJ, Sodhi JS, McGuffin LJ, Buxton BF, Jones DT. Prediction and functional analysis of native disorder in proteins from the three kingdoms of life. J Mol Biol 2004; 337(3):635–45.
[15] Dunker AK, Brown CJ, Lawson JD, Iakoucheva LM, Obradović Z. Intrinsic disorder and protein function. Biochemistry 2002;41(21):6573–82.
[16] Eisenmesser EZ, Millet O, Labeikovsky W, Korzhnev DM, Wolf-Watz M, Bosco DA, et al. Intrinsic dynamics of an enzyme underlies catalysis. Nature 2005;438 (7064):117–21.
[17] Agarwal PK. Role of protein dynamics in reaction rate enhancement by enzymes. J Am Chem Soc 2005;127(43):15248–56.
[18] Artymiuk PJ, Blake CCF, Grace DEP, Oatley SJ, Phillips DC, Sternberg MJE. Crystallographic studies of the dynamic properties of lysozyme. Nature 1979;280 (5723):563–8.
[19] Austin RH, Beeson KW, Eisenstein L, Frauenfelder H, Gunsalus IC. Dynamics of ligand binding to myoglobin. Biochemistry 1975;14(24):5355–73.
[20] Frauenfelder H, Petsko GA, Tsernoglou D. Temperature-dependent X-ray diffraction as a probe of protein structural dynamics. Nature 1979;280(5723):558–63.
[21] Krittanai C, Johnson Jr WC. The relative order of helical propensity of amino acids changes with solvent environment. Proteins 2000;39(2):132–41.
[22] Macdonald JR, Johnson Jr WC. Environmental features are important in determining protein secondary structure. Protein Sci 2001;10(6):1172–7.
[23] O'Brien EP, Brooks BR, Thirumalai D. Effects of pH on proteins: predictions for ensemble and single-molecule pulling experiments. J Am Chem Soc 2012;134(2): 979–87.
[24] Di Russo NV, Estrin DA, Martí MA, Roitberg AE. pH-dependent conformational changes in proteins and their effect on experimental pKas: the case of Nitrophorin 4. PLoS Comput Biol 2012;8(11), e1002761.
[25] Tilton RF, Dewan JC, Petsko GA. Effects of temperature on protein structure and dynamics: x-ray crystallographic studies of the protein ribonuclease-A at nine different temperatures from 98 to 320K. Biochemistry 1992;31(9):2469–81.
[26] Mehra R, Dehury B, Kepp KP. Cryo-temperature effects on membrane protein structure and dynamics. Phys Chem Chem Phys 2020;22(10):5427–38.
[27] Miyashita Y, Ohmae E, Nakasone K, Katayanagi K. Effects of salt on the structure, stability, and function of a halophilic dihydrofolate reductase from a hyperhalophilic archaeon, Haloarcula japonica strain TR-1. Extremophiles 2015;19(2):479–93.
[28] Mao Y-J, Sheng X-R, Pan X-M. The effects of NaCl concentration and pH on the stability of hyperthermophilic protein Ssh10b. BMC Biochem 2007;8(1):28.

[29] Sasidharan S, Nag N, Tripathi T, Saudagar P. Experimental methods to study the thermodynamics of protein–protein interactions. In: Tripathi T, Dubey VK, editors. Advances in protein molecular and structural biology methods. USA: Academic Press; 2022. p. 103–14.
[30] Sarkar M, Li C, Pielak GJ. Soft interactions and crowding. Biophys Rev 2013; 5(2):187–94.
[31] Hawe A, Sutter M, Jiskoot W. Extrinsic fluorescent dyes as tools for protein characterization. Pharm Res 2008;25(7):1487–99.
[32] Acharya P, Rao NM. Stability studies on a lipase from Bacillus subtilis in guanidinium chloride. J Protein Chem 2003;22(1):51–60.
[33] Goto Y, Fink AL. Conformational states of beta-lactamase: molten-globule states at acidic and alkaline pH with high salt. Biochemistry 1989;28(3):945–52.
[34] Lerner E, Cordes T, Ingargiola A, Alhadid Y, Chung S, Michalet X, et al. Toward dynamic structural biology: two decades of single-molecule Förster resonance energy transfer. Science 2018;359(6373):eaan1133.
[35] Lipman EA, Schuler B, Bakajin O, Eaton WA. Single-molecule measurement of protein folding kinetics. Science 2003;301(5637):1233–5.
[36] Margittai M, Widengren J, Schweinberger E, Schröder GF, Felekyan S, Haustein E, et al. Single-molecule fluorescence resonance energy transfer reveals a dynamic equilibrium between closed and open conformations of syntaxin 1. Proc Natl Acad Sci U S A 2003;100(26):15516–21.
[37] Mazal H, Haran G. Single-molecule FRET methods to study the dynamics of proteins at work. Curr Opin Biomed Eng 2019;12:8–17.
[38] Shi Z, Chen K, Liu Z, Kallenbach NR. Conformation of the backbone in unfolded proteins. Chem Rev 2006;106(5):1877–97.
[39] Ebbinghaus S, Dhar A, McDonald JD, Gruebele M. Protein folding stability and dynamics imaged in a living cell. Nat Methods 2010;7(4):319–23.
[40] Gruebele M. Chapter 6 Fast protein folding. In: Protein folding, misfolding and aggregation: classical themes and novel approaches. The Royal Society of Chemistry; 2008. p. 106–38.
[41] Suryawanshi VD, Walekar LS, Gore AH, Anbhule PV, Kolekar GB. Spectroscopic analysis on the binding interaction of biologically active pyrimidine derivative with bovine serum albumin. J Pharm Anal 2016;6(1):56–63.
[42] Cavalli A, Salvatella X, Dobson CM, Vendruscolo M. Protein structure determination from NMR chemical shifts. Proc Natl Acad Sci U S A 2007;104(23):9615.
[43] Purslow JA, Khatiwada B, Bayro MJ, Venditti V. NMR methods for structural characterization of protein-protein complexes. Front Mol Biosci 2020;7:9. https://doi.org/10.3389/fmolb.2020.00009.
[44] Hu Y, Cheng K, He L, Zhang X, Jiang B, Jiang L, et al. NMR-based methods for protein analysis. Anal Chem 2021;93(4):1866–79.
[45] Zhuravleva A, Korzhnev DM. Protein folding by NMR. Prog Nucl Magn Reson Spectrosc 2017;100:52–77.
[46] Tyler R, Pelczer I, Carey J, Copié V. Three-dimensional solution NMR structure of Apo-L75F-TrpR, a temperature-sensitive mutant of the tryptophan repressor protein. Biochemistry 2002;41(40):11954–62.
[47] Ghosh A, Ostrander JS, Zanni MT. Watching proteins wiggle: mapping structures with two-dimensional infrared spectroscopy. Chem Rev 2017;117(16):10726–59.
[48] Kottke T, Lórenz-Fonfría VA, Heberle J. The grateful infrared: sequential protein structural changes resolved by infrared difference spectroscopy. J Phys Chem B 2017;121(2):335–50.
[49] Barth A. Infrared spectroscopy of proteins. Biochim Biophys Acta 2007;1767(9): 1073–101.

[50] Tamm LK, Tatulian SA. Infrared spectroscopy of proteins and peptides in lipid bilayers. Q Rev Biophys 1997;30(4):365–429.
[51] Arkin IT. Isotope-edited IR spectroscopy for the study of membrane proteins. Curr Opin Chem Biol 2006;10(5):394–401.
[52] Arrondo JLR, Muga A, Castresana J, Goñi FM. Quantitative studies of the structure of proteins in solution by Fourier-transform infrared spectroscopy. Prog Biophys Mol Biol 1993;59(1):23–56.
[53] Siebert F. [20] Infrared spectroscopy applied to biochemical and biological problems. In: Methods in enzymology, vol. 246. Academic Press; 1995. p. 501–26.
[54] Fabian H, Mäntele W. Infrared spectroscopy of proteins. In: Handbook of vibrational spectroscopy. John Wiley & Sons, Ltd; 2006.
[55] Vogel R, Siebert F. Vibrational spectroscopy as a tool for probing protein function. Curr Opin Chem Biol 2000;4(5):518–23.
[56] Dioumaev A. Infrared methods for monitoring the protonation state of carboxylic amino acids in the photocycle of bacteriorhodopsin. Biochemistry (Moscow) 2001;66(11):1269–76.
[57] Fabian H, Mantsch HH, Schultz CP. Two-dimensional IR correlation spectroscopy: sequential events in the unfolding process of the λ Cro-V55C repressor protein. Proc Natl Acad Sci U S A 1999;96(23):13153–8.
[58] Dyer RB, Gai F, Woodruff WH, Gilmanshin R, Callender RH. Infrared studies of fast events in protein folding. Acc Chem Res 1998;31(11):709–16.
[59] Mäntele W. Reaction-induced infrared difference spectroscopy for the study of protein function and reaction mechanisms. Trends Biochem Sci 1993;18(6):197–202.
[60] Braiman MS, Rothschild KJ. Fourier transform infrared techniques for probing membrane protein structure. Annu Rev Biophys Biophys Chem 1988;17:541–70.
[61] Chalmers JM, Griffiths PR. Handbook of vibrational spectroscopy. J. Wiley; 2002.
[62] Radu I, Schleeger M, Bolwien C, Heberle J. Time-resolved methods in biophysics. 10. Time-resolved FT-IR difference spectroscopy and the application to membrane proteins. Photochem Photobiol Sci 2009;8(11):1517–28.
[63] Noguchi T. Fourier transform infrared difference and time-resolved infrared detection of the electron and proton transfer dynamics in photosynthetic water oxidation. Biochim Biophys Acta 2015;1847(1):35–45.
[64] Kratochvil HT, Carr JK, Matulef K, Annen AW, Li H, Maj M, et al. Instantaneous ion configurations in the K+ ion channel selectivity filter revealed by 2D IR spectroscopy. Science (New York, NY) 2016;353(6303):1040–4.
[65] Benevides JM, Overman SA, Thomas Jr GJ. Raman spectroscopy of proteins. Curr Protoc Protein Sci 2004;Chapter 17:1–35. Unit 17.8.
[66] Kitagawa T. Investigation of higher order structures of proteins by ultraviolet resonance Raman spectroscopy. Prog Biophys Mol Biol 1992;58(1):1–18.
[67] Asher SA. UV resonance Raman studies of molecular structure and dynamics: applications in physical and biophysical chemistry. Annu Rev Phys Chem 1988;39:537–88.
[68] Lednev IK, Karnoup AS, Sparrow MC, Asher SA. α-Helix peptide folding and unfolding activation barriers: a nanosecond UV resonance Raman study. J Am Chem Soc 1999;121(35):8074–86.
[69] Song S, Asher SA. UV resonance Raman studies of peptide conformation in poly (L-lysine), poly (L-glutamic acid), and model complexes: the basis for protein secondary structure determinations. J Am Chem Soc 1989;111(12):4295–305.
[70] Chi Z, Chen XG, Holtz JS, Asher SA. UV resonance Raman-selective amide vibrational enhancement: quantitative methodology for determining protein secondary structure. Biochemistry 1998;37(9):2854–64.

[71] Balakrishnan G, Hu Y, Case MA, Spiro TG. Microsecond melting of a folding intermediate in a coiled-coil peptide, monitored by T-jump/UV Raman spectroscopy. J Phys Chem B 2006;110(40):19877–83.
[72] Spiro TG, Smulevich G, Su C. Probing protein structure and dynamics with resonance Raman spectroscopy: cytochrome c peroxidase and hemoglobin. Biochemistry 1990; 29(19):4497–508.
[73] Kuhar N, Sil S, Umapathy S. Potential of Raman spectroscopic techniques to study proteins. Spectrochim Acta A Mol Biomol Spectrosc 2021;258, 119712.
[74] Saio T, Guan X, Rossi P, Economou A, Kalodimos CG. Structural basis for protein antiaggregation activity of the trigger factor chaperone. Science 2014;344 (6184):1250494.
[75] Müller R, Gräwert MA, Kern T, Madl T, Peschek J, Sattler M, et al. High-resolution structures of the IgM Fc domains reveal principles of its hexamer formation. Proc Natl Acad Sci U S A 2013;110(25):10183–8.
[76] Yu Z, Reid JC, Yang YP. Utilizing dynamic light scattering as a process analytical technology for protein folding and aggregation monitoring in vaccine manufacturing. J Pharm Sci 2013;102(12):4284–90.
[77] Lewis EN, Qi W, Kidder LH, Amin S, Kenyon SM, Blake S. Combined dynamic light scattering and Raman spectroscopy approach for characterizing the aggregation of therapeutic proteins. Molecules 2014;19(12):20888–905.
[78] Mohr BG, Dobson CM, Garman SC, Muthukumar M. Electrostatic origin of in vitro aggregation of human γ-crystallin. J Chem Phys 2013;139(12), 121914.
[79] Wu D, Minton AP. Quantitative characterization of the interaction between sucrose and native proteins via static light scattering. J Phys Chem B 2013;117(1):111–7.
[80] Roberts D, Keeling R, Tracka M, van der Walle CF, Uddin S, Warwicker J, et al. The role of electrostatics in protein-protein interactions of a monoclonal antibody. Mol Pharm 2014;11(7):2475–89.
[81] Li W, Persson BA, Morin M, Behrens MA, Lund M, Zackrisson Oskolkova M. - Charge-induced patchy attractions between proteins. J Phys Chem B 2015;119 (2):503–8.
[82] Blanco MA, Perevozchikova T, Martorana V, Manno M, Roberts CJ. Protein-protein interactions in dilute to concentrated solutions: α-chymotrypsinogen in acidic conditions. J Phys Chem B 2014;118(22):5817–31.
[83] Bremer A, Wolff M, Thalhammer A, Hincha DK. Folding of intrinsically disordered plant LEA proteins is driven by glycerol-induced crowding and the presence of membranes. FEBS J 2017;284(6):919–36.
[84] Hernandez P, Müller M, Appel RD. Automated protein identification by tandem mass spectrometry: issues and strategies. Mass Spectrom Rev 2006;25(2):235–54.
[85] Noor Z, Ahn SB, Baker MS, Ranganathan S, Mohamedali A. Mass spectrometry-based protein identification in proteomics—a review. Brief Bioinform 2021;22(2):1620–38.
[86] Salzano AM, Crescenzi M. Mass spectrometry for protein identification and the study of post translational modifications. Ann Ist Super Sanita 2005;41(4):443–50.
[87] Mann C, Griffin JH, Downard KM. Detection and evolution of SARS-CoV-2 coronavirus variants of concern with mass spectrometry. Anal Bioanal Chem 2021;413 (29):7241–9.
[88] Hillenkamp F, Karas M, Beavis RC, Chait BT. Matrix-assisted laser desorption/ionization mass spectrometry of biopolymers. Anal Chem 1991;63(24):1193a–203a.
[89] Brader ML, Baker EN, Dunn MF, Laue TM, Carpenter JF. Using X-ray crystallography to simplify and accelerate biologics drug development. J Pharm Sci 2017;106 (2):477–94.

[90] Ivanova J, Leitans J, Tanc M, Kazaks A, Zalubovskis R, Supuran CT, et al. X-ray crystallography-promoted drug design of carbonic anhydrase inhibitors. Chem Commun (Camb) 2015;51(33):7108–11.
[91] Blundell TL, Patel S. High-throughput X-ray crystallography for drug discovery. Curr Opin Pharmacol 2004;4(5):490–6.
[92] Maveyraud L, Mourey L. Protein X-ray crystallography and drug discovery. Molecules 2020;25(5):1030.
[93] Zhang J, Adrián FJ, Jahnke W, Cowan-Jacob SW, Li AG, Iacob RE, et al. Targeting Bcr–Abl by combining allosteric with ATP-binding-site inhibitors. Nature 2010;463 (7280):501–6.

# CHAPTER 2

# Fluorescence-based techniques to assess biomolecular structure and dynamics

**Jakub Sławski and Joanna Grzyb**
Department of Biophysics, Faculty of Biotechnology, University of Wrocław, Wroclaw, Poland

## 1. Introduction

Since the seminal determination of the myoglobin structure by John Kendrew in the late 1950s [1], X-ray crystallography along with nuclear magnetic resonance (NMR) spectroscopy and cryo-electron microscopy (cryo-EM) emerged as primary methods for structural biologists [2]. Although indispensable for the determination of complete structures of entire protein molecules, these techniques are bound by limitations such as a need for a large amount of protein, diffraction grade crystals for X-ray crystallography, low molecular weight limit for NMR, and sample preparation issues with cryo-EM [3,4]. Furthermore, they share one common drawback—they are predominantly restricted to static or time-averaged structures, usually in nonnative or nonphysiological conditions [5]. With its relative ease of use, fluorescence spectroscopy represents an important complementary method and sometimes reaches beyond standard structural techniques. It can be used to study the structural features of biomolecules and the kinetics of processes they are involved in. In the current chapter, we focus on proteins; however, it must be remembered that the assays may also be adapted to other biomolecules. Fluorescence helps track the localization and determine the kinetics of interaction between two or more biomolecules, in vitro or in vivo. For the dynamics studies, the calculation of the equilibrium dissociation constant ($K_d$) comes in handy. It may be obtained from an experimental dose-response curve based on the change in concentration of reactants ($A$, $B$) and a complex ($AB$):

$$xA + yB = A_xB_y \tag{1}$$


https://doi.org/10.1016/B978-0-323-99127-8.00007-6

$$K_d = \frac{[A]^x[B]^y}{[A_xB_y]} \tag{2}$$

Some experiments exploit the binding kinetics in more detail, which makes the calculation of the on-rate ($k_{on}$) and the off-rate ($k_{off}$) also possible. $K_d$ can then be obtained from the ratio of $k_{on}$ and $k_{off}$. The concentration of reactants may also be deduced based on fluorescence changes (discussed in Section 6).

For fluorescence examination, the biomolecule needs to have fluorescent properties or in other words be a fluorophore. The origin of fluorescence is schematically presented in the Jablonski diagram (Fig. 1). Perhaps the most noticeable feature of a fluorescent dye is the emission spectrum, typically of longer wavelengths than excitation (the so-called Stokes shift), which is a direct manifestation of energy loss due to transitions between electron vibrational energy levels [6]. The efficiency of the fluorescence process is characterized by the quantum yield, $\Phi$:

$$\Phi = \frac{\#\text{photons emitted}}{\#\text{photons absorbed}} = \frac{k_{em}}{k_{em} + \sum k_{nr}} \tag{3}$$

where $k_{em}$ is a rate constant of the emission process and $\sum k_{nr}$ is the sum of rate constants on nonradiative relaxation processes. The value of $\Phi$ varies significantly between fluorophores, such as it is 0.12 for tryptophan and 0.92 for AlexaFluor 488. Additionally, every fluorophore has its specific

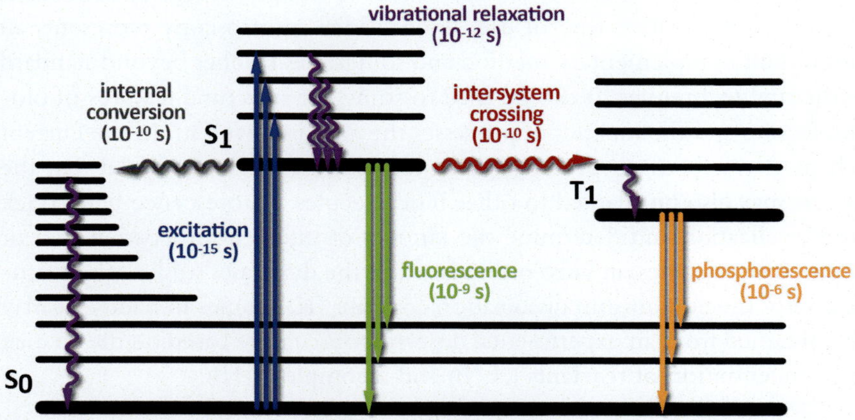


**Fig. 1** Jablonski diagram for basic fluorescence-related processes. Radiative *(straight lines)* and nonradiative *(waved lines)* ways of relaxation are shown with typical time ranges.

lifetime, $\tau$, which defines the average time a molecule spends in the excited state before emission.

Fluorescence polarization results from preferential absorption and emission of photons whose electric vectors have a defined orientation about the molecular axis of the absorbing/emitting molecule [7]. The polarization of fluorescence emission may be interchangeably expressed by fluorescence anisotropy, $r$, or fluorescence polarization, $P$:

$$r = \frac{I_{\parallel} - I_{\perp}}{I_{\parallel} + 2I_{\perp}} \qquad P = \frac{I_{\parallel} - I_{\perp}}{I_{\parallel} + I_{\perp}} \tag{4}$$

where $I_{\parallel}$ and $I_{\perp}$ are fluorescence intensities of the vertically and horizontally polarized emission, respectively, when excitation light is polarized vertically.

## 1.1 Types of fluorescence measurements

There are two general types of fluorescence measurements: steady-state and time-resolved. In steady-state fluorometry, excitation of a fluorophore is achieved by a constant illumination, and time-averaged emission is recorded as a function of wavelength. In time-resolved fluorometry, the sample is excited by a pulse-light source (primarily laser or light-emitting diode, LED), with the width of the pulse much shorter than the emission $\tau$. Following the excitation pulse, emission intensity is measured as a function of time and recorded as time-dependent intensity decay. Most of the present time-resolved measurements are performed using time-correlated single-photon counting (TCSPC) using photomultiplier tubes (PMTs) as detectors [8,9]. Intensity decay is usually described by the single- or multiexponential decay law:

$$I(t) = \sum\nolimits_{i=1}^{n} A_i e^{-t/\tau_i} \tag{5}$$

where intensity in the function of time $I(t)$ is a sum of $n$ exponential decays characterized by the amplitude $A_i$ (intensity in $t=0$) and lifetime $\tau_i$ (for single-exponential decay, $n=1$). Individual components of multiexponential decay are interpreted as a possibility of several emission processes with different lifetimes within one molecule (e.g., simultaneous emission from different conformational states [10]) or as a presence of fluorophore molecules in several different environments in the sample (e.g., regions of different polarity within protein molecule [11]). Alternatively, for the multiexponential decay, one can expect a continuous lifetime distribution rather than a limited number of discrete lifetime values. Examples include

Förster resonance energy transfer (FRET)-triggered donor's lifetime decrease in a flexible donor-acceptor system [12] or distribution of fluorophore molecules in a continuously changing environment [13]. In such cases, decay description has an integral form:

$$I(t) = \int_{\tau=0}^{\infty} \alpha(\tau) e^{-t/\tau}\, d\tau \tag{6}$$

where $\int \alpha(\tau) d\tau = 1$ and $\alpha(\tau)$ is the function describing the shape of the distribution, usually arbitrarily supposed as Gaussian or Lorentzian distribution [14,15].

Fitting algorithms used for extraction of lifetime parameters from emission decays include nonlinear least squares (NLLS, used widely for discrete lifetime analysis [8]) and the maximum entropy method (MEM, used for the analysis of lifetime distribution [12,13]).

## 1.2 Fluorophores

Besides temporal resolution of the fluorescence signal, most applications of fluorescence spectroscopy for protein structural studies require strictly determined localization of fluorophore within protein structure or sequence. In analogy to site-directed mutagenesis, defined as site-directed fluorescence [16], this approach may be realized by site-specific mutagenesis and specific labeling. The fluorescent toolbox used in the structural studies of proteins includes:

1. **Intrinsic tryptophan residues:** With extinction coefficient and quantum yield higher than other aromatic amino acids (tyrosine and phenylalanine), tryptophan is a valuable intrinsic protein fluorophore [6,17]. Single-tryptophan proteins are preferred to avoid complex or even insolvable analysis, although analysis of multiple-tryptophan protein fluorescence is also possible, e.g., by fractional quenching or time-resolved measurements [11,18].
2. **Fluorescent amino acid analogues:** They include tryptophan analogues or fluorophores chemically coupled to amino acids and exhibiting desirable brightness, emission spectrum, environment sensitivity, and other fluorescent properties [19,20]. These nonnatural amino acids may be incorporated into the peptide chain by solid-phase peptide synthesis or translational introduction via orthogonal tRNA/aminoacyl-tRNA synthetase pairs [21,22].

3. **Extrinsic fluorophores, attached covalently or noncovalently, with high affinity:** The method of choice in most structural studies of proteins is covalent labeling of cysteine residues mainly due to the relatively low abundance of cysteine in proteins and flexibility of thiol chemistry [21,23]. Targeting amino groups is not only less specific but is also used frequently. The range of fluorophores to be attached by this method is vast, varying from small molecules such as cyanine dyes to nanoparticles, like quantum dots. Other approaches include genetically encoded fluorescent tags (e.g., green fluorescent protein, GFP) or self-labeling enzymatic tags (e.g., HaloTag, SNAP-tag [24,25]); however, they may be of less use because of the large size of fluorescent groups.
4. **Intrinsic fluorescent cofactors:** They are less frequently used in structural studies, examples include chromophores of GFP-related and other fluorescent proteins [10], phycobilins [26], and porphyrin derivatives [27,28].

A particular fluorophore is chosen following specific requirements of apparatus, namely, available excitation source and detection range. Companies often provide interactive online fluorophore selection tools, providing excitation and emission spectra (e.g., see www.aatbio.com or www.thermofisher.com). The chemical structures of some commonly used fluorophores are shown in Fig. 2.

## 2. Environmental effects used in fluorescence structural studies

Protein studies frequently take advantage of the influence of polarity of the surrounding environment on the emission of a fluorophore. This effect is manifested mainly as a polarity-dependent emission spectrum shift accompanied by $\tau$ change. Excitation of fluorophore molecule typically increases its electric dipole moment ($\mu$), followed by reorientation of solvent molecule dipoles [29,30], thereby lowering the energy of fluorophore's excited state and consequently the energy of the emitted photon. In the subsequent subsections, we discuss the examples of environmental effects used for protein structure studies.

### 2.1 Environment-dependent shifts of fluorescence parameters

The dependence of basic fluorescence parameters on environment polarity was beautifully demonstrated by taking a simple example of Tim23 protein. Tim23 is composed of four transmembrane helices (TMS1–TMS4), the

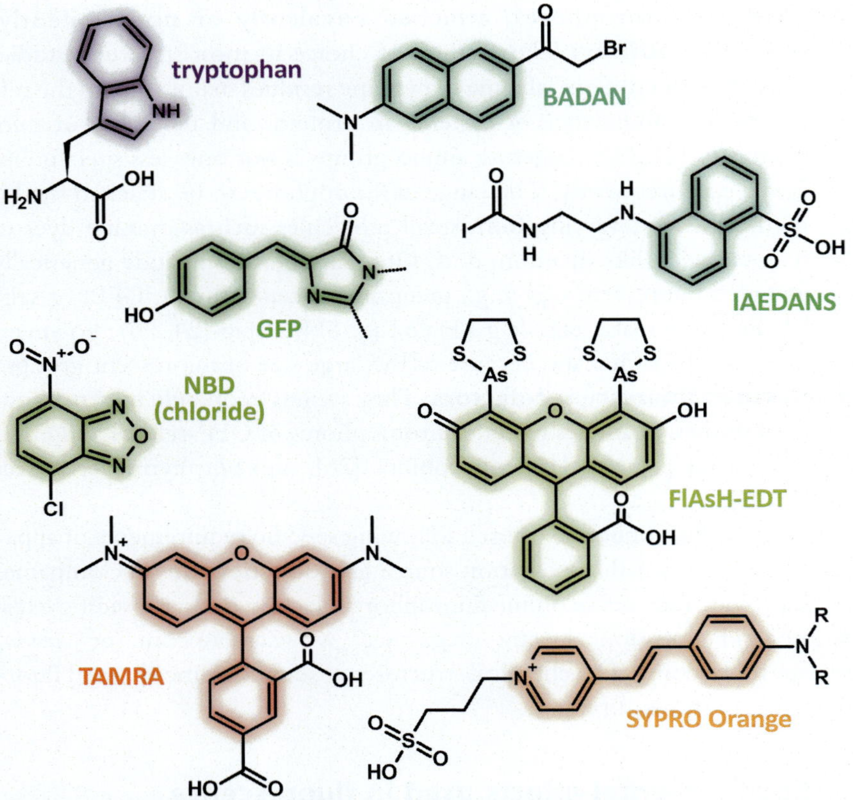


**Fig. 2** Chemical structure of examples of fluorophores used in protein structural studies. *BADAN*, 6-bromo-acetyl-2-(dimethylamino)-naphthalene; *FlAsH*, fluorescein arsenical hairpin (in ethanedithiol-chelated form); *GFP*, green fluorescence protein (chromophore shown); *IAEDANS*, 5-((((2-iodoacetyl)amino)ethyl)amino) naphthalene-1-sulfonic acid; *NBD*, 7-nitrobenz-2-oxa-1,3-diazolyl; *TAMRA*, 5-carboxytetramethylrhodamine.

central component of the TIM23 complex, importing proteins through the inner mitochondrial membrane [31]. To evaluate its amphipathic character, cysteine residues were introduced in different positions of the helix [32] by incorporating cysteine labeled using a 7-nitrobenz-2-oxa-1,3-diazolyl (NBD) probe in in vitro translation reaction. NBD-Tim23 was integrated into the inner membrane of isolated mitochondria. The steady-state and time-resolved emission of the probe exhibited alterations, clearly related to the position of the fluorophore on the studied helix, as it faced or hydrophobic environment of lipid bilayer (or nonpolar surface of other helix) or aqueous interior, forming a channel (Fig. 3).

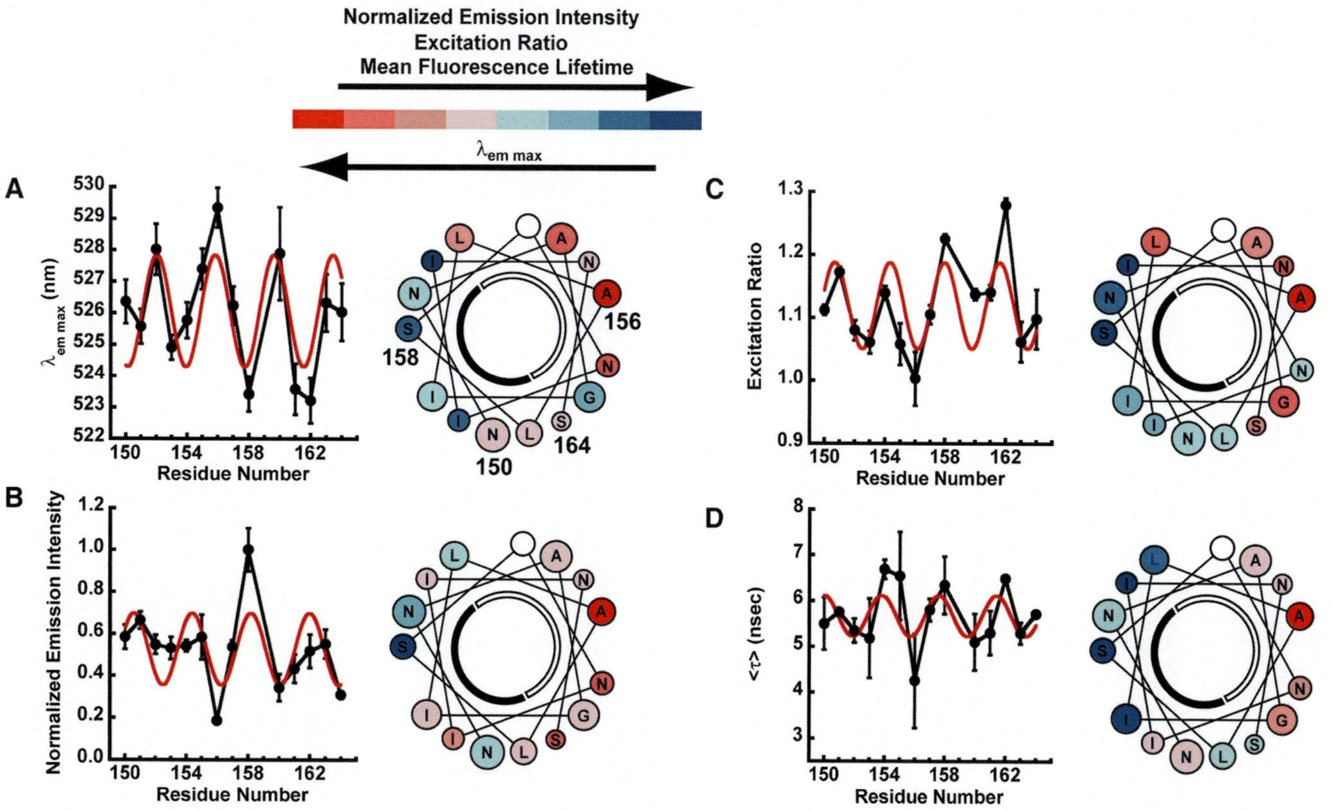


**Fig. 3** Dependence of NBD fluorescence parameters on position within Tim23 TMS2 α-helix. (A) Plots show values of maximum emission wavelength, (B) emission intensity, (C) excitation ratio, i.e., emission at 468 nm/emission at 491 nm, and (D) average lifetime. Sinusoidal functions are depicted in *red*. The position of amino acid residues within TMS2 is presented as a helical wheel projection. The *red and blue color* codes indicate hydrophilic and hydrophobic environments, respectively. *(Reprinted from Alder NN, Jensen RE, Johnson A.E. Fluorescence mapping of mitochondrial TIM23 complex reveals a water-facing, substrate-interacting helix surface. Cell 2008;134:439–50. https://doi.org/10.1016/j.cell.2008.06.007, with permission.)*

## 2.2 Time-resolved emission spectra

The dynamics of fluorescence environmental effects are frequently studied by time-resolved emission spectra (TRES) [33,34]. Obtaining TRES usually consists of measurement of the emission decays $I(\lambda, t)$ at several wavelengths $\lambda$ across the emission spectrum and their normalization with respect to the corresponding steady-state spectrum $F(\lambda)$:

$$I'(\lambda, t) = \frac{F(\lambda)\sum_{i=1}^{n} A_i(\lambda)e^{-t/\tau_i(\lambda)}}{\sum_{i=1}^{n} A_i(\lambda)\tau_i(\lambda)} \tag{7}$$

where $A_i(\lambda)$ and $\tau_i(\lambda)$ are amplitude and lifetime of component $i$ of multi-exponential decay for a given emission wavelength $\lambda$.

In a complex environment such as a protein interior or protein-membrane interface, solvent relaxation may be contributed by different processes (Fig. 4). TRES of water-soluble *N*-acetyl-L-tryptophanamide (NATA) and lipid-soluble DL-tryptophan octyl ester (TOE) have been used

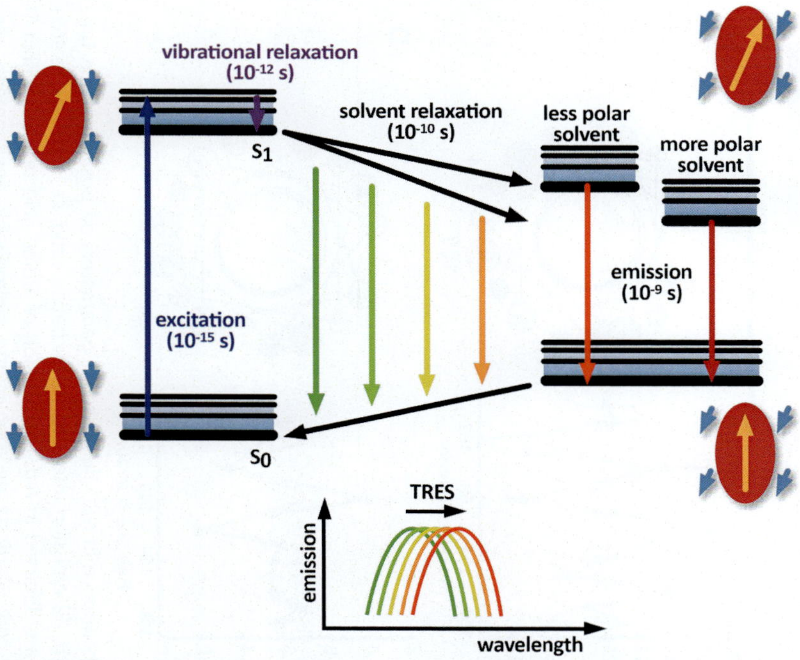


**Fig. 4** Jablonski diagram of solvent relaxation. Alignment and reorientation of fluorophore and solvent dipoles are depicted for ground and excited states. The principle of TRES formation is shown.

to resolve solvent relaxation of tryptophan of single tryptophan transmembrane protein mistic (Fig. 5) [33]. To extract solvent relaxation dynamics, spectral barycenter values $\nu$ were used to calculate the solvation correlation function $C(t)$:

$$C(t) = \frac{\nu(t) - \nu(\infty)}{\nu(0) - \nu(\infty)} \tag{8}$$

where $\nu$ (in $cm^{-1}$) values represent emission maximum in a given time $t$ and after extrapolation to $t=0$ and $t=\infty$. $C(t)$ itself may be fitted to a single- or multiexponential decay function to obtain relaxation times $\tau_r$.

The analysis revealed two components of solvent relaxation of tryptophan. Fast component, with $\tau_r$ smaller than the 50 ps time window accessible in the study, represented hydration of tryptophan by water molecules. Slow $\tau_r$ component (2.7–6.0 ns) was due to relaxation mediated by protein-lipid interior and was dependent on conformational freedom of mistic in a given hydrophobic environment.

When interpreted and related to particular processes, TRES dynamics may serve as an insightful tool for studying protein conformational dynamics. Soybean lipoxygenase (SLO) is a nonheme iron enzyme catalyzing the cleavage of the C—H bond within the linoleic acid alkene chain that employs a hydrogen tunneling mechanism [34]. Thermally modulated conformational motions of Cys-substituted and 6-bromo-acetyl-2-(dimethylamino)-naphthalene (BADAN)-labeled SLO mutants were monitored by TRES. These results showed that the reorientation of the protein hydration network plays a crucial role in the thermal activation of enzymatic reactions and associated conformational changes in the enzyme. We have also successfully employed TRES to characterize the origin of electrons in the photoinduced reduction of a ferredoxin and cytochrome *c* by quantum dots [35].

## 2.3 Ultrafast fluorescence upconversion spectroscopy

Kinetics of fluorescence shifts are accessible only if the time window of high enough resolution is comparable with the timescale of the monitored process. Therefore subpicosecond water-protein dynamics require time resolution higher than achievable by TCSPC. The most popular ultrafast technique is fluorescence upconversion [36,37]. It consists of excitation of the sample by laser pulse followed by a second delayed probe pulse. Fluorescence beam and probe pulse are cofocused into an upconverting optical crystal where sum frequency photons are generated with a time resolution of

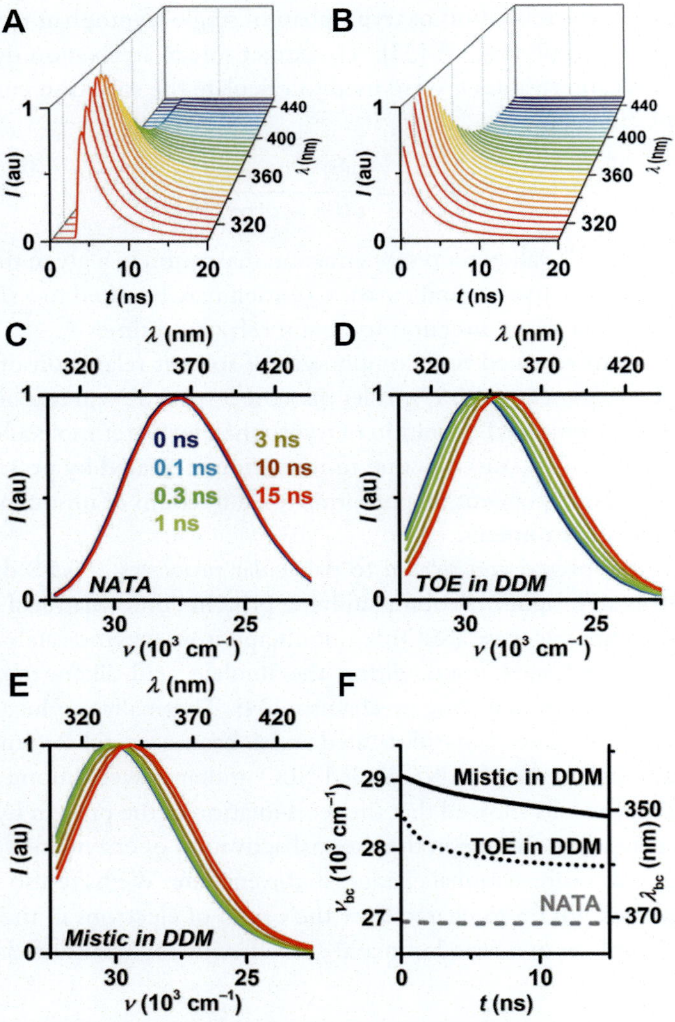


**Fig. 5** Time-resolved fluorescence to study membrane proteins. (A) Emission decays of mistic in DDM (dodecyl-β-D-maltopyranoside) micelles collected for different emission wavelengths and (B) triple exponential functions derived from this data set. (C) TRES of NATA in aqueous buffer, (D) TOE in DDM micelles, and (E) mistic in DDM micelles, (F) Spectral barycenters as a function of time, calculated from TRES. *(Reprinted from Frotscher E, Krainer G, Schlierf M, Keller S. Dissecting nanosecond dynamics in membrane proteins with dipolar relaxation upon tryptophan photoexcitation. J Phys Chem Lett 2018;9:2241–45. https://doi.org/10.1021/acs.jpclett.8b00834, with permission. Copyright 2018 American Chemical Society.)*

laser pulse width (~100 fs). Varying the time delay of the probe pulse allows for tracing emission signals throughout time.

The upconversion method was used to disentangle coupled water-protein dynamics on the surface of eye crystallin [38]. Tryptophan fluorescence scans were performed using crystallin mutants with tryptophan residues located in sites of different solvent exposure. $C(t)$ functions were then fitted with three $\tau_r$ components. Each $\tau_r$ corresponded with different protein-water relaxation processes related to bulk water motions or reorientation of hydration layer molecules. A similar approach was employed to probe water dynamics in the cavity of the *E. coli* GroEL chaperone [39].

## 2.4 Red-edge excitation shifts

In the examples mentioned previously, solvent relaxation times $\tau_r$ were several orders shorter in magnitude than emission decay times $\tau$. However, when $\tau_r$ is comparable or longer than $\tau$, red-edge excitation shift (REES) may occur by the red shift of the wavelength of the emission maximum when excited at the red edge of the excitation spectrum. Red-edge excited fraction of fluorophores in the sample has absorption at lower energy close to the solvent-relaxed state; thus it displays red-shifted emission. For REES to occur, high heterogeneity (for wide distribution of absorption energy) and motional restriction (to decrease $\tau_r$ value) of the environment are required.

REES of buried W76 tryptophan residue was used to reveal exceptionally ordered water molecules ($\tau_r$ ~1.4 ns) inside κ-casein, an intrinsically disordered protein (IDP) [40]. Treatment of κ-casein with two denaturants, sodium dodecyl sulfate (SDS) detergent and guanidinium chloride chaotropic salt, surprisingly resulted in opposite effects. An increase of REES in the presence of SDS was interpreted as the formation of a molten globule state by collapsed and even more compacted κ-casein chain. Although REES is a sensitive indicator of slow solvent relaxation, it does not offer time resolvability—for $\tau_r$ calculation, TRES were used.

The molten globule concept of proteins assumes the disturbance of protein structure by water molecules (wet molten globule). However, growing evidence indicates that breakage of side-chain interactions may occur without water penetration—resulting in a dry molten globule state. This process was visualized by REES of humans serum albumin (HSA) labeled with buried 5-((((2-iodoacetyl)amino)ethyl)amino)naphthalene-1-sulfonic acid (IAEDANS) [41]. Here, the increase of REES in the presence of urea denaturant was interpreted as an indication of the exceptionally slow relaxation of the IAEDANS probe inside the water-free interior of HSA dry molten globule, behaving like a dense, viscous liquid.

## 2.5 Voltage clamp fluorometry

Measurements of fluorescence environmental effects may be coupled with other techniques. An example is voltage-clamp fluorometry (VCF), where electrode voltage clamp equipment is linked with a fluorescence microscope for emission change detection [42,43]. VCF was used to monitor the voltage-gated human sodium glucose cotransporter (hSGLT1) cycle comprising several conformational states [44]. The cysteine residue, located on the internal surface of the hSGLT1 channel, was labeled with carboxytetramethylrhodamine (TAMRA). The conformational changes in response to gradual potential alteration could be monitored in ms timescale by VCF, providing information on activity and gating of hSGLT1 in various external conditions.

# 3. Förster resonance energy transfer

FRET is a phenomenon commonly explored in biochemical studies [45,46] to study biomolecular interaction and dynamics. It is a nonradiative energy transfer from an excited donor fluorophore to a ground-state acceptor. The efficiency of the process, $E$, is calculated from the intensities or lifetimes of donor with ($I_{DA}$, $\tau_{DA}$) and without ($I_D$, $I_D$) acceptor:

$$E = 1 - \frac{I_{DA}}{I_D} = 1 - \frac{\tau_{DA}}{\tau_D} = \frac{R_0^6}{R_0^6 + R^6} \tag{9}$$

where $R$ is the distance between donor and acceptor and $R_0$ (Förster distance) is constant for a given donor-acceptor pair in particular conditions and in the simplified form can be expressed as

$$R_0^6 = 0.021 \frac{J\kappa^2\Phi_D}{n_r^4} \left[\mathrm{nm}^6\right] \tag{10}$$

where $J$ is the spectral overlap of donor emission and acceptor absorption in $M^{-1}cm^{-1}nm^4$, $\kappa^2$ is dipole-dipole orientation factor, $\Phi_D$ is donor quantum yield, and $n_r$ is the refractive index of the medium. Typical $R_0$ values are in the range 30–60 Å, which is the most sensitive distance scale in FRET-based assays. The dependence of $E$ on $R$ allows the application of FRET as a molecular ruler:

$$R = R_0 \sqrt[6]{\frac{1}{E} - 1} \quad (11)$$

## 3.1 Transition-metal ion FRET

The dimensions of biomolecules over typical $1 < R < 10\,\text{nm}$ create a requirement to extend the FRET range. The application of transition-metal complexes as energy acceptors comprises in the range below 1 nm. Due to low extinction coefficients, metal ion complexes exhibit short $R_0$ values and thus act as FRET acceptors over distances of 5–20 Å. $Cu^{2+}$ and $Ni^{2+}$ complexed by di-histidine motif allowed to map the structure and conformational changes of maltose-binding protein (MBP) with an accuracy of 1.4–3.6 Å root-mean-square deviation from X-ray crystal structures [47]. In close proximity to the protein backbone, the location of metal complexes reflected protein structure more accurately than covalently linked fluorophores. Fluorescein-nickel FRET was also used to monitor the opened-closed conformation of the LeuT leucine transporter [48].

## 3.2 Lanthanide-based resonance energy transfer

Luminescent lanthanide ions complexes have several favorable properties as energy donors, such as large $R_0$ values and small size. Exceptionally, long (up to msec) lifetimes allow for time-gated detection of lanthanide emission, reducing fluorescent background [49]. These advantages were particularly useful for mapping the structure of voltage-gated channels by time-resolved lanthanide-based resonance energy transfer (LRET) measurements. A terbium ion was complexed by inserting specific amino acid sequence without labeling reaction in selected transmembrane regions and served as an LRET donor to single fluorescent dyes [50,51].

## 3.3 Determination of quaternary protein structure by FRET

General quaternary structure and stoichiometry of protein complexes may be determined by FRET without the need for site-specific labeling. Oligomeric assembly of G protein-coupled receptors (GPCRs) could be monitored in living cells by selective labeling of receptor subunits by the SNAP-tag method [25]. Time-resolved FRET signal of receptors in different combinations of labeled and unlabeled subunits allowed for the determination of the formation of dimers and dimers of dimers. Another example involved advanced and careful normalization of FRET efficiency by simultaneously measuring donor and acceptor emission in different acceptor:donor molar ratios [52]. It allowed the determination of stoichiometry of protein complexes and the calculation of apparent affinity constants.

### 3.4 Bioluminescence resonance energy transfer

The use of bioluminescent energy donors offers several advantages related to the absence of the exciting light source, such as no direct excitation of the acceptor, no autofluorescence in living cell assays, and no photobleaching of donor and acceptor dyes [53]. However, conjugation with enzymes converting fluorophore molecules may lack site-specificity. Nonetheless, bioluminescence resonance energy transfer (BRET)-based systems were able to detect subtle conformational changes in proteins of interest.

BRET conformational sensor was developed for angiotensin II AT1 receptor (AT1R), a member of the GPCR family [54]. The C-terminal intracellular tail of AT1R was conjugated with *Renilla* luciferase (Rluc) as an energy donor, and FlAsH (fluorescein arsenical hairpin that attaches to CCXXCC amino acid motif) was located in several different positions in a series of ATR1 mutant. A set of BRET measurements with FlAsH as an acceptor located in different sites in intracellular loops (referred to as FlAsHwalk) distinguished conformational changes induced by various AT1R agonists (Fig. 6). Another sensor was based on two nanobodies recognizing two different conformational states (active and inactive) of the κ-opioid receptor [55]. This system employing BRET from Rluc to mVenus-tagged nanobodies was transferable to other GPCRs.

## 4. Fluorescence quenching

Fluorescence quenching is a collective term for the decrease of fluorophore emission intensity in response to physical or chemical factors. It can be a result of various mechanisms, including photoinduced electron transfer (PET) [35,56,57] and intersystem crossing due to encounters with heavy atoms (e.g., halogens [58]). Quantitatively, quenching can be described in two general types: static and dynamic quenching [59]. Dynamic quenching may be considered as an additional nonradiative relaxation process, reducing both emission intensity and lifetime. Static quenching, which assumes complete turn-off of a fraction of the fluorophores in the sample (e.g., by the formation of a nonfluorescent complex), decreases emission intensity without change in the lifetime. A combined mechanism of both types of

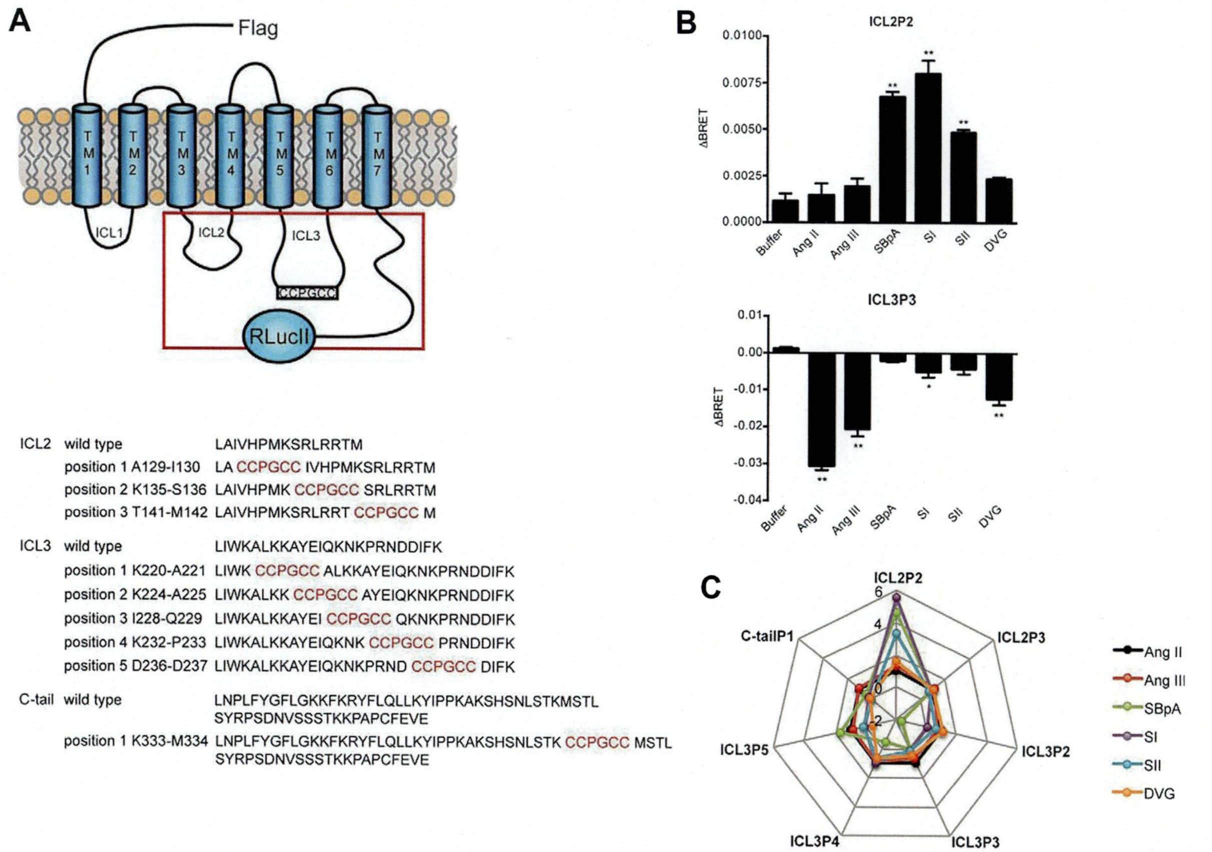


**Fig. 6** BRET conformational sensor of AT1R. (A) Schematic of AT1R structure conjugated with Rluc and containing FlAsH-binding sequence and sequences of AT1R mutants. (B) Example of calculated BRET responses of two AT1R sensors for different AT1R agonists. (C) Radar plot showing conformational profiling of AT1R for different agonists (C). *(Reprinted from Devost D, Sleno R, Pé Trin D, Zhang A, Shinjo Y, Okde R, Aoki J, Inoue A, Hébert TE. Conformational profiling of the AT1 angiotensin II receptor reflects biased agonism, G protein coupling, and cellular context. J Biol Chem 2017;292:5443–56. https://doi.org/10.1074/jbc.M116.763854, with permission.)*

quenching may also be observed [28]. Quantitatively, quenching may be described by the Stern-Volmer equation:

$$\frac{I_0}{I} = (1 + K_D[Q])(1 + K_S[Q]) \tag{12}$$

where $I_0$ is unquenched emission intensity, $I$ is intensity in the presence of [Q] quencher concentration, and $K_D$ and $K_S$ are dynamic and static quenching constants, respectively.

The standard application of fluorescence quenching is the determination of fluorophore exposure by water-soluble, nonpenetrating quenchers such as acrylamide or iodide ions [18,59]. The efficiency of quenching depends then on the accessibility of the fluorophore to the quencher and provides topographical information about the location of a fluorophore within the protein molecule. However, more complex approaches utilizing quenching have also been developed [18,59].

## 4.1 Photoinduced electron transfer as a molecular ruler

As mentioned earlier, distance dependence of FRET ($E \propto 1/R^6$) is commonly used in FRET-based structural studies. However, distance-dependent quenching processes may also serve as an alternative. An example is PET, a quenching mechanism that consists of the transfer of electrons from excited donor to acceptor. Requiring close contact ($E \propto e^{-R}$), PET allows for distance measurement in the range of 0–20 Å [60]. PET measurement by VCF setup was implemented to detect conformational changes of potassium BK channel in response to voltage jumps [56]. A tryptophan residue served as a PET donor for tetramethylrhodamine maleimide (TMRM) fluorophore conjugated by linkers of gradually increased length and controlled spatial orientation. Dependence of quenching efficiency on linker length allowed for the measurement of intramolecular distances between four transmembrane TMRM-labeled α-helices. A similar approach, using nitroxide-bearing quencher and oligo-PEG (polyethylene glycol) linkers, was employed for the *Shaker* channel [61].

## 4.2 Depth-dependent quenching in lipid bilayers

Distance-dependent quenching may determine the location of fluorophores in membranes. Unstructured in solution, apolipoprotein C-III (ApoCIII) forms amphipathic α-helix when associated with the surface of micelles or lipid vesicles [58]. Insertion depth of three ApoCIII tryptophan residues was measured by quenching experiments using brominated phospholipids.

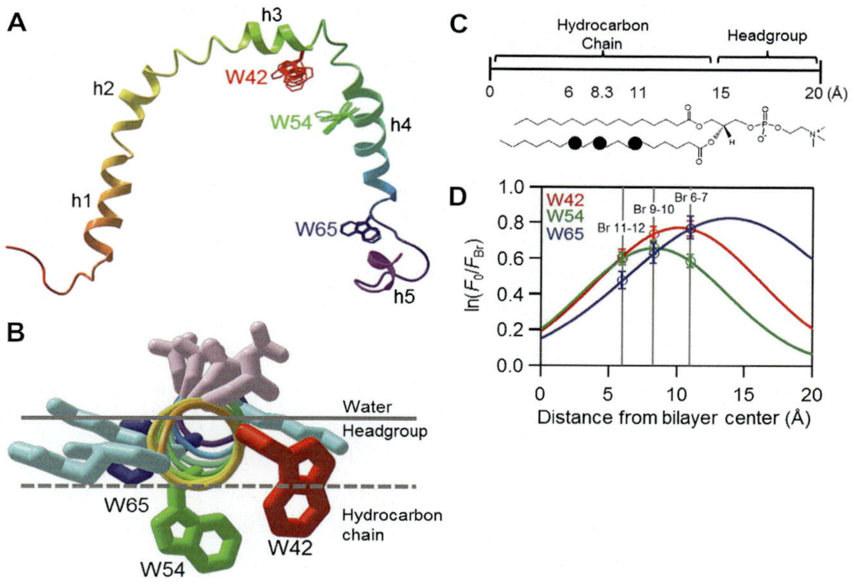


**Fig. 7** Depth-dependent quenching of ApoCIII. (A) Structure of ApoIIIC in SDS-micelles with three native tryptophan residues. (B) Helical wheel projection of ApoCIII associated with the lipid bilayer. (C) Structure of brominated phosphatidylcholine molecules used as quenchers for depth-dependence analysis. (D) Quenched tryptophan emission values fitted to Gaussian distance distribution. *(Reprinted from Pfefferkorn CM, Walker RL, He Y, Gruschus JM, Lee JC. Tryptophan probes reveal residue-specific phospholipid interactions of apolipoprotein C-III. Biochim Biophys Acta Biomembr 2015;1848:2821–28. https://doi.org/10.1016/j.bbamem.2015.08.018, with permission.)*

Bromine atoms, located at a different distance from the bilayer center, quenched tryptophan emission to different extents depending on the intermolecular distance. Fitting of $F_0/F$ data to Gaussian function provided distributions of depth distance for each tryptophan residue (Fig. 7). Other examples of depth-dependent quenching application include transmembrane α-helix of NBD-labeled Bcl-xL protein [62] and membrane-penetrating peptides [63].

## 4.3 Thermal shift assay

Thermal shift assay (TSA), also known as differential scanning fluorimetry (DFT) or ThermoFluor, is a procedure that allows the determination of the melting temperature ($T_m$) of proteins. The measurement is based on a change in fluorescence intensity resulting from the attachment or detachment of fluorophores to hydrophobic regions of proteins. $T_m$ is then related to changes in the protein conformation and subsequently the loss of tertiary and secondary structures. During the unfolding of a protein, the accessibility of hydrophobic patches usually increases, which leads to an increase in fluorescence intensity. The recorded development of fluorescence intensity as a function of temperature is presented as dose-response curves and then fitted. Sometimes, the first derivative of a function is given for straightforward identification of $T_m$. The TSA may employ virtually any fluorophore, binding selectively to folded and unfolded states of a protein; however, SyproOrange is frequently chosen due to its compatibility with regular RT-PCR machines. Occasionally, intrinsic tryptophan fluorescence is also used [64]. Ligands generally stabilize proteins, and in their presence, the Tm increases. Such molecules are called N-binders, opposite to destabilizers (U-binders), which decrease the $T_m$ upon binding. An example of a U-binder is the $Zn^{2+}$ ion, used in the case of a recombinant porcine growth hormone [65]. The destabilization was observed because the $Zn^{2+}$ ion has a higher affinity to the unfolded peptide than the folded peptide. Melting experiments performed for a series of protein:ligand ratios may be used for $K_d$ determination in a wide range of affinities, from subnanomolar to micromolar. To illustrate, $K_d$ of 3.2 μM was measured for maltose binding to MBP [64]), while for the system based on protein POT1 (protection of telomeres 1), $K_d$ as low as $10^{-11}$ M was estimated [66]. It should be remembered that the absolute value of $T_m$ may not be related to $K_d$. For reaction with negative enthalpies, the $T_m$ shift may be lower.

TSA was used for the analysis of elementary and more complex systems. A simple case was the demonstration of chloroplastic E3 Ubiquitin Ligase SPL2 stabilization by lanthanide ions but not by its native $Zn^{2+}$ partner ion [67]. A study involving a heterodimeric protein, soluble guanylyl cyclase (sGC) containing heme cofactor, was more sophisticated. Interestingly, only a combination of several factors (namely, $MnCl_2$, oligonucleotides, and activator) was found to significantly increase the $T_m$ of sGC [68]. To conclude, a broad library of potential ligands of different chemical nature may be tested, including lipid molecules. An example of the last is fenofibric acid, binding to intestinal fatty acid-binding protein [69]. However, no study using TSA is available that shows protein interaction with model membrane liposomes. This is because lipid membranes also undergo temperature-dependent transition, which might significantly influence the binding event.

TSA is very easily adapted for high-throughput studies, as it demands one-point measurement and can be done in a volume as small as 5 fL [70]. Thus a series of protein samples can be easily checked for stability. Also, multiple potential ligands can be screened at once, as shown using the example of chemoreceptors binders [71].

Cellular TSA (CeTSA) has been portrayed as an advancement in TSA; however, it should be noted that such a description is a simplification. What is common with TSA is a temperature challenge, here being done to whole cells. There might be samples pretreated with compounds of interest, as shown using inhibitors of the human mitogen-activated kinase 14 (MAPK14) [72]. The principle of CeTSA is that proteins denatured upon heating aggregate and then can be easily removed from cell lysate by centrifugation. Therefore what is left in a supernatant is a protein or a protein complex in its native conformation. However, there is no online tracking of denaturation, which is possible in TSA. A treated cell lysate should be analyzed for the concentration of biomolecules of interest. The detection can be done with antibodies, labeled fluorescently, or other techniques, such as mass spectrometry. With detection assays based on FRET [73] (see, e.g., PerkinElmer AlphaCeTSA [74]), a centrifugation step of CeTSA can be skipped. In this procedure, two antibodies (one with a donor and the second with an acceptor fluorophore) participate, targeting two native epitopes on the same protein. Therefore if a protein is in its native conformation, the distance between donor and acceptor is short enough, and an acceptor signal is measured. Otherwise, for a denatured protein, no binding occurs, and the distance is too high for a FRET.

## 5. Fluorescence anisotropy

Fundamental anisotropy $r_0$ of a sample arises from orientation-selective excitation of randomly oriented fluorophore molecules and displacement of their excitation and emission dipoles [7]. Further loss of anisotropy results from depolarizing processes such as rotational diffusion or energy transfer. Actual anisotropy r of rotating emitting molecule is described by Perrin equation [75,76]:

$$r = \frac{r_0}{1 + \left(\frac{\tau}{\theta}\right)} \tag{13}$$

where $\tau$ is emission lifetime and $\theta$ is rotational correlation time. Stokes-Einstein theory relates $\theta$ with temperature $T$ and viscosity $\eta$ of environment and volume of molecule $V$:

$$\theta = \frac{\eta V}{RT} \tag{14}$$

where $R$ is gas constant. With $V$ proportional to molecular weight, simple anisotropy measurements may determine approximate protein size, which we will discuss next.

## 5.1 Time-resolved anisotropy measurements

In analogy to fluorescence intensity, the time-dependence of anisotropy may be described as exponential decay, $r(t)$ [76]. In the case of labeled proteins, $r(t)$ is usually a sum of depolarization by slow overall tumbling of the protein molecule and fast localized rotational motions of side chains. Anisotropy decay single-tryptophan mutants of α-synuclein exhibited two $\theta$ correlation times: faster (~0.2 ns) of the local mobility of the indole side chain and slower (~1.4 ns), which was in the order of magnitude shorter than typical global tumbling rotational times [77]. Hence, the slower component was related to segmental motions of α-synuclein of exceptionally large amplitude, which were sufficient to completely depolarize emission before global rotation could reveal its influence. Moreover, faster $\theta$ was independent of tryptophan residue location within the α-synuclein chain, indicating the lack of steric constraints resulting from a structured environment. Slower $\theta$ of α-synuclein mutants was shown to increase (up to ~45 ns) after binding to lipid membrane was assigned to translational diffusion on the membrane surface [78].

## 5.2 Determination of homoFRET by anisotropy measurements

Homotransfer is a FRET process between two identical molecules of the same fluorophore. Because donor and acceptor are spectroscopically indistinguishable, anisotropy loss caused by energy migration is the only way to study homoFRET [79]. Seven single cysteine mutants of α-synuclein were labeled by fluorescein to cover its entire length, and fluorescent and unlabeled protein was mixed in a different molar ratio [80]. The formation of amyloid fibers by α-synuclein was monitored by anisotropy increase (slow $\theta$ of aggregate), and then homoFRET was manifested by determining the labeled:unlabeled ratio-dependent anisotropy loss (Fig. 8). The structure of aggregated α-synuclein was mapped by homoFRET efficiencies with $7 \times 7$ intermolecular distances. HomoFRET was also used to resolve conformational factors responsible for ion selectivity of the KcsA potassium channel [81]. Tryptophan-tryptophan homoFRET between four subunits of KcsA homotetramer was employed using a mathematical model for $r(t)$ in squared geometry to calculate intermolecular distances.

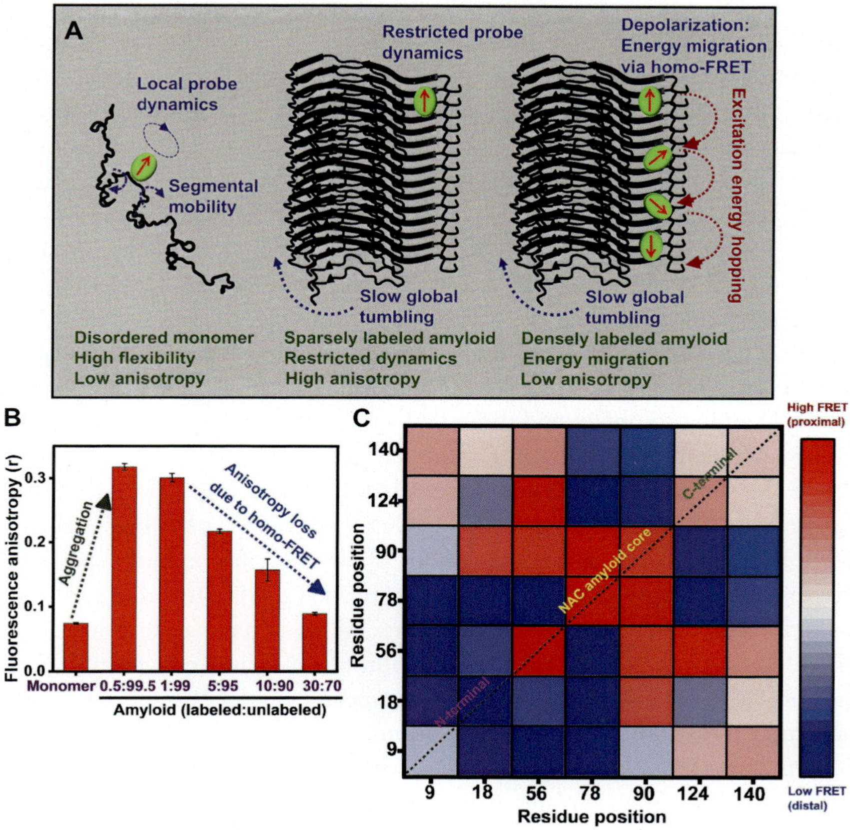


**Fig. 8** Determination of homoFRET by anisotropy measurements. Schematic of anisotropy changes due to rotational movement and energy transfer (A). Steady-state anisotropy of fluorescein-labeled α-synuclein in the function of labeled: unlabeled ratio (B). HomoFRET efficiency map of α-synuclein (C). *(Reprinted from Majumdar A, Das D, Madhu P, Avni A, Mukhopadhyay S. Excitation energy migration unveils fuzzy interfaces within the amyloid architecture. Biophys J 2020;118:2621–26. https://doi.org/10.1016/j.bpj.2020.04.015, with permission.)*

## 6. Diffusion-based estimation of biomolecular size and shape

### 6.1 Analytical ultracentrifugation

Analytical ultracentrifugation (AUC) supports the determination of biomolecular mass and, to some extent, shape [82]. There are two primary types of AUC experiments: sedimentation velocity (SV) and sedimentation equilibrium (SE). In SE, the final gradient reached after centrifugation is analyzed, while in SV, the evolution of the sedimentation process is the source of information. Analysis of sedimentation profiles leads to the calculation of sedimentation coefficient related to apparent molecular weight. In its typical application, AUC is nondestructive; therefore the samples may be reused (for the repetition of AUC or other methods). The online evaluation of the formed gradient is possible due to specially designed centrifuge units (cells) with transparent windows and a detection system mounted inside the centrifuge. Commonly, the spectrophotometers are exploited, as absorption is a universal characteristic of a biomolecule. Currently, the AUC instruments with fluorescence detection (FD-AUC) are used. The fluorescence signal allows for higher sensitivity and, therefore, a low concentration of an analyte is required. This comes in handy when biomolecules are challenging to obtain and those that aggregate in higher concentrations. A low concentration of compounds is also crucial for resolving lower $K_d$ [83]—similar to fluorescence correlation spectroscopy (FCS)/fluorescence crosscorrelation spectroscopy (FCCS), described in the previous subsection. The advantage of AUC is its application for extreme particles, comparable in size with its focal volume. In AUC, it is also easier to handle experiments under anaerobic conditions. With certain assumptions, one may determine the overall shape of the biomolecule based on a friction coefficient [84]. This may answer the "sphere versus cylinder" question for a single molecule and help discuss monomers' relative orientation in the oligomer.

Lamm equation (15), describing recorded sedimentation profiles, and Svedberg equation (16) are the basis for further analysis of sedimentation profiles:

$$\frac{\partial c}{\partial t} = D\left[\left(\frac{\partial^2 c}{\partial r^2}\right) + \frac{1}{r}\left(\frac{\partial c}{\partial r}\right)\right] - s\omega^2\left[r\left(\frac{\partial c}{\partial r}\right) + 2c\right] \tag{15}$$

$$s = \frac{v}{\omega^2 r} = \frac{M(1 - \rho\,\overline{v})}{N_{AV} f} \tag{16}$$

where $D$ is the diffusion coefficient, $s$ is the sedimentation coefficient, $r$ is the radial position (the radial distance from the axis), $c$ is the particle concentration along the radius at the given time $t$, $v$ is solute velocity, $\omega$ is the angular velocity, $M$ is molecular weight, $N_{AV}$ is Avogadro's number, $f$ is the frictional coefficient, and $\rho$ is the density of the medium.

From the equations, one may find that separation might be improved by optimizing the angular velocity (in simple words, speed of rotation).

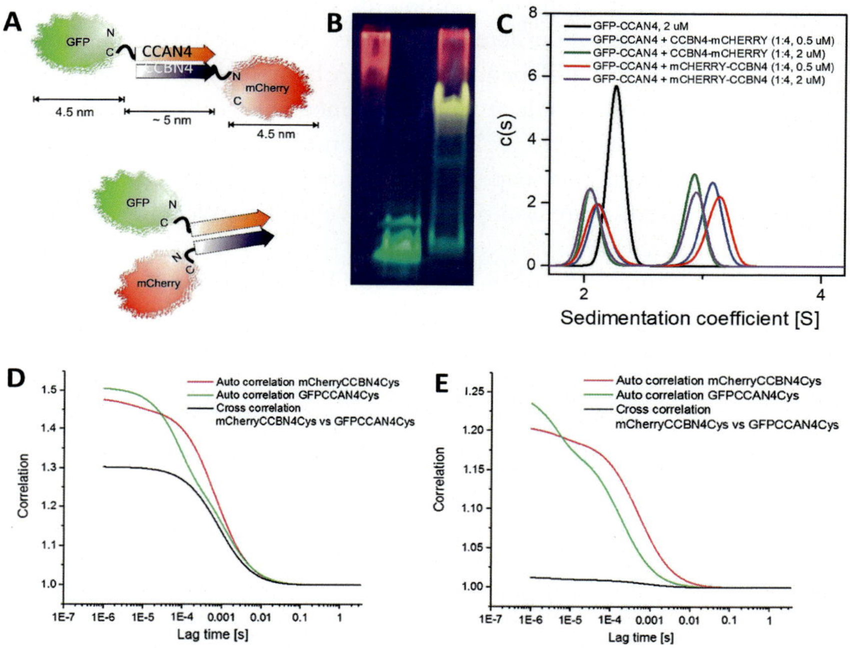


**Fig. 9** Fluorescent techniques applied to study the binding of GFP and mCherry. (A) Schematic representation of the possible organization of a dimer, in dependence of a helix position on N- or C-terminus of carrying protein. (B) Native electrophoresis of proteins alone and a mixture of them, visualized with UV-transilluminator (real fluorescence colors presented). (C) FD-AUC recorded for both possible orientations of heterodimers. (D) FCCS with significant autocorrelation recorded heterodimers stabilized by disulfide bridge and (E) the same mixture with added DTT, breaking the S—S bond. *(Adapted from Sztatelman O., Kopeć K., Pędziwiatr M., Trojnar M., Worch R., Wielgus-Kutrowska B., et al. Heterodimerizing helices as tools for nanoscale control of the organization of protein-protein and protein-quantum dots. Biochimie 2019;167:93–105. https://doi.org/10.1016/j.biochi.2019.09.015, with permission.)*

Components, which easily change medium density (e.g., glycerol), may also be used with care. Free software SEDFIT is a broadly known tool for solving the Lamm equation to analyze AUC data [85]. It offers the fit for both SV and SE experiments, with multiple models of binding [84,86].

We have successfully used FD-AUC to study the binding of engineered GFP and mCherry molecules (Fig. 9) [87]. Those fluorescent proteins bare helices (shortly named A and B), whose sequences were optimized for heterodimerization. The question to be resolved was stoichiometry. Therefore several mixtures were run in parallel with varied ratios between A and

B proteins. Concentrations of compounds were chosen to secure stable complexes, as $K_d$ was earlier determined by isothermal titration calorimetry (ITC). We found that independent of an A:B mixture ratio, there were always some monomers, indicating the presence of nonfunctional proteins. The experiment confirmed a 1:1 heterodimer stoichiometry of A and B proteins, although the shape of the formed complex differed. Some amount of B proteins were homodimers; however, they were broken during heterodimer formation.

A combination of absorption AUC and FD-AUC was adapted to study the folding and maturation of eGFP. Absorbance at 280 nm detected the total protein, while fluorescence signals indicated the presence of a correctly folded (mature) eGFP molecule [18]. The clear demonstration of FD-AUC to determine $K_d$ was described by Chaturvedi et al. [83] while studying the binding between eGFP to its antibodies. The $K_d$ as low as 20 pM was estimated. As for any other $K_d$ determination protocol, titration curves must be constructed. Fluorescence detection allows for a very low protein concentration (in this case, 8 pM eGFP); therefore some free proteins may still be present in the assay, which assures correct titration curves construction and analysis.

It should be remembered that the crucial point for any AUC experiment determining $K_d$ or oligomerization state is that the system needs to reach equilibrium before the start of the run. It may be a problem even for systems with low $K_d$ but relatively high $k_{on}$. For such situations, prolonged incubation may be a solution. If no crude estimation of that time is known, one may test two or more incubation periods.

## 6.2 Microscale thermophoresis

Microscale thermophoresis (MST) is a technique dedicated to the determination of dissociation constants. In the simplest description; it analyses the binding between two molecules based on their diffusion. Samples are placed in capillaries, which are optimized for fluorescence detection. A laser is used to excite a fluorescent sample. There is also a source of infrared radiation, heating capillaries, which leads to molecule's movement in capillaries. In the end, a difference in fluorescence intensity in the measurement point relevant to the velocity of molecules is observed. Smaller molecules (as nonbound cofactors) diffuse faster than the same compound attached to their protein partner. Accordingly, fluorescence decreases much stronger for smaller than for bigger molecules; the higher the size difference, the higher the sensitivity of the assay. Thus when performing MST, it is better to label the smaller of the two interacting molecules. The larger of the two interacting molecules can also be labeled; however, such an assay may not be advantageous in all cases. MST machines, available currently from NanoTemper, are equipped with one or more detection channels: blue, green, and red, which determine useful fluorophores. Unfortunately, mixing channels in one measurement is not possible currently. Besides, there are MST versions called "label-free," with optics optimized for the detection of intrinsic tryptophan fluorescence [88].

MST method, in general, analyses the mobility of proteins. Because hydrodynamic radius, and therefore diffusion rate, may change with conformational changes (induced by small ions $Ca^{2+}$ or $Mg^{2+}$, pH change, etc.) [89], the shape of the molecule may be further analyzed. Finally, one may use the label-free MST to analyze ligand-dependent tryptophan quenching, similar to the classical quenching experiment. The advantage is that a very small amount of protein is needed for assay compared to the cuvette measurement.

MST not only allows a 1:1 mode binding study but can also assess the situation of several binding places, simultaneous or sequential, with different affinities. More complicated assays may not be solved in a single experiment and may need several concentration ranges and nontypical concentration scales. Analyzing such data may require additional software, allowing other models than the basic ones proposed by the apparatus supplier. MST is a fast and easy method for standard systems, with high reproducibility, although not the cheapest. It is important to recognize that the binding curve is created in a single experiment with very little material. It is also possible to obtain results even for a cell lysate or denaturing conditions. For better results, one should avoid any aggregates in the samples. For labeling in a lysate, it is straightforward to use chemistry targeting a HisTag [90]. The problem for MST might be FRET between binding molecules (if both are fluorescent). In such a situation, it is better to aim the assay on acceptor fluorescence; however, one may also comprise for energy transfer effects in the analysis.

## 6.3 Fluorescence correlation spectroscopy

FCS recognizes fluorescence emitted by a molecule while its diffusion by a focal point of a confocal microscope [91,92]. A registered signal is, in principle, the noise of a classical cuvette measurement. The signal is averaged to obtain the final value for a point or a spectrum. In FCS, we observe an increase and decrease in fluorescence intensity caused by molecules diffusing in and out of measurement points. This trace is then analyzed with autocorrelation function, allowing the determination of time $\tau_D$ needed for a fluorophore to pass through confocal volume. $\tau_D$ is in relation to diffusion coefficient is as follows:

$$D = \frac{w_0^2}{4\tau_D} \tag{17}$$

where $w_0$ is the axial radius of the confocal volume.

With a fluorophore of a known $D$, one may precisely test the confocal volume. This may be thought of as a system calibration. The $D$ of unknown fluorophores may be measured, which can be used for the calculation of the hydrodynamic radius $r$ of the molecule using the Einstein-Stokes equation:

$$r = \frac{k_B T}{6\pi\eta D} \tag{18}$$

where $k_B$ is the Boltzmann constant, $T$ is the absolute temperature during measurement, and $\eta$ is the solvent viscosity at $T$.

FCS was used to analyze the protein size and the interaction between proteins and nanoparticles [28,57,93,94]. In all cases, Alexa Fluor488 was used as a standard. Other fluorophores, such as rhodamine, were also used successfully. It should be remembered that the standard for confocal volume scaling needs to be purely monomeric. In our experience with Alexa-Fluor488, nanomolar concentrations should be used, and the working solution should be freshly prepared from concentrated stocks.

To use FCS for the quantification of protein hydrodynamic radius, the protein needs to be fluorescent. This can be easily achieved for the GFP and its relatives. For other proteins, the fluorescent label must be added, which might be any fluorophore that can be excited by the available laser set. It should be remembered that a confocal volume depends on the laser wavelength, and it needs to be remeasured for changed $\lambda_{exc}$. A good laboratory practice is to measure it on different days, even if the same excitation is used. For the second mentioned application of FCS, the fluorescence of QDs was measured directly. When proteins attach to the QD surface, the hydrodynamic radius of the nanoparticle increase; in this way, we can prove the formation of the QD amphipathic shell by peptides [93,94]. We also analyzed complex formation between QDs and redox-active proteins, cytochrome *c*, and ferredoxins. Here, we did not calculate the $K_d$; however, the obtained data could be treated as the preliminary titration curve [28]. For the best results, samples should be homogenous; however, some small amount of unspecific aggregation, usually present in protein preparations, can be tolerated and excluded during analysis. It should be noted that the calculated $D$ is averaged for the molecules/particles in the solution. Therefore it is impossible to determine the distribution of sizes based on only one measurement.

Carrying FCS at once for two separate detection channels allows analysis of the comovement of two fluorophores. This method is called FCCS. Here, the determination of $D$ is not the aim; however, it still may be done as all necessary data are available. For two fluorophores moving together, the change in their fluorescence intensities in time will be identical. This measure is the crosscorrelation function, obtained by comparing the signal from two channels. Optimally, fluorophores analyzed by the FCCS should not form a FRET pair. This might be achieved by selecting appropriated dyes with insignificant spectral overlap or securing a proper distance between fluorophores, those reducing the probability of energy transfer. We measured crosscorrelation between GFP and mCherry proteins bearing heterodimerizing helices (Fig. 9C and D). Interestingly, although we proved otherwise that the $K_d$ values of complex formation were in the nanomolar range, the crosscorrelation was unclear. This was because of the low protein concentration (10 nM or lower) needed for an optimal FCS signal. One might argue that for 10 nM, a significant amount of complex should form, indicating the weakness of the method. In some cases, a crosscorrelation signal can hardly be used to construct titration curves. However, we had shown positive crosscorrelation for those heterodimerizing helices when the binding between them was stabilized with disulfide bridges. By adding dithiothreitol, we were able to break the interaction and demonstrate the loss of crosscorrelation. FCCS is a valuable tool for analyzing protein interactions in a living cell as it requires very low protein concentrations [92]. The $K_d$ for binding of NFκB p50/p65 heterodimer was found by the FCCS to be 0.46 μM in the cytoplasm and 1.06 μM in the nucleus [95]. The interaction between mRNA and CRDBP protein was also examined in single living cells [96]. In this case, the molecular beacon labeling technique provided fluorescent labels.

FCCS can also be combined with a FRET analysis. Using this combination, Wennmalm et al. [97] detected a small fraction of oligomers of an amyloid-beta (Aβ) peptide coexisting with monomers. Theoretically, FCS and FCCS may be measured in any solution, including a cell lysate; however, in such a case, the viscosity of the solute should also be checked, possibly by measuring a standard in a lysate without the protein of interest. Optimizing the FCS signal often demands dilutions; in such cases, special care should be taken to compensate for the viscosity and presence of components, which may influence diffusion. For these reasons, FCS is often used for isolated protein and molecule complexes.

Employing one detection channel but checking a *D* for different focal volumes is called a spot-variation FCS (svFCS) [98]. The experimental dependence between *D* and volume is called a diffusion law. It extracts information about fluorophore dynamics in cell membranes and other cell regions. A recent study using this technique has shown surfactin-dependent changes in the mobility of lipids and proteins in membrane nanodomains but not in lipid rafts [99]. This data was possible due to the availability of GFP-anchored membrane protein Thy-1 and rigidity-specific lipid probes. The svFCS demands the ability to vary the waste volume, which is accomplished by a diaphragm introduction into an optical path [98]. It was proposed that similar information might be extracted from the classic FCS measurement by postprocessing the data. However, the svFCS does not allow tracking of a single molecule, though this is manageable with certain high-resolution confocal microscopy techniques.

## 7. Biomolecules in situ

Although all the techniques discussed earlier provide detailed knowledge about the shape and interactions of biomolecules, i.e., information about an isolated case. This is what a researcher requires for most situations, as the isolated systems allow strict conditions and identification of factors important to the studied process. However, in the end, there is always a question about the actual behavior of a biomolecule of interest in vivo. The development of fluorescent microscopic techniques provides insight into the dynamics of biomolecules in vivo. Here, we refer primarily to confocal microscopy, which has a higher resolution and better focus than conventional fluorescent microscopy. The basic version of confocal laser scanning microscopy (CLSM) provides a static picture of the specimen (Fig. 10). Frequently, the sample needs to be fixed and stained due to the lack of natural, unique fluorophores in the biomolecule under investigation. For proteins expressed as a fusion with a fluorescent tag (such as with GFP), the observation can be performed in vivo. A few classic examples are real-time studies of a cytoskeleton dynamic during the morphogenesis of

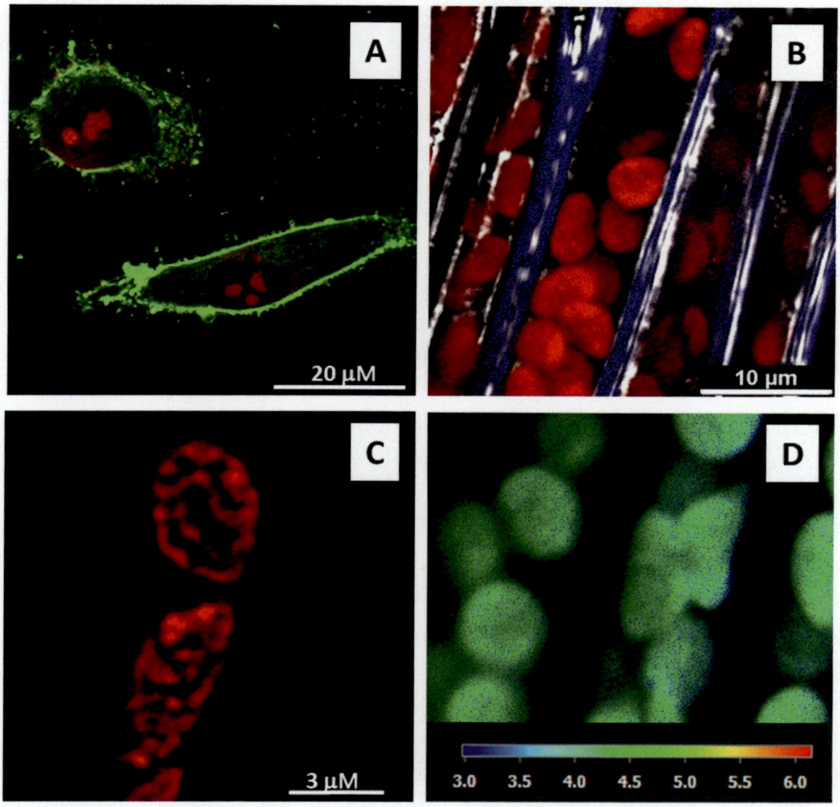


**Fig. 10** CLSM application to study biomolecules. (A) Hela cells fixed with 2% glutaraldehyde, with a Laurdan stained membrane (*green*, excitation 405 nm, detection 450–500 nm) and a propidine iodide stained nucleus (*red*, excitation 550 nm, detection 600–700 nm). (B–D) Living cells of *Pleurozium schreberi* (a common middle Europe moss species): (B) an autofluorescence of a cell wall (*blue*, excitation 405 nm, detection 450–500 nm), overlaid with chlorophyll autofluorescence (*red*, excitation 488 nm, detection 650–700 nm) and a transmission image (greyscale); (C) a superresolution ($SIM^2$) image of chloroplasts (a chlorophyll autofluorescence signal analyzed); (D) a FLIM image of chloroplasts. Images obtained by authors. For Hela cells, the participation of Ms. Karolina Wójtowicz is cordially acknowledged.

*Drosophila* cells [100] and microtubule reorganization during stomatal closure in *Arabidopsis thaliana* leaves [101].

## 7.1 CLSM and its auxiliary techniques

A static visualization of cell components may permit the localization of a specific biomolecule. In the simplest case, it gives the place with an organelle accuracy. The basic information on biomolecular dynamics that CLSM can offer is the fluctuation of the signal between organelles, corresponding to cell physiological state. Using this method, a GFP-tagged bZIP28, an ER membrane-associated transcription factor in *A. thaliana* was shown to translocate from the nucleus to the Golgi apparatus [102]. However, if tagging is not possible, the identification might be made on cells treated with specific antibodies.

The resolution of CLSM is limited by the wavelength used for fluorophore excitation. It also depends on the characteristics of the objective lens (numeric aperture, NA) and immersion fluid refraction index (see the Abby equation and other details in [103]). For example, the calculated lateral resolution limit for objective 63x (NA = 1.4, oil immersion) is 227 nm for AlexaFluor 488 and 301 nm for AlexaFluor 600. The smallest spot will likely depend on the quality of the other part of microscope optics. Additionally, the pixel size is dependent on the strictly mechanical properties of the detecting system and the precision of measuring spot movement. Maximal accuracy is often higher than spot size defined by laser wavelength. One should then decide on the picture parameters, as the compromise between resolution, sensitivity, and measuring time, which could lead to photobleaching of the fluorophore. The CLSM also supports recording several focal planes and reconstructing 3D images of fluorescent features. It is worth mentioning that CLSM detection may be done using filters (a bandpass or a long pass type) or monochromators. Both configurations result in the same image, but the monochromator offers additional, often underestimated advantages. It allows recording the emission spectrum and, therefore, can distinguish between separate fluorophores emitting light in the same filter-defined channels. Modifying the detection range with a monochromator makes it easier to minimize background and cross-talk, improving image quality.

The CLSM supports the visualization of biomolecules inside the cells but offers other advantages using auxiliary procedures [104], as mentioned earlier, such as fluorescence lifetime image (FLIM) and FCCS. Among other benefits are the fluorescence recovery after the photobleaching (FRAP) technique, the fluorescence loss in photobleaching (FLIP), and the fluorescence localization after photobleaching (FLAP) [104]. These are based on applying a high-power laser pulse to a sample that irreversibly photobleaches some fluorophores. As soon as the laser pulse is irradiated, the place of the impact appears as a black, nonfluorescent patch, surrounded by a region that actively emits fluorophores. The imaging is then done in real time to observe the appearance and rebuild fluorescence intensity in the photobleached spot. These techniques are particularly useful for studying diffusion in membranes. The Mullineaux group first demonstrated the method on phycobilisomes, a pigment-protein complex attached peripherally to photosynthetic membranes of the cyanobacteria [105]. The FRAP method showed that phycobilisomes work as antennae for photosynthetic reaction centers and can rearrange themselves. Within the time scale, the diffusion rate can be calculated based on intensity variations. Notably, this can be done for lateral diffusion of molecules in the membranes by using fluorescent lipid analogues [106].

## 7.2 FLIM complements localization data

The addition of time-resolved detection to the CLSM brings about the FLIM (Fig. 10D). First, FLIM adds additional data, which may serve fluorophore discrimination in the same specific spectral range. Second, it provides valuable information about the state of the native pigment-protein complexes during in vivo measurements. The measurement is similar to the fluorescence lifetime determination in a cuvette via photon counting; the decay curve is recorded, and then the exponential fit is performed. Usually, the software does the fitting automatically; however, a good practice is to inspect the fit. Routinely, the obtained image is colored by artificial colors, representing $\tau$ or $\tau$ average for multiple component decays. As $\tau$ strongly depends on the environment (pH, viscosity, crowding), FLIM measurement complements the localization data.

For native chlorophyll molecules bound in the photosynthetic apparatus, $\tau$ represents the functional state of photosynthesis. In a highly efficient photosynthetic apparatus, fluorescence is strongly quenched due to its coupling to photosynthetic phosphorylation. There is always some basic level of emission (a result of imperfections in membranes), which may be recorded. Compared to the characteristic of free chlorophylls, $\tau$ is also shortened significantly. In some conditions, e.g., in strong light causing a disturbance in an energy or electron transfer, the intensity of fluorescence increases due to uncoupling between the following elements in photosynthetic membranes. This is known as the Kautsky effect. Mostly, light-harvesting complexes II (LHCII) fluorescence is observed, as the energy is not accepted by photosystem II (PSII). At that moment, $\tau$ also increases. After some time, the photosynthetic apparatus reorganizes, and the fluorescence intensity again goes down. Measurement of the fluorescence intensity is performed in instruments like pulse-amplitude modulated (PAM) chlorophyll fluorimeter or the FluorCam. The details can be found elsewhere [107,108]. A CLSM is not a perfect system for comparing fluorescence intensities. However, $\tau$ may still be registered and compared, and therefore it is a valuable tool, e.g., to obtain information about the functional state of chloroplasts [109].

FLIM applied in combination with other methods is a powerful tool. A study of FLIM with Raman scattering on the human retina revealed carotenoid isomerization acting as molecular blinds that dynamically control the amount of light reaching photoreceptors [110]. Lipid translocation and cholesterol redistribution in cell membranes might also be investigated by FLIM linked with fluorescence anisotropy recording [111]. FLIM with FRET provided in vivo insight into a plant receptor kinases complex formation over time [112].

## 7.3 Superresolution and single-molecule studies

An increase of a resolution below the lambda limit may be achieved by additional optical elements plus mathematical analysis of the data. The STED (stimulated emission depletion) microscope is probably the most known optics improvement due to its inventor, Prof. Stefan Hell (Nobel prize in Chemistry, 2014). The resolution of STED is estimated to be about 30 nm. In practice, STED use is often limited due to specific demands of specific fluorophores. The fluorophore fluorescence is to be quenched by STED laser, which can be done only when this laser emission is in the range of the emission spectrum of a fluorophore. The second, less remembered condition is that the STED laser should not be in the range of the fluorophore absorption spectrum. For chlorophyll, whose absorption spectrum is very broad, it is impossible to fulfill the second condition with available STED lasers. In practice, a good STED experiment needs STED-optimized fluorophores.

Alternative methods do not provide such high resolution but still increase it below the lambda limit. There is so-called structure illuminated microscopy (SIM and $SIM^2$) imaging (Fig. 10C). Here, the specimen is illuminated using special grids of different patterns. The set of images is registered with a CCD camera and then analyzed, providing a resolution of 60–120 nm. This might be good enough to study localization inside subcellular organelles, such as mitochondrion or chloroplast, and distinguish localization in the membrane or the matrix/lumen [113]. Newer techniques, using switchable fluorophores, are another approach leading to superresolution. These include photoactivated localization microscopy (PALM) and closely related stochastic optical reconstruction microscopy (STORM). The principle of both techniques is the same, and the difference is the way to achieve photoactivation. For PALM, there are photoswitchable proteins, while STORM uses dyes and specific buffers, affecting fluorescence by redox or other mechanisms.

However, even without superresolution, single molecule imaging is accomplishable. For this, fluorescent molecules should be diluted to secure a distance higher than the Abby limit. This approach came in handy, e.g., in studying interactions between fluorescent pigment-protein complexes and nanomaterials, enhancing or quenching fluorescence [114–116].

## 8. Conclusions

As discussed, fluorescence-based methods offer a plethora of possibilities to study the structure and conformational changes of proteins and other biomolecules. It also allows analysis of biomolecules in vivo, starting from identification of the localization, by real-time tracking of the dynamics of the molecules. Studies of binding are of particular interest, as high sensitivity for fluorescent signal favors high affinities pairs, not accessed by other methods. It is worth mentioning that fluorescence-based techniques usually have low sample requirements, thereby reducing time and material consumption. Some methods are also available as high throughput for fast and reliable analysis of a series of compounds.

## References

[1] Kendrew JC, Bodo G, Dintzis HM, Parrish RG, Wyckoff H, Phillips DC. A three-dimensional model of the myoglobin molecule obtained by x-ray analysis. Nature 1958;181:662–6. https://doi.org/10.1038/181662a0.

[2] Renaud JP, Chari A, Ciferri C, Liu WT, Rémigy HW, Stark H, Wiesmann C. Cryo-EM in drug discovery: achievements, limitations and prospects. Nat Rev Drug Discov 2018;17:471–92. https://doi.org/10.1038/nrd.2018.77.

[3] Shi Y. A glimpse of structural biology through X-ray crystallography. Cell 2014;159:995–1014. https://doi.org/10.1016/j.cell.2014.10.051.

[4] Murata K, Wolf M. Cryo-electron microscopy for structural analysis of dynamic biological macromolecules. Biochim Biophys Acta Gen Subj 2018;1862:324–34. https://doi.org/10.1016/j.bbagen.2017.07.020.

[5] Miller MD, Phillips GN. Moving beyond static snapshots: protein dynamics and the Protein Data Bank. J Biol Chem 2021;296. https://doi.org/10.1016/j.jbc.2021.100749, 100749.

[6] Lakowicz JR. Principles of fluorescence spectroscopy. 3rd ed. US, Boston, MA: Springer; 2006. https://doi.org/10.1007/978-0-387-46312-4.

[7] Weber G. Polarization of the fluorescence of macromolecules. 1. Theory and experimental method. Biochem J 1952;51:145–55. https://doi.org/10.1042/bj0510145.

[8] Birch DJS, Imhof RE. Time-domain fluorescence spectroscopy using time-correlated single-photon counting. In: Lakowicz JR, editor. Top. fluoresc. spectrosc. Boston, MA: Springer US; 2006. p. 1–95. https://doi.org/10.1007/0-306-47057-8_1.

[9] Becker W. Introduction to multi-dimensional TCSPC. In: Becker W, editor. Adv. time-correlated single phot. count. appl. 1st ed. Cham: Springer Nature Switzerland; 2015. p. 1–63. https://doi.org/10.1007/978-3-319-14929-5_1.

[10] Yoon E, Konold PE, Lee J, Joo T, Jimenez R. Far-red emission of mPlum fluorescent protein results from excited-state interconversion between chromophore hydrogen-bonding states. J Phys Chem Lett 2016;7:2170–4. https://doi.org/10.1021/acs.jpclett.6b00823.

[11] Tcherkasskaya O, Bychkova VE, Uversky VN, Gronenborn AM. Multisite fluorescence in proteins with multiple tryptophan residues: apomyoglobin natural variants and site-directed mutants. J Biol Chem 2000;275:36285–94. https://doi.org/10.1074/jbc.M003008200.

[12] Connolly M, Arra A, Zvoda V, Steinbach PJ, Rice PA, Ansari A. Static kinks or flexible hinges: multiple conformations of bent DNA bound to integration host factor revealed by fluorescence lifetime measurements. J Phys Chem B 2018;122:11519–34. https://doi.org/10.1021/acs.jpcb.8b07405.

[13] Das A, Raghuraman H. Conformational heterogeneity of the voltage sensor loop of KvAP in micelles and membranes: a fluorescence approach. Biochim Biophys Acta Biomembr 2021;1863. https://doi.org/10.1016/j.bbamem.2021.183568, 183568.

[14] Alcala JR, Gratton E, Prendergast FG. Resolvability of fluorescence lifetime distributions using phase fluorometry. Biophys J 1987;51:587–96. https://doi.org/10.1016/S0006-3495(87)83383-0.

[15] Lakowicz JR, Cherek H, Gryczynski I, Joshi N, Johnson ML. Analysis of fluorescence decay kinetics measured in the frequency domain using distributions of decay times. Biophys Chem 1987;28:35–50. https://doi.org/10.1016/0301-4622(87)80073-X.

[16] Raghuraman H, Chatterjee S, Das A. Site-directed fluorescence approaches for dynamic structural biology of membrane peptides and proteins. Front Mol Biosci 2019;6:1–25. https://doi.org/10.3389/fmolb.2019.00096.

[17] Teale FWJ, Weber G. Ultraviolet fluorescence of the aromatic amino acids. Biochem J 1957;65:476–82. https://doi.org/10.1042/bj0650476.

[18] Glandières J-M, Twist C, Haouz A, Zentz C, Alpert B. Resolved fluorescence of the two tryptophan residues in horse apomyoglobin. Photochem Photobiol 2000;71:382. https://doi.org/10.1562/0031-8655(2000)071<0382:rfottt>2.0.co;2.

[19] Krueger AT, Imperiali B. Fluorescent amino acids: modular building blocks for the assembly of new tools for chemical biology. Chembiochem 2013;14:788–99. https://doi.org/10.1002/cbic.201300079.

[20] Cheng Z, Kuru E, Sachdeva A, Vendrell M. Fluorescent amino acids as versatile building blocks for chemical biology. Nat Rev Chem 2020;4:275–90. https://doi.org/10.1038/s41570-020-0186-z.

[21] Braun N, Sheikh ZP, Pless SA. The current chemical biology tool box for studying ion channels. J Physiol 2020;598:4455–71. https://doi.org/10.1113/JP276695.

[22] de la Torre D, Chin JW. Reprogramming the genetic code. Nat Rev Genet 2021;22:169–84. https://doi.org/10.1038/s41576-020-00307-7.

[23] Toseland CP. Fluorescent labeling and modification of proteins. J Chem Biol 2013;6:85–95. https://doi.org/10.1007/s12154-013-0094-5.

[24] Los GV, Encell LP, McDougall MG, Hartzell DD, Karassina N, Zimprich C, Wood MG, Learish R, Ohana RF, Urh M, Simpson D, Mendez J, Zimmerman K, Otto P, Vidugiris G, Zhu J, Darzins A, Klaubert DH, Bulleit RF, Wood KV. HaloTag: a novel

protein labeling technology for cell imaging and protein analysis. ACS Chem Biol 2008;3:373–82. https://doi.org/10.1021/cb800025k.

[25] Maurel D, Comps-Agrar L, Brock C, Rives ML, Bourrier E, Ayoub MA, Bazin H, Tinel N, Durroux T, Prézeau L, Trinquet E, Pin JP. Cell-surface protein-protein interaction analysis with time-resolved FRET and snap-tag technologies: application to GPCR oligomerization. Nat Methods 2008;5:561–7. https://doi.org/10.1038/nmeth.1213.

[26] Squires AH, Moerner WE. Direct single-molecule measurements of phycocyanobilin photophysics in monomeric C-phycocyanin. Proc Natl Acad Sci U S A 2017; 114:9779–84. https://doi.org/10.1073/pnas.1705435114.

[27] Monni R, Al Haddad A, Van Mourik F, Auböck G, Chergui M. Tryptophan-to-heme electron transfer in ferrous myoglobins. Proc Natl Acad Sci U S A 2015; 112:5602–6. https://doi.org/10.1073/pnas.1423186112.

[28] Sławski J, Białek R, Burdziński G, Gibasiewicz K, Worch R, Grzyb J. Competition between photoinduced electron transfer and resonance energy transfer in an example of substituted cytochrome c–quantum dot systems. J Phys Chem B 2021;125:3307–20.

[29] Stratt RM, Maroncelli M. Nonreactive dynamics in solution: the emerging molecular view of solvation dynamics and vibrational relaxation. J Phys Chem 1996;100: 12981–96. https://doi.org/10.1021/jp9608483.

[30] Jurkiewicz P, Sýkora J, Olzyńska A, Humpolíčková J, Hof M. Solvent relaxation in phospholipid bilayers: principles and recent applications. J Fluoresc 2005; 15:883–94. https://doi.org/10.1007/s10895-005-0013-4.

[31] Mokranjac D, Neupert W. The many faces of the mitochondrial TIM23 complex. Biochim Biophys Acta Bioenerg 2010;1797:1045–54. https://doi.org/10.1016/j.bbabio.2010.01.026.

[32] Alder NN, Jensen RE, Johnson AE. Fluorescence mapping of mitochondrial TIM23 complex reveals a water-facing, substrate-interacting helix surface. Cell 2008;134:439–50. https://doi.org/10.1016/j.cell.2008.06.007.

[33] Frotscher E, Krainer G, Schlierf M, Keller S. Dissecting nanosecond dynamics in membrane proteins with dipolar relaxation upon tryptophan photoexcitation. J Phys Chem Lett 2018;9:2241–5. https://doi.org/10.1021/acs.jpclett.8b00834.

[34] Zaragoza JPT, Nguy A, Minnetian N, Deng Z, Iavarone AT, Offenbacher AR, Klinman JP. Detecting and characterizing the kinetic activation of thermal networks in proteins: thermal transfer from a distal, solvent-exposed loop to the active site in soybean lipoxygenase. J Phys Chem B 2019;123:8662–74. https://doi.org/10.1021/acs.jpcb.9b07228.

[35] Darżynkiewicz ZM, Pędziwiatr M, Grzyb J. Quantum dots use both LUMO and surface trap electrons in photoreduction process. JOL 2017;183. https://doi.org/10.1016/j.jlumin.2016.11.070.

[36] Xu J, Knutson JR. Chapter 8 ultrafast fluorescence spectroscopy via upconversion. Applications to biophysics. In: Brand L, Johnson ML, editors. Methods enzymol. Elsevier; 2008. p. 159–83. https://doi.org/10.1016/S0076-6879(08)03408-3.

[37] Chen K, Gallaher JK, Barker AJ, Hodgkiss JM. Transient grating photoluminescence spectroscopy: an ultrafast method of gating broadband spectra. J Phys Chem Lett 2014;5:1732–7. https://doi.org/10.1021/jz5006362.

[38] Houston P, Macro N, Kang M, Chen L, Yang J, Wang L, Wu Z, Zhong D. Ultrafast dynamics of water-protein coupled motions around the surface of eye crystallin. J Am Chem Soc 2020;142:3997–4007. https://doi.org/10.1021/jacs.9b13506.

[39] Macro N, Chen L, Yang Y, Mondal T, Wang L, Horovitz A, Zhong D. Slowdown of water dynamics from the top to the bottom of the GroEL cavity. J Phys Chem Lett 2021;12:5723–30. https://doi.org/10.1021/acs.jpclett.1c01216.

[40] Arya S, Mukhopadhyay S. Ordered water within the collapsed globules of an amyloidogenic intrinsically disordered protein. J Phys Chem B 2014;118:9191–8. https://doi.org/10.1021/jp504076a.
[41] Mishra P, Jha SK. Slow motion protein dance visualized using red-edge excitation shift of a buried fluorophore. J Phys Chem B 2019;123:1256–64. https://doi.org/10.1021/acs.jpcb.8b11151.
[42] Patti M, Fenollar-Ferrer C, Werner A, Forrest LR, Forster IC. Cation interactions and membrane potential induce conformational changes in NaPi-IIb. Biophys J 2016;111:973–88. https://doi.org/10.1016/j.bpj.2016.07.025.
[43] Wulf M, Pless SA. High-sensitivity fluorometry to resolve ion channel conformational dynamics. Cell Rep 2018;22:1615–26. https://doi.org/10.1016/j.celrep.2018.01.029.
[44] Gorraitz E, Hirayama BA, Paz A, Wright EM, Loo DDF. Active site voltage clamp fluorometry of the sodium glucose cotransporter hSGLT1. Proc Natl Acad Sci U S A 2017;114:E9980–8. https://doi.org/10.1073/pnas.1713899114.
[45] Algar WR, Hildebrandt N, Vogel SS, Medintz IL. FRET as a biomolecular research tool—understanding its potential while avoiding pitfalls. Nat Methods 2019;16:815–29. https://doi.org/10.1038/s41592-019-0530-8.
[46] Sobakinskaya E, Schmidt am Busch M, Renger T. Theory of FRET "spectroscopic ruler" for short distances: application to polyproline. J Phys Chem B 2018;122:54–67. https://doi.org/10.1021/acs.jpcb.7b09535.
[47] Yu X, Wu X, Bermejo GA, Brooks BR, Taraska JW. Accurate high-throughput structure mapping and prediction with transition metal ion FRET. Structure 2013;21:9–19. https://doi.org/10.1016/j.str.2012.11.013.
[48] Billesbølle CB, Mortensen JS, Sohail A, Schmidt SG, Shi L, Sitte HH, Gether U, Loland CJ. Transition metal ion FRET uncovers K+ regulation of a neurotransmitter/sodium symporter. Nat Commun 2016;7:12755. https://doi.org/10.1038/ncomms12755.
[49] Bhattacharya K, Bernasconi L, Picard D. Luminescence resonance energy transfer between genetically encoded donor and acceptor for protein-protein interaction studies in the molecular chaperone HSP70/HSP90 complexes. Sci Rep 2018;8:2801. https://doi.org/10.1038/s41598-018-21210-6.
[50] Castillo JP, Sánchez-Rodríguez JE, Hyde HC, Zaelzer CA, Aguayo D, Sepúlveda RV, Luk LYP, Kent SBH, Gonzalez-Nilo FD, Bezanilla F, Latorre R. β1-subunit-induced structural rearrangements of the $Ca^{2+}$– and voltage-activated $K^+$ (BK) channel. Proc Natl Acad Sci U S A 2016;113:E3231–9. https://doi.org/10.1073/pnas.1606381113.
[51] Kubota T, Durek T, Dang B, Finol-Urdaneta RK, Craik DJ, Kent SBH, French RJ, Bezanilla F, Correa AM. Mapping of voltage sensor positions in resting and inactivated mammalian sodium channels by LRET. Proc Natl Acad Sci U S A 2017;114:E1857–65. https://doi.org/10.1073/pnas.1700453114.
[52] Hochreiter B, Kunze M, Moser B, Schmid JA. Advanced FRET normalization allows quantitative analysis of protein interactions including stoichiometries and relative affinities in living cells. Sci Rep 2019;9:8233. https://doi.org/10.1038/s41598-019-44650-0.
[53] Picard LP, Schönegge AM, Lohse MJ, Bouvier M. Bioluminescence resonance energy transfer-based biosensors allow monitoring of ligand- and transducer-mediated GPCR conformational changes. Commun Biol 2018;1:106. https://doi.org/10.1038/s42003-018-0101-z.
[54] Devost D, Sleno R, Pé Trin D, Zhang A, Shinjo Y, Okde R, Aoki J, Inoue A, Hébert TE. Conformational profiling of the AT1 angiotensin II receptor reflects biased agonism, G protein coupling, and cellular context. J Biol Chem 2017;292:5443–56. https://doi.org/10.1074/jbc.M116.763854.
[55] Che T, English J, Krumm BE, Kim K, Pardon E, Olsen RHJ, Wang S, Zhang S, Diberto JF, Sciaky N, Carroll FI, Steyaert J, Wacker D, Roth BL. Nanobody-enabled

monitoring of kappa opioid receptor states. Nat Commun 2020;11:1145. https://doi.org/10.1038/s41467-020-14889-7.
[56] Pantazis A, Westerberg K, Althoff T, Abramson J, Olcese R. Harnessing photoinduced electron transfer to optically determine protein sub-nanoscale atomic distances. Nat Commun 2018;9:4738. https://doi.org/10.1038/s41467-018-07218-6.
[57] Grzyb J, Kalwarczyk E, Worch R. Photoreduction of natural redox proteins by CdTe quantum dots is size-tunable and conjugation-independent. RSC Adv 2015;5. https://doi.org/10.1039/c5ra02900g.
[58] Pfefferkorn CM, Walker RL, He Y, Gruschus JM, Lee JC. Tryptophan probes reveal residue-specific phospholipid interactions of apolipoprotein C-III. Biochim Biophys Acta Biomembr 2015;1848:2821–8. https://doi.org/10.1016/j.bbamem.2015.08.018.
[59] Eftink MR. Fluorescence quenching: theory and applications. In: Lakowicz JR, editor. Top. fluoresc. spectrosc. 1st ed. Boston, MA: Springer US; 2006. p. 53–126. https://doi.org/10.1007/0-306-47058-6_2.
[60] Soranno A, Holla A, Dingfelder F, Nettels D, Makarov DE, Schuler B. Integrated view of internal friction in unfolded proteins from single-molecule FRET, contact quenching, theory, and simulations. Proc Natl Acad Sci U S A 2017;114: E1833–9. https://doi.org/10.1073/pnas.1616672114.
[61] Jarecki BW, Zheng S, Zhang L, Li X, Zhou X, Cui Q, Tang W, Chanda B. Tethered spectroscopic probes estimate dynamic distances with subnanometer resolution in voltage-dependent potassium channels. Biophys J 2013;105:2724–32. https://doi.org/10.1016/j.bpj.2013.11.010.
[62] Vasquez-Montes V, Vargas-Uribe M, Pandey NK, Rodnin MV, Langen R, Ladokhin AS. Lipid-modulation of membrane insertion and refolding of the apoptotic inhibitor Bcl-xL. Biochim Biophys Acta Proteins Proteomics 2019;1867:691–700. https://doi.org/10.1016/j.bbapap.2019.04.006.
[63] Rabe M, Aisenbrey C, Pluhackova K, de Wert V, Boyle AL, Bruggeman DF, Kirsch SA, Böckmann RA, Kros A, Raap J, Bechinger B. A coiled-coil peptide shaping lipid bilayers upon fusion. Biophys J 2016;111:2162–75. https://doi.org/10.1016/j.bpj.2016.10.010.
[64] Bai N, Roder H, Dickson A, Karanicolas J. Isothermal analysis of ThermoFluor data can readily provide quantitative binding affinities. Sci Rep 2019;9:2650. https://doi.org/10.1038/s41598-018-37072-x.
[65] Cimmperman P, Baranauskiene L, Jachimovičiute S, Jachno J, Torresan J, Michailoviene V, Matuliene J, Sereikaite J, Bumelis V, Matulis D. A quantitative model of thermal stabilization and destabilization of proteins by ligands. Biophys J 2008; 95:3222–31. https://doi.org/10.1529/biophysj.108.134973.
[66] DeLeeuw LW, Monsen RC, Petrauskas V, Gray RD, Baranauskiene L, Matulis D, Trent JO, Chaires JB. POT1 stability and binding measured by fluorescence thermal shift assays. PLoS One 2021;16. https://doi.org/10.1371/journal.pone.0245675, e0245675.
[67] Tracz M, Górniak I, Szczepaniak A, Białek W. E3 ubiquitin ligase SPL2 is a lanthanide-binding protein. Int J Mol Sci 2021;22:5712. https://doi.org/10.3390/ijms22115712.
[68] Elgert C, Rühle A, Sandner P, Behrends S. Thermal shift assay: strengths and weaknesses of the method to investigate the ligand-induced thermostabilization of soluble guanylyl cyclase. J Pharm Biomed Anal 2020;181. https://doi.org/10.1016/j.jpba.2019.113065, 113065.
[69] Velkov T, Lim MLR, Horne J, Simpson JS, Porter CJH, Scanlon MJ. Characterization of lipophilic drug binding to rat intestinal fatty acid binding protein. Mol Cell Biochem 2009;326:87–95. https://doi.org/10.1007/s11010-008-0009-x.
[70] Liu WW, Zhu Y, Fang Q. Femtomole-scale high-throughput screening of protein ligands with droplet-based thermal shift assay. Anal Chem 2017;89:6678–85. https://doi.org/10.1021/acs.analchem.7b00899.

[71] Ehrhardt MKG, Warring SL, Gerth ML. Screening chemoreceptor–ligand interactions by high-throughput thermal-shift assays. In: Manson M, editor. Bact. chemosensing. methods mol. biol. New York, NY: Humana Press; 2018. p. 281–90. https://doi.org/10.1007/978-1-4939-7577-8_22.
[72] Seashore-Ludlow B, Axelsson H, Almqvist H, Dahlgren B, Jonsson M, Lundbäck T. Quantitative interpretation of intracellular drug binding and kinetics using the cellular thermal shift assay. Biochemistry 2018;57:6715–25. https://doi.org/10.1021/acs.biochem.8b01057.
[73] Cho EJ, Dalby KN. Luminescence energy transfer–based screening and target engagement approaches for chemical biology and drug discovery. SLAS Discov 2021;26:984–94. https://doi.org/10.1177/24725552211036056.
[74] PerkinElmer. Technical data sheet: alpha CETSA® research reagents; n.d. Available at: https://www.perkinelmer.com/pl/lab-products-and-services/application-support-knowledgebase/alpha-cetsa-knowledgebase.html.
[75] Jameson DM, Ross JA. Fluorescence polarization/anisotropy in diagnostics and imaging. Chem Rev 2010;110:2685–708. https://doi.org/10.1021/cr900267p.
[76] Vogel SS, Nguyen TA, Blank PS, Wieb Van Der Meer B. An introduction to interpreting time resolved fluorescence anisotropy curves. In: Becker W, editor. Adv. time-correlated single phot. count. appl. Cham: Springer Nature Switzerland; 2015. p. 385–406. https://doi.org/10.1007/978-3-319-14929-5_12.
[77] Jain N, Narang D, Bhasne K, Dalal V, Arya S, Bhattacharya M, Mukhopadhyay S. Direct observation of the intrinsic backbone torsional mobility of disordered proteins. Biophys J 2016;111:768–74. https://doi.org/10.1016/j.bpj.2016.07.023.
[78] Bhasne K, Jain N, Karnawat R, Arya S, Majumdar A, Singh A, Mukhopadhyay S. Discerning dynamic signatures of membrane-bound α-synuclein using site-specific fluorescence depolarization kinetics. J Phys Chem B 2020;124:708–17. https://doi.org/10.1021/acs.jpcb.9b09118.
[79] Chan FTS, Kaminski CF, Schierle GSK. HomoFRET fluorescence anisotropy imaging as a tool to study molecular self-assembly in live cells. ChemPhysChem 2011;12:500–9. https://doi.org/10.1002/cphc.201000833.
[80] Majumdar A, Das D, Madhu P, Avni A, Mukhopadhyay S. Excitation energy migration unveils fuzzy interfaces within the amyloid architecture. Biophys J 2020;118:2621–6. https://doi.org/10.1016/j.bpj.2020.04.015.
[81] Renart ML, Giudici AM, Poveda JA, Fedorov A, Berberan-Santos MN, Prieto M, Díaz-García C, González-Ros JM, Coutinho A. Conformational plasticity in the KcsA potassium channel pore helix revealed by homo-FRET studies. Sci Rep 2019;9:6215. https://doi.org/10.1038/s41598-019-42405-5.
[82] Edwards GB, Muthurajan UM, Bowerman S, Luger K. Analytical ultracentrifugation (AUC): an overview of the application of fluorescence and absorbance AUC to the study of biological macromolecules. Curr Protoc Mol Biol 2020;133. https://doi.org/10.1002/cpmb.131.
[83] Chaturvedi SK, Ma J, Zhao H, Schuck P. Use of fluorescence-detected sedimentation velocity to study high-affinity protein interactions. Nat Protoc 2017;12:1777–91. https://doi.org/10.1038/nprot.2017.064.
[84] Brown PH, Schuck P. Macromolecular size-and-shape distributions by sedimentation velocity analytical ultracentrifugation. Biophys J 2006;90:4651–61.
[85] Schuck P, Perugini MA, Gonzales NR, Howlett GJ, Schubert D. Size-distribution analysis of proteins by analytical ultracentrifugation: strategies and application to model systems. Biophys J 2002;82:1096–111.
[86] Chaturvedi SK, Ma J, Brown PH, Zhao H, Schuck P. Measuring macromolecular size distributions and interactions at high concentrations by sedimentation velocity. Nat Commun 2018;9:1–9.

[87] Sztatelman O, Kopeć K, Pędziwiatr M, Trojnar M, Worch R, Wielgus-Kutrowska B, Jemioła-Rzemińska M, Bzowska A, Grzyb J. Heterodimerizing helices as tools for nanoscale control of the organization of protein-protein and protein-quantum dots. Biochimie 2019;167:93–105. https://doi.org/10.1016/j.biochi.2019.09.015.
[88] Seidel SAI, Wienken CJ, Geissler S, Jerabek-Willemsen M, Duhr S, Reiter A, Trauner D, Braun D, Baaske P. Label-free microscale thermophoresis discriminates sites and affinity of protein–ligand binding. Angew Chem Int Ed 2012;51:10656–9.
[89] Seidel SAI, Dijkman PM, Lea WA, van den Bogaart G, Jerabek-Willemsen M, Lazic A, Joseph JS, Srinivasan P, Baaske P, Simeonov A. Microscale thermophoresis quantifies biomolecular interactions under previously challenging conditions. Methods 2013;59:301–15.
[90] Lai Y-T, Chang Y-Y, Hu L, Yang Y, Chao A, Du Z-Y, Tanner JA, Chye M-L, Qian C, Ng K-M. Rapid labeling of intracellular His-tagged proteins in living cells. Proc Natl Acad Sci U S A 2015;112:2948–53.
[91] Petrov EP, Schwille P. State of the art and novel trends in fluorescence correlation spectroscopy. In: Standardization and quality assurance in fluorescence measurements II. Springer series on fluorescence, vol. 6. Berlin, Heidelberg: Springer; 2008. p. 145–97. https://doi.org/10.1007/4243_2008_032.
[92] Bacia K, Kim SA, Schwille P. Fluorescence cross-correlation spectroscopy in living cells. Nat Methods 2006;3:83–9. https://doi.org/10.1038/NMETH822.
[93] Kopeć K, Pędziwiatr M, Gront D, Sztatelman O, Sławski J, Łazicka M, Worch R, Zawada K, Makarova K, Nyk M, Grzyb J. Comparison of α-helix and β-sheet structure adaptation to a quantum dot geometry: toward the identification of an optimal motif for a protein nanoparticle cover. ACS Omega 2019;4:13086–99. https://doi.org/10.1021/acsomega.9b00505.
[94] Dabrowska A, Nyk M, Worch R, Grzyb J. Hydrophilic colloidal quantum dots with long peptide chain coats. Colloids Surf B Biointerfaces 2016;145. https://doi.org/10.1016/j.colsurfb.2016.05.081.
[95] Tiwari M, Mikuni S, Muto H, Kinjo M. Determination of dissociation constant of the NFκB p50/p65 heterodimer using fluorescence cross-correlation spectroscopy in the living cell. Biochem Biophys Res Commun 2013;436:430–5.
[96] Yu S, Li F, Huang X, Dong C, Ren J. In situ study of interactions between endogenous c-myc mRNA with CRDBP in a single living cell by combining fluorescence cross-correlation spectroscopy with molecular beacons. Anal Chem 2020;92:2988–96.
[97] Wennmalm S, Chmyrov V, Widengren J, Tjernberg L. Highly sensitive FRET-FCS detects amyloid β-peptide oligomers in solution at physiological concentrations. Anal Chem 2015;87:11700–5.
[98] Mailfert S, Wojtowicz K, Brustlein S, Blaszczak E, Bertaux N, Łukaszewicz M, et al. Spot variation fluorescence correlation spectroscopy for analysis of molecular diffusion at the plasma membrane of living cells. J Vis Exp 2020;(165):e61823. https://doi.org/10.3791/61823.
[99] Wójtowicz K, Czogalla A, Trombik T, Łukaszewicz M. Surfactin cyclic lipopeptides change the plasma membrane composition and lateral organization in mammalian cells. Biochim Biophys Acta Biomembr 2021;, 183730.
[100] Edwards KA, Demsky M, Montague RA, Weymouth N, Kiehart DP. GFP-Moesin illuminates actin cytoskeleton dynamics in living tissue and demonstrates cell shape changes during morphogenesis in Drosophila. Dev Biol 1997;191:103–17.
[101] Biel A, Moser M, Meier I. Arabidopsis KASH proteins SINE1 and SINE2 are involved in microtubule reorganization during ABA-induced stomatal closure. Front Plant Sci 2020;11.
[102] Srivastava R, Chen Y, Deng Y, Brandizzi F, Howell SH. Elements proximal to and within the transmembrane domain mediate the organelle-to-organelle movement of bZIP28 under ER stress conditions. Plant J 2012;70:1033–42.

[103] Pawley J. Handbook of biological confocal microscopy. Springer Science & Business Media; 2006.

[104] Ishikawa-Ankerhold HC, Ankerhold R, Drummen GPC. Advanced fluorescence microscopy techniques—Frap, Flip, Flap, Fret and flim. Molecules 2012;17:4047–132.

[105] Aspinwall CL, Sarcina M, Mullineaux CW. Phycobilisome mobility in the cyanobacterium Synechococcus sp. PCC7942 is influenced by the trimerisation of photosystem I. Photosynth Res 2004;79:179–87.

[106] Klein C, Pillot T, Chambaz J, Drouet B. Determination of plasma membrane fluidity with a fluorescent analogue of sphingomyelin by FRAP measurement using a standard confocal microscope. Brain Res Protoc 2003;11:46–51.

[107] Schreiber U. Pulse-amplitude-modulation (PAM) fluorometry and saturation pulse method: an overview. In: Papageorgiou GC, Govindjee, editors. Chlorophyll a fluorescence. Advances in photosynthesis and respiration, vol. 19. Dordrecht: Springer; 2004. p. 279–319. https://doi.org/10.1007/978-1-4020-3218-9_11.

[108] Murchie EH, Lawson T. Chlorophyll fluorescence analysis: a guide to good practice and understanding some new applications. J Exp Bot 2013;64:3983–98.

[109] Janik E, Bednarska J, Zubik M, Luchowski R, Mazur R, Sowinski K, Grudzinski W, Garstka M, Gruszecki WI. A chloroplast "wake up" mechanism: illumination with weak light activates the photosynthetic antenna function in dark-adapted plants. J Plant Physiol 2017;210:1–8.

[110] Luchowski R, Grudzinski W, Welc R, Mendes Pinto MM, Sek A, Ostrowski J, Nierzwicki L, Chodnicki P, Wieczor M, Sowinski K. Light-modulated sunscreen mechanism in the retina of the human eye. J Phys Chem B 2021;125:6090–102.

[111] Wu A, Grela E, Wójtowicz K, Filipczak N, Hamon Y, Luchowski R, Grudziński W, Raducka-Jaszul O, Gagoś M, Szczepaniak A. ABCA1 transporter reduces amphotericin B cytotoxicity in mammalian cells. Cell Mol Life Sci 2019;76:4979–94.

[112] Weidtkamp-Peters S, Stahl Y. The use of FRET/FLIM to study proteins interacting with plant receptor kinases. In: Plant recept. kinases. Springer; 2017. p. 163–75.

[113] Iwai M, Roth MS, Niyogi KK. Subdiffraction-resolution live-cell imaging for visualizing thylakoid membranes. Plant J 2018;96:233–43.

[114] Wiwatowski K, Podlas P, Twardowska M, Maćkowski S. Fluorescence studies of the interplay between metal-enhanced fluorescence and graphene-induced quenching. Materials (Basel) 2018;11:1916.

[115] Mackowski S, Wörmke S, Maier AJ, Brotosudarmo THP, Harutyunyan H, Hartschuh A, Govorov AO, Scheer H, Bräuchle C. Metal-enhanced fluorescence of chlorophylls in single light-harvesting complexes. Nano Lett 2008;8:558–64.

[116] Kowalska D, Szalkowski M, Sulowska K, Buczynska D, Niedziolka-Jonsson J, Jonsson-Niedziolka M, Kargul J, Lokstein H, Mackowski S. Silver island film for enhancing light harvesting in natural photosynthetic proteins. Int J Mol Sci 2020;21:2451.

# CHAPTER 3

# Structural analysis of biomacromolecules using circular dichroism spectroscopy

**Xue Zhao, Yuxuan Wang, and Di Zhao**
Key Laboratory of Meat Processing, MOA, Key Laboratory of Meat Processing and Quality Control, MOE, Jiang Synergetic Innovation Center of Meat Production, Processing and Quality Control, Nanjing Agricultural University, Nanjing, PR China

## 1. Introduction

A plane polarized light can be regarded as a superposition of left and right circularly polarized light of the same frequency and amplitude. A chiral molecule has different absorption coefficients to the left- and right-handed circularly polarized lights. Its circular dichroism (CD) spectrum can be obtained by recording the changes in absorption coefficients with the change in the wavelength of polarized light. CD is a well-established spectroscopic technique extensively used to investigate structures of proteins, polysaccharides, and nucleic acids and also provides important information on ligand binding. The advantages of CD tools could be briefly accounted to three main aspects [1,2].

1. **Low requirement for protein concentration**: Around $\leq$20 μg of samples in solution buffers are required to analyze secondary structures. Wide range of concentrations could meet the requirement for CD measurements. The secondary and tertiary structures of varieties of proteins have been revealed even at 0.001–250 mg/mL protein concentrations [3].
2. **Simple pretreatment**: Unlike other spectroscopic methods such as Fourier transform infrared (FTIR) spectroscopy or Raman spectroscopy, where samples often need to be prepared in deuterium oxide or made as KBr tablet, or derivatization/radiolabeling as in nuclear magnetic resonance (NMR), the CD does not require specific preparation of the sample.
3. **Fast and efficient data collection:** A single CD measurement including far and near UV regions (240–180 and 320–260 nm, respectively) could be accomplished within minutes [4]. Although technologies like

*Advanced Spectroscopic Methods to Study Biomolecular Structure and Dynamics*
https://doi.org/10.1016/B978-0-323-99127-8.00013-1

NMR spectroscopy, X-ray crystallography, and electron microscopy could provide more specific details for structures of biomolecules, their operation protocol and analysis are much more time consuming and rather complex than a CD experiment.

Usually, some prerequirement should be satisfied to better utilize CD. First, the samples should be well dissolved in water (or other solvent buffers) because the insoluble biomolecules or suspended particles (such as lipids) would induce light scattering and result in absorption flattening effects, thereby providing inaccurate absorbance values, especially at low wavelengths (below 210 nm). Second, the buffer, cell path length, and cell material should be carefully chosen to avoid too high absorbance at the near and far-ultraviolet (UV) wavelength region. Due to the difference in optical components, CD values obtained from two distinctive instruments may vary by a certain level. In addition, high-quality CD data require a precise concentration determination and choice of proper datasets. This chapter presents an overview of analytical methods of biomolecules, including proteins, polysaccharides, and nucleic acid, using a CD spectrometer.

## 2. Applications of CD in protein studies

### 2.1 CD in the analysis of protein structure

In the last 40 years, many reviews have thoroughly explained the basics of the CD technique, where the readers can get valuable insight into the underlined principle [4–6]. A CD spectrum originates due to differential absorption of the left- and right-handed circularly light components of the plane polarized light by the chiral protein structure. Therefore the CD could be calculated in two ways: (i) the electric field margin ($\Delta E$) between the absorbance of two circularly polarized light by an asymmetric molecule or (ii) ellipticity degrees ($[\theta]$, the molar ellipticity in deg. $cm^2$/dmol), which reflects the ratio of the long and short axis of the ellipse derived from the rotated plane of a light wave. The CD technique is often complemented with other methods to generate more details about protein structure, such as fluorescence spectroscopy [7], cryo-electron microscopy, FTIR spectroscopy, small-angle scattering, and molecular dynamics (MD) simulations [8].

### 2.2 CD spectra of proteins

Since all amino acids (except glycine) are asymmetric and CD signal active, CD spectroscopy is an excellent tool to determine the secondary structure of proteins [2]. In protein molecules, the changes in the array of chromophores residues lead to optical transitions and therefore exhibit different CD spectra, which not only reflect protein conformation but also the conformational changes as affected by environmental conditions such as temperature, pH, ionic strength, concentrations. Although some aromatic amino acid residues (such as Phe, Tyr, and Trp residues) in asymmetric environments and disulfide bonds also exhibit CD response, their bands are located in the near UV region (250–320 nm). Secondary structure analysis by CD is an empirical method based on reference datasets of spectra (especially around the far-UV region) derived from proteins with known structures. The listed examples include some well-studied phytochemical proteins, such as soy protein, rice protein, corn zein, and animal-sourced proteins, such as ovalbumin, bovine serum albumin (BSA), myoglobin, lysozyme, myofibrillar, and gelatin. The characteristics of CD spectra are also discussed in detail. The secondary and tertiary structures and the thermodynamics associated with the folding process can be easily studied using CD spectroscopy [9–12].

For soy protein, two typical peaks are shown, a positive peak at around 195 nm and a negative peak at around 208 nm [13,14]. The former peak is indicative of the presence of β-sheet structures, and the latter peak is caused by the Cotton effect of α-helix structures [13,14]. A typical CD spectrum of rice protein shows two strong positive peaks at 192 and 195 nm and a negative peak at around 208 nm [15]. Another widely studied plant protein is zein from corn, which is always fabricated as a stabilizer in pickering emulsion. Since it is not soluble in water, the structure of zein is especially solvent dependent. Many reports mentioned the CD spectra of zein having two negative peaks at around 208 and 222 nm; both of which are characteristic of α-helix structures [16,17]. Similarly, there are three apparent peaks in CD spectra of myofibrillar protein, ovalbumin, lysozyme, BSA, and myoglobin, i.e., a positive peak at around 192 nm and two negative peaks at 208 and 222 nm. These three peaks are indicative of α-helix structures, suggesting that they are all α-helix dominated proteins [18–22]. Gelatin is a type of protein with a unique conformation. It consists of three chains that are entwined together to form a triple helix structure. Each chain is assembled by the poly-L-proline and forms an extended helical structure. All the bonds in these chain subunits are *trans* to each other. Therefore the pattern for gelatin is distinct from other proteins. The spectra of gelation consisted of a positive peak at 220–230 nm and a negative peak at around 200 nm, indicating the existence of triple-helical conformation [23].

## 2.3 Calculation of secondary structural by CD spectrogram

Since the electronic transitions (mainly $n \rightarrow \pi^*$ transition and parallel/perpendicular $\pi \rightarrow \pi^*$ transition) of polypeptides are sourced from changed dihedral angels ($\Phi$, $\Psi$) within adjacent residues, the signals of CD indicate various types of secondary structures. The collection of these secondary structure elements is correlated with the CD spectrum at the far UV region in a linear manner. Many different methods have been adopted to calculate secondary structures of protein from CD spectrum data; the two most commonly used approaches are discussed here.

### 2.3.1 Method I—Approximate estimation

The α-helix structure shows typical strong negative bands at 222 and 208 nm. In contrast, the β-strand and random coil structures have very low ellipticity, approximately −4000 deg. $cm^2$/dmol at 208 nm. Therefore the fraction of α-helix content ($f_\alpha$) can be reliably estimated at $[\theta]_{208}$ using the following equation:

$$f_\alpha = -\left([\theta]_{208} - 4000\right)/29,000$$

To further clarify, this is just a rough estimation with less accuracy because of two reasons: (i) it ignores the spectra of other secondary structures and (ii) it is found that the $[\theta]$ of α-helix is dependent on the chain length [24].

### 2.3.2 Method II—Deconvolution method based on reference databases

Based on the reference dataset of known proteins, many deconvolution methods have been developed to quantitatively calculate the secondary structure information from a CD spectrum. This protocol has been thoroughly reviewed by Miles et al. recently, and the information is briefly summarized here [6]. The deconvolution methods include parameterized fits, simple least-squares algorithms, ridge regression, singular value deconvolution, neural network training, etc., which match the reference database produced from proteins with known structures. The currently available database is built in different servers like DichroWeb, BeStSel [25], and K2D (recently upgraded with K2D3 servers) [26]. The last two methods improved the quality of secondary structure prediction, mostly for beta-sheet-rich proteins. For the distinctive type of proteins, the most suitable database is varied based on its structure (chiefly globular or flexible) and solubility (soluble, denatured, or membrane protein). A larger size of datasets often gives rise to a more accurate estimation.

## 2.4 Recent advances in predicting secondary structure using CD

The traditional algorithms are more accurate in estimating proteins with a high level of helix structures because of the well-defined dihedral angles of the stretched polypeptide that generate intense CD signals. Due to the varied direction (parallel or antiparallel) and twist level of β-strands, the β-sheet-rich structure is more complicated to be rightfully predicted. In recent years, one of the advances in CD technique is the development and upgradation of databases and algorithms more suitable for β-sheet dominated protein structure [25–27].

Another advancement is the emerging applications of synchrotron radiation CD (SRCD) in recent years. Compared to the conventional CD spectrum, the SRCD (namely, coupled with high-flux SR beamlines) possesses many advantages, such as it could provide information at low UV regions (the flux kept relatively constant down to 140 nm), improve signal-to-noise levels, and inhibit some components interference like salts. The SRCD was first well applied in structure estimation by Wallace [28] and then gradually expanded worldwide. Before 2010, only ~100 publications used SRCD for structure studies [29]. In 2017, Kumagai et al. (2017) pointed out that owing to the shortage of SRCD beamline operations in America, the development of this technique in Latin America was largely limited [30]. Currently, the SRCD technology is not limited to life science, biological studies, and drug design but is also extensively utilized in the biopolymer science field to reveal amphiphilicity-induced structure changes, structural transformation during processing, or taste mechanism [22,31–33]. The most favored merit of SRCD is its ability to measure structures of insoluble or membrane proteins, which brings in opportunities for a more detailed and direct examination of the conformational changes occurring during denaturation or autocatalysis [34].

## 2.5 Changes in CD spectra by environmental factors

### *2.5.1 pH*

It is noticed that around the pH of the protein, the CD spectra lose strong signals due to scattered polarized light induced by turbid aggregates formation. Other than that, the soy glycinin exhibits somewhat similar spectra with minor changes at different pH [35]. While for chicken myofibrillar protein, it seems that the CD spectra pattern is affected by alkaline pH, exhibiting a less negative peak around 208 and 222 nm, which could be interpreted as hugely lost α-helix architecture [36].

### 2.5.2 Temperature

CD has been used to investigate the unfolding pathway of proteins such as casein, whey protein, and zein as a function of temperature to reveal their thermal stability. Based on CD results, in most of the proteins (i.e., quinoa protein isolate, soy protein, mantle protein) that were confronted with denaturation during heating, a loss of α-helix/β-sheet or both were observed [37–39]. Ma et al. (2020) claimed that the lower protein concentration or higher pH resulted in a higher level of β-sheets loss during heating [37]. It was also found the 37% α-helix structures in the mixture of mantle actomyosin/paramyosin turned to 100% β-sheets during heating [39].

### 2.5.3 Chemical modifications

The modification of amino acid residues (e.g., acetylation, glycation, oxidation, and phosphorylation) change the secondary structure of proteins and is widely investigated by CD tools.

1. **Acetylation**: CD results confirmed that the acetylation significantly modified pea protein structure by decreasing the content of α-helix while increasing the content of β-turn [40]. It was found that acetylation gave rise to an unordered structure of soy protein, exhibiting increased band intensity of far-UV CD spectra and shifted peaks at 210 nm toward a shorter wavelength [41].
2. **Glycation**: By using far-UV CD, Song et al. (2018) suggested after glycosylated with oligochitosan, soy protein lost some α-helix and β-sheet while obtaining some random coil curvatures [42]. Differently, it indicates that the glucose-based glycosylation did not contribute to secondary structure changes for soy protein isolate [43]. Similar results were observed in lactose-glycated whey protein samples, reflecting similar CD spectra after glycation [44].
3. **Oxidation**: It is determined by CD that the oxidation of myofibrillar proteins would first convert α-helix to β-sheet and then to a disordered β-turn afterward [45]. For soy protein, oxidation would lead to decreased intensity of negative peaks at 208, 218, and 222 nm and a positive peak at 194 nm, implying a loss of α-helix and β-sheet structures [13].

4. **Phosphorylation**: As studied by CD spectroscopy, the secondary structure of soy protein would not be changed by phosphorylation [43]. However, for myofibrillar protein, it is reported that the phosphorylated protein at pH 5.0 exhibited an essentially distinctive helical pattern, denoting a significant loss of α-helices compared to untreated protein sample (from 28.3% to 17.2%) [46]. Hu et al. (2019) have concluded that the phosphorylation at around acid or neutral pH would only slightly affect the CD spectra, while the treatment at alkaline pH led to significant changes in the secondary structure of rice bran protein [47].

## 2.6 Changes in CD spectra by protein-related interactions

### 2.6.1 Protein-protein interaction

The protein-protein interaction could be noticed from different ellipticity patterns against the respective spectrum [48]. By comparing the CD spectra between coprecipitated and blended pea/whey protein complex, it was found that the blended samples exhibited prominent double peaks with lower intensity. In contrast, the coprecipitated complex showed a whey protein dominated structure, indicating a varied assembly behavior of pea and whey proteins among these two processes [49]. Tan et al. (2019) used near-UV CD spectra to study the interaction between tilapia and soybean proteins during coprecipitation. The molar ellipticity changes and peaks migrating both indicated a strong interaction between aromatic residues within two protein chains [50]. From similar CD spectra, it was concluded that the interaction between rice glutelin and casein did not change their secondary structure [51].

### 2.6.2 Protein-ligand interaction

As the far-UV signal often indicates secondary structure changes, in the near-UV region, the tertiary structure changes could be probed by ligand chromophore moieties, which unveiled protein-ligand interactions [52,53]. Besides, some prosthetic groups like flavonoids, carotenoids, and beams could be examined in the visible regions (400–800 nm). To study protein-ligand interactions, the CD could provide unique and direct information that was otherwise unattainable through other tools like fluorescence, isothermal titration calorimetry, or surface plasmon resonance. Therefore the CD tools are always accompanied by these measurements and other spectroscopic methods (i.e., UV-vis, fluorescence, Raman spectrum, and FTIR) to provide more comprehensive information. The protocol of CD in studying protein-ligand interactions has been previously reviewed by Rodger et al. (2005) [54]. There are mainly two aspects of protein-ligand interactions that can be deduced from CD spectra:

1. **Using far-UV CD spectra**: Many studies have widely investigated how ligand binding affects the secondary structure changes of the macromolecular protein. It was observed that the binding of cyanidin-3-O-glucoside with various thermal treated soy proteins resulted in negative effects occurring at 205, 210, and 220 nm, indicating the apparent change in protein structure [55]. Zhao et al. (2020) also found that the complex formation between betalain and soy protein helped stabilize β-sheet structures [56]. By binding epigallocatechin gallate (EGCG) and EGCG-Cu upon BSA, the content of α-helix could be decreased because of induced partial unfolding [57]. Using CD to examine the binding properties between BSA and SDS, it was found that the changes in secondary structure induced by thermal and SDS ratio indicated increased stability after the saturation of the first binding site of proteins [58]. As a cryoprotectant, the ligand lactobionic acid (LBA) and BSA interactions were studied by CD. The result revealed the protective role of LBA toward the conformation of BSA from freeze-thaw damage [59].
2. **Using near-UV CD spectra**: In the case of the transitions of the aromatic side chains, tertiary structures, and ligands with chromophores binding toward protein backbones, the near-UV CD spectra would be changed by extrinsic CD bands [60]. These changes can be used to trace binding details. The interactions between EGCG and serum albumin (SA) proteins were investigated by CD, and effects appeared around 325 and 280 nm. These effects are attributed to interactions between protein-EGCG and conformational changes of EGCG and protein [61]. Similar results have been reported in the EGC-BSA complex, in which the EGC exhibited positive peaks at 280 nm and negative peaks at 320 nm toward BSA in a dose-dependent manner [62].

## 3. Applications of CD in studying polysaccharides

### 3.1 Basic knowledge of CD spectrum of polysaccharides

The ideal aim of CD analysis for polysaccharides should be to determine the absolute conformations, distinguish coils from helices, distinguish flexible from rigid structures, and detect the linkage information [63]. However, in contrast to proteins, the structures of polysaccharides usually have no simple secondary structures that describe proteins, including α-helices, β-sheets, β-turns, and random coils. In solutions, polysaccharides can usually show disordered, extended, collapsed, and helical structures [64]. Whether disordered, extended, collapsed, or helical, the structure of polysaccharides can be determined by the multidimensional potential energy surfaces. Current CD analysis of polysaccharides can be divided into vacuum UV region (<200 nm) and induced CD region (>200 nm).

## 3.2 Polysaccharides in the vacuum UV region

Linkages of monosaccharides show wavelengths shorter than 200 nm in CD spectra, and the CD of the linkage ether chromophore is directly related to the linkage conformation of polysaccharides [65]. In addition, different ether chromophores in sugar rings give rise to overlapped CD spectra. If the electronic transitions of the two ether groups are strongly coupled, the entire acetal group should be considered as the chromophore [66]. Arndt and Stevens [67] measured seven types of pyranoside thin films by circular dichromatic chromatography (140–200 nm) and obtained curves that correspond microscopically to the types of electron level transitions in the vacuum UV region. Cziner et al. [68] proposed an empirical rule, the quadrant rule, for the interpretation of CD in the 170–175 nm region. Later, Arndt and Stevens further improved this theory in the 140–200 nm region, called the n-3$p_z$ and n-3$p_y$ quadrant rules [67]. As shown in Fig. 1, the rules indicated that the acetal two ether oxygen chromophores independently contributed to the CD. Each oxygen ether bond formed by the plane and its symmetry plane divides the space into four quadrants, and spatial adjacent oxygen atoms are called "perturbers," depending on their position in the space.

Therefore positive or negative CD spectra of polysaccharides are produced between 160 and 180 nm. Generally speaking, only two cases need to be considered. First, the chromophore and the perturbator are in a side position (gauche) by the C—C bond. Second, the chromophore is correlated with the perturbator in the one and three axial directions, in which

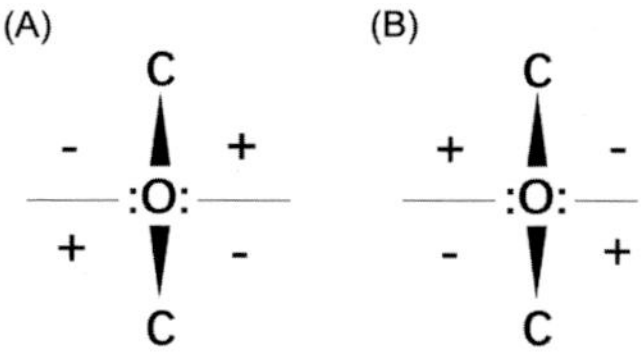


**Fig. 1** Quadrant role for the n-3pz (A) and n-3py (B) transition of acetal oxygen chromophores in saccharides.

**Table 1** CD spectra of native polysaccharides.

| Polysaccharides | CD spectra characteristics | Ref. |
|---|---|---|
| Amylose | Strong positive peak at 166 nm and negative peak at 182 nm | [63] |
| Cellulose | Strong negative peak at 158 nm | [64] |
| β-Galactan | Strong negative band at 171 nm (film CD) | [65] |
| Carrageenan | Strong negative band at 172 nm (film CD) | [60] |
| Agarose | Strong negative band at 155 nm and positive band at 172 nm | [60,66] |
| chitosan | Strong negative band at 211 nm | [62] |
| Hyaluronic acid | Strong band at 210 nm and weak band at 190 nm | [59] |
| Chondroitin sulfate | Strong band at 210 nm and weak band at 190 nm | [59] |
| Dermatan sulfate | Weak band at 210 nm and strong band at 190 nm | [59] |
| Heparin | Weak band at 210 nm and strong band at 190 nm | [59] |
| Heparan sulfate | Strong band at 210 nm and strong band at 190 nm | [59] |
| Keratan sulfate | Strong band at 210 nm and strong band at 190 nm | [59] |

both O [1] and O [3] are axial directions. This quadrant rule can readily explain the shape of CD in the region of 160–170 and 170–175 nm. CD spectra of native polysaccharides in the vacuum UV region, including amylose, cellulose, β-galactan, carrageenan, and agarose, have been well established, as shown in Table 1 [67,69–73].

## 3.3 Polysaccharides in the induced CD region

The CD spectrum with a wavelength range larger than 200 cm is also called induced CD, which is usually applied to identify the primary structure of oligosaccharides/polysaccharides with π electrons. Wiesler et al. determined the composition and sequence of oligosaccharides through a series of derivative reactions and dichroism measurements [74]. In addition, some polysaccharides with π electrons donators, including carboxylate, amide, or carboxymethyl groups, also present CD spectra in the induced CD region, as shown in Table 2 [66,75–77]. A general example is a glycosaminoglycan, which is a group of polysaccharides that include dermatan sulfate, hyaluronic acid, heparin, chondroitin sulfate, keratan sulfate, and heparan sulfate. Glycosaminoglycan generally exhibits CD bands around 210 and 190 nm [66]. These CD bands can be concluded as the n-π and π-π* amide bands. Both the changes in the band at 210 and 190 nm are closely related to the intersaccharide linkages, substituent position, and conformation of the

**Table 2** Orbital designation for CD bands in polysaccharides.

| λ (nm) | Orbital designation | Ref. |
|---|---|---|
| 160 | $n_O(2p_x)[b_1]$-$\sigma^{*\prime}[a_1]$ | [56] |
| 168 | $n_O(2p_x)[b_1]$-$3p_y[a_1]$ | [56] |
| 171 | $n_O(2p_x)[b_1]$-$3p_z[a_1]$ | [56] |
| 183 | $n_O(2p_x)[b_1]$-$\sigma^*/3s[a_1]$ | [56] |
| 190 | π–π* amide transition | [59] |
| 196–200 | n–π carboxymethyl transition | [59,68] |
| 210 | n–π amide transition | [59] |
| 210–220 | n–π carboxylate/carboxylate ester transition | [69,70] |

polymers [66]. Hyaluronic acid and chondroitin sulfate show relatively stronger intensity at 210 nm and weaker intensity at 190 nm, whereas dermatan sulfate, heparin, heparan sulfate, and keratan sulfate show relative stronger intensity at 190 nm. Based on their different CD spectra, heparin can be easily discriminated from chondroitin sulfate in adulteration detection [78].

A mixture of glucuronic acid and *N*-acetylglucosamine showed one negative peak at 210 nm (rotational strengths of n–π*). In contrast, the combination of *N*-acetylglucosamine and glucuronic acid into disaccharide drastically enhanced the peak at 190 cm (π–π* amide transition). By elongating the oligosaccharides of glycosaminoglycans from 2 to 12 monosaccharide units, the intensity at 190 nm gradually increased and the intensity at 210 nm gradually decreased [66]. Further elongation in the chain of glycosaminoglycans induced greater changes in the ratio between intensity at 190 and 210 nm [66]. In addition, 1, 3-linked amino sugars could show a negative CD and 1,4-linked could indicate a positive one at the π–π* amide transition region (210 nm) [79]. These phenomena clearly indicate the complex contribution of the glycosidic bond to the CD spectra of glycosaminoglycans. In analyzing the configuration and sequence of oligosaccharides, induced CD spectra were created by introducing chromophores [80]. However, this measurement may be more appropriate in the analysis of pyranose. CD spectra of glycosaminoglycans in the induced CD region have been well established, as shown in Table 2.

## 3.4 Polysaccharides CD spectra affected by environmental parameters

Even though the polysaccharides CD database is far from perfect, the CD is a very efficient way to detect the conformation changes in polysaccharides.

CD is very sensitive to conformational changes of polysaccharides when the environment parameters are changed [73,81,82]. Influences of environmental factors including solvent, pH, metal ion, temperature, ultrasonic treatment, and chemical modification on the CD spectra of polysaccharides have been widely reported, illustrated in Table 3. The thermal stability of xanthan and agarose was easily investigated using CD scans [73,81]. Here, the CD spectrometer detected the denaturation temperatures of polysaccharides similar to a differential scanning calorimeter (DSC). In detecting chemical modification of polysaccharides such as sulfated, acetylated, and carboxymethylated modifications, the CD can be used to detect the stem grafting groups in a similar way to infrared and Raman spectroscopy [75,76,83–85]. Ion-polysaccharides interaction also changed the Cotton effect of the original polysaccharides, indicating the formation of ion-polysaccharides [82,86–89]. A case refers to the "egg-box" model found in the polysaccharide-based complex with $Ca^{2+}$, $Cu^{2+}$, and $Fe^{3+}$ ions. Qualitative analysis of *N*-acetyl in polysaccharides using CD has been established, and measurement of other groups may be a promising scope in future CD applications in polysaccharides analysis. In addition, the effect of pH, ultrasonic treatment, and the addition of flavonoids on CD spectra of polysaccharides have also been studied [42,86,90]. Changes in the morphology of polysaccharides (order vs. disordered, rigid vs. flexible, helix) induced by environmental factors have been widely reported in CD measurements, where most conclusions were obtained with the assistance of AFM and X-ray diffraction analyses [77,91,92]. The use of Congo red during CD measurements proved to be a reliable indicator for the existence of triple helix structure in polysaccharides [65,93].

Therefore the structure of carbohydrates determines the strength and shape of CD, and the amount of conformational information obtained from CD depends on the complexity of the sample structure. CD can provide more reliable spatial structure information for those carbohydrate compounds with simple repeating series. For the structure of complex polysaccharides, even if their absolute conformation cannot be obtained, some empirical rules can be applied. In addition, the CD can be used to detect the trivial conformational changes of polysaccharides, such as solution-gel and disorder-orderly transition and formation of the polysaccharide-based complex. However, the universality of the n-3$p_z$ quadrant rule and the empirical rule need to be further verified. In addition, the CD database of polysaccharides is far from perfect, limiting the applications of CD in detecting the structure of unknown polysaccharides. However, the CD technique

**Table 3** CD spectra of polysaccharide affected by environmental parameters and chemical modifications.

| Origin of the polysaccharide | Influencing factor | Changes in CD spectra | Possible structural changes | Ref. |
|---|---|---|---|---|
| Macroalga *Ulva* | Acid treatment | Resulted in two isodichroic points at 230 and 206 nm | Indicator of an equilibrium between the protonated and deprotonated uronic units | [79] |
| | $Cu^{2+}$ | Red shift of the n-$\pi^*$ transition of carboxylate | Binding of $Cu^{2+}$ with negatively charged groups | |
| | Heating | Reveal no isodichroic point | Equilibrium between different conformers | |
| *Inonotus obliquus* | Ultrasonic treatment | The n-$\pi^*$ region changed drastically | Possible degradation and aggregation | [83] |
| Alginate | $Ca^{2+}$ | The n-$\pi^*$ region changed drastically | Sol $\rightarrow$ gel transition | [75] |
| *Hericium erinaceus* | $Bi^{3+}$ | Positive cotton effect at 222 nm | Ordered complex | [80] |
| Gonad of pacific abalone | $Fe^{3+}$ | Negative Cotton effect at 208 nm became weaker | Chelation reaction through Fe—O chemical bond | [81] |
| *Hohenbuehelia serotina* | Selenium modified | Strong negative cotton effect presented at 205 nm | The n $\rightarrow$ $\pi$ transition of selenate group | [76] |
| Xanthan | Heating | Blue shift of the n-$\pi^*$ transition of carboxylate | Reversible denaturation of double helical structure | [74] |
| Agarose | Heating | Along with heated, band at 172 nm gradually shifted to 185 nm | Helix-helix aggregation | [66] |

*Continued*

**Table 3** CD spectra of polysaccharide affected by environmental parameters and chemical modifications—cont'd

| Origin of the polysaccharide | Influencing factor | Changes in CD spectra | Possible structural changes | Ref. |
|---|---|---|---|---|
| *Artemisia sphaerocephala* | Sulfation | Red shift of the n-$\pi^*$ transition of carboxylate or carboxylate ester | Toward a more rigid conformation | [77,78] |
| Corn silk | Acetylation | A negative peak at 198 nm | The $n \rightarrow \pi$ transition of acetyl group | [68] |
| Mushroom *Inonotus obliquus* | Carboxymethylation | A negative peak at 200 nm | The $n \rightarrow \pi$ transition of carboxymethyl group | [69] |
| Corn silk (*Maydis stigma*) | Flavonoids | Strong negative Cotton effect appeared in 202 nm | Ordered helical structure of the complex | [84] |

is adequate to indicate structural changes when environmental factors are modulated. In addition, new CD techniques such as second-harmonic generation circular dichroism (SHG-CD) and SRCD are promising tools to provide a more accurate analysis of polysaccharides structure [94,95].

# 4. Applications of CD in nucleic acid measurements

## 4.1 Basic understanding of CD spectrum of nucleic acids

Theoretically, the CD spectrum of nucleic acid can be calculated according to the principle of quantum mechanics. However, such calculations cannot reliably and quantitatively explain the measured CD spectrum due to the diversity of influencing factors in the actual measurement [96]. Many assays of nucleic acids using CD have been carried out in the UV region, whereas structural information is also obtained in the vacuum UV region [97]. These experiments are more easily carried out in SRCD that provide large photon fluxes. CD signal intensities of nucleic acids in the vacuum UV region are much more significant than those in the UV region, and the electronic coupling between bases can also increase with excitation energy [98].

CD spectroscopy is especially helpful in detecting the level of electronic coupling in a strand, such as the delocalization of excited-state electronic wavefunctions. The character of the excited state is related to base compositions, sequence, and DNA folding motifs [99–102]. The latter is related to environmental parameters, including pH, temperature, solvent, and ionic strength. In neutral solutions, DNA usually exists in a B form. RNA can also assume various conformations, depending on environmental parameters, whereas their conformation in neutral solutions usually exists in an A-form [103]. The CD spectra of DNA/RNA depend on the base composition and sugar type. Interestingly, it has been shown that the features of the CD spectra of RNA are opposite to those of DNA [103]. The delocalized π-electron of the four DNA bases accounts for their optical transitions in the UV region [98]. The bases alone are not chiral, and the CD activity is generated by the asymmetric environment induced by sugars [104]. pH has a trivial effect on the absorption but essentially changes the CD spectra since most of the CD active bands are generated by n-π* transitions other than π-π* that have stronger absorption [104].

## 4.2 CD spectra of nucleic acids: From simple sequence to native DNA

A simple model to establish a CD signal for a DNA strand is the sum of the individual base signals. This, if applicable, implies that the dimer spectrum should be similar to the spectrum of the monomer mixture, which is usually uncommon to observe since other factors, including the nearest-neighbor interactions and next-to-nearest neighbor interactions, also count [97]. For the double-strand DNA, both the pairing of the two strands and the base stacking participate in the CD signal. The hybridization signal is revealed by comparing the CD spectra of individual strands and the duplex. Single strands of adenine, which have a negative band at 178 nm and a positive band at 190 nm in the vacuum UV region, can shape the single helical structures of B-type DNA [105,106]. The signal for the dA dimer is quite different from the monomer, thus demonstrating the contribution of interactions between stacked bases to the CD spectra of nucleic acids. When the strand length of adenine increases from 2 to 20, the signals for the bands at 190, 251, and 218 nm increase linearly [97]. This indicates that the excitation is delocalized when the number of bases is larger than two. For the strand of thymine, the signals for the bands at 182, 251, and 275 nm increase linearly along with the elongation in length [106]. The CD spectra for poly (dC) have a negative band at 266 nm and a positive band at 288 nm, which can be assigned to the signal of the hemiprotonated base pairing [107,108]. The two bands for the hemiprotonated complex are related to the signal for the protonated base and interbase Hoogsteen pairing (266 nm). The signals for the 190, 200, and 275 nm bands increase linearly along with the elongation of poly (dG) [109]. Spectra of dAdT and dTdA are quite similar in the UV region but are different in the vacuum UV region [110]. Therefore the order of the bases also largely contributes to the CD signal of nucleic acids.

Native nucleic acids are less recorded by CD spectra. The spectrum of the three strands, c-mycI (CCCCACCCTCCC), c-mycII (GGGAGG-GTGGGG), and c-mycIII (GGGGTGGGTGGG), have been recorded [111]. These three sequences can form antiparallel triplexes. The spectra for the c-mycI:c-mycII duplex, the c-mycI:c-mycII:c-mycII triplex, and the c-mycI:c-mycII:c-mycIII triplex are also measured. The duplex spectrum is quite similar to that for $(dG)_{10}$:$(dC)_{10}$, and the triplex residual spectra are very similar to that of $(dG)_{10}$:($(dG)_{10}$:$(dC)_{10}$) in the UV region. In contrast, discrepancies in intensity and location of bands are found in the vacuum UV region. In addition, calf thymus DNA is another well-characterized nucleic acid in plenty of studies, which has a characteristic negative peak at 245 nm and a positive peak at 275 nm [112–114].

## 4.3 Binding of drugs with DNA

As shown in Table 4, even though the structure of native nucleic acids is less studied by CD, the interaction between DNA and antitumor drugs has been of interest to pharmaceuticals since CD can be used as an auxiliary screening method to find new DNA-targeted drugs, such as antitumor drugs, antibacterial drugs, antiviral drugs [111–129]. CD spectrum is extremely sensitive to conformational changes of DNA, especially in the range of 180–320 nm. There are three typical ways through which drugs interact with DNA: embedding, groove binding, and alkylation/metallization. The first two ways are noncovalent, and the third is covalent [115,116,123]. Planar aromatic hydrocarbon molecules, such as valacyclovir, imatinib, 2-methyl-9-hydroxyellipticinium acetate, ethidium bromide, acridone-based derivatives, and ellipticine, can be embedded into the lamellar base pairs of DNA double-helix molecules, inducing unwinding and elongation of DNA [111,113,117–120]. Some molecules with flexible structures, such as metformin, adriamycin, calicheamicin, thiabendazole, piperine, methimazole, and dapsone, can bind to DNA grooves [112,121,122,124–128]. Epirubicin, an anticancer drug, interacted with G-quadruplex in a mixed model [128]. In alkylation/metallization drugs (nitrogen mustard and platinum antitumor drugs), covalent interactions occur between drugs and DNA molecules that can cause cross-linking within or between DNA chains and affect its replication, transcription, and other biological functions [129]. The conformation of some drugs with no chirality and optical activity is changed into a spatial arrangement with chirality after interacting with DNA. Therefore CD signals can also be generated by coupling with the electron transition dipole moment of bases in DNA. Monnot et al. reported that antitumor drug 2-methyl-9-hydroxyellipticinium could interact with the 8 bp oligonucleotide sequence of CAMP-reaction element and induced dose-dependent positive band at 330 nm [117]. CD studies that screen the interaction between antineoplastic drugs and compounds with DNA are generally consistent with those reported in the literature. Therefore CD spectroscopy can be used as an additional screening tool in the research of new drugs.

# 5. Conclusions and outlook

CD has become an indispensable tool in the structural analysis of biological macromolecules due to several advantages, including its fast test speed, simple sample pretreatment, data collection, and analysis [130]. CD is an

**Table 4** CD spectra of drug-DNA complexes.

| Type of drugs | DNA | Combination type | Changes in CD spectra | Ref. |
|---|---|---|---|---|
| Valacyclovir | CT DNA | Intercalative binding | Red shift and increased intensity of bands at 245 and 270 nm | [106] |
| Metformin | CT DNA | Groove binding | Increased intensity of bands at 245 and 275 nm | [107] |
| Imatinib | CT DNA | Intercalative binding | Red shift of bands at 275 nm | [108] |
| 2-Methyl-9-hydroxyellipticinium acetate | Octanucleotide | Intercalative binding | Decreased intensity of bands at 312 and 335 nm | [112] |
| Ethidium bromide | CT DNA | Intercalative binding | Increased intensity of bands at 245 and 275 nm | [113] |
| Epirubicin | G-quadruplex $[d\text{-}(TTAGGGT)]_4$ | Mixed binding | Nonlinear decrease in band at 264 nm | [115] |
| Adriamycin | G-quadruplex $[d\text{-}(TTAGGGT)]_4$ | Groove binding | Linear decrease in band at 264 nm | [116] |
| Calicheamicin | 5'-CCCGGTCCTAAG-3′/3'-GGGCCAGGATTC-5′ dodecamer | Groove binding | Decreased intensity of bands at 270 and 315 nm | [117] |
| Thiabendazole | CT DNA | Groove binding | Slight decrement in the negative DNA band at 250 nm | [118] |
| Piperine | CT DNA | Groove binding | Red shift of bands at 245 nm | [119] |

| | | | | |
|---|---|---|---|---|
| Methimazole | CT DNA | Groove binding | Decreased intensity of bands at 245 nm and increased intensity of bands at 275 nm | [120] |
| Acridone-based derivatives | CT DNA | Intercalative binding | Increased intensity of bands at 245 and 275 nm | [114] |
| Pregabalin | CT DNA | Groove binding | Increased intensity of bands at 245 and 275 nm | [121] |
| Dapsone | CT DNA | Groove binding | Decreased intensity of bands at 245 and increased intensity of bands at 275 nm | [122] |
| Ellipticine | CT DNA | Intercalative binding | Decreased intensity of bands at 245 and 275 nm, and increased intensity of band at 315 nm | [123] |

efficient scanner to detect conformation alterations when the environmental parameters of biomolecules are changed. CD databases and algorithms for proteins have been well established, whereas those for polysaccharides and nucleic acid are still far from perfect, making the interpretation of the CD spectra of complex polysaccharides and native nucleic acid a challenge. Quantum mechanical calculations could be of vital assistance in predicting CD spectra of nucleic acids, which may help establish the CD database of nucleic acids. Studying insoluble proteins using CD is still challenging, and SRCD could emerge as an excellent choice to measure insoluble or membrane proteins. In addition, high-throughput computational techniques could be further combined with CD measurements to provide more accessible information for three-dimensional biological systems. These developments may extend the applications of CD beyond the fields of drug discovery and structural biology to new areas such as the analysis of nanomaterials, disease diagnosis, etc.

## References

[1] Wang K, Sun D-W, Pu H, Wei Q. Principles and applications of spectroscopic techniques for evaluating food protein conformational changes: a review. Trends Food Sci Technol 2017;67:207–19.

[2] Tripathi T. Calculation of thermodynamic parameters of protein unfolding using far-ultraviolet circular dichroism. J Proteins Proteomics 2013;4(2):85–91.

[3] Li C, Arakawa T. Feasibility of circular dichroism to study protein structure at extreme concentrations. Int J Biol Macromol 2019;132:1290–5.

[4] Kelly SM, Price NC. The use of circular dichroism in the investigation of protein structure and function. Curr Protein Pept Sci 2000;1(4):349–84.

[5] Greenfield NJ. Using circular dichroism spectra to estimate protein secondary structure. Nat Protoc 2006;1(6):2876–90.

[6] Miles AJ, Janes RW, Wallace BA. Tools and methods for circular dichroism spectroscopy of proteins: a tutorial review. Chem Soc Rev 2021;50:8400–13.

[7] Honisch C, Donadello V, Hussain R, Peterle D, De Filippis V, Arrigoni G, et al. Application of circular dichroism and fluorescence spectroscopies to assess photostability of water-soluble porcine lens proteins. ACS Omega 2020;5(8):4293–301.

[8] Shukla R, Tripathi T. Molecular dynamics simulation of protein and protein-ligand complexes. In: Singh DB, editor. Computer-aided drug design. Singapore: Springer Nature; 2020. p. 133–61.

[9] Kalita P, Shukla H, Shukla R, Tripathi T. Biochemical and thermodynamic comparison of the selenocysteine containing and non-containing thioredoxin glutathione reductase of Fasciola gigantica. Biochim Biophys Acta Gen Subj 2018;1862(6):1306–16.

[10] Sonkar A, Shukla H, Shukla R, Kalita J, Pandey T, Tripathi T. UDP-N-acetylglucosamine enolpyruvyl transferase (MurA) of *Acinetobacter baumannii* (AbMurA):structural and functional properties. Int J Biol Macromol 2017;97:106–14.

[11] Tripathi T, Na BK, Sohn WM, Becker K, Bhakuni V. Structural, functional and unfolding characteristics of glutathione S-transferase of *Plasmodium vivax*. Arch Biochem Biophys 2009;487(2):115–22.
[12] Tripathi T, Röseler A, Rahlfs S, Becker K, Bhakuni V. Conformational stability and energetics of *Plasmodium falciparum* glutaredoxin. Biochimie 2010;92(3):284–91.
[13] Wu W, Zhang C, Kong X, Hua Y. Oxidative modification of soy protein by peroxyl radicals. Food Chem 2009;116(1):295–301.
[14] Liu Q, Geng R, Zhao J, Chen Q, Kong B. Structural and gel textural properties of soy protein isolate when subjected to extreme acid pH-shifting and mild heating processes. J Agric Food Chem 2015;63(19):4853–61.
[15] Li S, Yang X, Zhang Y, Ma H, Liang Q, Qu W, et al. Effects of ultrasound and ultrasound assisted alkaline pretreatments on the enzymolysis and structural characteristics of rice protein. Ultrason Sonochem 2016;31:20–8.
[16] Erickson DP, Ozturk OK, Selling G, Chen F, Campanella OH, Hamaker BR. Corn zein undergoes conformational changes to higher β-sheet content during its self-assembly in an increasingly hydrophilic solvent. Int J Biol Macromol 2020;157: 232–9.
[17] Wang Y, Padua GW. Nanoscale characterization of zein self-assembly. Langmuir 2012;28(5):2429–35.
[18] Dai H, Sun Y, Xia W, Ma L, Li L, Wang Q, et al. Effect of phospholipids on the physicochemical properties of myofibrillar proteins solution mediated by NaCl concentration. LWT 2021;141, 110895.
[19] Lv M, Mei K, Zhang H, Xu D, Yang W. Effects of electron beam irradiation on the biochemical properties and structure of myofibrillar protein from *Tegillarca granosa* meat. Food Chem 2018;254:64–9.
[20] Sheng L, Huang M, Wang J, Xu Q, Hammad HHM, Ma M. A study of storage impact on ovalbumin structure of chicken egg. J Food Eng 2018;219:1–7.
[21] Sheng L, Ye S, Han K, Zhu G, Ma M, Cai Z. Consequences of phosphorylation on the structural and foaming properties of ovalbumin under wet-heating conditions. Food Hydrocoll 2019;91:166–73.
[22] Day L, Zhai J, Xu M, Jones NC, Hoffmann SV, Wooster TJ. Conformational changes of globular proteins adsorbed at oil-in-water emulsion interfaces examined by synchrotron radiation circular dichroism. Food Hydrocoll 2014;34:78–87.
[23] Nikoo M, Benjakul S, Ocen D, Yang N, Xu B, Zhang L, et al. Physical and chemical properties of gelatin from the skin of cultured Amur sturgeon (*Acipenser schrenckii*). J Appl Ichthyol 2013;29(5):943–50.
[24] Greenfield NJ, Fasman GD. Computed circular dichroism spectra for the evaluation of protein conformation. Biochemistry 1969;8(10):4108–16.
[25] Micsonai A, Wien F, Bulyáki É, Kun J, Moussong É, Lee Y-H, et al. BeStSel: a web server for accurate protein secondary structure prediction and fold recognition from the circular dichroism spectra. Nucleic Acids Res 2018;46(W1):W315–22.
[26] Louis-Jeune C, Andrade-Navarro MA, Perez-Iratxeta C. Prediction of protein secondary structure from circular dichroism using theoretically derived spectra. Proteins Struct Funct Bioinf 2012;80(2):374–81.
[27] Miles AJ, Ramalli SG, Wallace BA. DichroWeb, a website for calculating protein secondary structure from circular dichroism spectroscopic data. Protein Sci 2022;31(1): 37–46.
[28] Wallace BA, Janes RW. Synchrotron radiation circular dichroism spectroscopy of proteins: secondary structure, fold recognition and structural genomics. Curr Opin Chem Biol 2001;5(5):567–71.
[29] Wallace BA. Protein characterisation by synchrotron radiation circular dichroism spectroscopy. Q Rev Biophys 2009;42(4):317–70.

[30] Kumagai PS, Araujo APU, Lopes JLS. Going deep into protein secondary structure with synchrotron radiation circular dichroism spectroscopy. Biophys Rev 2017;9: 517–27.
[31] García-Moreno PJ, Yang J, Gregersen S, Jones NC, Berton-Carabin CC, Sagis LMC, et al. The structure, viscoelasticity and charge of potato peptides adsorbed at the oil-water interface determine the physicochemical stability of fish oil-in-water emulsions. Food Hydrocoll 2021;115, 106605.
[32] Wong BT, Zhai J, Hoffmann SV, Aguilar M-I, Augustin M, Wooster TJ, et al. Conformational changes to deamidated wheat gliadins and β-casein upon adsorption to oil-water emulsion interfaces. Food Hydrocoll 2012;27(1):91–101.
[33] García-Estévez I, Ramos-Pineda AM, Escribano-Bailón MT. Interactions between wine phenolic compounds and human saliva in astringency perception. Food Funct 2018;9(3):1294–309.
[34] Théron L, Bonifacie A, Delabre J, Sayd T, Aubry L, Gatellier P, et al. Investigation by synchrotron radiation circular dichroism of the secondary structure evolution of pepsin under oxidative environment. Foods 2021;10(5):998.
[35] Kim KS, Kim S, Yang HJ, Kwon DY. Changes of glycinin conformation due to pH, heat and salt determined by differential scanning calorimetry and circular dichroism. Int J Food Sci Technol 2004;39(4):385–93.
[36] Li L, Zhao X, Xu X. Trace the difference driven by unfolding-refolding pathway of myofibrillar protein: emphasizing the changes on structural and emulsion properties. Food Chem 2022;367, 130688.
[37] Ma W, Wang T, Wang J, Wu D, Wu C, Du M. Enhancing the thermal stability of soy proteins by preheat treatment at lower protein concentration. Food Chem 2020;306, 125593.
[38] Mir NA, Riar CS, Singh S. Improvement in the functional properties of quinoa (*Chenopodium quinoa*) protein isolates after the application of controlled heat-treatment: effect on structural properties. Food Struct 2021;28, 100189.
[39] Tolano-Villaverde IJ, Santacruz-Ortega H, Rivero-Espejel IA, Torres-Arreola W, Suárez-Jiménez GM, Márquez-Ríos E. Effect of temperature on the actomyosin-paramyosin structure from giant squid mantle (*Dosidicus gigas*). J Sci Food Agric 2019;99(12):5377–83.
[40] Shen Y, Li Y. Acylation modification and/or guar gum conjugation enhanced functional properties of pea protein isolate. Food Hydrocoll 2021;117, 106686.
[41] Wan Y, Liu J, Guo S. Effects of succinylation on the structure and thermal aggregation of soy protein isolate. Food Chem 2018;245:542–50.
[42] Song C-L, Ren J, Chen J-P, Sun X-H, Kopparapu N-K, Xue Y-G. Effect of glycosylation and limited hydrolysis on structural and functional properties of soybean protein isolate. J Food Meas Charact 2018;12(4):2946–54.
[43] Liu J, Wan Y, Ren L, Li M, Lv Y, Guo S, et al. Physical-chemical properties and in vitro digestibility of phosphorylated and glycosylated soy protein isolate. LWT 2021;152, 112380.
[44] A'Yun Q, Demicheli P, De Neve L, Wu J, Balcaen M, Setiowati AD, et al. Dry heat induced whey protein-lactose conjugates largely improve the heat stability of O/W emulsions. Int Dairy J 2020;108, 104736.
[45] Zhang Z, Xiong Z, Lu S, Walayat N, Hu C, Xiong H. Effects of oxidative modification on the functional, conformational and gelling properties of myofibrillar proteins from *Culter alburnus*. Int J Biol Macromol 2020;162:1442–52.
[46] Chen J, Ren Y, Zhang K, Qu J, Hu F, Yan Y. Phosphorylation modification of myofibrillar proteins by sodium pyrophosphate affects emulsion gel formation and oxidative stability under different pH conditions. Food Funct 2019;10(10):6568–81.

[47] Hu Z, Qiu L, Sun Y, Xiong H, Ogra Y. Improvement of the solubility and emulsifying properties of rice bran protein by phosphorylation with sodium trimetaphosphate. Food Hydrocoll 2019;96:288–99.
[48] Wang T, Xu P, Chen Z, Wang R. Mechanism of structural interplay between rice proteins and soy protein isolates to design novel protein hydrocolloids. Food Hydrocoll 2018;84:361–7.
[49] Kristensen HT, Christensen M, Hansen MS, Hammershøj M, Dalsgaard TK. Protein-protein interactions of a whey-pea protein co-precipitate. Int J Food Sci Technol 2021;56(11):5777–90.
[50] Tan L, Hong P, Yang P, Zhou C, Xiao D, Zhong T. Correlation between the water solubility and secondary structure of tilapia-soybean protein co-precipitates. Molecules 2019;24(23):4337.
[51] He C, Hu Y, Liu Z, Woo MW, Xiong H, Zhao Q. Interaction between casein and rice glutelin: binding mechanisms and molecular assembly behaviours. Food Hydrocoll 2020;107, 105967.
[52] Hussain R, Hughes CS, Siligardi G. Enzyme-ligand interaction monitored by synchrotron radiation circular dichroism. In: Targeting enzymes for pharmaceutical development. Springer; 2020. p. 87–118.
[53] Singh DB, Tripathi T. Frontiers in protein structure, function, and dynamics. Singapore Springer Nature; 2020.
[54] Rodger A, Marrington R, Roper D, Windsor S. Circular dichroism spectroscopy for the study of protein-ligand interactions. In: Protein-ligand interactions. Springer; 2005. p. 343–63.
[55] Chen Z, Wang C, Gao X, Chen Y, Santhanam RK, Wang C, et al. Interaction characterization of preheated soy protein isolate with cyanidin-3-O-glucoside and their effects on the stability of black soybean seed coat anthocyanins extracts. Food Chem 2019;271:266–73.
[56] Zhao H-S, Ma Z, Jing P. Interaction of soy protein isolate fibrils with betalain from red beetroots: morphology, spectroscopic characteristics and thermal stability. Food Res Int 2020;135, 109289.
[57] Zhang L, Liu Y, Wang Y. Interaction between an (−)-epigallocatechin-3-gallate-copper complex and bovine serum albumin: fluorescence, circular dichroism, HPLC, and docking studies. Food Chem 2019;301, 125294.
[58] Tesmar A, Kogut MM, Żamojć K, Grabowska O, Chmur K, Samsonov SA, et al. Physicochemical nature of sodium dodecyl sulfate interactions with bovine serum albumin revealed by interdisciplinary approaches. J Mol Liq 2021;340, 117185.
[59] Misugi CT, Savi LK, Iwankiw PK, Masson ML, de Oliveira MAS, Igarashi-Mafra L, et al. Effects of freezing and the cryoprotectant lactobionic acid in the structure of GlnK protein evaluated by circular dichroism (CD) and isothermal titration calorimetry (ITC). J Food Sci Technol 2017;54(1):236–43.
[60] Segatta F, Rogers DM, Dyer NT, Guest EE, Li Z, Do H, et al. Near-ultraviolet circular dichroism and two-dimensional spectroscopy of polypeptides. Molecules 2021;26(2):396.
[61] Nozaki A, Hori M, Kimura T, Ito H, Hatano T. Interaction of polyphenols with proteins: binding of (−)-epigallocatechin gallate to serum albumin, estimated by induced circular dichroism. Chem Pharm Bull 2009;57(2):224–8.
[62] Li M, Hagerman AE. Role of the flavan-3-ol and galloyl moieties in the interaction of (−)-epigallocatechin gallate with serum albumin. J Agric Food Chem 2014;62(17): 3768–75.
[63] Fasman GD. Circular dichroism and the conformational analysis of biomolecules. Plenum; 1996.

[64] Ren Y, Bai Y, Zhang Z, Cai W, Del Rio FA. The preparation and structure analysis methods of natural polysaccharides of plants and fungi: a review of recent development. Molecules 2019;24(17):3122.
[65] Song Q-y, Zhu Z-Y, Wang X-T, Chen L-T, Wang D-Y. Effects of solution behavior on polysaccharide structure and inhibitory of α-glucosidase activity from *Cordyceps militaris*. J Mol Struct 2019;1178:630–8.
[66] Chakrabarti B. Carboxyl and amide transitions in the circular dichroism of glycosaminoglycans. ACS Publications; 1981.
[67] Arndt ER, Stevens ES. Anhydro sugar and linkage contributions to circular dichroism of agarose and carrageenan, with conformational implications. Carbohydr Res 1997;303(1):73–8.
[68] Cziner DG, Stevens ES, Morris ER, Rees DA. Vacuum ultraviolet circular dichroism of dermatan sulfate: iduronate ring geometry in solution and solid state. J Am Chem Soc 1986;108(13):3790–5.
[69] Domard A. Determination of *N*-acetyl content in chitosan samples by cd measurements. Int J Biol Macromol 1987;9(6):333–6.
[70] Lewis DG, Johnson Jr WC. Optical properties of sugars. VI. Circular dichroism of amylose and glucose oligomers. Biopolymers 1978;17(6):1439–49.
[71] Stipanovic AJ, Stevens ES, editors. Vacuum ultraviolet circular dichroism of cellulose and cellulose acetates. Binghamton: State Univ. of New York; 1983.
[72] Duda CA, Stevens ES. Solution conformation of laminaribioside and (1 → 3)-β-D-glucan from optical rotation. Biopolymers 1991;31(12):1379–85.
[73] Liang JN, Stevens ES, Morris ER, Rees DA. Spectroscopic origin of conformation-sensitive contributions to polysaccharide optical activity: vacuum-ultraviolet circular dichroism of agarose. Biopolymers 1979;18(2):327–33.
[74] Wiesler WT, Berova N, Ojika M, Meyers HV, Chang M, Zhou P, et al. A CD-spectroscopic alternative to methylation analysis of oligosaccharides: reference spectra for identification of chromophoric glycopyranoside derivatives. Helv Chim Acta 1990;73(3):509–51.
[75] Chen S, Chen H, Tian J, Wang Y, Xing L, Wang J. Chemical modification, antioxidant and α-amylase inhibitory activities of corn silk polysaccharides. Carbohydr Polym 2013;98(1):428–37.
[76] Ma L, Chen H, Zhang Y, Zhang N, Fu L. Chemical modification and antioxidant activities of polysaccharide from mushroom *Inonotus obliquus*. Carbohydr Polym 2012;89(2):371–8.
[77] Liu S, Xie L, Shen M, Xiao Y, Yu Q, Chen Y, et al. Dual modifications on the gelatinization, textural, and morphology properties of pea starch by sodium carbonate and Mesona chinensis polysaccharide. Food Hydrocoll 2020;102, 105601.
[78] Stanley FE, Stalcup AM. The use of circular dichroism as a simple heparin-screening strategy. Anal Bioanal Chem 2011;399(2):701–6.
[79] Stone AL. Optical rotary dispersion of mucopolysaccharides III. Ultraviolet circular dichroism and conformation al specificity in amide groups. Biopolymers 1971;10(4):739–51.
[80] Engle AR, Purdie N, Hyatt JA. Induced circular dichroism study of the aqueous solution complexation of cello-oligosaccharides and related polysaccharides with aromatic dyes. Carbohydr Res 1994;265(2):181–95.
[81] Matsuda Y, Biyajima Y, Sato T. Thermal denaturation, renaturation, and aggregation of a double-helical polysaccharide xanthan in aqueous solution. Polym J 2009;41(7):526–32.
[82] Morris ER, Rees DA, Thom D. Characterization of polysaccharide structure and interactions by circular dichroism: order-disorder transition in the calcium alginate system. J Chem Soc Chem Commun 1973;(7):245–6.

[83] Wang L, Li X, Wang B. Synthesis, characterization and antioxidant activity of selenium modified polysaccharides from *Hohenbuehelia serotina*. Int J Biol Macromol 2018;120:
1362–8.
[84] Wang J, Niu S, Zhao B, Luo T, Liu D, Zhang J. Catalytic synthesis of sulfated polysaccharides. II: comparative studies of solution conformation and antioxidant activities. Carbohydr Polym 2014;107:221–31.
[85] Wang J, Yang W, Tang Y, Xu Q, Huang S, Yao J, et al. Regioselective sulfation of *Artemisia sphaerocephala* polysaccharide: solution conformation and antioxidant activities in vitro. Carbohydr Polym 2016;136:527–36.
[86] Paradossi G, Cavalieri F, Pizzoferrato L, Liquori AM. A physico-chemical study on the polysaccharide ulvan from hot water extraction of the macroalga Ulva. Int J Biol Macromol 1999;25(4):309–15.
[87] Zhu Y, Chen Y, Li Q, Zhao T, Zhang M, Feng W, et al. Preparation, characterization, and anti-helicobacter pylori activity of Bi3+-Hericium erinaceus polysaccharide complex. Carbohydr Polym 2014;110:231–7.
[88] Wang L, Wang L, Su C, Wen C, Gong Y, You Y, et al. Characterization and digestion features of a novel polysaccharide-Fe (III) complex as an iron supplement. Carbohydr Polym 2020;249, 116812.
[89] Wang J, Chen H, Wang Y, Xing L. Synthesis and characterization of a new *Inonotus obliquus* polysaccharide-iron (III) complex. Int J Biol Macromol 2015;75:210–7.
[90] Guo Q, Ma Q, Xue Z, Gao X, Chen H. Studies on the binding characteristics of three polysaccharides with different molecular weight and flavonoids from corn silk (*Maydis stigma*). Carbohydr Polym 2018;198:581–8.
[91] Milas M, Rinaudo M. Conformational investigation on the bacterial polysaccharide xanthan. Carbohydr Res 1979;76(1):189–96.
[92] Zhu Z-y, Dong F, Liu X, Lv Q, Liu F, Chen L, et al. Effects of extraction methods on the yield, chemical structure and anti-tumor activity of polysaccharides from *Cordyceps gunnii* mycelia. Carbohydr Polym 2016;140:461–71.
[93] Gao Y, Peng B, Xu Y, Yang J-n, Song L-y, Bi S-x, et al. Structural characterization and immunoregulatory activity of a new polysaccharide from *Citrus medica* L. var. sarcodactylis. RSC Adv 2019;9(12):6603–12.
[94] Phillips-Jones MK, Harding SE. Tapping into synchrotron and benchtop circular dichroism spectroscopy for expanding studies of complex polysaccharides and their interactions in anoxic archaeological wood. Heritage 2019;2(1):121–34.
[95] Zhuo GY, Lee H, Hsu KJ, Huttunen MJ, Kauranen M, Lin YY, et al. Three-dimensional structural imaging of starch granules by second-harmonic generation circular dichroism. J Microsc 2014;253(3):183–90.
[96] Padula D, Jurinovich S, Di Bari L, Mennucci B. Simulation of electronic circular dichroism of nucleic acids: from the structure to the spectrum. Chem A Eur J 2016;22(47):17011–9.
[97] Holm AIS, Nielsen LM, Hoffmann SV, Nielsen SB. Vacuum-ultraviolet circular dichroism spectroscopy of DNA: a valuable tool to elucidate topology and electronic coupling in DNA. Phys Chem Chem Phys 2010;12(33):9581–96.
[98] Gekko K. Synchrotron-radiation vacuum-ultraviolet circular dichroism spectroscopy in structural biology: an overview. Biophys Physicobiol 2019;16:41–58.
[99] Vorlíčková M, Kejnovská I, Bednářová K, Renčiuk D, Kypr J. Circular dichroism spectroscopy of DNA: from duplexes to quadruplexes. Chirality 2012;24(9):691–8.
[100] Ivanov VI, Minchenkova LE, Schyolkina AK, Poletayev AI. Different conformations of double-stranded nucleic acid in solution as revealed by circular dichroism. Biopolymers 1973;12(1):89–110.

[101] Miyahara T, Nakatsuji H, Sugiyama H. Helical structure and circular dichroism spectra of DNA: a theoretical study. J Phys Chem A 2013;117(1):42–55.
[102] Vorlíčková M, Kejnovská I, Sagi J, Renčiuk D, Bednářová K, Motlová J, et al. Circular dichroism and guanine quadruplexes. Methods 2012;57(1):64–75.
[103] Miyahara T, Nakatsuji H, Sugiyama H. Similarities and differences between RNA and DNA double-helical structures in circular dichroism spectroscopy: a SAC-CI study. J Phys Chem A 2016;120(45):9008–18.
[104] Sprecher CA, Johnson Jr WC. Circular dichroism of the nucleic acid monomers. Biopolymers 1977;16(10):2243–64.
[105] Dewey TG, Turner DH. Laser temperature-jump study of stacking in adenylic acid polymers. Biochemistry 1979;18(26):5757–62.
[106] Johnson KH, Gray DM, Sutherland JC. Vacuum UV CD spectra of homopolymer duplexes and triplexes containing AT or AU base pairs. Nucleic Acids Res 1991;19(9):2275–80.
[107] Inman RB. Transitions of DNA homopolymers. J Mol Biol 1964;9(3):624–37.
[108] Antao VP, Gray DM. CD spectral comparisons of the acid-induced structures of poly [d (A)] poly [r (A)], poly [d (C)], and poly [r (C)]. J Biomol Struct Dyn 1993;10(5):819–39.
[109] Holm AIS, Kohler B, Hoffmann SV, Brøndsted NS. Synchrotron radiation circular dichroism of various G-quadruplex structures. Biopolymers 2010;93(5):429–33.
[110] Munksgaard Nielsen L, Holm AIS, Varsano D, Kadhane U, Hoffmann SV, Di Felice R, et al. Fingerprints of bonding motifs in DNA duplexes of adenine and thymine revealed from circular dichroism: synchrotron radiation experiments and TDDFT calculations. J Phys Chem B 2009;113(28):9614–9.
[111] Johnson KH, Durland RH, Hogan ME. The vacuum UV CD spectra of GGC triplexes. Nucleic Acids Res 1992;20(15):3859–64.
[112] Shahabadi N, Fatahi N, Mahdavi M, Nejad ZK, Pourfoulad M. Multispectroscopic studies of the interaction of calf thymus DNA with the anti-viral drug, valacyclovir. Spectrochim Acta A Mol Biomol Spectrosc 2011;83(1):420–4.
[113] Shahabadi N, Heidari L. Binding studies of the antidiabetic drug, metformin to calf thymus DNA using multispectroscopic methods. Spectrochim Acta A Mol Biomol Spectrosc 2012;97:406–10.
[114] Hegde AH, Seetharamappa J. Fluorescence and circular dichroism studies on binding and conformational aspects of an anti-leukemic drug with DNA. Mol Biol Rep 2014;41(1):67–71.
[115] Nordén B, Kurucsev T. Analysing DNA complexes by circular and linear dichroism. J Mol Recognit 1994;7(2):141–55.
[116] Rodger A. Circular and linear dichroism of drug-DNA systems. In: Drug-DNA interaction protocols. Springer; 2010. p. 37–54.
[117] Monnot M, Mauffret O, Lescot E, Fermandjian S. Probing intercalation and conformational effects of the anticancer drug 2-methyl-9-hydroxyellipticinium acetate in DNA fragments with circular dichroism. Eur J Biochem 1992;204(3):1035–9.
[118] Kundu P, Das S, Chattopadhyay N. Switching from endogenous to exogenous delivery of a model drug to DNA through micellar engineering. J Photochem Photobiol B Biol 2020;203, 111765.
[119] Thimmaiah K, Ugarkar AG, Martis EF, Shaikh MS, Coutinho EC, Yergeri MC. Drug-DNA interaction studies of acridone-based derivatives. Nucleosides Nucleotides Nucleic Acids 2015;34(5):309–31.
[120] Koninti RK, Palvai S, Satpathi S, Basu S, Hazra P. Loading of an anti-cancer drug into mesoporous silica nano-channels and its subsequent release to DNA. Nanoscale 2016;8(43):18436–45.

[121] Tariq Z, Barthwal R. Binding of anticancer drug daunomycin to parallel G-quadruplex DNA [d-(TTGGGGT)] 4 leads to thermal stabilization: a multispectroscopic investigation. Int J Biol Macromol 2018;120:1965–74.
[122] Proni G, Tami K, Berova N, Ellestad GA. Circular dichroism analysis of the calicheamicin-DNA interaction revisited. J Pharm Biomed Anal 2017;144:1–5.
[123] Kypr J, Kejnovská I, Renčiuk D, Vorlíčková M. Circular dichroism and conformational polymorphism of DNA. Nucleic Acids Res 2009;37(6):1713–25.
[124] Jalali F, Dorraji PS. Interaction of anthelmintic drug (thiabendazole) with DNA: spectroscopic and molecular modeling studies. Arab J Chem 2017;10:S3947–54.
[125] Maurya N, Parray ZA, Maurya JK, Islam A, Patel R. Ionic liquid green assembly-mediated migration of piperine from calf-thymus DNA: a new possibility of the tunable drug delivery system. ACS Omega 2019;4(25):21005–17.
[126] Dorraji PS, Jalali F. Spectral, electrochemical, and molecular docking evaluation of the interaction of the anti-hyperthyroid drug methimazole with DNA. Can J Chem 2015;93(10):1132–9.
[127] Shahabadi N, Amiri S. Spectroscopic and computational studies on the interaction of DNA with pregabalin drug. Spectrochim Acta A Mol Biomol Spectrosc 2015;138:840–5.
[128] Chakraborty A, Panda AK, Ghosh R, Biswas A. DNA minor groove binding of a well known anti-mycobacterial drug dapsone: a spectroscopic, viscometric and molecular docking study. Arch Biochem Biophys 2019;665:107–13.
[129] Brabec V, Kasparkova J. Modifications of DNA by platinum complexes: relation to resistance of tumors to platinum antitumor drugs. Drug Resist Updat 2005;8(3): 131–46.
[130] Tripathi T, Dubey VK. Advances in protein molecular and structural biology methods. 1st ed. Cambridge, MA, USA: Academic Press; 2022.

CHAPTER 4

# Nuclear magnetic resonance spectroscopy for protein structure, folding, and dynamics

**Sourya Bhattacharya**[a], **Sebanti Gupta**[c], **and Saugata Hazra**[a,b]
[a]Department of Biotechnology, Indian Institute of Technology Roorkee, Roorkee, India
[b]Centre of Nanotechnology, Indian Institute of Technology Roorkee, Roorkee, India
[c]Yenepoya Research Centre, Yenepoya (Deemed to be University), Mangalore, India

## 1. Introduction

Biological macromolecules may be seen as statistical ensembles of conformational states with Boltzmann distribution-based equilibrium populations [1]. A Boltzmann distribution (Gibbs distribution) is known as a probability distribution or probability measure in statistical mechanics or mathematics that describes the chance that a system will be in a particular state with respect to the state's energy and temperature [2]. The distribution is given as

$$p_i = e^{-\varepsilon i}/kT$$

Here, $p_i$ is the probability of the system in state $i$, $\varepsilon_i$ is the energy of that particular state, and $kT$ is a constant term of the probability distribution, i.e., the multiplication of the Boltzmann's constant $(k)$ and temperature $T$.

The native (N) state or the populous arrangement of a biological macromolecule is the main focus of structural biology. The structure-function relationship of biological macromolecules states that the main functions of different macromolecules come entirely from the native state conformation. In this context, determining the three-dimensional (3D) structure of the native state using high-resolution techniques such as nuclear magnetic resonance (NMR) spectroscopy and X-ray crystallography were thought to provide detailed information about the molecular mechanisms behind the function of these biomolecules [3,4] (Fig. 1).

Many biologically important processes like substrate binding, allosteric modification and regulation, catalysis, pathogenic misbehavior, and aggregation-associated diseases are controlled by excited (E) states as suggested by extensive molecular dynamics (MD) simulation studies. Subsequently, there

*Advanced Spectroscopic Methods to Study Biomolecular Structure and Dynamics*
https://doi.org/10.1016/B978-0-323-99127-8.00004-0

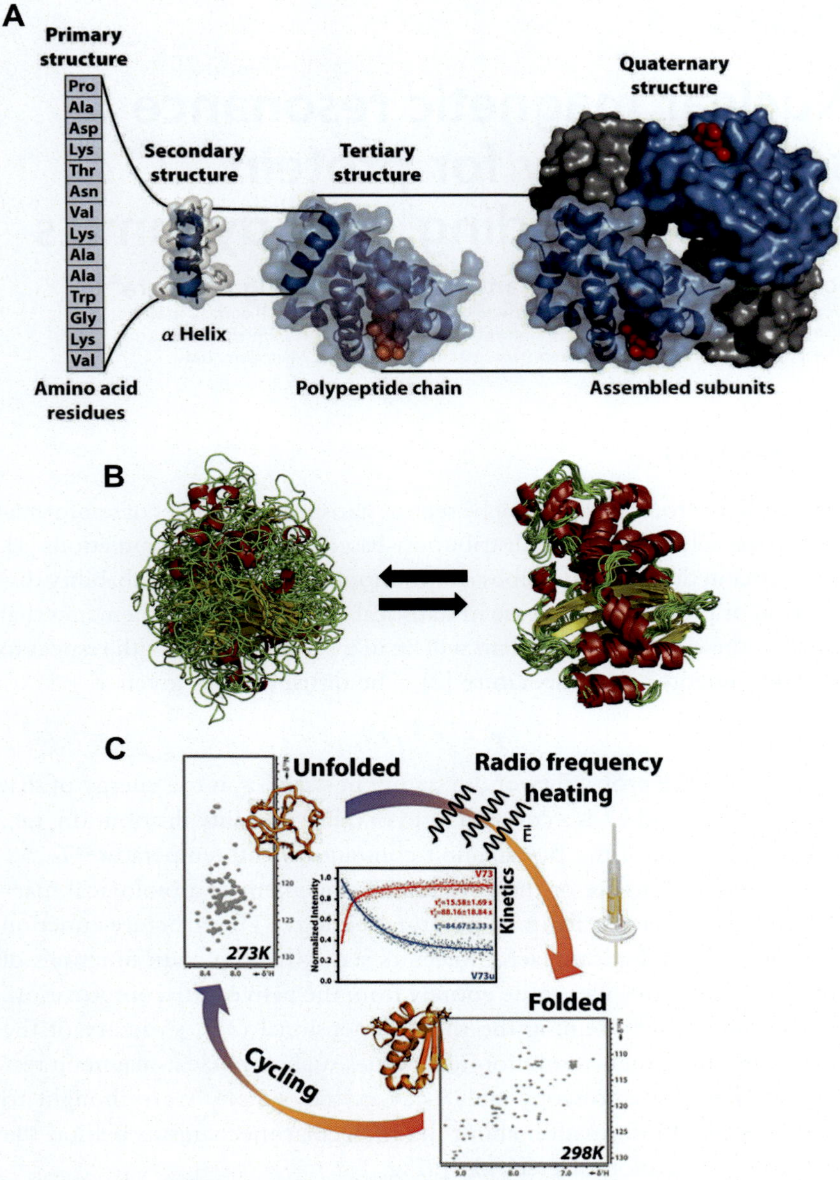


**Fig. 1** Use of NMR spectroscopy in studying protein folding and dynamics. (A) The four levels of protein structural architecture. (B) Protein folding/unfolding dynamics. (C) Overall steps for understanding protein folding and dynamics using NMR spectroscopy. *(Figure reproduced with permission from Rashid MA, Khatib F, Sattar A. Protein preliminaries and structure prediction fundamentals for computer scientists.*

is a requirement for different experimental procedures to analyze the role of the excited-state conformation at the subatomic level (Fig. 2A). NMR spectroscopy has emerged as one of the most used test procedures for analyzing the structural dynamics of biomolecules at the nuclear level on account of systemic advancements accomplished during the last decade. Currently, different NMR strategies are accessible, covering different timescales of subatomic elements. The quick timescale (submilliseconds) dynamics, emerging from small energy boundaries among the interchangeable states, primarily provide information about the nearby conformational variances in the particle; the significant structural fluctuations contribute to slow time scale dynamics. Two critical NMR experiments, like Carr Purcell-Meiboom-Gill (CPMG)-type relaxation-dispersion and rotating-frame spin-relaxation experiments, can accurately analyze dynamic behavior coming from the millisecond (ms) to the subsecond timeframe. The excited-state (E) molecule, which is sought to be short lived, and the native state (N) molecules are thought to be separated by a significant energy barrier of several kcal/mol. Notably, the interconversion of the two states generally takes place within the timescale of 1 s and can be studied with the help of real-time NMR [5,6] (Fig. 2B).

## 2. NMR spectroscopy approaches to study protein folding and dynamics

### 2.1 Real-time NMR spectroscopy

According to the previously described Boltzmann distribution, biological macromolecules invariably maintain populous conformational states depending on their respective energy states. The native state of a statistical ensemble is the most populous and lower in energy and is considered the most functional entity. This native state is the main focus of many structure determination techniques [1].

**Fig. 1, cont'd** 2015 Oct 9 [cited 2022 Apr 9]; Available from: http://arxiv.org/abs/1510.02775. Bhattacharya S, Junghare V, Pandey NK, Ghosh D, Patra H, Hazra S. An insight into the complete biophysical and biochemical characterization of novel class A beta-lactamase (Bla1) from *Bacillus anthracis*. Int J Biol Macromol 2020;145:510–26. New method to study protein folding by real-time NMR spectroscopy, Wiley Analytical Science, 2020 [cited 2022 Apr 9]. Available from: https://analyticalscience.wiley.com/do/10.1002/was.00080152.)

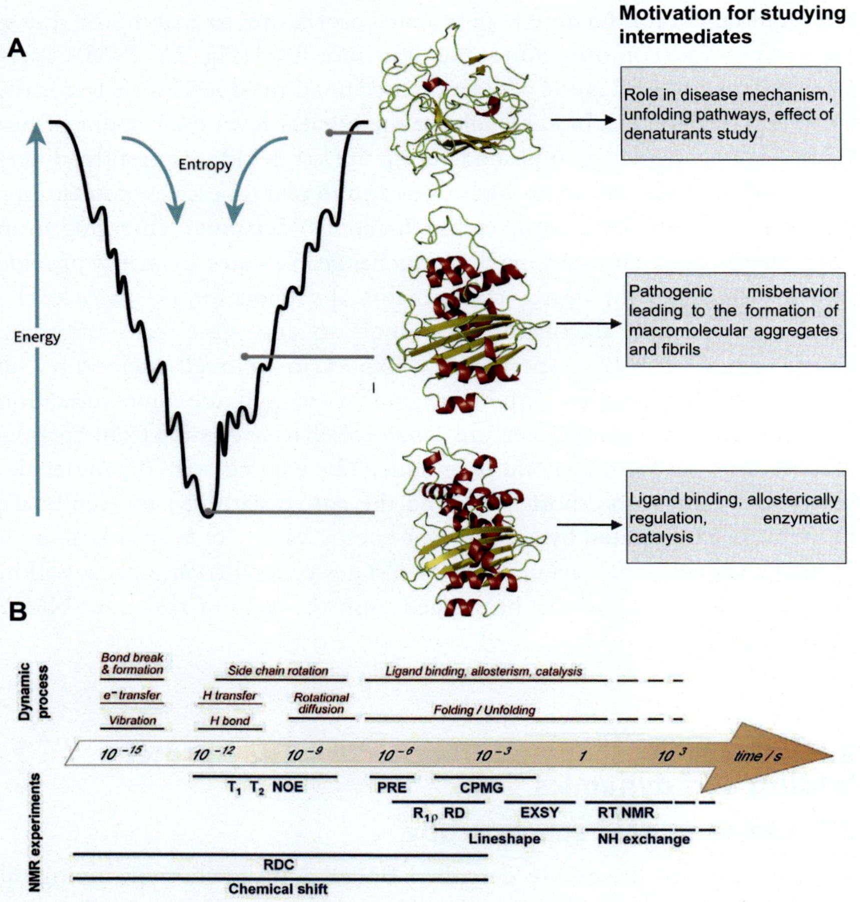


**Fig. 2** (A) Role of different energy state molecules in the folding funnel. (B) Timescale resolution of different NMR techniques to understand different dynamic processes. *(Figure reproduced with permission from Ortega G, Pons M, Millet O. Protein functional dynamics in multiple timescales as studied by NMR spectroscopy. Adv Protein Chem Struct Biol 2013;92:219–51.)*

Off-equilibrium real-time NMR spectroscopy monitors the conformational shift from starting unfolded state (U) to the last native state (N). The high-energy intermediate state population and the folding from U to N may be transiently boosted by fast shifting the sample circumstances from, for instance, a denaturing condition to a native environment. Changing different physicochemical conditions like temperature, pH, or denaturants by a quick mixing mechanism, or removing a photolabile protective group, brings about an adjustment of test conditions inside the NMR spectrometer. During the kinetic refolding process toward the native form, several or multidimensional NMR spectra can be acquired to explore high-energy states. Conformational states with high energy are set of briefly populous states in the course of refolding and are commonly named intermediate states or I state. The main utilization of ongoing NMR analyses of proteins or nucleic acids depended on recording a one-dimensional (1D) $^1$H NMR spectrum progression. It is a delicate and quick method [7–11].

There are two reasonably different methodologies to perform multidimensional real-time NMR tests. The methodology relies upon obtaining a single-spectrum TnD and kinetic rate constant *k*, depending on whether the folding happens on a period size of minutes (2D) or hours (3D). During the acquisition time of the data collection, it is thought to be clear that different conformation at different states is not a constant entity [12,13]. Thus the collapsing energy influence the line states of the NMR signals. Sharp lines possessing negative horizontal wings are noticed for the resonances of the conformational states that develop through the course of kinetics. Interestingly, wide lines result in the case of the resonances of the decaying species. In the current chapter, we discuss recent improvements in multidimensional real-time NMR and different NMR strategies presented during the last 10 years to show and understand the additional opportunities these methods represent to get definite data on briefly populous conditions proteins [14] (Fig. 3).

## 2.2 Multidimensional real-time NMR spectroscopy

Real-time NMR is a method to get quick multidimensional NMR data required to gain information about the short-lived, transiently populous conformational states at the subatomic level. Several rapid NMR methodologies have recently been developed to investigate biomolecular structure and dynamics. These rapid acquisition strategies have opened up new interesting opportunities for the exploration in real time of short-lived conformational states of proteins and nucleic acids formerly unreachable by traditional multidimensional NMR methods [15,16].

## 2.3 Polarization-enhanced fast-pulsing techniques

Fast-pulsing methods are particularly appealing for real-time applications since they dramatically reduce experimental time and improve gross experimental sensitivity. This particular technique aims to reduce the gap between succeeding pulse sequence (scans) repeats to obtain faster rates of repetition

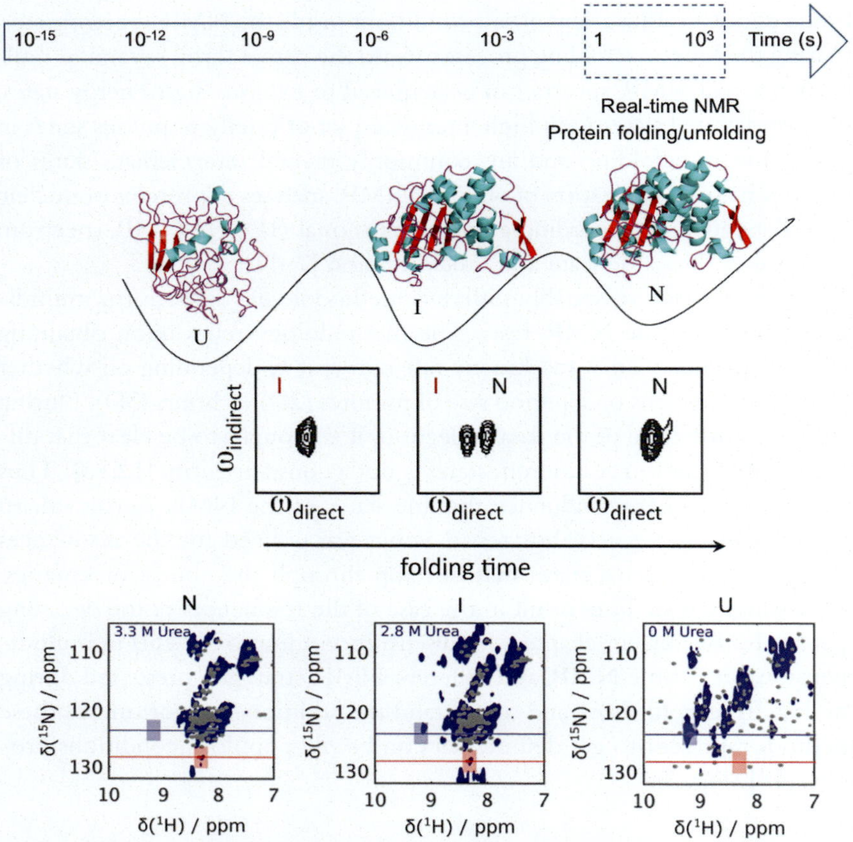


**Fig. 3** Real-time NMR timescale and use of the real-time NMR method to understand the protein folding/unfolding process. *(Figures reproduced with permission from Rennella E, Brutscher B. Fast real-time NMR methods for characterizing short-lived molecular states. ChemPhysChem 2013 [cited 2022 Apr 9];14(13):3059–70. Available from: https://onlinelibrary.wiley.com/doi/full/10.1002/cphc.201300339 and Kim J, Mandal R, Hilty C. 2D NMR spectroscopy of refolding RNase Sa using polarization transfer from hyperpolarized water. J Magn Reson 2021;326:106942.)*

and collect similar scans in a limited time. In the case of multiscan studies or at the equilibrium, the spin polarization at the steady state available at the start of the pulse sequence determines the sensitivity of an NMR experiment.

## 2.4 2D real-time NMR spectroscopy

Because it gives the maximum sensitivity in the shortest amount of time, SOFAST-HMQC has emerged as the most used approach for 2D real-time

NMR. It helps in the identification of the folding pathways and the conformational states. In the past, it was utilized to monitor particular amine and side chains, mainly $CH_3$ groups in proteins and the imino groups in DNA. 2D SOFAST-HMQC record spectra under 5 s on different molecular systems (like monomeric globular protein to multimeric assembly) in the favorable experimental condition of the instrumental setup and macromolecular conditions (magnetic field strength of the instrument, proper cryoprobe, standard protein concentration for NMR studies). This low-time requirement can be further reduced by using sparse nonuniform data sampling. The ultraSOFAST technology, which uses gradient-assisted spatial encoding of NMR frequencies to increase the repetition of 10 spectra per second, is particularly useful [17].

## 2.5 Chemical shifts

The NMR spectroscopy is a technique that largely determines the structure and dynamics of biomacromolecules. Chemical shifts are the most effectively and precisely estimated NMR parameters and address the conformities of local and nonnative conditions of proteins with high explicitness. Chemical shifts are extremely sensitive molecular structural probes. This property gives them their one-of-a-kind worth in testing the properties of frameworks to analyze the subatomic details of natural and inorganic mixtures to complex organic macromolecules. It permits the determination of specific signals coming from chemically identical groups in various regions of the environment like local and global maxima. Chemical shifts are commonly employed in structural biology to anticipate secondary structure regions in native and nonnative protein states, help refine complicated structures, identify unfolding pathways, and significant changes in conformation due to unfolding and ligand binding. Chemical shift combined with other probes can be used to infer the inter proton distances using the nuclear Overhauser effect (NOE) and the orientation of different NMR active nuclei in the structure of a macromolecule. This process also assists in identifying the tertiary structure of proteins.

Chemical shifts are the main NMR attributes that could be gathered on specified conditions of a protein with any level of fulfillment in numerous conditions to examine how well these qualities are accustomed to distinguish high-resolution structures. In the case of a low-molecular-weight molecule, chemical shifts intrinsically convey enough information to determine the structure at high resolution, based on the distinctive fingerprints supplied by their NMR spectra. On the other hand, it also provides structural data that is very different than NOEs. (NOE outlines pairwise distances between explicit protons and can provide unequivocal data concerning general spatial areas of various residual architectures in a protein arrangement.) The chemical shift related to a particular molecule is a summation of many contributing variables, making solid distinguishing proof of interaction connection, although they can be fundamentally affected by contacts between residues at totally different areas in the protein, for example, hydrogen bonding and aromatic rings. If such effects could be fully understood, they would allow the precise environment of nearly every atom in the structure to be characterized and lead to the development of a distinctive gross conformation belonging to a similar environment. Recent developments in the study of chemical shifts have made it possible to utilize their values to gain information on various properties of protein conformations, most notably dihedral angles, with great accuracy. Because even tiny inaccuracies in dihedral angles lead to gradual inaccuracy accumulation, these quantifications do not allow the high-resolution structures of proteins to be characterized without the use of considerable supplementary information, like the one given by NOEs or RDC, as described later. Chemical shift constraints can be considered ensemble averages as specified for other NMR observables because they are sensitive to dynamics on the microsecond timescale. This method allows for the description of structural and dynamical features of individual proteins under a range of situations. Furthermore, recent studies show that new approaches can enable the acquisition and assignment of NMR spectra for systems previously thought to be out of reach to this spectroscopic strategy, such as enormous or transient multimolecular congregations, low-populous states associated with enzymatic catalysis, allostery of differently populous entities, and folding/unfolding conversion of proteins. Using this method, the capacity to define precise structures from chemical changes might be critical in tackling the structural issues that such systems provide and, hence, play an increasingly essential and distinctive role in structural and molecular biology [18] (Fig. 4).

## 2.6 Spin-lattice (T1) and spin-spin (T2) and nuclear Overhauser effect

Using spin relaxation studies, NMR spectroscopy has emerged as a potent tool for examining bimolecular dynamics. Backbone $^{15}N$ NMR dynamics analysis of a protein aims to discover and discriminate mobile versus restricted residues in the main sequence and connect this information to

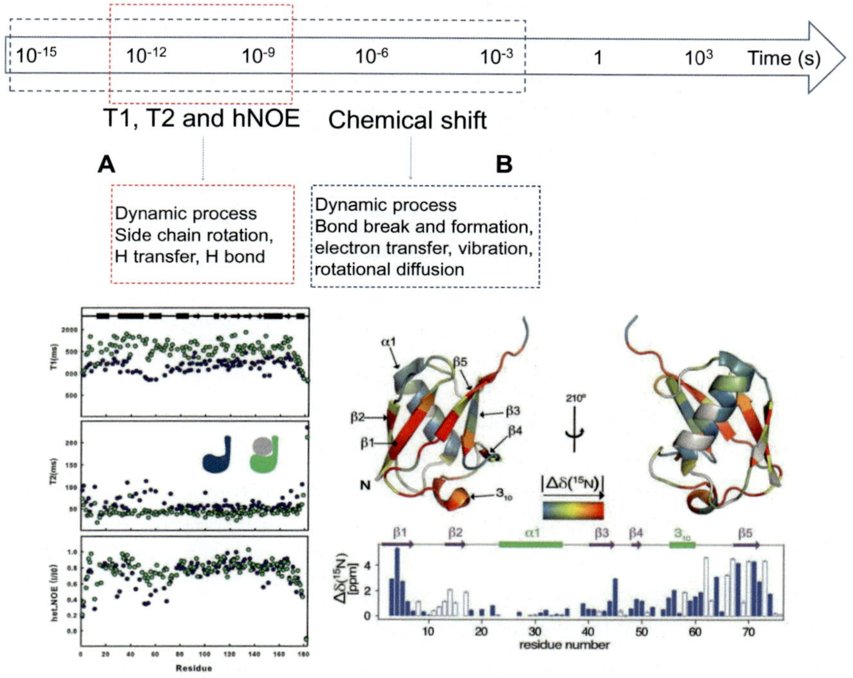


**Fig. 4** (A) Time scale and application of T1, T2, and hNOE to understand protein folding and dynamics. The figure indicates a comparison of the T1, T2, and hNOE experiments of native $^{15}N$ (blue) and its monoubiquitinated form. (B) Time scale and application of chemical shift to understand $^{15}N$ chemical shift differences, $|\Delta\delta(^{15}N)| = |\delta I - \delta F|$, between the metastable intermediate and natively folded VA2-ubiquitin. *((A) Figure reproduced with permission from Faggiano S, Menon RP, Kelly GP, Todi SV, Scaglione KM, Konarev PV, et al. Allosteric regulation of deubiquitylase activity through ubiquitination. Front Mol Biosci 2015;2(2). (B) Figure reproduced with permission from Charlier C, Courtney JM, Alderson TR, Anfinrud P, Bax A. Monitoring $^{15}N$ chemical shifts during protein folding by pressure-jump NMR. J Am Chem Soc 2018;140:8096–9. https://pubs.acs.org/doi/10.1021/jacs.8b04833.)*

predict the structure, function, and relationship. Several significant discoveries have resulted from such information. NMR spin relaxation investigations offer detailed observations on bond vector movements in the picosecond to the nanosecond timeframe. The initial characteristics derived from standard $^{15}N$ or $^{13}C$ NMR relaxation investigations are spin-lattice (T1) and spin-spin (T2) relaxation periods, as well as the steady-state NOE.

To better understand the microscopic causes of solution-state NMR spin relaxation, the completely relaxed thermal equilibrium state is a straightforward method to comprehend the NMR spin relaxation process. At thermal equilibrium, there are two key requirements: (i) the Boltzmann distribution that gives the spin state of each and every population and (ii) the loss of all the coherence between these spins. Spin excitation using a radiofrequency (RF) resonance pulse can disrupt the equilibrium and cause one of these criteria. The key interaction between excited spin and the neighboring environment, referred to as the lattice for historical reasons, causes the reestablishment of the Boltzmann distribution or spin-lattice relaxation (defined by a time constant, i.e., T1). Random orientation of the molecule in liquid takes place when the given spin experiences a local magnetic field. The relative dipole orientation is the cause of direct dipole-dipole coupling against surrounding spins induced by the magnetic field. The molecular electronic current produces a magnetic field, which depends on the molecular orientation of the static magnetic field. These two are exclusive components of the macromolecular study using NMR. By causing spin-state conversion, these fluctuations, especially field fluctuations at the Larmor frequency of the nucleus, induce spin–lattice relaxation. Spin-spin relaxation (T2) occurs due to individual variations in the Larmor frequency (described by the rate of precession of the magnetic moment of the proton when the proton experiences an external magnetic field). Unlike spin-lattice relaxation, the spin-spin relaxation of a designated nuclear spin would not need a spin-state conversion. Nonetheless, field fluctuations depend on nearby nuclei's spin-state transitions.

In NMR, the heteronuclear NOE is regularly used to increase magnetism proportionate to the level of the thermal equilibrium for different dynamics-related studies. The relative gyromagnetic proportion of individual spins and the dipole-dipole distance determines the NOE.

Spin relaxation is intuitively responsive to dynamics. For example, it is the bond vector motion that gives intuitive motional control. S2 is a motion amplitude parameter that spans 0 (for an unfettered bond vector) and 1 (for complete motion limitation). S2 is insufficient for determining the route or model of the motion or dynamics of a bond vector. Internal bond vector dynamics and overall molecule dynamics are assumed to be separable in this technique [19] (Fig. 4).

## 2.7 Relaxation dispersion (RD) techniques

Protein folding takes anywhere from 1 s to seconds or even longer, with small proteins and a single domain containing globular proteins folding in milliseconds. Proteins and modular domains are at their lowest energy native state under physiological conditions. Fully folded modular domain proteins set out on protein folding and convert from meagerly populated unfolded and intermediate state conformer (in local minima of folding landscape) to a native conformation (in global minima with lesser energy). These unfolded conformers or intermediates, also known as denatured states, have varying structural contents and are involved in folding, misfolding, and aggregation. The saturation transfer (ST) and relaxation dispersion (RD) techniques are NMR approaches for analyzing s to ms protein dynamics, investigating protein folding-related structural associations of minor intermediate states on the folding energy landscape, and folding kinetics. The two main classes of RD experiments are (i) CPMG methods, which use a sequence of evenly spaced refocusing pulses to modulate Rex and (ii) rotating-frame R1 relaxation experiments, which use an on- or off-resonance continuous wave (CW) RF field to modulate Rex.

During μs to ms timescale, RD spectroscopy provides exact exchangeable kinetic observation between native species and intermediate species, covering significant chemical shift variations between the states. Significant line broadening and RD would not happen unless substantial chemical shift variations exist between the states. Relaxation dispersion measurements and kinetic rate constants produce chemical shifts for intermediate or low populous species, which can be utilized to get an essential structural understanding of lowly populous intermediate states.

## 2.8 Rotating frame relaxation

During the last decade, NMR approaches, especially solution-state NMR, for detecting the spin relaxation rate constants in proteins and other biomacromolecules, have proven particularly critical for studying ps to ns and μs to ms timescale dynamics. For studying exchange mechanisms that occur on the μs to ms timescale and involve different populations, R1F experiments have been the preferred method. The timescale of the process that CPMG and R1F approaches analyze is governed by the available range of effective magnetic field strengths. The functional field strengths used in the R1F relaxation assay are generally 1–6 kHz. As a result, R1F experiments are frequently employed to study chemical exchange mechanisms at the μs timescale. Chemical exchange, a common occurrence in NMR spectroscopy, changes line morphologies in free-precession NMR spectra and, more importantly, comes up with the magnetization spin-locked relaxation in the rotating frame of reference. Chemical exchange is a concept that relates to intramolecular or intermolecular kinetic processes that exchange nuclear spins between environments with varying localized magnetic fields. R1F and R2F are two distinct relaxation rate constants, R1F comes from magnetization locked in the rotating frame along the direction of the effective field, and R2F comes from magnetization orthogonal to the direction of the effective field. Rh 1 and Rh 2 are found in the rotating-frame relaxation rate constants due to the change from the laboratory to the rotating frame. This dependency may be broken by testing Rh 1 and Rh 2 separately (Fig. 5A).

Depending on the size of the pulse sequences in R1F, the estimation can be named "off-resonance" or "on-resonance" (also dubbed "near-resonance"), reliant on the size of $\Omega$, or as "strong field" and "weak field" reliant on the magnitude of $\omega 1$. The rotating-frame relaxation rate R1ρ, is estimated as a component of strength, $\omega 1$, and/or offset, $\Omega$, of the applied RF field (spin-lock) for magnetization aligned along the direction of the effective field $\omega e = (\omega 1,0,\Omega)$ [20,21]. Getting the most reliable and exact R1F estimations requires cautious advancement of designed experimentation, monitoring temperature, water concealment, and adjustment of $\omega 1$ rf fields.

Chemical exchange processes occur based on conformational alternations or chemical reactions on the timescale of the chemical shift that alter a particular spin's magnetic surroundings and stochastically regulate its isotropic chemical shift. R1F spin relaxation in the rotating frame is a collection of NMR techniques for characterizing chemical exchange processes in the μs to ms timescale. The thermodynamics and kinetic features of the conformational alternation or chemical processes that cause transition widening may be determined using R1F RD data and the isotropic shifts changes between various structural configurations or chemical entities. The structural information regarding ordinarily unobservable small molecular or chemical species is included in the differences in isotropic induces chemical shift associated values commonly denoted as, ij, acquired from dispersion measurements. Chemical shifts, on the other hand, are influenced by many conformational factors, including the architecture of hydrogen bond, torsion angle coming from protein side chain, backbone dihedral angle, and other intramolecular interactions like electrostatic interaction, etc. Due to these factors, a conformational alternation can affect the chemical shifts of distinct nuclei. These intricacies make interpreting the differences in chemical shifts more difficult for any finite collection of nuclear spins. As a result, quantification of R1F dispersion data for various nuclei (e.g., the protein

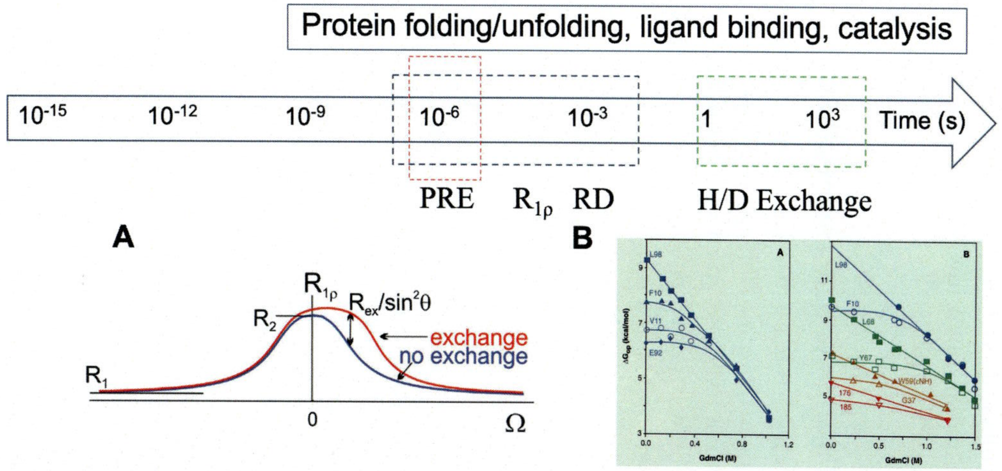


**Fig. 5** (A) Rotating-frame R1ρ pulse sequences for an isolated nucleus and offset dependence of the rotating-frame R1ρ rates to understand chemical exchange to understand protein folding dynamics. (B) Hydrogen exchange method to understand the hydrogen exchange phenomenon in unfolding condition c. *((A) Figure reproduced with permission from Zhuravleva A, Korzhnev DM. Protein folding by NMR. Prog Nucl Magn Reson Spectrosc. 2017 [cited 2022 Mar 14];100:52–77. Available from: https://pubmed.ncbi.nlm.nih.gov/28552172/. (B) Figure reproduced with permission from Bai Y, Sosnick TR, Mayne L, Englander SW. Protein folding intermediates: native-state hydrogen exchange. Science 1995 [cited 2022 Apr 9];269(5221):193–7. Available from: https://pubmed.ncbi.nlm.nih.gov/7618079/.)*

backbone, $^{15}N$, $^{13}CO$, and $^{13}CR$) offers a more thorough diagram of the transition process. Under equilibrium native state circumstances, R1F techniques can quantify folding kinetics and offer structural observations on the assembly of unfolded structures. There is enough population of unfolded states to offer a widening process. For R1F studies to be practical, the population of unfolded molecules must typically be $>0.005$. Studies that match RD data with a two-site model implicitly imply that kinetic conversion within the unfolded entities is quick.

## 2.9 Paramagnetic relaxation enhancement (PRE)

Other NMR measures, such as PRE, can be used to explore temporary nonnative protein states. In general, profoundly high populous states are found inside the area of global minimum (lesser in energy) of the free energy landscape. However, only a little is understood about lowly populous states, which cannot trap in the landscape and remain undetected by standard structural approaches of structural biology and biophysics. Many biological activities, such as molecular recognition, molecular binding, enzyme catalytic efficiency, enzyme allosterically behavior, and self-assembly, occur via these low populous intermediates involving quick but occasional transitions between the global minimum and local minimum states. $^{2}H$, $^{13}C$, and $^{15}N$ relaxation measurements have offered a plethora of information on the ps to ns timescale for NMR dynamics research. Recent improvements in RD spectroscopy have revealed information on operation in the μs to ms range. RD spectroscopy and PRE are methods of choice that may be utilized to analyze equilibrated low populous states in the energy landscape. [22].

In recent years, paramagnetic NMR has arisen as an alternate and effective method for studying many dynamic processes involving macromolecules. Because of its distance dependence between the paramagnetic center and the particular nucleus, the PRE resulting from unpaired electrons that possess a significant isotropic g factor (it is referred to as a scaling factor between the orbital and spin angular momentum of the respective unpaired electron) has proven to be a remarkably versatile tool for providing several structural information is in a dynamic process. The PRE arises from a magnetic dipolar interaction between the paramagnetic core's unpaired electron and a nucleus. This interaction causes nuclear relaxation rates and is known as the PRE effect. The PRE effect is quite significant; the effect results in a significant magnetic moment of an unpaired electron and the exact magnetic moment we observe in NOE, but in the case of NOE, where the effects are tiny and hence are confined to short-range interproton interaction. The PRE rates in a fast interconversion of major and minor species are a fingerprint measure to understand low populous states. As a result, if the minor species' proton and paramagnetic center distance are lesser than the significant species' proton and paramagnetic center, the minor species' imprints could be visible in the reported PRE rates, allowing minor species' structural information.

PRE has also been used to study the unfolded or partially unfolded forms of proteins. The modestly populous intermediate states and other nonnative states usually help in many critical biological processes. One of the major drawbacks of basic biophysical and biochemical approaches is that these methods could not provide information on the different structural changes associated with these intermediates. PRE emerges as a versatile approach that recognizes and describes these intermediates in a fast-folding mechanism. The same idea may be used for nonspecific protein-DNA interactions, interdomain migrations, and transitory protein connections, among other dynamic processes.

## 2.10 ZZ-exchange

Longitudinal exchange studies allow quantifying interconversion rates and longitudinal relaxation rates between exchanging species by examining the cross-peak exchange phenomenon and the direct correlation of time dependence. The exchange of heteronuclear longitudinal magnetization of product operator terms such as IzSz or Sz to quantitate changes between particular conformities of biomolecules for isolated resonances has been observed in exchange experiments. Since they permit admittance to practically essential dynamic processes occurring on time scales going from 0.5 to $50\,s^{-1}$, these experiments significantly influence NMR dynamics examinations of organic frameworks from that point onward. This technique also quantifies the interconversion on a per-residue basis without disrupting the system's thermodynamic equilibrium. This property makes NMR spectroscopy a must-have tool for studying macromolecular dynamics.

Longitudinal exchange experiments measure the transfer of magnetization at the mixing time between two species of exchanging conformation. In the standard ZZ exchange experimental (frequency-labeled) setup, four spectra or peaks are noticed for each spin: two spectra or peaks (diagonal) coming from direct correlation and two spectra or peaks coming from exchange cross. During the mixing time or interconversion between several species, a progression of spectra is recorded to observe the interconversion rates between the exchanging conformers. The four spectra or peaks generated from the ZZ-exchange experiment from direct correlation and exchange are quantified by applying the peak's magnitude to appropriate theoretical curves.

The principles from $^{1}$H EXSY and $^{1}$H-X (X= $^{13}$C or 15 N) HMQC or HSQC heteronuclear correlation studies were used to design the first heteronuclear experiments to assess chemical exchange in biomolecules. These ZZ-exchange experiments, like $^{1}$H EXSY, are suitable in slow chemical exchange, which is set on resonances for particular chemical states are seen. Montelione and Wagner's first studies employed 2IzSz longitudinal two-spin order or Sz magnetization ($I$ = $^{1}$H, S= $^{13}$C or $_{15}$N) instead of Iz magnetization, among other $^{1}$H EXSY during the mixing period. Wüthrich et al. and Kay et al. went on to design an established pulse sequence for heteronuclear ZZ-exchange spectroscopy that uses Sz magnetization on interconversion or mixing time and HSQC-style frequency labelling [23].

## 2.11 Chemical exchange saturation transfer (CEST)

CEST is a novel MRI contrast approach. Replaceable protons or interchangeable atoms in a compound (either exogenous or endogenous) are precisely saturated and distinguished by implication through the water signal with increased sensitivity. The requirements for utilizing an agent (mainly for generating contrast) in MRI are similar to those agents that are frequently used in other imaging techniques. They generate the appropriate contrast (by the effect of that agent) while using the lowest feasible agent concentration to escape disrupting the physiological milieu and cut down toxicity. Most (super) paramagnetic metals that aid relaxation are hazardous when not covered or chelated. The use of exchangeable protons for MRI contrast allows diamagnetic substances to be added to the list of probable MR contrast agents. "Chemical exchange saturation transfer" was termed after the ability of a unique mechanism of "on and off" of the particular contrast by utilizing the proton of interest's particular RF (radiofrequency) exchange saturation. Replaceable solute protons (*s*) with a distinct resonate frequency than bulk water protons (*w*) are precisely saturated using RF irradiation, according to the basic principles of CEST imaging. During the exchange procedure of replaceable solute, protons, and bulk water protons (exchange rate *ksw*), the generated saturation is transmitted to bulk water, and the water signal is significantly muted [24].

## 2.12 Hydrogen-deuterium exchange (H-D or H/D exchange)

According to the Boltzmann distribution, a tiny percentage of protein entities (different conformational states) occupies different potential energy in the folding funnel (Fig. 2A) that includes protein population in the fully unfolded state as well as protein entity with a partially folded form. Even while these partially unfolded forms (intermediates) are generally opaque to measurement, studying them can reveal the essentially cooperative character of protein structure and identify the folding/unfolding kinetics and mechanisms. Denaturants and temperature can modify the energy levels and occupancy of these conformationally excited states. Hydrogen-deuterium exchange (H-D or H/D exchange) experiments may estimate the amount of hydrogen exposed in unfolded or partially folded (higher energy form) conformations, their solvent exchange kinetics, and sensitivity to the perturbation. Accordingly, we may deduce the protein structure, surface occupancy or exposure, and associated free energy. When replaceable amide hydrogens (NH) in hydrogen-bonded structures are exposed to solvent transiently, the replaceable amide hydrogens can exchange with solvent hydrogens (Fig. 5B) [25,26].

# 3. Conclusion and future prospects

NMR has emerged as the essential exploratory instrument for portraying the effect of various nonnative states (low crowded states) on the protein folding landscape or protein-energy scene. It provides the primary and valuable experiences played by these intermediates that play various significant functions in protein collapsing, misfolding, accumulation, and binding [27,28]. We discussed different NMR strategies by comparing their overall capacity and test experiences during the process. The main advantage of using NMR spectroscopy to study protein folding is to analyze the folding mechanism of isolated protein nearly in its thermodynamic equilibrium. One can study the dynamic changes associated with the help of chemical or physical denaturants that we commonly referred as reversible renaturation/denaturation. We discussed several NMR techniques to understand protein folding dynamics. Real-time NMR measurement detects real-time amide exchange procedures to understand intermediate states' equilibrium. Other techniques like RD, chemical shift, and residual dipole coupling are introduced to describe the transition of excited states and structural information associated with intermediated states. One of the significant disadvantages of NMR techniques to study the dynamic process of protein include it can only measure isolated protein in varying physicochemical denaturant conditions; however, in the real thing, the folding mechanism in a cell is not in general equilibrium. Several factors like posttranslational modification are associated with it, and proteolysis cleavage is also a factor that can fold a protein in a cell. These things are not considered in the NMR techniques and are its limitation. Solution NMR, alone or related to other primary and biophysical approaches (such as cryo-electron microscopy, X-ray crystallography, and computational analysis), is relied upon to assume an undeniably significant part in understanding protein folding/unfolding in vivo and in vitro.

## References

[1] Rennella E, Brutscher B. Fast real-time NMR methods for characterizing short-lived molecular states. ChemPhysChem 2013;14(13):3059–70.

[2] Landau LD, Lifshitz EM, Landau LD, Lifschitz EM. Vol. 9—Statistical physics part 2. pdf. ZAMM Zeitschrift f 1980;61:603.

[3] Baldwin AJ, Kay LE, et al. Nat Chem Biol 2009;5(11):808–14. [cited 2022 Mar 5]. Available from: https://pubmed.ncbi.nlm.nih.gov/19841630/.

[4] Dobson CM. Principles of protein folding, misfolding and aggregation. Semin Cell Dev Biol 2004;15(1):3–16. [cited 2022 Mar 5]. Available from: https://pubmed.ncbi.nlm.nih.gov/15036202/.

[5] Korzhnev DM, Kay LE, et al. Acc Chem Res 2008;41(3):442–51. [cited 2022 Mar 5]. Available from: https://pubs.acs.org/doi/abs/10.1021/ar700189y.

[6] Co &, Creighton TE; WH, Freeman & Co, Wennerström H, Deverell C, Morgan RE, Strange JH, et al. Monitoring macromolecular motions on microsecond to millisecond time scales by $R1\rho - R1$ constant relaxation time NMR spectroscopy. J Am Chem Soc 1996;118(4):911–2. [cited 2022 Mar 5]. Available from: https://pubs.acs.org/doi/full/10.1021/ja953503r.

[7] Kautz RA, Fox RO. NMR analysis of staphylococcal nuclease thermal quench refolding kinetics. Protein Sci 1993;2(5):851–8. [cited 2022 Mar 5]. Available from: https://onlinelibrary.wiley.com/doi/full/10.1002/pro.5560020514.

[8] Frieden C, Hoeltzli SD, Ropson IJ. NMR and protein folding: equilibrium and stopped-flow studies. Protein Sci 1993;2(12):2007. [cited 2022 Mar 5]. Available from: /pmc/articles/PMC2142323/?report=abstract.

[9] Koide S, Dyson HJ, Wright PE. Characterization of a folding intermediate of apoplastocyanin trapped by proline isomerization 1. Biochemistry 1993;32:12299–310. [cited 2022 Mar 5]. Available from: https://pubs.acs.org/sharingguidelines.

[10] Kiefhaber T, Labhardt AM, Baldwin RL. Direct NMR evidence for an intermediate preceding the rate-limiting step in the unfolding of ribonuclease A. Nature 1995;375 (6531):513–5. [cited 2022 Mar 5]. Available from: https://www.nature.com/articles/375513a0.

[11] Segel DJ, Bachmann A, Hofrichter J, Hodgson KO, Doniach S, Kiefhaber T, et al. Compaction during protein folding studied by real-time NMR diffusion experiments. J Am Chem Soc 2000;122(24):5887–8. [cited 2022 Mar 5]. Available from: https://pubs.acs.org/doi/full/10.1021/ja994514d.

[12] Liu X, Siegel DL, Fan P, Brodsky B, Baum J. Direct NMR measurement of the folding kinetics of a trimeric peptide. Biochemistry 1996;35(14):4306–13. [cited 2022 Mar 5]. Available from: https://pubs.acs.org/doi/abs/10.1021/bi952270d.

[13] Roy M, Jennings PA. Real-time NMR kinetic studies provide global and residue-specific information on the non-cooperative unfolding of the β-trefoil protein, interleukin-1β. J Mol Biol 2003;328(3):693–703.

[14] Balbach J, Forge V, Lau WS, Van Nuland NAJ, Brew K, Dobson CM. Protein folding monitored at individual residues during a two-dimensional NMR experiment. Science 1996;274(5290):1161–3. [cited 2022 Mar 5]. Available from: https://www.science.org/doi/abs/10.1126/science.274.5290.1161.

[15] Bermel W, Bertini I, Felli IC, Pierattelli R. Speeding up $^{13}C$ direct detection biomolecular NMR spectroscopy. J Am Chem Soc 2009;131(42):15339–45. https://doi.org/10.1021/ja9058525.

[16] Felli IC, Brutscher B. Recent advances in solution NMR: fast methods and heteronuclear direct detection. ChemPhysChem 2009;10(9–10):1356–68. [cited 2022 Mar 5]. Available from: https://onlinelibrary.wiley.com/doi/full/10.1002/cphc.200900133.

[17] Schanda P, Forge V, Brutscher B. Protein folding and unfolding studied at atomic resolution by fast two-dimensional NMR spectroscopy. Proc Natl Acad Sci U S A 2007;104(27):11257–62.

[18] Cavalli A, Salvatella X, Dobson CM, Vendruscolo M. Protein structure determination from NMR chemical shifts. Proc Natl Acad Sci U S A 2007;104(23):9615–20.

[19] Reddy T, Rainey JK. Interpretation of biomolecular NMR spin relaxation parameters. Biochem Cell Biol 2010;88(2):131–42.

[20] Zhuravleva A, Korzhnev DM. Protein folding by NMR. Prog Nucl Magn Reson Spectrosc 2017;100:52–77. [cited 2022 Mar 14]. Available from: https://pubmed.ncbi.nlm.nih.gov/28552172/.

[21] Williamson G, Clifford MN. Role of the small intestine, colon and microbiota in determining the metabolic fate of polyphenols. Biochem Pharmacol 2017;139:24–39. https://doi.org/10.1016/j.bcp.2017.03.012. Epub 2017 Mar 18. PMID: 28322745.

[22] Marius Clore G, Iwahara J. Theory, practice, and applications of paramagnetic relaxation enhancement for the characterization of transient low-population states of biological macromolecules and their complexes. Chem Rev 2009;109(9):4108–39. [cited 2022 Mar 10]. Available from: https://pubmed.ncbi.nlm.nih.gov/19522502/.

[23] Kloiber K, Spitzer R, Grutsch S, Kreutz C, Tollinger M. Longitudinal exchange: an alternative strategy towards quantification of dynamics parameters in ZZ exchange spectroscopy. J Biomol NMR 2011;51(1–2):123–9.

[24] Van Zijl PCM, Yadav NN. Chemical exchange saturation transfer (CEST): what is in a name and what isn't? Magn Reson Med 2011;65(4):927–48.

[25] Bai Y, Sosnick TR, Mayne L, Englander SW. Protein folding intermediates: native-state hydrogen exchange. Science 1995;269(5221):193–7. [cited 2022 Mar 10]. Available from: https://pubmed.ncbi.nlm.nih.gov/7618079/.

[26] Schmitt & Segert. 基因的改变NIH public access. Bone 2008;23(1):1–7.

[27] Singh D, Tripathi T, editors. Frontiers in protein structure, function, and dynamics. 1st ed. Singapore: Springer; 2020. p. 1–458.

[28] Tripathi T, Dubey V, editors. Advances in protein molecular and structural biology methods. 1st ed. USA: Academic Press; 2022. p. 1–714.

# CHAPTER 5

# Advanced NMR spectroscopy methods to study protein structure and dynamics

**Ashish A. Kawale**[a,b] **and Björn M. Burmann**[a,b]
[a]Wallenberg Centre for Molecular and Translational Medicine, University of Gothenburg, Gothenburg, Sweden
[b]Department of Chemistry and Molecular Biology, University of Gothenburg, Gothenburg, Sweden

## 1. Introduction

In layman's terms, life at the cellular level can be depicted as a soup of a large and diverse pool of biomolecules with complex structural and chemical compositions. These biomolecules must interact with each other to orchestrate various biochemical pathways crucial for the survival of organisms. Proteins are one of the most important biomolecules and display remarkable structural as well as functional diversity. Proteins are best described by the beads on a string model, i.e., a series of amino acids joined via a peptide bond, subsequently folded into a distinct three-dimensional (3D) tertiary structure. A protein's 3D structure plays an essential role in its functionality. In the past decades, numerous studies have highlighted that many misfolded proteins, which lack native 3D structures, either get degraded naturally or create a threat to life in the form of prions and have implications for several diseases and aging processes (reviewed in Refs. [1–3]). Hence, a thorough understanding of the structure-function relationships of proteins through biophysical and structural studies becomes indispensable for learning various molecular mechanisms underlying fundamental biochemical pathways and thus can provide the basis for developing novel therapeutic approaches in diseased conditions.

Structural biology focuses on elucidating the structural properties of biomolecules by depicting the relative arrangement of atomic coordinates in the macromolecules and correlating their functional repertoire [4]. X-ray crystallography has been phenomenal in elucidating the structures of protein macromolecular complexes, as seen by the enormous number of structures deposited in the protein databank (PDB) every year [5,6].

*Advanced Spectroscopic Methods to Study Biomolecular Structure and Dynamics*
https://doi.org/10.1016/B978-0-323-99127-8.00010-6

Similarly, the recent "resolution revolution" in cryo-electron microscopy (cryo-EM) has fueled rapid structure determination of large macromolecular complexes at near-atomic resolution [6–8]. Together with nuclear magnetic resonance (NMR) spectroscopy, these techniques have resulted in a wealth of information about protein macromolecular structures at near-atomic resolution in *apo* and *holo* form, revealing ab-initio information about their biological inner workings.

Notably, it is now widely accepted that proteins are not LEGO-type rigid entities; instead, they undergo a series of structural and conformational fluctuations, essentially acting as molecular machines [9]. These fluctuations are guided by the intrinsic dynamic properties of proteins that impart them the remarkable ability to fine-tune their 3D structures in response to the diverse and complex array of cellular signals that include environmental triggers such as modulations in the pH, temperature, ligand binding, etc. [10–12]. The physical changes exerted by these signals prompt subtle adaptations to their 3D structures, crucial for their functional niche. Out of an ensemble of diverse possible conformations, time-averaged static structures representing the lowest energy state, as illustrated by the standard structural biology approaches, such as X-ray crystallography and cryo-EM, provide limited and inadequate information to corroborate multifaceted protein functions [13,14]. Standard structural biology approaches often fail to detect these low populated, short-lived, intermediate high-energy conformational states pivotal in the respective reaction pathways such as signal transduction, conformational selection, enzyme catalysis, protein folding, and aggregation [13].

Rapid progress in the NMR methodology and hardware over the last few decades has resulted in the widespread application of NMR spectroscopy in the field of life sciences, and it has emerged as an indispensable tool in deciphering atomic-level information [15]. NMR spectroscopy stands out in comparison to the other structural biology techniques as, in addition to serving as a structure determination tool, it also offers several methodologies to dissect macromolecular dynamics as well as protein-ligand interactions at an atomic level under near-physiological conditions (reviewed in Refs. [16–18]). Moreover, NMR spectroscopy has been the method of choice for the proteins that are challenging to crystalize, intrinsically disordered proteins (IDPs), and offers the possibility to study proteins in living cells via *in-cell* NMR [19–21]. Especially, *in-cell* NMR, employing electroporation as a transfer of isotopically labeled into living mammalian cells has lately shown great potential to study IDPs in vivo at atomic resolutions [22–25].

Although NMR is applicable for the characterization of various biomolecules and is classified into solution and solid-state NMR spectroscopy, for practical purposes, this chapter aims to provide our perspective on the principal steps of solution NMR spectroscopy of proteins with a particular focus on the dynamics of large protein molecules.

## 2. Traditional NMR spectroscopy approaches for small-medium-sized proteins

NMR utilizes the nuclear magnetic properties of NMR-active atomic nuclei acting as bar magnets [26]. Protein NMR spectroscopy exploits the NMR-active nuclei such as the $^{1}H$, $^{13}C$, and $^{15}N$ composed of spin half isotopes showing magnetic dipole moments as well as $^{2}H$ manifesting both magnetic dipole and quadrupole moments [27]. Protein molecules enriched with these NMR active nuclei, when placed under the influence of a strong magnetic field, magnetic moments of these nuclei align with the field. Subsequently, when the radiofrequency (*RF*) pulses are applied, as a result of the resonance condition at a characteristic frequency, each NMR-active nucleus gives rise to a characteristic NMR signal, referred to as peak, as the individual protein residues experience a different chemical environment in a protein structure. Thus with the help of sophisticated NMR pulse sequences, the structural and dynamic properties of the proteins can be extracted at an atomic level.

The commonly used approach for performing NMR studies aimed at protein structure determination using traditional NMR spectroscopic approaches is illustrated in Fig. 1. It is critical to use biochemically pure, stable, and isotopically labeled protein samples. Generally, considerable efforts are put into protein sample preparation by screening various conditions during sample production, such as protein domain boundaries, purification steps, buffer conditions, stability parameters, etc., before starting NMR measurements to ensure the best outcomes [28]. Typically, a 0.1–0.6 mL protein sample of 0.1–2 mM concentration loaded in a suitable NMR tube is loaded in a higher magnetic field magnet ($\geq$500 MHz) in solution NMR spectroscopy, and the NMR data are commonly recorded at ambient temperature (25°C). Using buffers mimicking physiological conditions makes the data measurement under near-physiological conditions feasible. Also, provided the stability of the samples at ambient temperatures, the application of NMR spectroscopy does not destroy the sample or alter its composition, making it a nondestructive technique where samples remain reusable for several months to years.

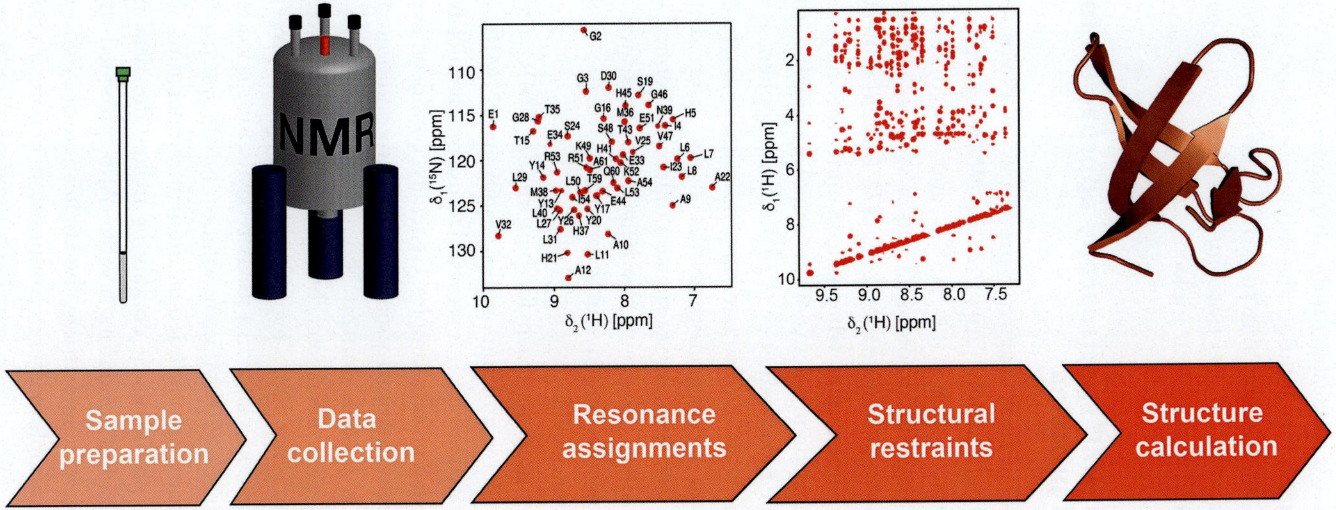


**Fig. 1** Schematic representation of the steps involved in traditional structure determination by NMR spectroscopy. Essential steps performed for the solution NMR structure determination using conventional NMR spectroscopic approaches limited by the protein size (~2025 kDa).

Though it is feasible to utilize the natural abundance of NMR-active $^{1}H$ nuclei from the proteins by applying one-dimensional (1D) NMR spectroscopy on unlabeled proteins or peptides with very small sizes, the severe signal overlap poses a serious challenge for the interpretation of the 1D NMR spectra as the protein size increases and makes its application unfeasible because of its restricted resolving capacity. The development of multidimensional NMR spectroscopy approaches circumvented this problem, paving the way for the application of the NMR spectroscopy to the larger peptides and small-medium-sized proteins. Using multidimensional NMR, it became feasible to understand whether two nuclei are connected via a chemical bond (through-bond, scalar, or *J*-couplings) or are simply close in space (through space, NOE) [29]. This information is crucial for the resonance assignments and subsequent structure determination of proteins.

Barring $^{1}H$, the other most important constituent of proteins, such as nitrogen, carbon, and oxygen, are naturally composed of NMR inactive isotopes ($^{12}C$, $^{14}N$, and $^{16}O$, respectively). Hence, for the successful application of heteronuclear multidimensional NMR spectroscopy approaches, protein samples need to be enriched with $^{15}N$ and $^{13}C$ isotopes comprised of NMR-active spin half nuclei. This is achieved via recombinantly producing proteins in bacteria grown under minimal growth media such as M9, supplemented with $^{15}N$-ammonium chloride and $^{13}C$-glucose as the sole source for nitrogen and carbon, respectively [30].

The practical application of protein NMR spectroscopy usually starts with recording the two-dimensional [$^{15}N$, $^{1}H$] heteronuclear single quantum coherence (HSQC) experiment [31], also referred to as a "fingerprint" experiment. This experiment involves the magnetization transfer from $^{1}H$ spins to the directly attached $^{15}N$ spins via through-bond scalar couplings, evolving the chemical shift on $^{15}N$ with subsequent magnetization transfer back to the $^{1}H$ spins for the detection. As a result, all the protein backbone amide groups where $^{15}N$ nuclei are directly bonded to $^{1}H$ nuclei give rise to one backbone amide NH peak for each amino acid, with the sole exception of proline for the uniformly $^{15}N$ labeled (either [*U*-$^{15}N$] or [*U*-$^{13}C$,$^{15}N$]) proteins. The success of this experiment determines the general feasibility of performing solution NMR spectroscopy studies with the protein of interest. The presence of the high-quality, well-dispersed resonances within the 2D [$^{15}N$, $^{1}H$]-HSQC spectrum indicates the folded protein and typically serves as a reference spectrum for the backbone resonance assignment.

A sequence-specific resonance assignment provides the solid foundation for interpreting NMR data at atomic resolution, which involves assigning each resonance frequency in the NMR spectrum to its origin nuclei from the protein. Backbone resonance assignment refers to the process of correlating the resonances from the 2D [$^{15}$N, $^{1}$H]-HSQC spectrum to the respective amino acid from the protein sequence [28]. The distinctive chemical shifts of the amino acids and application of specific triple resonance experiments allow the identification of amino acids by linking resonances from adjacent residues. Triple resonance experiments such as HNCACB, HNCA, HNCO, and HN(*CA*)CO are typically recorded to perform backbone resonance assignments [32]. In these experiments, magnetization is transferred from $^{1}$H to $^{15}$N and subsequently to the connected $^{13}$C atoms. Details of the NMR resonance assignment strategies are reviewed by Frueh [28]. TOCSY-based triple resonance experiments are recorded for the side-chain chemical shift assignment. After the resonance assignments of $^{1}$H, $^{15}$N, and $^{13}$C, NMR signals from the protein of interest distance and angle restraints are obtained for the solution NMR structure determination. The $^{15}$N and $^{13}$C edited 3D heteronuclear NOESY spectra provide pairwise interproton distance restraints, and the torsion angle restraints are derived from the backbone chemical shifts to determine the solution NMR structure [33].

Resonance assignment allows the interpretation of the dynamics (see Sections 3 and 5) and protein-ligand interactions at an atomic level. NMR spectroscopy offers the remarkable possibility to qualitatively and quantitatively characterize biomolecular interactions via chemical shift mapping. Typically, the 2D [$^{15}$N, $^{1}$H]-HSQC (for backbone amides) or [$^{13}$C, $^{1}$H]-HMQC (for methyl groups) experiment is employed to study protein-ligand interactions. Upon serial addition of the interaction partner, changes in the chemical shifts of the protein, which are sensitive to its chemical environment, are monitored. NMR has been employed to study numerous protein-protein, protein-ligand interactions under physiological conditions at the atomic level, even if they are transient and weak (reviewed in Refs. [34–39]).

## 3. Protein backbone dynamics

Almost all proteins show inherent flexibility at an ambient temperature. Protein dynamics refers to the collective motions displayed by the protein molecule, which can span a diverse range of amplitudes and time scales, resulting in a variety of protein movements such as overall tumbling in solution, bond vibrations, torsion angle rotation, side-chain rotations, internal mobility, protein folding-unfolding transitions, enzyme catalysis, ligand binding, domain motions, conformational changes, and diffusion. These dynamic fluctuations governed by the conformational changes are characterized by diverse spatial and temporal features and are associated with distinct energy barriers. A thorough depiction of these fascinating yet complex motions embedded in the protein can be characterized by studying protein dynamics. NMR spectroscopy offers a diverse experimental toolbox to site-specifically quantify dynamic processes occurring over a broad range of time scales as well as amplitudes of motions (Fig. 2; reviewed in Refs. [16,17,40,41]). These methods are highly effective in investigating protein motions ranging from fast (ps-ns) to slow (μs-h) time scales, discerning intricate details of protein conformational motions controlling protein structural rearrangement and functions.

The most common approach for studying protein motions using NMR spectroscopy is to perform backbone dynamics characterization by measuring the $^{15}N$ (attached to $^{1}H$ nuclei) relaxation rates of the backbone amide nitrogen. This is achieved by the measurement of the longitudinal relaxation rate ($R_1$), the transverse relaxation rate ($R_2$), and the heteronuclear NOE (hetNOE) values, Fig. 3 [42]. These experiments are built on the 2D [$^{15}N$,$^{1}H$] HSQC experiment as its foundation and can be readily employed once the sequence-specific backbone assignment of the protein of interest is available. These experiments allow the monitoring of the intensity changes of backbone amide peaks as a function of varying relaxation delays [27]. The $R_1$, $R_2$, and hetNOE values are then analyzed using model-free formalism [43,44], providing site-specific information about the time scales as well as the amplitudes of dynamic motions and together discern important information regarding the protein's inherent dynamics, overall size, correlation time ($\tau_c$) as well as diffusion properties [42,45]. In a recent application, a compare-and-contrast approach based on the inherent backbone dynamics revealed subtle functional differences, which were already deducible based on subtle modulations of their inherent dynamical properties despite only minute structural differences [46,47].

## 3.1 The heteronuclear NOE

The heteronuclear NOE experiment is an excellent probe to characterize fast time scale protein motions. The experiment detects the motion of the N—H bond vectors of the protein via through-space magnetization transfer as a means of dipolar coupling of directly bonded $^{1}H$ and $^{15}N$ nuclei

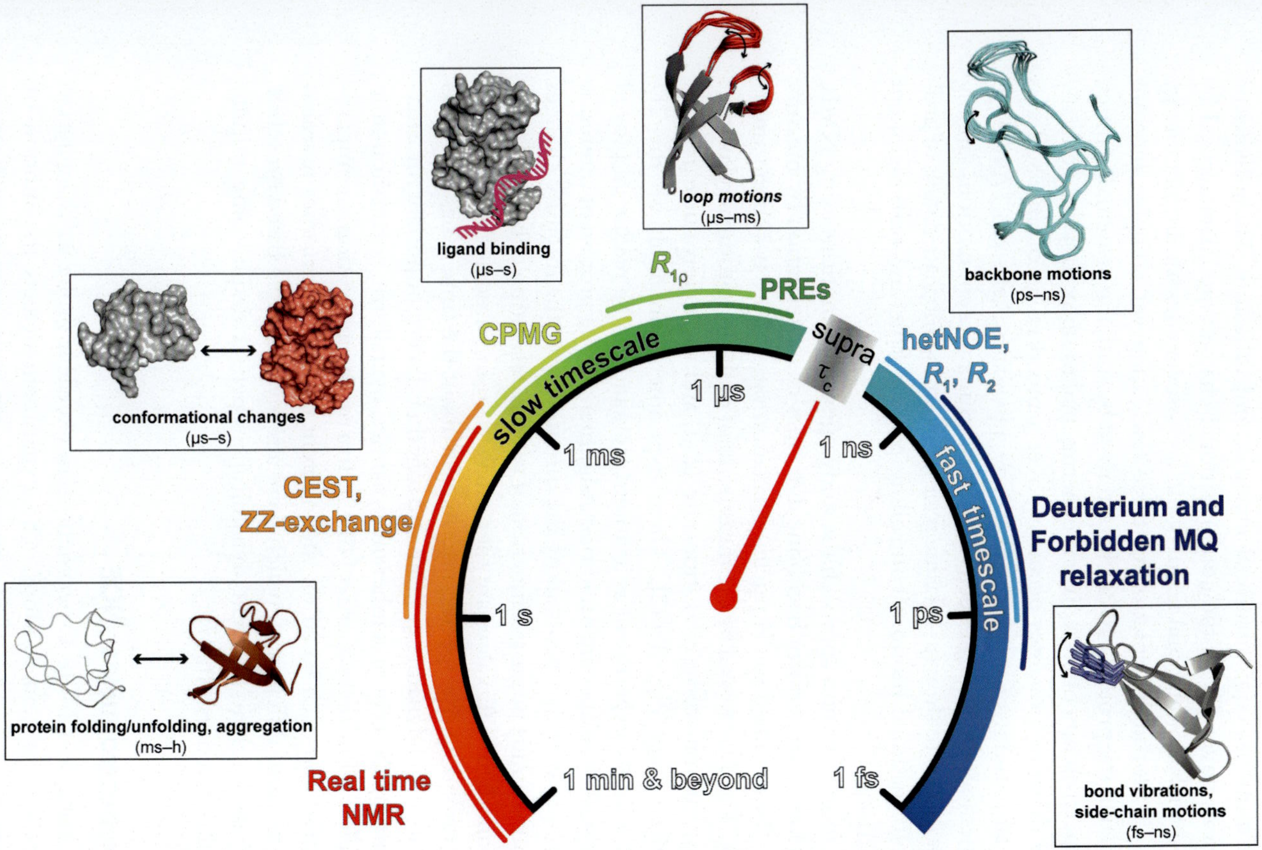


**Fig. 2** Representation of NMR methods for the characterization of protein dynamics. Dynamic fluctuations occurring over a broad range of time scales and amplitudes are coupled to the protein functions. Solution NMR spectroscopy offers a range of methods to probe these motions in detail. Two distinct regimes can be distinguished for the dynamics accessible, the fast and slow time scale motions [40]. These two regimes are separated by the so-called "supra-$\tau_c$" gap that is technically not (yet) accessible for NMR relaxation experiments. This gap spans the range between the rotational correlation time ($\sim$ 100 ns) and about the length of the shortest spin-refocusing pulses that are technically possible (a few μs). The time scale for the motions is represented by the color gradient. Typical NMR experiments for the characterization of motions falling into the respective time scale regimes are indicated above the time range; these are operational.

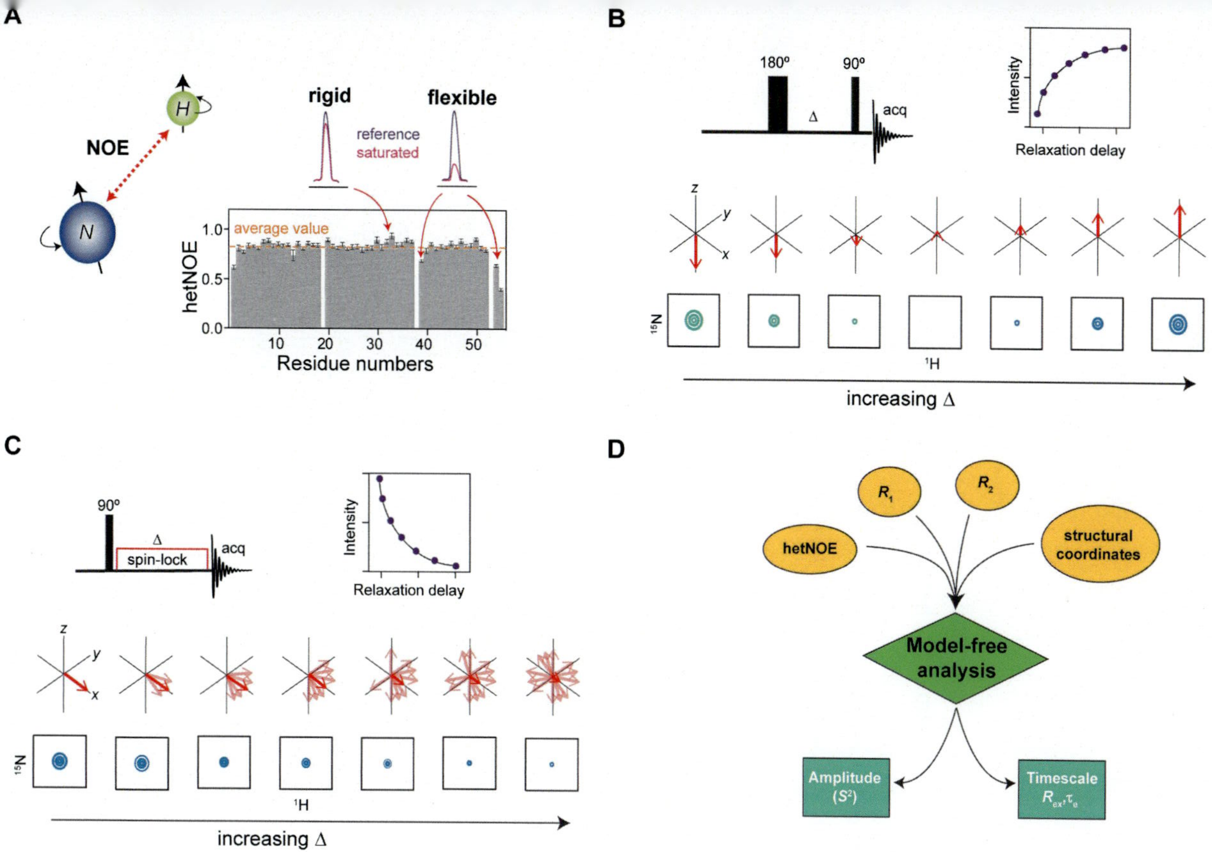

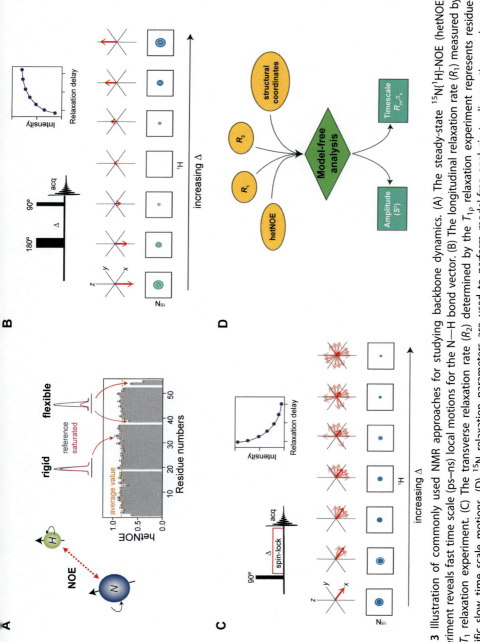

**Fig. 3** Illustration of commonly used NMR approaches for studying backbone dynamics. (A) The steady-state $^{15}N\{^{1}H\}$-NOE (hetNOE) experiment reveals fast time scale (ps–ns) local motions for the N—H bond vector. (B) The longitudinal relaxation rate ($R_1$) measured by the $T_1$ relaxation experiment. (C) The transverse relaxation rate ($R_2$) determined by the $T_{1\rho}$ relaxation experiment represents residue-specific slow time scale motions. (D) $^{15}N$ relaxation parameters are used to perform model-free analysis to discern time scales and amplitudes of motions experienced by each N—H bond vector. Higher-order parameters ($S^2 > 0.8$) indicate that the residues are structurally rigid. Residues undergoing fast time scale as well as slow time scale motions are represented by the observable $\tau_e$ and $R_{ex}$ terms, respectively.

(Fig. 3A). The steady-state $^{15}N\{^{1}H\}$-NOE experiment [48,49] is recorded to determine the hetNOE values by measuring two almost identical experiments in an interleaved fashion with (saturated) and without (reference) saturation of the coupled $^{15}N$ nucleus. The ratio between the intensities of the resulting peaks in these two spectra corresponds to the hetNOE value for each residue.

$$\text{hetNOE} = \frac{I_{\text{saturated}}}{I_{\text{reference}}}$$

The residues undergoing motions on the pico-nanosecond time frame display a considerable drop in the hetNOE values with respect to the average hetNOE values observed for the protein molecule. Commonly, a more dramatic effect is observed for the flexible amino- and carboxy-terminal residues, which undergo fast time scale motions and sometimes even display negative hetNOE values.

## 3.2 $T_1$ relaxation

The $T_1$ relaxation experiment is performed to quantitate the longitudinal relaxation (also called spin-lattice relaxation or $T_1$), which is the relaxation process by which the excited NMR spins relax back to the thermal equilibrium state dictated by the Boltzmann distribution (Fig. 3B). The time constant describing the recovery of the net magnetization to the equilibrium is termed $T_1$ and is the simple inverse of the longitudinal relaxation rate ($R_1$).

The $T_1$ relaxation experiment is usually performed by employing the inversion recovery scheme, where a first 180° pulse is applied to transfer the magnetization from $z$ to $z$-axis, a variable delay ($\Delta$) is applied to allow the recovery of magnetization to the $z$-axis followed by the application of a 90° pulse for final detection of magnetization at the $xy$ plane. This experiment is serially repeated with different variable delays, and subsequently, the quantification of time-dependent recovery of the intensity is performed for each backbone amide peak to calculate the longitudinal relaxation rate ($R_1$), which is the simple inverse of $T_1$ by using the following equation.

$$I_{(t)} = I_{\infty} - I_0 * e^{(-R_1 * t)}$$

where $-I_0$ is the intensity of the negative net magnetization at the start, $I_{\infty}$ is the final magnetization with positive intensity, $t$ is the delay time ($\Delta$), and $R_1$ is the longitudinal relaxation rate.

## 3.3 $T_2$ relaxation

The $T_2$ relaxation experiment provides information about the transverse relaxation time constant (also called spin-spin relaxation or $T_2$), reporting protein motions occurring on the μs-ms time scale (Fig. 3C). $T_2$ relaxation is associated with the loss of coherence of excited net magnetization due to the dephasing of magnetization in the transverse (*xy*) plane, perpendicular to the applied magnetic field ($B_0$). Experiments for directly measuring the transverse relaxation rate ($R_2$, the simple inverse of $T_2$) are often unreliable due to the phase inhomogeneity. The $R_2$ relaxation rate is therefore often determined by an indirect approach by quantifying the transverse relaxation rate in the rotating frame ($R_{1\rho}$) using the $T_{1\rho}$ experiment [50], where transverse magnetization is locked in the rotating frame by introducing spin-lock conditions.

In this experiment, first, a 90° pulse is applied to excite the spins. A variable spin-lock period (Δ) is applied under the spin-lock conditions to control the dephasing of the spins by generating a rotating magnetic field in the transverse (*xy*) plane. This experiment is iterated with varying spin-lock periods to quantify the $R_{1\rho}$ rate using the following equation.

$$I_{(t)} = I_0 * e^{\left(-R_{1\rho} * t\right)}$$

where $I_0$ is the net magnetization at the start, *t* is the delay time, and $R_{1\rho}$ is the transverse relaxation rate in the rotating frame.

$R_2$ relaxation rate is subsequently derived from the $R_{1\rho}$ and $R_1$ rates using the following relationship:

$$R_{1\rho} = R_1 \cos^2\theta + R_2 \sin^2\theta$$

where $\theta = \tan^{-1}(\nu_1/\Delta\nu)$ and $\Delta\nu$ is the offset of the *RF* field to the resonance [50].

## 3.4 Model-free analysis

The experimentally determined steady-state NOE, $R_1$, and $R_2$ relaxation data can subsequently be interpreted in terms of physically meaningful quantitative motional parameters, which provide information about the time scale and amplitudes of dynamic motions by using the so-called model-free formalism (Fig. 3D; [43,44]). The output parameter, the *generalized order parameter* ($S^2$), and the product of $S_f^2$ and $S_s^2$ (the squares of the order parameters for fast and slow time scale internal motions) reflect the rigidity of each N—H bond vector. $S^2$ values range from 0 (full disorder) to 1 (total rigidity), reporting on the amplitude of motions experienced by each backbone amide group. Time scales of the motions are reported by the model-free output parameter, the effective correlation time for internal motions ($\tau_e$) attributing to the fast time scale motions from the individual N—H vector faster than the overall tumbling time ($\tau_c$) and the conformational exchange term ($R_{ex}$) reflecting the slow (μs-ms) time scale motions via chemical exchange.

## 4. NMR spectroscopy for large proteins

NMR spectroscopy has been tremendously successful in deciphering the structures as well as dynamical information of small up to medium-sized proteins (~2025 kDa). The proteins larger than this size pose significant challenges to the traditional NMR spectroscopy approach, such as extensive signal overlap and signal broadening caused by the longer rotational correlation times and faster transverse relaxation ($T_2$), resulting in rapid decay of the detectable signal intensity. Incidentally, the majority of protein molecules are rather larger than the range that traditional NMR spectroscopy can access, posing a significant hurdle or size limitation for the application of NMR spectroscopy for the characterization of large macromolecules or complexes. With the advent of advanced isotope labeling techniques such as deuteration and selective methyl labeling, transverse relaxation optimized spectroscopy (TROSY) schemes, the evolution of high-field magnets and cryogenic probes, as well as novel coherence transfer strategies, have contributed significantly to pushing the size limit for the larger proteins accessible to the NMR spectroscopy as high as 1 MDa [17,41,51].

### 4.1 Deuteration

Protein deuteration approaches such as perdeuteration (referring to the complete deuteration of the amino acid side chains) and selective deuteration result in a dramatic simplification of the NMR spectra of large protein molecules. Deuteration reduces the proton density within the protein allowing the suppression of the undesired spin diffusion, resulting in the subsequent reduction of relaxation rates of $^{15}N$ and $^{13}C$ nuclear spins and, thus, leads to a significant increase in the signal-to-noise ratio for large proteins [52]. Deuteration is achieved by recombinantly producing proteins in minimal media, where $^{2}H_2O$ ($D_2O$) is used as the solvent instead of $H_2O$ and deuterated $^{13}C$ glucose if ~99% deuteration is desired. Backbone amide protons are then subsequently reintroduced using an $H_2O$-based buffer, allowing back exchange of labile protons, such as the amide protons of the protein backbone.

## 4.2 Selective methyl labeling

Methyl groups stand out as an excellent probe for studying structural and dynamic processes for the large proteins because of their higher sensitivity (as three methyl protons contribute to the same signal), strong occurrence in the protein core, and interfaces in macromolecular protein-protein, protein-ligand complexes as well as catalytic active sites [41,51]. Methyl groups are also located toward the end of the amino acid side chain, imparting faster tumbling than the rest of the protein molecule, resulting in shaper resonance peaks. In addition, methyl groups are susceptible to the TROSY approach, which further enhances the spectral sensitivity. As a result of this improved sensitivity, even with the minute changes in the protein environment, the structural and dynamical changes occurring to the large protein molecules can be successfully extracted. Selective methyl labeling refers to the recombinantly producing protein with methyl groups labeled as $^{13}CH_3$ in an otherwise highly deuterated environment, commonly exploiting the biochemical amino acid pathways of *E. coli* [53]. This is achieved by using specific amino acid precursors, which allows the selective enrichment of $^{1}H$ and $^{13}C$ of methyl groups improving the resolution and sensitivity of NMR spectra [54]. Various strategies used for preparing samples labeled selectively with different methyl groups are reviewed by Schütz and Sprangers [51].

## 4.3 TROSY

The application of TROSY (transverse relaxation optimized spectroscopy) has tremendously increased the sensitivity of NMR spectroscopy for larger protein molecules, pushing the size range for NMR spectroscopy for macromolecules further. The TROSY effect is observed at higher magnetic fields, where the $T_2$ relaxation mechanism is contributed by the dipole-dipole mechanism as well as the chemical shift anisotropy. The TROSY principle utilizes the relaxation interferences to achieve the suppression of the $T_2$ relaxation mechanism by selecting the component where these relaxation mechanisms almost cancel each other out, resulting in a sharp single peak leading to the increased sensitivity of the signals for large proteins [55]. Application of TROSY-based assignment experiments [56] and relaxation experiments [57,58] have significantly contributed to the successful application of NMR spectroscopy on larger proteins.

## 5. Methods for probing protein dynamics of large proteins

Traditional amide-based ($^{15}N$-$^{1}H$) relaxation methods have been pivotal in the dynamical characterization of proteins but are limited by the protein size (~20–25 kDa), as described in Section 3. In the past few years, side-chain methyl groups have emerged as sensitive probes for extracting meaningful information about the protein dynamics of large proteins and their complexes (reviewed in Refs. [36,41,51,59]). By using methyl groups as a probe, it is now possible to quantify protein dynamics for such large proteins that were not discernable by employing traditional backbone dynamics approaches. NMR methods such as relaxation dispersion, multiple quantum (MQ) relaxation, chemical-exchange saturation transfer (CEST)/dark state excitation saturation transfer (DEST), and ZZ-exchange spectroscopy were initially developed for assessing $^{15}N$-based dynamics. These methods were adapted over the course of time to exploit the methyl groups transforming them into a powerful probe to study dynamics at a large diversity of biologically relevant time scales [41]. Once the selectively methyl labeled sample and resonance assignment data for the protein of interest are available, these methods can be readily applied to probe the dynamics of larger proteins by exploiting favorable spectral properties of methyl groups providing insights into the biologically relevant dynamic events.

### 5.1 Fast (ps-ns) time scale motions

Protein molecules experience a multitude of motions on the fast (ps-ns) time scales, such as bond vibrations, side-chain rotations, and loop motions. The most commonly used approach to describe the fast time scale motions experienced using the methyl group as a probe is to determine the order parameter ($S^2_{axis}$) [41]. The $S^2_{axis}$ value provides the measure of the amplitude of motion experienced by the methyl group about its threefold symmetry axis. Similar to the backbone-based *generalized order parameter* ($S^2$), the $S^2_{axis}$ value ranges from 0 (totally flexible) to 1 (totally rigid) methyl group bond vector. Another approach to assess fast time scale motions exhibited by the methyl group is to determine the picosecond time scale lifetime ($\tau_e$) for the local reorientation of the side-chain methyl group, though this approach is not widely used [41].

### 5.1.1 Deuterium relaxation

Deuterium relaxation was first implemented by the Kay group in their pioneering study to quantify fast time scale (ps-ns) dynamics of the methyl groups on uniformly $^{13}C$ and fractionally $^{2}H$-labeled methyl groups, i.e., $^{13}CH_2D$ and $^{13}CHD_2$ (Fig. 4A; [60]). The deuterium relaxation process is driven by the quadrupolar interaction, which is insensitive to the exchange contributions as opposed to the dipole-dipole interaction and chemical shift anisotropy driven $^{1}H$ and $^{13}C$ relaxation processes [60]. The much higher efficiency of quadrupolar interaction results in a robust measurement and the interpretation of fast time scale motions via deuterium relaxation. Deuterium relaxation allows the possibility to measure five unique relaxation rates per deuteron and nine relaxation rates per methyl group [61,62]. Moreover, with the help of a prior estimated rotation correlation time, $\tau_c$, deuterium relaxation facilitates the determination of the order parameter of the methyl group ($S^2_{axis}$) as well as correlation times of the methyl groups, making it an excellent probe to study side-chain dynamics at a fast time scale [61].

### 5.1.2 "Forbidden" multiple quantum transitions

*Multiple quantum* relaxation violated coherence transfer NMR spectroscopy offers another approach to discern the fast time scale motions of the methyl groups, which can directly be applied to the $^{13}CH_3$ groups without the need for incorporation of deuterons into methyl groups, unlike deuterium relaxation [63,64]. In this approach, methyl proton MQ transitions, also called "forbidden" transitions, are used to provide quantitative estimation of order parameters for the most routinely prepared protein samples containing uniformly $^{13}C$ labeled protonated methyl groups ($^{13}CH_3$) in the otherwise highly deuterated background (Fig. 4B; [63–66]).

Within the methyl group, $^{1}H$-$^{1}H$ dipolar cross-relaxation rates relax with substantially different rates and provide information about side-chain motions. Selection of "forbidden" $^{1}H$ *Double Quantum* (DQ) or *Triple Quantum* (TQ) proton transitions via the relaxation violated coherence transfer scheme paves the way for the measurements of $^{1}H$-$^{1}H$ dipolar cross-correlation rates $\eta$. The time dependence of the build-up of the $^{1}H$ magnetization used to calculate $\eta$ is subsequently used to extract the methyl order parameters $S^2_{axis}$, if the protein's overall rotational correlation time ($\tau_c$) is known. It should be noted that the TQ version is ~50% more sensitive than the DQ version but is only applicable for the larger protein molecules as it requires longer macromolecular tumbling times, e.g., larger $t_c$ [65].

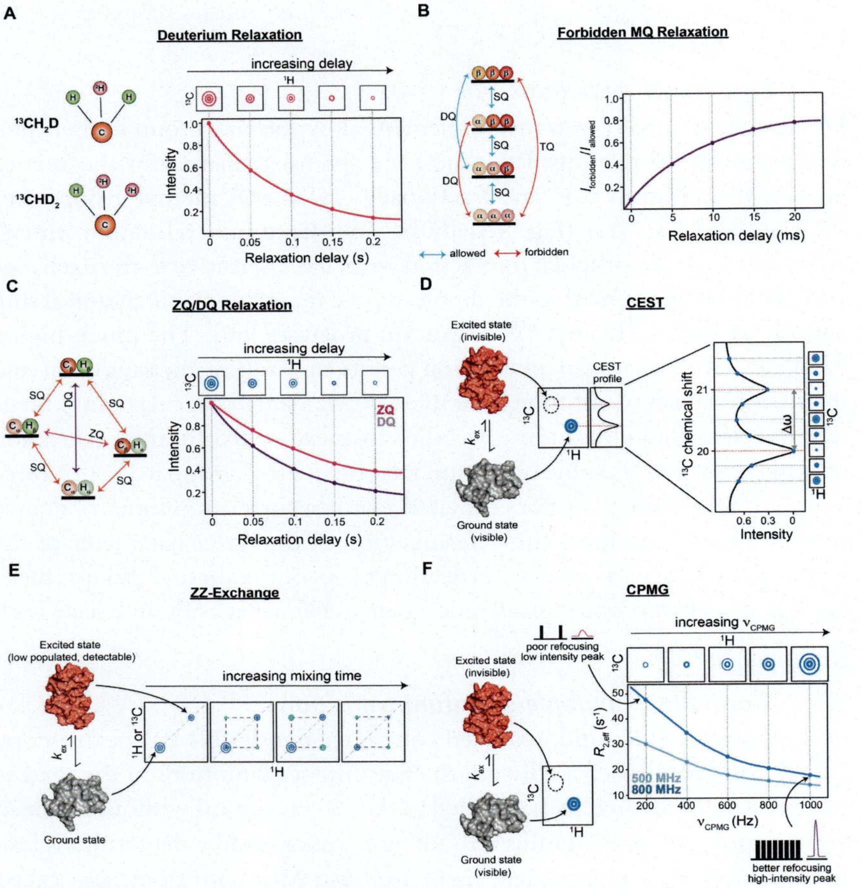


**Fig. 4** NMR approaches for studying protein dynamics using methyl groups as probes. (A) Deuterium relaxation (applicable for fractionally $^{2}H$ labeled methyl groups, i.e., $^{13}CH_2D$, $^{13}CHD_2$, left panel) and (B) the relaxation-violated or "forbidden" MQ relaxation (quantifies the allowed (single-quantum (SQ) coherences) and the traditionally forbidden (either double quantum (DQ) or TQ coherences of intramethyl proton transitions illustrated in the left panel) methods can be used to determine the methyl group order parameters describing the fast (ps-ns) time scale motions of the side chains. (C) The ZQ and DQ coherences for the $^{1}H$-$^{13}C$ spin system (left panel) are quantified by the ZQ and DQ relaxation experiments (right panel) to study chemical exchange occurring over a slow (μs–ms) time scale. (D) The CEST (chemical exchange saturation transfer) experiment can be used to quantify chemical exchange by applying a weak *RF* field at the resonance frequency of excited invisible conformation, which results in loss of the intensity of the ground-state NMR peak if two conformations are in an exchange over a slow time scale. (E) ZZ-exchange is applicable to study the slowly exchanging conformations, where the signals of the lowly populated excited conformation are also detected in the NMR spectrum along with the visible conformation. This experiment employs the variable mixing time to determine exchange rates and populations of protein conformations. (F) CPMG experiment employs suppression of the exchange broadening of the visible ground-state population via application of CPMG pulse trains, allowing the quantification of chemical exchange between the individual conformations.

## 5.2 Slow (μs-ms) time scale motions

Important biological processes such as protein folding-unfolding, conformational changes, allostery, and protein-ligand binding, as well as enzyme catalysis, are characterized by the dynamic motions takings place at the slow time scales such as μs, ms, or s and are associated with many cases involving intermediate protein conformations [13]. These sparsely populated, transient, and high-energy minor conformations of the protein often result in weak undetectable NMR signals, which are in exchange with the low-energy, long-lived, and highly populated ground state of the protein [67,68]. The slow time scale motions associated with these exchanges are characterized by the line broadening of NMR resonances via a chemical exchange process, where the nucleus experiences the transition from one state to another ascribed to distinct chemical environments and chemical shifts. NMR spectroscopy offers numerous methods such as CEST), Carr-Purcell-Meiboom-Gill (CPMG) relaxation dispersion, $R_{1\rho}$, ZZ-exchange spectroscopy, or paramagnetic relaxation enhancement (PRE). These different methods can be applied to extract slow time scale motions at a diverse range of time scales provided that motions fall under a slow or intermediate exchange regime on the NMR time scale, i.e., the exchange rates of these motions are slower in comparison to the chemical shift difference between two states [69].

### *5.2.1 Zero quantum (ZQ) and double quantum (DQ) relaxation*

ZQ and DQ experiments are used to study the slow time scale dynamics and are considered as an alternative for the $R_{1\rho}$ experiments (see Section 3.3), which are deemed to be highly challenging because of dipolar cross-correlation within uniformly labeled methyl groups ($^{13}CH_3$) [70]. ZQ and DQ experiments rely on the distinct relaxation rates of the ZQ and DQ terms and employ the difference in the relaxation rates for DQ and ZQ coherences, i.e., $\Delta R_{MQ} = R_{DQ} – R_{ZQ}$ and the relaxation rate for MQ coherence ($R_{MQ}$), to quantitate the underlying chemical exchange (Fig. 4C; [71–74]).

### *5.2.2 Chemical exchange saturation transfer (CEST) and dark state excitation saturation transfer (DEST)*

CEST is applied to derive chemical exchange occurring on a slow time scale and allows the detection of minor states which are populated $\geq 0.5\%$ [68]. CEST reports on the lowly populated invisible minor or excited conformations undergoing chemical exchange with visible (major) states at slow time scales ranging from ~3 to 100 ms provided that both states have distinct chemical shifts [67,75–77]. CEST employs the saturation transfer method to selectively saturate invisible protein conformation by implementing a saturation field with an offset from the visible state frequency to saturate invisible protein conformation while the visible protein conformation is kept unsaturated (Fig. 4D). Chemical exchange between two conformational states allows the transfer of saturation from the invisible (minor) to the visible (major) state and can be detected by the loss of the intensity of the visible state. Chemical exchange parameters, as well as the invisible and visible state populations, are derived from the relaxation profile of the visible state intensity with respect to the offset of the saturation field using Bloch-McConnell equations [67,78].

Dark state excitation saturation transfer (DEST), pioneered by the Clore group, employs a similar principle as CEST and is suitable for detecting slow time scale motions similar to CEST. The major difference between CEST and DEST is that DEST exploits distinct transverse relaxation rates ($R_2$) between the visible and invisible states as opposed to CEST, which requires significant chemical shift differences between the individual states [79,80]. Thus DEST is applicable to the sparsely populated invisible state characterized by the large molecular size (>1 MDa) in a chemical exchange with the visible state in the time scale in the range of 0.5–1000 ms [13,79–81]. The differences in line-width of these two states are exploited to saturate the invisible state by applying a saturation field, which is then transferred to the visible state during a chemical exchange and is characterized by a decrease in the intensity similar to the CEST.

### 5.2.3 ZZ-exchange spectroscopy

ZZ-exchange spectroscopy, also known as Exchange Spectroscopy (EXSY), is a suitable method to quantify the chemical exchange occurring in the 10–5000 ms time window, when the minor state has a significant detectable population and is in slow chemical exchange with the major state on the NMR time scale, e.g., ms to s, resulting in two distinct resonances for the same nuclei (Fig. 4E; [69,82–84]). The ZZ-Exchange experiment is usually recorded as a 2D NMR experiment, wherein a $^{1}H$-$^{13}C$ HMQC version upon indirect dimension $^{13}C$ frequency-labeling during $t_1$, magnetization is restricted to be in phase $^{13}C$ Z magnetization and is permitted to exchange (mixing time). Subsequently, the magnetization is transferred back to the proton and detected. A shorter mixing time results in an intense on-diagonal signal for both major (ground) and minor (excited) states with weak or absent off-diagonal exchange cross-peaks. With an increase in the mixing times, the system is allowed to exchange longer, resulting in an increase in the intensities of the cross-peaks. After surpassing a certain mixing time, both on-diagonal and off-diagonal peaks experience the loss of intensities due to longitudinal relaxation. Measurement of ZZ-exchange spectra with varying mixing time yields the dependences of peak intensities as a function of mixing time, from which the population of the minor state as well as chemical exchange can be quantified [82].

The simplest ZZ-exchange experiment, which uses homonuclear NOESY as a template, results in spectral crowding because of overlapping extensive overlap along the diagonal of the spectrum. The use of heteronuclear correlation experiments such as $^{1}H$–$^{13}C$ HMQC alleviates this problem. Similarly, with an increase in the protein size, crowded regions such as the region containing the signals for isoleucine, leucine, and valines result in extensive overlap for the major and minor populations, complicating the analysis [41]. In such cases, methionine resonances that have strikingly distinct chemical shifts and thus have greater resolution can be analyzed to successfully apply ZZ-exchange experiments on larger protein molecules.

### 5.2.4 CPMG relaxation dispersion experiments

CPMG relaxation dispersion experiments monitor chemical exchange processes falling in the slow time scale range (~500 μs–10 ms) and provide the kinetics (rate constants), thermodynamic (populations of both states), and structural (magnitudes of the chemical shifts difference) information for the protein dynamic motions [17,85]. CPMG experiments are similar to the $T_{1\rho}$ experiments (see Section 3.3) but monitor the dynamic processes occurring at a slower time scale window providing a valuable tool to investigate sparsely populated "excited" states which are not directly observable in the NMR spectrum [86–88].

When conformational changes occur, the chemical exchange influences the transverse relaxation rate ($R_2$), which results in the additional loss of the signal when the magnetization is in the $xy$ plane. As a consequence, the effective transverse relaxation rate experienced by the magnetization at the $xy$ plane can be described as $R_{2,\text{eff}} = R_2 + R_{ex}$. CPMG experiments involve the determination of the effective transverse relaxation rates ($R_{2,\text{eff}}$), which are proportional to the line width of the NMR signal (Fig. 4F). CPMG relaxation dispersion experiments employ the suppression of the dephasing magnetization associated with the chemical exchange process by applying spin-echo sequences with varying the so-called CPMG frequency ($\nu_{CPMG}$), which results in the chemical shift refocusing of the 180° pulses during a fixed relaxation delay. The increase in $\nu_{CPMG}$ results in the gain in the detectable magnetization as the dephasing of the resonance signals due to chemical exchange is reduced, which is then used to determine the effective transverse relaxation rates ($R_{2,\mathrm{eff}}$) with respect to the $\nu_{CPMG}$. For the reliable application of $\nu_{CPMG}$, it is crucial to repeat CPMG experiments at two different static magnetic field strengths, which results in higher accuracy of the extracted parameters.

## 6. Methods for simultaneous study of the structure and dynamics of proteins

### 6.1 Paramagnetic relaxation enhancement

The PRE effect is caused by the enhancement of the relaxation rates of the nucleus upon interaction with a spin label carrying an unpaired electron due to a through-space magnetic dipolar interaction [89]. Thus PREs result in the enhancement of site-specific nuclear relaxation rates $R_1$ and $R_2$ in a distance-dependent manner, which is inversely proportional to the sixth power of the distance of the paramagnetic center of the spin-label and nucleus under observation, providing long-range distance restraints (up to 35 Å). PREs are also used to probe the dynamic motions occurring at the μs time scales and are extremely sensitive to rapidly exchanging conformational states populated as low as 1% within the PRE observable distances [13]. This allows the revelation of dynamic transient states that undergo a fast time scale interconversion with the ground-state structure, hence used as a probe for the protein dynamics [89]. To observe the PRE effect, proteins are either introduced with the nitroxide label or with paramagnetic metal ions in either hydrated in the solvent or attached to the protein. Relaxation rates of the paramagnetic samples are first measured, followed by the measurements of diamagnetic protein samples, which can be achieved by either reducing the nitroxide tag in the case of nitroxide labels or exchanging the bound paramagnetic metal ion with a diamagnetic ion. The difference in relaxation rates is used to derive the PRE rates. Lately, PRE-based approaches have been used in a variety of different implementations, such as characterizing protein domains previously inaccessible due to extensive line broadening as for the bacterial trigger factor chaperone, the movements of transcription complexes along with DNA, as well as the spatial orientation and stability of chaperone:client complexes [90–93].

## 6.2 Residual dipolar couplings

Under the influence of a strong magnetic field, the interaction of two magnetically active nuclei results in the residual dipolar couplings (RDCs) [94]. Generally, RDCs are averaged to zero due to the isotropic tumbling of protein molecules in the solution. By introducing an anisotropic media, it is possible to restrict the movements of the protein molecules. This results in a partial alignment, leading to the anisotropic tumbling of proteins in the solution resulting in an incomplete averaging of anisotropic dipolar couplings, which can be measured in the form of RDCs [95,96]. One of the most important applications of RDCs is the comparison of protein X-ray crystallographic structures with solution conformations. Common alignment media for RDCs include stretched polyacrylamide gels, phages, liquid crystalline media, etc. [16,94,96]. RDCs improve the precision of solution NMR structures significantly. RDCs have been used to discern protein dynamic motions occurring on the slow (μs) time scale [97]. RDCs provide long-range orientational restraints, which are extremely useful for improving the accuracy of the NMR structure calculation [96,98].

# 7. Conclusions

The biological functions of proteins are controlled largely by their structural and dynamical variations[99,100]. In this chapter, we have outlined the application of a large diversity of NMR spectroscopic methods to characterize the protein structure and broad range of dynamic motions, which remain poorly understood in most cases. The recent advancements in methyl-based NMR strategies have been instrumental in pushing the size barrier traditional NMR spectroscopy faced. These advancements present an exciting trajectory for the application of NMR to large proteins and their dynamic motions over a wide range of time scales, as well as reveal boundless opportunities to observe proteins in either near physiological or native conditions, which will significantly contribute to increasing our understanding of biomolecular pathways and their associations in diseases at unprecedented details. Employing such sophisticated NMR approaches as outlined previously already, many important studies in recent years increased tremendously the understanding of the functionality of protein kinases [101–104], the evolutionary origin of their functional features [105–107], the basis of how molecular chaperones interact with their client proteins [92,108–118], and finally have culminated to rich novel insights into how multimeric proteases are regulated and employ their proteolytic function [118–125].

## Acknowledgments

B.M.B. gratefully acknowledges funding from the Swedish Research Council (Starting Grant 2016-04721; Consolidator Grant 2020-00466), the Swedish Cancer Foundation (2019-0415), and the Knut och Alice Wallenberg Foundation through a Wallenberg Academy Fellowship (2016.0163) as well as through the Wallenberg Centre for Molecular and Translational Medicine, Göteborgs Universitet, Sweden. The Swedish NMR Centre of the Göteborgs Universitet is acknowledged for spectrometer time.

## References

[1] Kaushik S, Cuervo AM. Proteostasis and aging. Nat Med 2015;21(12):1406–15.
[2] Lee C, Yu MH. Protein folding and diseases. J Biochem Mol Biol 2005;38(3):275–80.
[3] Hartl FU, Bracher A, Hayer-Hartl M. Molecular chaperones in protein folding and proteostasis. Nature 2011;475(7356):324–32.
[4] Berman HM, Lawson CL, Vallat B, Gabanyi MJ. Anticipating innovations in structural biology. Q Rev Biophys 2018;51, e8.
[5] Burley SK. Integrative/hybrid methods structural biology: role of macromolecular crystallography. Adv Exp Med Biol 2018;1105:11–8.
[6] Berman HM, Vallat B, Lawson CL. The data universe of structural biology. IUCrJ 2020;7(Pt 4):630–8.
[7] Danev R, Yanagisawa H, Kikkawa M. Cryo-electron microscopy methodology: current aspects and future directions. Trends Biochem Sci 2019;44(10):837–48.
[8] Cheng Y. Single-particle cryo-EM-how did it get here and where will it go. Science 2018;361(6405):876–80.
[9] Levy Y. Protein assembly and building blocks: beyond the limits of the LEGO brick metaphor. Biochemistry 2017;56(38):5040–8.
[10] Marsh JA, Teichmann SA. Structure, dynamics, assembly, and evolution of protein complexes. Annu Rev Biochem 2015;84:551–75.
[11] Skjaerven L, Reuter N, Martinez A. Dynamics, flexibility and ligand-induced conformational changes in biological macromolecules: a computational approach. Future Med Chem 2011;3(16):2079–100.
[12] Grutsch S, Brüschweiler S, Tollinger M. NMR methods to study dynamic allostery. PLoS Comput Biol 2016;12(3), e1004620.
[13] Anthis NJ, Clore GM. Visualizing transient dark states by NMR spectroscopy. Q Rev Biophys 2015;48(1):35–116.
[14] Schiro A, Carlon A, Parigi G, Murshudov G, Calderone V, Ravera E, Luchinat C. On the complementarity of X-ray and NMR data. J Struct Biol X 2020;4, 100019.
[15] Bax A, Clore GM. Protein NMR: boundless opportunities. J Magn Reson 2019;306:187–91.
[16] Markwick PR, Malliavin T, Nilges M. Structural biology by NMR: structure, dynamics, and interactions. PLoS Comput Biol 2008;4(9), e1000168.
[17] Kovermann M, Rogne P, Wolf-Watz M. Protein dynamics and function from solution state NMR spectroscopy. Q Rev Biophys 2016;49, e6.
[18] Sekhar A, Kay LE. An NMR view of protein dynamics in health and disease. Annu Rev Biophys 2019;48:297–319.

[19] Kang C. Applications of in-cell NMR in structural biology and drug discovery. Int J Mol Sci 2019;20(1):139.
[20] Sciolino N, Burz DS, Shekhtman A. In-cell NMR spectroscopy of intrinsically disordered proteins. Proteomics 2019;19(6), e1800055.
[21] Grudziaz K, Zawadzka-Kazimierczuk A, Kozminski W. High-dimensional NMR methods for intrinsically disordered proteins studies. Methods 2018;148:81–7.
[22] Gerez JA, Prymaczok NC, Riek R. In-cell NMR of intrinsically disordered proteins in mammalian cells. Methods Mol Biol 2020;2141:873–93.
[23] Matečko-Burmann I, Burmann BM. Recording in-cell NMR-spectra in living mammalian cells. Methods Mol Biol 2020;2141:857–71.
[24] Theillet FX, Binolfi A, Bekei B, Martorana A, Rose HM, Stuiver M, Verzini S, Lorenz D, van Rossum M, Goldfarb D, Selenko P. Structural disorder of monomeric α-synuclein persists in mammalian cells. Nature 2016;530(7588):45–50.
[25] Burmann BM, Gerez JA, Matečko-Burmann I, Campioni S, Kumari P, Ghosh D, Mazur A, Aspholm EE, Sulskis D, Wawrzyniuk M, Bock T, Schmidt A, Rüdiger SGD, Riek R, Hiller S. Regulation of α-synuclein by chaperones in mammalian cells. Nature 2020;577(7788):127–32.
[26] Cavanagh J, Fairbrother WJ, Palmer 3rd AG, Rance M, Skelton NJ. Protein NMR spectroscopy. 2nd ed. Burlington: Academic Press; 2007.
[27] Kleckner IR, Foster MP. An introduction to NMR-based approaches for measuring protein dynamics. Biochim Biophys Acta Proteins Proteomics 2011;1814(8): 942–68.
[28] Frueh DP. Practical aspects of NMR signal assignment in larger and challenging proteins. Prog Nucl Magn Reson Spectrosc 2014;78:47–75.
[29] Howard MJ. Protein NMR spectroscopy. Curr Biol 1998;8(10):R331–3.
[30] Sambrook J, Russell DW. Molecular cloning: a laboratory manual. 3rd ed. New York: Cold Spring Harbor Laboratory Press; 2001.
[31] Bodenhausen G, Ruben DJ. Natural abundance nitrogen-15 NMR by enhanced heteronuclear spectroscopy. Chem Phys Lett 1980;69(1):185–9.
[32] Sattler M, Schleucher J, Griesinger C. Heteronuclear multidimensional NMR experiments for the structure determination of proteins in solution employing pulsed field gradients. Prog Nucl Magn Reson Spectrosc 1999;34(2):93–158.
[33] Clore GM, Gronenborn AM. New methods of structure refinement for macromolecular structure determination by NMR. Proc Natl Acad Sci U S A 1998;95(11):5891–8.
[34] Takeuchi K, Wagner G. NMR studies of protein interactions. Curr Opin Struct Biol 2006;16(1):109–17.
[35] Bonvin AM, Boelens R, Kaptein R. NMR analysis of protein interactions. Curr Opin Chem Biol 2005;9(5):501–8.
[36] Wiesner S, Sprangers R. Methyl groups as NMR probes for biomolecular interactions. Curr Opin Struct Biol 2015;35:60–7.
[37] Williamson MP. Using chemical shift perturbation to characterise ligand binding. Prog Nucl Magn Reson Spectrosc 2013;73:1–16.
[38] Furukawa A, Konuma T, Yanaka S, Sugase K. Quantitative analysis of protein-ligand interactions by NMR. Prog Nucl Magn Reson Spectrosc 2016;96:47–57.
[39] Becker W, Bhattiprolu KC, Gubensak N, Zangger K. Investigating protein-ligand interactions by solution nuclear magnetic resonance spectroscopy. ChemPhysChem 2018;19(8):895–906.
[40] Palmer 3rd AG. NMR probes of molecular dynamics: overview and comparison with other techniques. Annu Rev Biophys Biomol Struct 2001;30:129–55.
[41] Boswell ZK, Latham MP. Methyl-based NMR spectroscopy methods for uncovering structural dynamics in large proteins and protein complexes. Biochemistry 2019;58 (3):144–55.

[42] Kay LE, Torchia DA, Bax A. Backbone dynamics of proteins as studied by $^{15}N$ inverse detected heteronuclear NMR spectroscopy: application to staphylococcal nuclease. Biochemistry 1989;28(23):8972–9.
[43] Lipari G, Szabo A. Model-free approach to the interpretation of nuclear magnetic-resonance relaxation in macromolecules. 2. Analysis of experimental results. J Am Chem Soc 1982;104(17):4559–70.
[44] Lipari G, Szabo A. Model-free approach to the interpretation of nuclear magnetic-resonance relaxation in macromolecules. 1. Theory and range of validity. J Am Chem Soc 1982;104(17):4546–59.
[45] Xia J, Deng NJ, Levy RM. NMR relaxation in proteins with fast internal motions and slow conformational exchange: model-free framework and Markov state simulations. J Phys Chem B 2013;117(22):6625–34.
[46] Kawale AA, Burmann BM. UvrD helicase-RNA polymerase interactions are governed by UvrD's carboxy-terminal Tudor domain. Commun Biol 2020;3(1):607.
[47] Kawale AA, Burmann BM. Inherent backbone dynamics fine-tune the functional plasticity of Tudor domains. Structure 2021;29(11):1253–1265.e4.
[48] Renner C, Schleicher M, Moroder L, Holak TA. Practical aspects of the 2D $^{15}N$-[1H]-NOE experiment. J Biomol NMR 2002;23(1):23–33.
[49] Kharchenko V, Nowakowski M, Jaremko M, Ejchart A, Jaremko L. Dynamic $^{15}N$ $\{^1H\}$ NOE measurements: a tool for studying protein dynamics. J Biomol NMR 2020;74(12):707–16.
[50] Massi F, Johnson E, Wang C, Rance M, Palmer 3rd AG. NMR $R_{1\rho}$ rotating-frame relaxation with weak radio frequency fields. J Am Chem Soc 2004;126(7):2247–56.
[51] Schütz S, Sprangers R. Methyl TROSY spectroscopy: a versatile NMR approach to study challenging biological systems. Prog Nucl Magn Reson Spectrosc 2020;116:56–84.
[52] Sattler M, Fesik SW. Use of deuterium labeling in NMR: overcoming a sizeable problem. Structure 1996;4(11):1245–9.
[53] Kerfah R, Plevin MJ, Sounier R, Gans P, Boisbouvier J. Methyl-specific isotopic labeling: a molecular tool box for solution NMR studies of large proteins. Curr Opin Struct Biol 2015;32:113–22.
[54] Kay LE. NMR studies of protein structure and dynamics. J Magn Reson 2005;173 (2):193–207.
[55] Pervushin K, Riek R, Wider G, Wüthrich K. Attenuated $T_2$ relaxation by mutual cancellation of dipole-dipole coupling and chemical shift anisotropy indicates an avenue to NMR structures of very large biological macromolecules in solution. Proc Natl Acad Sci U S A 1997;94(23):12366–71.
[56] Salzmann M, Pervushin K, Wider G, Senn H, Wüthrich K. TROSY in triple-resonance experiments: new perspectives for sequential NMR assignment of large proteins. Proc Natl Acad Sci U S A 1998;95(23):13585–90.
[57] Lakomek NA, Ying J, Bax A. Measurement of $^{15}N$ relaxation rates in perdeuterated proteins by TROSY-based methods. J Biomol NMR 2012;53(3):209–21.
[58] Zhu G, Xia Y, Nicholson LK, Sze KH. Protein dynamics measurements by TROSY-based NMR experiments. J Magn Reson 2000;143(2):423–6.
[59] Tugarinov V, Kay LE. Methyl groups as probes of structure and dynamics in NMR studies of high-molecular-weight proteins. Chembiochem 2005;6(9):1567–77.
[60] Muhandiram DR, Yamazaki T, Sykes BD, Kay LE. Measurement of $^2H$ $T_1$ and $T_{1\rho}$. Relaxation times in uniformly $^{13}C$-labeled and fractionally $^2H$-labeled proteins in solution. J Am Chem Soc 1995;117(46):11536–44.
[61] Millet O, Muhandiram DR, Skrynnikov NR, Kay LE. Deuterium spin probes of side-chain dynamics in proteins. 1. Measurement of five relaxation rates per deuteron in $^{13}C$-labeled and fractionally $^2H$-enriched proteins in solution. J Am Chem Soc 2002;124(22):6439–48.

[62] Liao X, Long D, Li DW, Brüschweiler R, Tugarinov V. Probing side-chain dynamics in proteins by the measurement of nine deuterium relaxation rates per methyl group. J Phys Chem B 2012;116(1):606–20.
[63] Tugarinov V, Sprangers R, Kay LE. Probing side-chain dynamics in the proteasome by relaxation violated coherence transfer NMR spectroscopy. J Am Chem Soc 2007;129(6):1743–50.
[64] Sun H, Godoy-Ruiz R, Tugarinov V. Estimating side-chain order in methyl-protonated, perdeuterated proteins via multiple-quantum relaxation violated coherence transfer NMR spectroscopy. J Biomol NMR 2012;52(3):233–43.
[65] Sun H, Kay LE, Tugarinov V. An optimized relaxation-based coherence transfer NMR experiment for the measurement of side-chain order in methyl-protonated, highly deuterated proteins. J Phys Chem B 2011;115(49):14878–84.
[66] Kay LE, Prestegard JH. Methyl-group dynamics from relaxation of double quantum filtered NMR signals—application to deoxycholate. J Am Chem Soc 1987;109 (13):3829–35.
[67] Vallurupalli P, Bouvignies G, Kay LE. Studying "invisible" excited protein states in slow exchange with a major state conformation. J Am Chem Soc 2012;134 (19):8148–61.
[68] Vallurupalli P, Sekhar A, Yuwen T, Kay LE. Probing conformational dynamics in biomolecules via chemical exchange saturation transfer: a primer. J Biomol NMR 2017;67(4):243–71.
[69] Palmer 3rd AG. Chemical exchange in biomacromolecules: past, present, and future. J Magn Reson 2014;241:3–17.
[70] Ishima R, Louis JM, Torchia DA. Transverse C-13 relaxation of $CHD_2$ methyl isotopmers to detect slow conformational changes of protein side chains. J Am Chem Soc 1999;121(49):11589–90.
[71] Gill ML, Palmer 3rd AG. Multiplet-filtered and gradient-selected zero-quantum TROSY experiments for $^{13}C_1H_3$ methyl groups in proteins. J Biomol NMR 2011;51(3):245–51.
[72] Tugarinov V, Kay LE. $^1H$,$^{13}C$-$^1H$,$^1H$ dipolar cross-correlated spin relaxation in methyl groups. J Biomol NMR 2004;29(3):369–76.
[73] Toyama Y, Osawa M, Yokogawa M, Shimada I. NMR method for characterizing microsecond-to-millisecond chemical exchanges utilizing differential multiple-quantum relaxation in high molecular weight proteins. J Am Chem Soc 2016;138 (7):2302–11.
[74] Toyama Y, Kano H, Mase Y, Yokogawa M, Osawa M, Shimada I. Dynamic regulation of GDP binding to G proteins revealed by magnetic field-dependent NMR relaxation analyses. Nat Commun 2017;8:14523.
[75] Yuwen T, Huang R, Kay LE. Probing slow timescale dynamics in proteins using methyl $^1H$ CEST. J Biomol NMR 2017;68(3):215–24.
[76] Rangadurai A, Shi H, Al-Hashimi HM. Extending the sensitivity of CEST NMR spectroscopy to micro-to-millisecond dynamics in nucleic acids using high-power radio-frequency fields. Angew Chem Int Ed Eng 2020;59(28):11262–6.
[77] Tiwari VP, Vallurupalli P. A CEST NMR experiment to obtain glycine $^1H\alpha$ chemical shifts in 'invisible' minor states of proteins. J Biomol NMR 2020;74(8–9):443–55.
[78] McConnell HM. Reaction rates by nuclear magnetic resonance. J Chem Phys 1958;28 (3):430–1.
[79] Fawzi NL, Ying J, Torchia DA, Clore GM. Probing exchange kinetics and atomic resolution dynamics in high-molecular-weight complexes using dark-state exchange saturation transfer NMR spectroscopy. Nat Protoc 2012;7(8):1523–33.
[80] Fawzi NL, Ying J, Ghirlando R, Torchia DA, Clore GM. Atomic-resolution dynamics on the surface of amyloid-beta protofibrils probed by solution NMR. Nature 2011;480(7376):268–72.

[81] Tugarinov V, Clore GM. Exchange saturation transfer and associated NMR techniques for studies of protein interactions involving high-molecular-weight systems. J Biomol NMR 2019;73(8–9):461–9.
[82] Li Y, Palmer 3rd AG. TROSY-selected ZZ-exchange experiment for characterizing slow chemical exchange in large proteins. J Biomol NMR 2009;45(4):357–60.
[83] Jeener J, Meier BH, Bachmann P, Ernst RR. Investigation of exchange processes by 2-dimensional NMR-spectroscopy. J Chem Phys 1979;71(11):4546–53.
[84] Palmer 3rd AG, Kroenke CD, Loria JP. Nuclear magnetic resonance methods for quantifying microsecond-to-millisecond motions in biological macromolecules. Methods Enzymol 2001;339:204–38.
[85] Koss H, Rance M, Palmer 3rd AG. General expressions for Carr-Purcell-Meiboom-Gill relaxation dispersion for N-site chemical exchange. Biochemistry 2018; 57(31):4753–63.
[86] Sekhar A, Kay LE. NMR paves the way for atomic level descriptions of sparsely populated, transiently formed biomolecular conformers. Proc Natl Acad Sci U S A 2013;110(32):12867–74.
[87] Hansen DF, Vallurupalli P, Kay LE. Using relaxation dispersion NMR spectroscopy to determine structures of excited, invisible protein states. J Biomol NMR 2008;41 (3):113–20.
[88] Vallurupalli P, Hansen DF, Kay LE. Structures of invisible, excited protein states by relaxation dispersion NMR spectroscopy. Proc Natl Acad Sci U S A 2008;105 (33):11766–71.
[89] Clore GM, Iwahara J. Theory, practice, and applications of paramagnetic relaxation enhancement for the characterization of transient low-population states of biological macromolecules and their complexes. Chem Rev 2009;109(9):4108–39.
[90] Morgado L, Burmann BM, Sharpe T, Mazur A, Hiller S. The dynamic dimer structure of the chaperone trigger factor. Nat Commun 2017;8(1):1992.
[91] Takayama Y, Clore GM. Intra- and intermolecular translocation of the bi-domain transcription factor Oct1 characterized by liquid crystal and paramagnetic NMR. Proc Natl Acad Sci U S A 2011;108(22):E169–76.
[92] Burmann BM, Wang C, Hiller S. Conformation and dynamics of the periplasmic membrane-protein-chaperone complexes OmpX-Skp and tOmpA-Skp. Nat Struct Mol Biol 2013;20(11):1265–72.
[93] Thoma J, Burmann BM, Hiller S, Müller DJ. Impact of holdase chaperones Skp and SurA on the folding of β-barrel outer-membrane proteins. Nat Struct Mol Biol 2015;22(10):795–802.
[94] Bax A, Kontaxis G, Tjandra N. Dipolar couplings in macromolecular structure determination. Methods Enzymol 2001;339:127–74.
[95] Prestegard JH, Bougault CM, Kishore AI. Residual dipolar couplings in structure determination of biomolecules. Chem Rev 2004;104(8):3519–40.
[96] Chen K, Tjandra N. The use of residual dipolar coupling in studying proteins by NMR. Top Curr Chem 2012;326:47–67.
[97] Lange OF, Lakomek NA, Fares C, Schröder GF, Walter KF, Becker S, Meiler J, Grubmüller H, Griesinger C, de Groot BL. Recognition dynamics up to microseconds revealed from an RDC-derived ubiquitin ensemble in solution. Science 2008;320(5882):1471–5.
[98] Tjandra N, Marquardt J, Clore GM. Direct refinement against proton-proton dipolar couplings in NMR structure determination of macromolecules. J Magn Reson 2000;142(2):393–6.
[99] Singh D, Tripathi T, editors. Frontiers in protein structure, function, and dynamics. 1st ed. Singapore: Springer; 2020. p. 1–458.
[100] Tripathi T, Dubey V, editors. Advances in protein molecular and structural biology methods. 1st ed. USA: Academic Press; 2022. p. 1–714.

[101] Sonti R, Hertel-Hering I, Lamontanara AJ, Hantschel O, Grzesiek S. ATP site ligands determine the assembly state of the abelson kinase regulatory core via the activation loop conformation. J Am Chem Soc 2018;140(5):1863–9.
[102] Xie T, Saleh T, Rossi P, Kalodimos CG. Conformational states dynamically populated by a kinase determine its function. Science 2020;370(6513), eabc2754.
[103] Saleh T, Rossi P, Kalodimos CG. Atomic view of the energy landscape in the allosteric regulation of Abl kinase. Nat Struct Mol Biol 2017;24(11):893–901.
[104] Kumar GS, Clarkson MW, Kunze MBA, Granata D, Wand AJ, Lindorff-Larsen K, Page R, Peti W. Dynamic activation and regulation of the mitogen-activated protein kinase p38. Proc Natl Acad Sci U S A 2018;115(18):4655–60.
[105] Otten R, Padua RAP, Bunzel HA, Nguyen V, Pitsawong W, Patterson M, Sui S, Perry SL, Cohen AE, Hilvert D, Kern D. How directed evolution reshapes the energy landscape in an enzyme to boost catalysis. Science 2020;370(6523):1442–6.
[106] Hadzipasic A, Wilson C, Nguyen V, Kern N, Kim C, Pitsawong W, Villali J, Zheng Y, Kern D. Ancient origins of allosteric activation in a Ser-Thr kinase. Science 2020;367 (6480):912–7.
[107] Rogne P, Rosselin M, Grundstrom C, Hedberg C, Sauer UH, Wolf-Watz M. Molecular mechanism of ATP versus GTP selectivity of adenylate kinase. Proc Natl Acad Sci U S A 2018;115(12):3012–7.
[108] Saio T, Guan X, Rossi P, Economou A, Kalodimos CG. Structural basis for protein antiaggregation activity of the trigger factor chaperone. Science 2014;344 (6184):1250494.
[109] Huang C, Rossi P, Saio T, Kalodimos CG. Structural basis for the antifolding activity of a molecular chaperone. Nature 2016;537(7619):202–6.
[110] Jiang Y, Rossi P, Kalodimos CG. Structural basis for client recognition and activity of Hsp40 chaperones. Science 2019;365(6459):1313–9.
[111] Mas G, Burmann BM, Sharpe T, Claudi B, Bumann D, Hiller S. Regulation of chaperone function by coupled folding and oligomerization. Sci Adv 2020;6(43), abc5822.
[112] Callon M, Burmann BM, Hiller S. Structural mapping of a chaperone-substrate interaction surface. Angew Chem Int Ed Eng 2014;53(20):5069–72.
[113] Sekhar A, Velyvis A, Zoltsman G, Rosenzweig R, Bouvignies G, Kay LE. Conserved conformational selection mechanism of Hsp70 chaperone-substrate interactions. elife 2018;7, e32764.
[114] Sekhar A, Rosenzweig R, Bouvignies G, Kay LE. Hsp70 biases the folding pathways of client proteins. Proc Natl Acad Sci U S A 2016;113(20):E2794–801.
[115] Sekhar A, Rosenzweig R, Bouvignies G, Kay LE. Mapping the conformation of a client protein through the Hsp70 functional cycle. Proc Natl Acad Sci U S A 2015;112 (33):10395–400.
[116] Mas G, Guan JY, Crublet E, Debled EC, Moriscot C, Gans P, Schoehn G, Maček P, Schanda P, Boisbouvier J. Structural investigation of a chaperonin in action reveals how nucleotide binding regulates the functional cycle. Sci Adv 2018;4(9), eaau4196.
[117] Sučec I, Wang Y, Dakhlaoui O, Weinhäupl K, Jores T, Costa D, Hessel A, Brennich M, Rapaport D, Lindorff-Larsen K, Bersch B, Schanda P. Structural basis of client specificity in mitochondrial membrane-protein chaperones. Sci Adv 2020;6(51).
[118] Felix J, Weinhäupl K, Chipot C, Dehez F, Hessel A, Gauto DF, Morlot C, Abian O, Gutsche I, Velazquez-Campoy A, Schanda P, Fraga H. Mechanism of the allosteric activation of the ClpP protease machinery by substrates and active-site inhibitors. Sci Adv 2019;5(9), eaaw3818.
[119] Šulskis D, Thoma J, Burmann BM. Structural basis of DegP-protease temperature-dependent activation. Sci Adv 2021;7(50):eabj1816.
[120] Toyama Y, Harkness RW, Kay LE. Dissecting the role of interprotomer cooperativity in the activation of oligomeric high-temperature requirement A2 protein. Proc Natl Acad Sci U S A 2021;118(35), e2111257118.

[121] Harkness RW, Toyama Y, Ripstein ZA, Zhao H, Sever AIM, Luan Q, Brady JP, Clark PL, Schuck P, Kay LE. Competing stress-dependent oligomerization pathways regulate self-assembly of the periplasmic protease-chaperone DegP. Proc Natl Acad Sci U S A 2021;118(32), e2109732118.

[122] Toyama Y, Harkness RW, Lee TYT, Maynes JT, Kay LE. Oligomeric assembly regulating mitochondrial HtrA2 function as examined by methyl-TROSY NMR. Proc Natl Acad Sci U S A 2021;118(11), e2025022118.

[123] Vahidi S, Ripstein ZA, Juravsky JB, Rennella E, Goldberg AL, Mittermaier AK, Rubinstein JL, Kay LE. An allosteric switch regulates *Mycobacterium tuberculosis* ClpP1P2 protease function as established by cryo-EM and methyl-TROSY NMR. Proc Natl Acad Sci U S A 2020;117(11):5895–906.

[124] Ripstein ZA, Vahidi S, Houry WA, Rubinstein JL, Kay LE. A processive rotary mechanism couples substrate unfolding and proteolysis in the ClpXP degradation machinery. elife 2020;9, e52158.

[125] Maček P, Kerfah R, Boeri Erba E, Crublet E, Moriscot C, Schoehn G, Amero C, Boisbouvier J. Unraveling self-assembly pathways of the 468-kDa proteolytic machine TET2. Sci Adv 2017;3(4), e1601601.

# CHAPTER 6

# Applications of infrared spectroscopy to study proteins

**Riya Sahu[a], Banesh Sooram[a], Santanu Sasidharan[a], Niharika Nag[b], Timir Tripathi[b,c], and Prakash Saudagar[a]**
[a]Department of Biotechnology, National Institute of Technology Warangal, Warangal, India
[b]Molecular and Structural Biophysics Laboratory, Department of Biochemistry, North-Eastern Hill University, Shillong, India
[c]Regional Director's Office, Indira Gandhi National Open University (IGNOU), Regional Center Kohima, Kohima, India

## 1. Introduction

Owing to its sensitivity to chemical makeup and structure, one of the most used approaches to analyze the structure of small molecules is infrared (IR) spectroscopy. In addition to its sensitivity, IR spectroscopy provides extensive data to study protein structure [1–11], biological systems, molecular and chemical processes [1,2,9,12–31], and mechanisms of protein misfolding, folding, and unfolding [8,9,11,23,28,32–38]. The use of information received from the IR spectrum is not only limited to the analysis of proteins but also used to explore other biological systems such as the classification of microbes and the identification of bacterial strains. IR spectroscopy has a wide range of applications ranging from small soluble proteins to large membrane proteins. The method requires less effort, is economical, and gives high time resolution within a short span of measurement time and with a low amount of sample requirement. This article discusses the applicability of the middle region of the IR spectra for studying protein reactions utilizing reaction-driven IR difference spectroscopy.

When IR radiation is absorbed by a molecule, it undergoes vibrational transitions. This is usually the case with the mid- and far-IR spectral regions when the light frequencies and vibration are equal and the dipole moment of the molecule changes. The strength and the polarity of the vibrating bonds are dependent on the frequency of the vibration and the probability of absorption. However, this is also influenced by inter- and intramolecular effects. The underlying factors such as vibrating molecules, bond nature, and the position of electron movement's effect on intra- and intermolecular interactions aid in the determination of the approximate position of the IR


https://doi.org/10.1016/B978-0-323-99127-8.00005-2

absorption band. The absorption increases with the increase in polarity, indicating that polar bonds are the key to IR absorption. The advantage of IR is that it is a label/marker-free method for biomolecule detection as almost all the biomolecules absorb IR light. On the other hand, the overlapping bands of large molecules in the spectrum can hide information within broad and no feature bands. The fact that only biomolecules absorb IR got a turn when CN encoded genetically, that is, a tRNA capable of encoding para-cyanophenylalanine at the amber nonsense code, was utilized for probing the binding of the ligand to myoglobin, in which a site-specific IR absorbing probe was used [39].

## 2. Infrared spectrum

The IR spectrum is displayed as transmittance versus wavenumber, with the unit of $cm^{-1}$ (inverse of wavelength). The wavenumber is proportional to the transition energy, and as per the International Union for Pure and Applied Chemistry (IUPAC), the horizontal coordinate of the spectrum is arranged to form high wavenumbers to the low wavenumbers. In other words, it is equivalent to the ascending wavelength order as that in UV-vis spectroscopy. This convention by IUPAC requires attention, particularly in the near-IR spectral range, where certain spectra are plotted against wavelength and others against wavenumber.

The IR spectral range, which extends from 0.78 μm to around 1000 μm (close to the visible spectral zone), can further be divided into the near-IR region (0.78–2.5 μm), the mid-IR region (2.5–50 μm), and the far-IR region (50–1000 μm). The far-IR region is also known as the terahertz frequency regime. The mid-IR usually corresponds to 4000–200 $cm^{-1}$, which corresponds to $10^{13}$ to $10^{14}$ Hz frequencies. At normal room temperature, thermal energy $kT$ equals 200 $cm^{-1}$, meaning that absorption in the mid-IR spectral band is primarily from the vibrational ground state, that is, the first excited vibrational state. In IR, each vibrational level is associated with several closely placed rotational levels. Therefore, IR spectroscopy is sometimes also called vibrational-rotational spectroscopy. Vibrational spectroscopy is based on the absorption of light by molecules and the wavelength of the incident light. Different types of vibrational spectroscopy include mid-IR spectroscopy (photoacoustic IR), IR spectroscopy, and Raman spectroscopy [40–48].

# 3. Infrared spectrum for the structural characterization of proteins

IR spectroscopy is a tool that provides valuable information regarding the structural aspects of proteins. The following information can be ascertained from the IR spectroscopy of the proteins.

## 3.1 Alteration in the chemical structure of proteins

The proteins contain several overlapping groups, and thus, it is not possible to determine the structure of proteins using IR spectroscopy [49]. However, it is possible to determine the alteration in the chemical structure of vibrating groups, such as the protonation of side chains, phosphorylation, glycosylation, and other such modifications in proteins [26,50,51]. The method has been successfully employed to study how proteins work in proton pumps, such as bacteriorhodopsin [31].

## 3.2 Understanding the redox state and bonding in proteins

The energy absorption could affect the electron density of the molecules, which can be utilized to study the redox state of proteins. IR spectra have been used to determine the contribution of quinone reduction using the spectral difference of various states of photosynthetic reaction centers [52–54]. During redox reactions, the functional groups in proteins can absorb energy which affects the vibrational state of the molecules. Moreover, IR spectroscopy and its variants are sensitive techniques for detecting bond distortions [55]. Any bond distortions smaller than 1 pm in an enzymatic catalytic reaction can be easily detected with very high accuracy. The method has been exploited in enzymatic studies of chymotrypsin [56], $Ca^{2+}$-ATPase [57], Ras [58], myosin [59], and subtilisin [56].

## 3.3 Conformational aspects and hydrogen bonding

Vibrational coupling in molecules can be used to study the conformational status of the molecules [60]. The molecular geometry of molecules is dependent on the vibration coupling of the molecules. For instance, the vibrational coupling of two CO groups has been used to study $Ca^{2+}$-binding proteins [61]. Similarly, $PO^{2-}$ stretching and $VO_4^{3-}$ stretching were used to study myosin and Ras proteins [58,59]. It has been shown that the vibrational coupling of helices of bacteriorhodopsin influences the IR absorption of the protein [62]. The method also provides information on the conformational status of the backbone, the side chain's protonation state, and internal water molecules [14].

Hydrogen bonding in molecules is crucial as it stabilizes the structures. The protein structure is stabilized by a large number of hydrogen bonds, which are required for the function of proteins [63]. Vibrational spectroscopy is one of the few approaches that directly report on the strength of hydrogen bonds [64]. In general, hydrogen bonding reduces the frequency of stretching vibrations, whereas it raises the frequency of bending vibrations. With a single hydrogen bond contributing 16 $cm^{-1}$ in a nitrogen matrix, hydrogen bonding to $PO^{2-}$ groups lowers the observed band position of the symmetric stretching vibration by 3–20 $cm^{-1}$ and the antisymmetric stretching vibration by 20–34 $cm^{-1}$ [65–67]. Badger and Bauer's relationship is used for quantitative evaluation of the hydrogen bond induced band shifts [68].

## 3.4 Conformational freedom and electric fields

IR spectroscopy also provides bandwidth information for molecules that can be used to estimate conformational freedom [69]. The technique provides information about the population of conformers of given molecules because of its very narrow time scale. Because the band position for a given vibration is different for every conformer, the heterogenous broadening of bands can be observed as a consequence [70]. As a result, the rigid structures will have narrow bandwidths, whereas flexible ones will have broader bandwidths. Bandwidth also provides entropy information, and it can be used to quantify the thermodynamic parameters of the enzyme catalysis [44]. The binding of molecules to proteins restricts the conformational freedom by two folds [12]. The aspect of conformational freedom has been used to study the binding of the substrate to lactate dehydrogenase and binding of nucleotides to Ras protein [44,71]. In addition, IR spectroscopy also provides insights into the electric field of the molecules.

# 4. Infrared spectrophotometers

There are two types of IR spectrophotometers, namely, Fourier-transform IR (FT-IR) and dispersive IR spectrophotometers.

## 4.1 Fourier-transform IR (FT-IR) spectrophotometer

Most IR spectrophotometers used today are FT-IR spectrophotometers. In these spectrophotometers, the Fourier transform is used to produce the final spectrum, hence the name. The crucial component of the FT-IR spectrophotometer is the interferometer with a movable mirror, which generates a variable optical path difference. Initially, the light generated from the source reaches the beam splitter. About 50% of the light reaches the fixed mirror and the other half reaches the movable mirror. The light then passes through the sample and is finally recombined. The two lights interfere either constructively or destructively. The detector measures the light intensity in relation to the movable mirror. The spectrum generated in this case is called a Fourier-transformed spectrum. The main advantage of FT-IR spectrophotometers is that they are highly sensitive, and the data acquisition is rapid [72]. Schematics of common FT-IR spectrophotometers are given in Fig. 1.

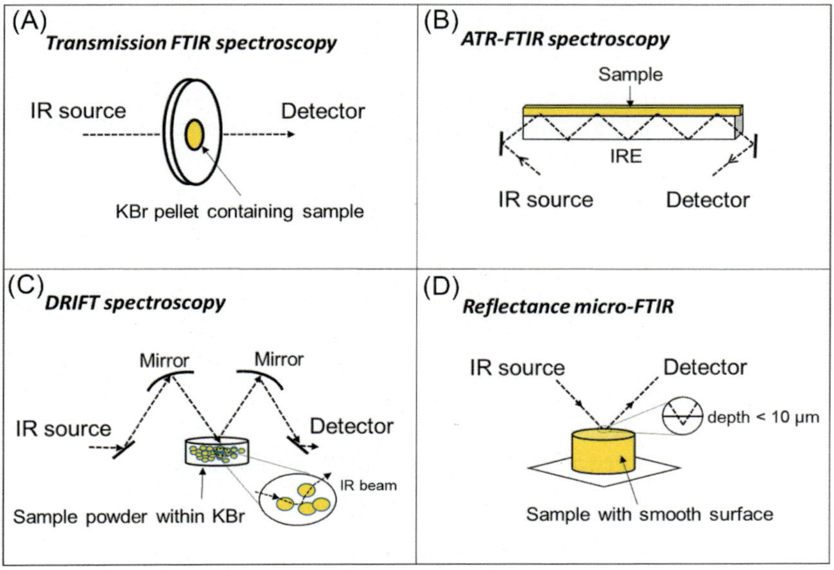


**Fig. 1** Common FT-IR spectroscopy analysis modes and their schematics. (A) Transmission FT-IR (B), attenuated total reflectance FT-IR (C), diffuse reflectance IR Fourier transform (D), and reflectance micro-FT-IR [73].

## 4.2 Dispersive IR spectrophotometers

Dispersive IR spectrophotometers are used for special measurements such as time-resolved spectral calculations. They consist of a simple light source, monochromator, and detector. Here, the radiation heat released from the sample can be avoided; however, the light that reaches the detector is very low. Because of this, the instrument has a low signal-to-noise ratio that makes the measurements unreliable [74].

# 5. Types of IR measurements

## 5.1 Transmission measurements

In transmission measurements, the sample incidence light is sensed by the detector [75]. The light absorption by the sample follows Beer-Lambert's law, given the sample is in a homogeneous form.

$$\text{Absorbance } (A) = \log\left(\frac{I_0}{I}\right) = \varepsilon \cdot c \cdot l$$

where $I_0$ is the incidence light intensity, $I$ is the reflected light intensity, $c$ is the concentration of the sample, $\varepsilon$ is the molar absorption coefficient, and $l$ is the path length of the cuvette [76].

Zinc, barium, and calcium fluoride are used as cuvette materials to facilitate the use of UV light as a light source. The path length of the cuvette is crucial as it decides the sample quantity and absorption. For a high signal-to-noise ratio, a sample concentration of 0.1–1 mM is needed for proteins, whereas a 1–100 mM concentration is required for small molecules. A variant of IR spectroscopy called surface-enhanced IR can be used to get an optimum signal even at lower concentrations of the material. The main drawback of the IR method is the interference water signal at $\sim$1645 $\text{cm}^{-1}$, which overlaps with the amide I signal [77].

## 5.2 Attenuated total reflectance (ATR) measurements

In attenuated total reflectance (ATR) measurements, the sample is placed on the crystal, and the light is focused on the crystal. The light that passes through the sample is detected and measured in the spectrum [78]. ATR measurements do not require a short path length as in transmission measurements. For protein and liquid samples, a thin film is prepared using compatible materials such as potassium bromide.

## 6. IR absorption and detection of amino acid side chains

Proteins are made up of amino acids, and the side chain of amino acids is the principal component that provides molecular functions. Therefore, information about the side chains is vital because they are involved in the mechanism. The primary use of IR to detect amino acid side chains is to identify the catalytically active residues and gain an understanding of the environment, their structural changes, protonation states, coordination of cations, and hydrogen bonding.

An overview of the IR absorption ranges of amino acid side chains is available in a classical review by Barth et al. [79]. The authors provide the strongest bands that are usually recorded in an IR spectrum of amino acids and others that are free of spectral overlap between groups. There are certain drawbacks, such as deviation in bands when the IR absorption of a protein is recorded in solution and in crystals. The influence of the environment, in such cases, is enormous as it changes the protonation, strength, and polarity of the bonds. There have been cases where the $pK_a$ of the residues influences the bands that are measured, such as Asp93 of bacteriorhodopsin [80].

## 7. IR absorption and detection of the protein backbone

The IR analysis of a protein backbone started with *N*-methyl acetamide (NMA), which is the smallest molecule in a trans peptide group. Therefore, it was considered the starting point of the normal mode analysis. When the $CH_3$ groups were considered, the number of normal modes was 12 (because the number of atoms in NMA is 6). Out of the 12, we now discuss the modes with the highest frequency in the following sections.

### 7.1 Vibrations of NH stretching (~3300 cm$^{-1}$ for amide A and ~3070 cm$^{-1}$ for amide B)

The stretching of the NH bond gives rise to two bands. The initial band can be observed between 3310 and 3270 cm$^{-1}$, and this corresponds to the amide A band. The band is specific to the NH groups, and it does not vary upon changes in the polypeptide backbone. However, the frequency is based on the strength of the hydrogen bond. Due to the Fermi resonance doublet, there is a second component in amide A that absorbs weakly between 3100 and 3030 cm$^{-1}$.

## 7.2 Vibrations of amide I (~1650 cm$^{-1}$)

The backbone structure determines the extent to which the internal coordinates influence the amide I normal mode. The CO bond stretching vibrations result in an absorption band near 1650 cm$^{-1}$. Apart from the CO bond, there are contributions from the out-of-phase CN stretching vibrations, NH in-plane bend, and CCN deformation. The side chains do not affect the amide I vibration and are mainly dependent on the secondary structure. Because of this, the amide I vibration is used for studying the secondary structure of proteins [81]. The amide I bands are usually sensitive to the secondary structure, and therefore, they are of interest in deciphering the structure of the protein. The relation of structure to spectrum can now be explained with the help of the transition dipole coupling [81–85], through-bond coupling [86–90], and hydrogen bonding [91–97]. There are various ways to calculate the amide I band of a protein:

**i.** The symmetry of the infinite secondary structure can reduce the vibrational mode that can be observed, and the transition dipole coupling can be measured from it [81,82,98].

**ii.** A "floating oscillator model" can be used to calculate the amide I band by considering the amide I oscillator [99,100].

## 7.3 Vibrations of amide II (~1550 cm$^{-1}$)

The amide II band results from the out-of-phase combination of the plane bend in NH and the stretching vibration of CN. There is also a contribution from the plane bend of CO and the stretching vibrations of CC and NC. Similar to the amide I vibration, amide II is also least affected by the side chain stretching vibrations. However, unlike amide I, there exists no straightforward relation between the frequency and the secondary structure of the protein. Structural analysis and the secondary structure prediction can be carried out from the amide II band independently [101]. The bending of NH is primarily modified by hydrogen bonding, and this results in modification in the amide II mode. Therefore, the influence of the environment and the backbone conformation can be understood from the amide II mode [102].

## 7.4 Vibrations of amide III (1400–1200 cm$^{-1}$)

The *in-phase* combination of the bending of NH and the stretching vibration of CN results in the amide III mode. There is also a contribution from the plane bending of the CO *in-phase* and the stretching vibrations of the CC bond. The mode is complex in polypeptides as it depends on the structure of the side chain, and several modes can be observed in 1400–1200 $cm^{-1}$ because of the NH bending. The secondary structure can be predicted using this mode despite the contributions from the side chains. The $N^2H$ bending, which is the bend between the nitrogen and deuterated hydrogen, can be separated off by N-deuteration, and the other coordinates can be redistributed [103–106].

# 8. IR spectroscopy for studying proteins

## 8.1 Understanding the protein secondary structure

IR spectroscopy has been extensively used to determine the secondary structure of proteins. This is exclusively conducted using the amide I, amide II, and amide III bands. In addition, the near-IR region has been useful for the determination of the secondary structure. The tool has been important in the current era of proteomics, considering the vast number of proteins that remain structurally uncharacterized. We discussed that the sensitivity of the amide I band is because of transition dipole coupling, and the resulting bands are observed predominantly in the amide I region (Table 1). There are two common methods to fit the spectrum to the structure:

**i.** The amide I bands are fitted into their component bands by band narrowing techniques [107,108].

**ii.** Amide I band decomposition can be performed by calibrating the set of spectra calculated from protein with a known structure [109–114].

## 8.2 Flexibility of proteins

The flexibility of proteins can be probed with the help of $^1H/^2H$ exchange experiments, where the amide II bands that are sensitive to the $^1H/^2H$

**Table 1** Amide I band positions of secondary structures [30].

| Secondary structure | Band position in $^1H_2O/cm^{-1}$ | | Band position in $^2H_2O/cm^{-1}$ | |
|---|---|---|---|---|
| | Average | Extremes | Average | Extremes |
| α-Helix | 1654 | 1648–1657 | 1652 | 1642–1660 |
| β-Sheet | 1633, 1684 | 1623–1641, 1674–1695 | 1630, 1679 | 1615–1638, 1672–1694 |
| Turn | 1672 | 1662–1686 | 1671 | 1653–1691 |
| Disordered | 1654 | 1642–1657 | 1645 | 1639–1654 |

exchange are checked. The major contribution to these bands is the bending vibration of the NH bond. The exchange rate of the protons in the amide is dependent on the solvent accessibility and the hydrogen bond strength [115,116]. These protons of the amide that are hydrogen-bonded are protected from the exchange when they are solvent-exposed, and this has been proved for small molecules where the exchange rate is in six orders of magnitude. Therefore, there will be local or global unfolding as a result of hydrogen bond breakage so that the $^{1}H/^{2}H$ exchange occurs at a significant rate [115].

## 8.3 Function of proteins

The mechanism of a protein function can be elucidated at the molecular level using IR spectroscopy. This is possible because of the high time resolution, applicability of all molecules from small peptides to large membrane proteins, and the large flow of information along with sensitivity. The changes in the environment can also be detected in most cases. The absorption spectrum, therefore, gives the information of a protein reaction directly, and in some instances, the associated changes in the IR absorption are monitored closely. Usually, this is carried out by subtracting or normalizing the protein after the reaction with the protein state before the reaction [117–124]. The changes in absorbance are so minimal to detect by absorption that the reaction has to be initiated in the cuvette directly to obtain the data. This is referred to as reaction-induced IR difference spectroscopy. There are other spectroscopic methods to study protein-ligand interactions, including Raman spectroscopy, which is detailed in these reviews [43,47].

## 8.4 Measuring enzyme activity

The enzyme activity is usually measured as a result of UV/vis absorbance of the substrate or the product and sometimes by fluorescence molecules. However, IR spectroscopy can be used to measure the "online" activity of the enzyme and monitor enzymatic reactions. Examples of such activity determination include deacetylation of cinnamoyl-chymotrypsin [125], adenosine triphosphate (ATP) hydrolysis of $Ca^{2+}$ [126], the synthesis of derivatives of hydroxamic acid by amides [127], and so on. This is achieved as the bands of the products of the enzymatic reactions are different from the initial band spectra.

## 8.5 Water and hydrated proton in proteins

Water is one of the essential solvents linked to protein structure and function, especially when biochemical reactions are considered. There are many reviews on IR spectroscopy of water and its interactions with proteins [128–131]. Examples of the study of protein interactions with water involve the hydration of bovine serum albumin (BSA) and lysozyme. The hydration involves (i) hydrogen bond formation between water and NH or CO groups present in the protein backbone, (ii) the protonation of the carboxylate groups, and (iii) the insertion of the water molecules without disrupting the hydrogen bond network. Although both the proteins are dry, a considerable amount of water remains with the proteins. A quarter of the NH and CO bonds are without hydrogen bonds when the proteins are in the dry state. About 50% of the CO and the NH bonds in the protein backbone are hydrogens bonded to water in the hydrated lysozyme [132,133].

# 9. Studying proteins with IR spectroscopy: Case studies

The $Ca^{2+}$-ATPases are enzymes of the P-type ATPases family. They are actively involved in biological membrane transport. The mechanism involves pumping two $Ca^{2+}$ for hydrolysis of one ATP molecule. This was the first protein studied using the photolytically induced $Ca^{2+}$ concentration jump [134]. For this, both caged ATP and $Ca^{2+}$ have been used [135–137]. A 65 ms time resolution in a rapid scan technique was used to study the kinetics of the pump before and after ATP release [138].

IR spectroscopy has also been used to study intermediates of enzymes, and one such example is the ATPase phosphoenzyme intermediates (E2P). The study revealed the high hydrolysis rates and also observed the exchange of the oxygen isotope at the phosphate groups when catalyzed by ATPase. The experiment provided the atomic-resolution details of the crowded regions that cause the three stretching vibrations of the phosphate group that is transiently bound. The spectrum of the intermediate is shown in Fig. 2. The terminal P—O stretching vibration of the phosphate group that is unlabeled was observed at 1194, 1115, and 1137 $cm^{-1}$. The data were confirmed by relating the P—O frequency with the P—O bond valence [57].

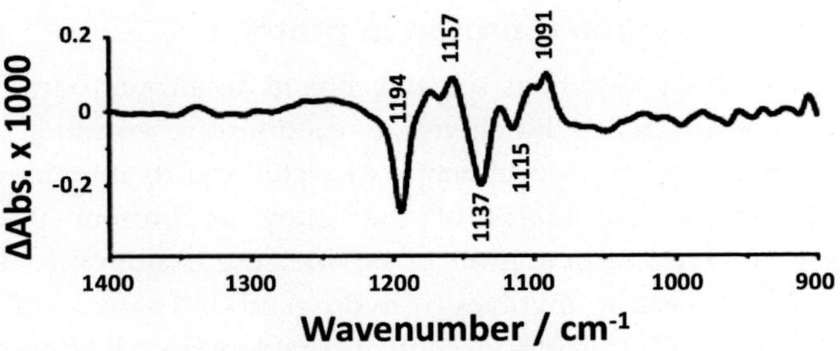


**Fig. 2** IR difference spectrum of the phosphate group of the E2 $P^{16}O_3$→E2 $P^{18}O_3$ isotope [57].

## 10. Conclusion and future perspectives

IR spectroscopy is a tool that can be utilized to determine the secondary structure composition of a protein. Although X-ray crystallography, cryo-EM, and NMR spectroscopy provide more detailed information on the structure of the protein, including the individual atomic positions, the cost of such experiments is very high, and therefore, the structure of all the proteins cannot be solved by these methods [139,140]. In addition, sometimes, the structure cannot be related to the dynamic properties of the protein in the solution. IR-based protein structure determination has now become more easily accessible and accurate and has huge potential in the field of protein biochemistry. However, FT-IR has drawbacks such as spectral interference of water and amide I/II bonds as well as sample preparation and handling. The discovery of the microfluidic IR spectroscopy system has simplified the prediction of protein structure by IR with the help of powerful lasers and microfluidic transmission cells. Although the IR-based equipment has complex operating conditions, they have progressed in the field of protein structure by leaps and bounds. The information is not limited to the structure alone but also to the dynamics and aggregation. This information is essential in order to understand the biological mechanism and function of macromolecules such as proteins. In the future, with the help of machine learning and cloud computing, the field can help strategize and revolutionize the secondary structure prediction of proteins.

## References

[1] Uversky VN, Permiakov EA. Methods in protein structure and stability analysis: vibrational spectroscopy. Nova Biomedical Books; 2007.

[2] Siebert F. Infrared spectroscopy applied to biochemical and biological problems. Methods Enzymol 1995;246:501–26. Elsevier.

[3] Jackson M, Mantsch HH. The use and misuse of FTIR spectroscopy in the determination of protein structure. Crit Rev Biochem Mol Biol 1995;30(2):95–120.
[4] Haris PI, Chapman D. Does Fourier-transform infrared spectroscopy provide useful information on protein structures? Trends Biochem Sci 1992;17(9):328–33.
[5] Haris PI, Chapman D. Analysis of polypeptide and protein structures using Fourier transform infrared spectroscopy. In: Microscopy, optical spectroscopy, and macroscopic techniques. Springer; 1994. p. 183–202.
[6] Goormaghtigh E. Determination of soluble and membrane protein structure by Fourier transform infrared spectroscopy. II. Experimental aspects, side chain structure, and H/D exchange. J Subcell Biochem 1994;23:363–403.
[7] Goormaghtigh E, Cabiaux V, Ruysschaert J-M. Determination of soluble and membrane protein structure by Fourier transform infrared spectroscopy. Subcell Biochem 1994;405–50.
[8] Fabian H, Mantsch HH, Schultz CP. Two-dimensional IR correlation spectroscopy: sequential events in the unfolding process of the λ Cro-V55C repressor protein. Proc Natl Acad Sci 1999;96(23):13153–8.
[9] Fabian H, Mäntele W. Infrared spectroscopy of proteins. In: Handbook of vibrational spectroscopy. John Wiley & Sons; 2006.
[10] Arrondo JLR, Muga A, Castresana J, Goñi FM. Quantitative studies of the structure of proteins in solution by Fourier-transform infrared spectroscopy. Prog Biophys Mol Biol 1993;59(1):23–56.
[11] Arrondo JLR, Goñi FM. Structure and dynamics of membrane proteins as studied by infrared spectroscopy. Prog Biophys Mol Biol 1999;72(4):367–405.
[12] Wharton CW. Infrared spectroscopy of enzyme reaction intermediates. Nat Prod Rep 2000;17(5):447–53.
[13] Vogel R, Siebert F. Vibrational spectroscopy as a tool for probing protein function. Curr Opin Chem Biol 2000;4(5):518–23.
[14] Rothschild KJ. FTIR difference spectroscopy of bacteriorhodopsin: toward a molecular model. J Bioenerg Biomembr 1992;24(2):147–67.
[15] Noguchi T. Light-induced FTIR difference spectroscopy as a powerful tool toward understanding the molecular mechanism of photosynthetic oxygen evolution. Photosynth Res 2007;91(1):59–69.
[16] Nabedryk E. Light-induced Fourier transform infrared difference spectroscopy of the primary electron donor in photosynthetic reaction centers. In: Infrared spectroscopy of biomolecules. John Wiley & Sons; 1996. p. 39–81.
[17] Mäntele W. Reaction-induced infrared difference spectroscopy for the study of protein function and reaction mechanisms. Trends Biochem Sci 1993;18(6):197–202.
[18] Mantele W. Infrared vibrational spectroscopy of the photosynthetic reaction center. vol. 2. Photosynthetic Reaction Center; 2013. p. 239.
[19] Mäntele W. Infrared and Fourier-transform infrared spectroscopy. In: Biophysical techniques in photosynthesis. Springer; 1996. p. 137–60.
[20] Mäntele W. Infrared vibrational spectroscopy of reaction centers. Anoxygenic photosynthetic bacteria. Springer; 1995. p. 627–47.
[21] Maeda A. Application of FTIR spectroscopy to the structural study on the function of bacteriorhodopsin. Isr J Chem 1995;35(3-4):387–400.
[22] Jung C. Insight into protein structure and protein-ligand recognition by Fourier transform infrared spectroscopy. J Mol Recognit 2000;13(6):325–51.
[23] Heberle J. Time-resolved ATR/FT-IR spectroscopy of membrane proteins. Rec Res Dev Appl Spectr 1999;2:147–59.
[24] Heberle J. Proton transfer reactions across bacteriorhodopsin and along the membrane. Biochim Biophys Acta Bioenerg 2000;1458(1):135–47.

[25] Goeldner M, Givens R. Dynamic studies in biology: phototriggers, photoswitches and caged biomolecules. John Wiley & Sons; 2006.

[26] Gerwert K. Molecular reaction mechanisms of proteins as monitored by time-resolved FTIR spectroscopy. Curr Opin Struct Biol 1993;3(5):769–73.

[27] Fahmy K. Application of ATR-FTIR spectroscopy for studies of biomolecular interactions. In: Recent research developments in biophysical chemistry. Inist-CNRS; 2001. p. 1–17.

[28] Dioumaev A. Infrared methods for monitoring the protonation state of carboxylic amino acids in the photocycle of bacteriorhodopsin. Biochemistry 2001;66(11): 1269–76.

[29] Breton J. Fourier transform infrared spectroscopy of primary electron donors in type I photosynthetic reaction centers. Biochim Biophys Acta Bioenerg 2001;1507 (1-3):180–93.

[30] Barth A, Zscherp C. What vibrations tell about proteins. Q Rev Biophys 2002;35 (4):369–430.

[31] Gerwert K. Molecular reaction mechanisms of proteins monitored by time-resolved FTIR-spectroscopy. Biol Chem 1999;380(7–8):931–5. https://doi.org/10.1515/BC.1999.115.

[32] Williams S, Causgrove TP, Gilmanshin R, Fang KS, Callender RH, Woodruff WH, et al. Fast events in protein folding: helix melting and formation in a small peptide. Biochemistry 1996;35(3):691–7.

[33] Troullier A, Reinstädler D, Dupont Y, Naumann D, Forge V. Transient non-native secondary structures during the refolding of α-lactalbumin detected by infrared spectroscopy. Nat Struct Biol 2000;7(1):78–86.

[34] Schultz CP. Illuminating folding intermediates. Nat Struct Biol 2000;7(1):7–10.

[35] Pozo Ramajo A, Petty SA, Starzyk A, Decatur SM, Volk M. The α-helix folds more rapidly at the C-terminus than at the N-terminus. J Am Chem Soc 2005;127 (40):13784–5.

[36] Kauffmann E, Darnton NC, Austin RH, Batt C, Gerwert K. Lifetimes of intermediates in the β-sheet to α-helix transition of β-lactoglobulin by using a diffusional IR mixer. Proc Natl Acad Sci 2001;98(12):6646–9.

[37] Gilmanshin R, Williams S, Callender RH, Woodruff WH, Dyer RB. Fast events in protein folding: relaxation dynamics and structure of the I form of apomyoglobin. Biochemistry 1997;36(48):15006–12.

[38] Dyer RB, Gai F, Woodruff WH, Gilmanshin R, Callender RH. Infrared studies of fast events in protein folding. Acc Chem Res 1998;31(11):709–16.

[39] Schultz KC, Supekova L, Ryu Y, Xie J, Perera R, Schultz PG. A genetically encoded infrared probe. J Am Chem Soc 2006;128(43):13984–5.

[40] Thomas Jr GJ. New structural insights from Raman spectroscopy of proteins and their assemblies. Biopolymers 2002;67(4-5):214–25.

[41] Spiro TG, Gaber BP. Laser Raman scattering as a probe of protein structure. Annu Rev Biochem 1977;46(1):553–70.

[42] Robert B. Resonance Raman studies in photosynthesis—chlorophyll and carotenoid molecules. In: Biophysical techniques in photosynthesis. Springer; 1996. p. 161–76.

[43] Deng H, Callender R. Vibrational studies of enzymatic catalysis. In: Infrared Raman spectroscopy of biological materials. CRC Press; 2001. p. 477–515.

[44] Deng H, Callender R. Raman spectroscopic studies of the structures, energetics and bond distortions of substrates bound to enzymes. Methods Enzymol 1999; 308:176–201. Elsevier.

[45] Carey PR. Raman spectroscopy, the sleeping giant in structural biology, awakes. J Biol Chem 1999;274(38):26625–8.

[46] Carey PR. Raman spectroscopy in enzymology: the first 25 years. J Raman Spectrosc 1998;29(1):7–14.
[47] Carey PR, Tonge PJ. Unlocking the secrets of enzyme power using Raman spectroscopy. Acc Chem Res 1995;28(1):8–13.
[48] Callender R, Deng H. Nonresonance Raman difference spectroscopy: a general probe of protein structure, ligand binding, enzymatic catalysis, and the structures of other biomacromolecules. Annu Rev Biophys Biomol Struct 1994;23(1):215–45.
[49] Singh BR. Basic aspects of the technique and applications of infrared spectroscopy of peptides and proteins. ACS Publications; 2000.
[50] Barth A, Mäntele W. ATP-induced phosphorylation of the sarcoplasmic reticulum $Ca^{2+}$ ATPase: molecular interpretation of infrared difference spectra. Biophys J 1998;75(1):538–44.
[51] Barth A. Phosphoenzyme conversion of the sarcoplasmic reticulum $Ca^{2+}$-ATPase: molecular interpretation of infrared difference spectra. J Biol Chem 1999;274 (32):22170–5.
[52] Mäntele W, Wollenweber A, Nabedryk E, Breton J. Infrared spectroelectrochemistry of bacteriochlorophylls and bacteriopheophytins: Implications for the binding of the pigments in the reaction center from photosynthetic bacteria. Proc Natl Acad Sci 1988;85(22):8468–72.
[53] Nabedryk E, Leonhard M, Mäntele W, Breton J. Fourier transform infrared difference spectroscopy shows no evidence for an enolization of chlorophyll a upon cation formation either in vitro or during P700 photooxidation. Biochemistry 1990;29(13):3242–7.
[54] Bauscher M, Mäntele W. Electrochemical and infrared-spectroscopic characterization of redox reactions of p-quinones. J Phys Chem 1992;96(26):11101–8.
[55] Czernuszewicz RS, Macor KA, Johnson MK, Gewirth A, Spiro TG. Vibrational mode structure and symmetry in proteins and analogs containing Fe4S4 clusters: resonance Raman evidence that HiPIP is tetrahedral while ferredoxin undergoes a D2d distortion. J Am Chem Soc 1987;109(23):7178–87.
[56] Tonge PJ, Carey PR. Length of the acyl carbonyl bond in acyl-serine proteases correlates with reactivity. Biochemistry 1990;29(48):10723–7.
[57] Barth A, Bezlyepkina N. P—O bond destabilization accelerates phosphoenzyme hydrolysis of sarcoplasmic reticulum $Ca^{2+}$-ATPase. J Biol Chem 2004;279(50): 51888–96.
[58] Cheng H, Sukal S, Deng H, Leyh TS, Callender R. Vibrational structure of GDP and GTP bound to RAS: an isotope-edited FTIR study. Biochemistry 2001;40 (13):4035–43.
[59] Deng H, Wang J, Callender RH, Grammer JC, Yount RG. Raman difference spectroscopic studies of the myosin S1⊙ MgADP⊙ vanadate complex. Biochemistry 1998;37(31):10972–9.
[60] Nie B, Stutzman J, Xie A. A vibrational spectral maker for probing the hydrogen-bonding status of protonated Asp and Glu residues. Biophys J 2005;88(4):2833–47.
[61] Tackett JE. FT-IR characterization of metal acetates in aqueous solution. Appl Spectrosc 1989;43(3):483–9.
[62] Karjalainen E-L, Barth A. Vibrational coupling between helices influences the amide i infrared absorption of proteins: application to bacteriorhodopsin and rhodopsin. J Phys Chem B 2012;116(15):4448–56.
[63] Hubbard RE, Haider MK. Hydrogen bonds in proteins: role and strength. In: eLS. John Wiley & Sons; 2010.
[64] Gorman M. The evidence from infrared spectroscopy for hydrogen bonding: a case history of the correlation and interpretation of data. ACS Publications; 1957.

[65] Brown EB, Peticolas WL. Conformational geometry and vibrational frequencies of nucleic acid chains. Biopolymers 1975;14(6):1259–71.
[66] Pohle W, Bohl M, Böhlig H. Interpretation of the influence of hydrogen bonding on the stretching vibrations of the $PO^{2-}$ moiety. J Mol Struct 1991;242:333–42.
[67] George L, Sankaran K, Viswanathan K, Mathews C. Matrix-isolation infrared spectroscopy of hydrogen-bonded complexes of triethyl phosphate with $H_2O$, $D_2O$, and methanol. Appl Spectrosc 1994;48(7):801–7.
[68] Badger RM, Bauer SH. Spectroscopic studies of the hydrogen bond. II. The shift of the O–H vibrational frequency in the formation of the hydrogen bond. J Chem Phys 1937;5(11):839–51.
[69] Woutersen S, Pfister R, Hamm P, Mu Y, Kosov DS, Stock G. Peptide conformational heterogeneity revealed from nonlinear vibrational spectroscopy and molecular-dynamics simulations. J Chem Phys 2002;117(14):6833–40.
[70] Nodland E, Libnau FO, Kvalheim OM, Luinge H-J, Klæboe P. Influence and correction of peak shift and band broadening observed by rank analysis on vibrational bands from variable-temperature measurements. Vib Spectrosc 1996;10(2):105–23.
[71] Cepus V, Scheidig A, Goody RS, Gerwert K. Time-resolved FTIR studies of the GTPase reaction of H-R as P21 reveal a key role for the β-phosphate. Biochemistry 1998;37(28):10263–71.
[72] Günzler H, Gremlich H-U. IR spectroscopy. An introduction. John Wiley & Sons; 2002.
[73] Chen Y, Zou C, Mastalerz M, Hu S, Gasaway C, Tao X. Applications of micro-fourier transform infrared spectroscopy (FTIR) in the geological sciences—a review. Int J Mol Sci 2015;16(12):30223–50.
[74] Diem M, Roberts G, Lee O, Barlow A. Design and performance of an optimized dispersive infrared dichrograph. Appl Spectrosc 1988;42(1):20–7.
[75] Hinsmann P, Haberkorn M, Frank J, Svasek P, Harasek M, Lendl B. Time-resolved FT-IR spectroscopy of chemical reactions in solution by fast diffusion-based mixing in a micromachined flow cell. Appl Spectrosc 2001;55(3):241–51.
[76] Mitschele J. Beer-Lambert law. J Chem Educ 1996;73(11):A260.
[77] Masuch R, Moss DA. Stopped flow system for FTIR difference spectroscopy of biological macromolecules. In: Greve J, Puppels GJ, Otto C, editors. Spectroscopy of biological molecules: new directions: 8th European conference on the spectroscopy of biological molecules, 29 August–2 September 1999, Enschede, The Netherlands. Dordrecht: Springer Netherlands; 1999. p. 689–90.
[78] Fringeli UP. In situ infrared attenuated total reflection membrane spectroscopy. In: Internal reflection spectroscopy. CRC Press; 1992. p. 255–324.
[79] Barth A. Infrared spectroscopy of proteins. Biochim Biophys Acta Bioenerg 2007;1767(9):1073–101.
[80] Zscherp C, Schlesinger R, Tittor J, Oesterhelt D, Heberle J. In situ determination of transient pKa changes of internal amino acids of bacteriorhodopsin by using time-resolved attenuated total reflection Fourier-transform infrared spectroscopy. Proc Natl Acad Sci 1999;96(10):5498–503.
[81] Krimm S, Abe Y. Intermolecular interaction effects in the amide I vibrations of β polypeptides. Proc Natl Acad Sci 1972;69(10):2788–92.
[82] Chirgadze YN, Nevskaya N. Infrared spectra and resonance interaction of amide-I vibration of the antiparallel-chain pleated sheet. Biopolymers 1976;15(4):607–25.
[83] Schweitzer-Stenner R. Advances in vibrational spectroscopy as a sensitive probe of peptide and protein structure: a critical review. Vib Spectrosc 2006;42(1):98–117.
[84] Chirgadze YN, Shestopalov B, Venyaminov SY. Intensities and other spectral parameters of infrared amide bands of polypeptides in the β-and random forms. Biopolymers 1973;12(6):1337–51.

[85] Iconomidou V, Chryssikos D, Gionis V, Pavlidis M, Paipetis A, Hamodrakas S. Secondary structure of chorion proteins of the teleostean fish Dentex dentex by ATR FT-IR and FT-Raman spectroscopy. J Struct Biol 2000;132(2):112–22.
[86] Hamm P, Lim M, DeGrado WF, Hochstrasser RM. The two-dimensional IR nonlinear spectroscopy of a cyclic penta-peptide in relation to its three-dimensional structure. Proc Natl Acad Sci 1999;96(5):2036–41.
[87] Torii H, Tasumi M. Ab initio molecular orbital study of the amide I vibrational interactions between the peptide groups in di-and tripeptides and considerations on the conformation of the extended helix. J Raman Spectrosc 1998;29(1):81–6.
[88] Antony J, Schmidt B, Schütte C. Nonadiabatic effects on peptide vibrational dynamics induced by conformational changes. J Chem Phys 2005;122(1), 014309.
[89] Gorbunov RD, Stock G. Ab initio based building block model of amide I vibrations in peptides. Chem Phys Lett 2007;437(4-6):272–6.
[90] Brauner JW, Dugan C, Mendelsohn R. 13C isotope labeling of hydrophobic peptides. Origin of the anomalous intensity distribution in the infrared Amide I spectral region of β-sheet structures. J Am Chem Soc 2000;122(4):677–83.
[91] Torii H, Tatsumi T, Tasumi M. Effects of hydration on the structure, vibrational wavenumbers, vibrational force field and resonance raman intensities of N-methylacetamide. J Raman Spectrosc 1998;29(6):537–46.
[92] Torii H, Tatsumi T, Kanazawa T, Tasumi M. Effects of intermolecular hydrogen-bonding interactions on the amide I mode of N-methylacetamide: matrix-isolation infrared studies and ab initio molecular orbital calculations. J Phys Chem B 1998;102(1):309–14.
[93] Mennucci B, Martínez JM. How to model solvation of peptides? Insights from a quantum-mechanical and molecular dynamics study of N-methylacetamide. 1. Geometries, infrared, and ultraviolet spectra in water. J Phys Chem B 2005;109 (19):9818–29.
[94] Kubelka J, Keiderling TA. Ab initio calculation of amide carbonyl stretch vibrational frequencies in solution with modified basis sets. 1. *N*-methyl acetamide. J Phys Chem A 2001;105(48):10922–8.
[95] Reisdorf WC, Krimm S. Infrared amide I 'band of the coiled coil. Biochemistry 1996;35(5):1383–6.
[96] Parrish JRJR, Blout ER. The conformation fo poly-L-alanine in hexafluoroisopropanol. Biopolymers 1972;11(5):1001–20.
[97] Manas ES, Getahun Z, Wright WW, DeGrado WF, Vanderkooi JM. Infrared spectra of amide groups in α-helical proteins: evidence for hydrogen bonding between helices and water. J Am Chem Soc 2000;122(41):9883–90.
[98] Moore W, Krimm S. Transition dipole coupling in Amide I modes of β-polypeptides. Proc Natl Acad Sci 1975;72(12):4933–5.
[99] Torii H, Tasumi M. Model calculations on the amide-I infrared bands of globular proteins. J Chem Phys 1992;96(5):3379–87.
[100] Torii H. Theoretical analyses of the amide I infrared bands of globular proteins. In: Infrared spectroscopy of biomolecules. John Wiley & Sons; 1996.
[101] Oberg KA, Ruysschaert JM, Goormaghtigh E. The optimization of protein secondary structure determination with infrared and circular dichroism spectra. Eur J Biochem 2004;271(14):2937–48.
[102] Krimm S, Bandekar J. Vibrational spectroscopy and conformation of peptides, polypeptides, and proteins. Adv Protein Chem 1986;38:181–364.
[103] Yuan L. Studies on amide IV infrared bands for the secondary structure determination of proteins. Chem J Chin Univ 2003;24(2):226–31.
[104] Fu F-N, Deoliveira DB, Trumble WR, Sarkar HK, Singh BR. Secondary structure estimation of proteins using the amide III region of Fourier transform infrared

spectroscopy: application to analyze calcium-binding-induced structural changes in calsequestrin. Appl Spectrosc 1994;48(11):1432–41.
[105] Cai S, Singh BR. Identification of β-turn and random coil amide III infrared bands for secondary structure estimation of proteins. Biophys Chem 1999;80(1):7–20.
[106] Cai S, Singh BR. A distinct utility of the amide III infrared band for secondary structure estimation of aqueous protein solutions using partial least squares methods. Biochemistry 2004;43(9):2541–9.
[107] Byler DM, Susi H. Examination of the secondary structure of proteins by deconvolved FTIR spectra. Biopolymers 1986;25(3):469–87.
[108] Susi H, Byler DM. Fourier transform infrared study of proteins with parallel β-chains. Arch Biochem Biophys 1987;258(2):465–9.
[109] Sarver Jr RW, Krueger WC. Protein secondary structure from Fourier transform infrared spectroscopy: a data base analysis. Anal Biochem 1991;194(1):89–100.
[110] Pribic R, Vanstokkum I, Chapman D, Haris PI, Bloemendal M. Protein secondary structure from Fourier transform infrared and/or circular dichroism spectra. Anal Biochem 1993;214(2):366–78.
[111] Kalnin N, Baikalov I, Venyaminov SY. Quantitative IR spectrophotometry of peptide compounds in water ($H_2O$) solutions. III. Estimation of the protein secondary structure. Biopolymers 1990;30(13-14):1273–80.
[112] Dousseau F, Pezolet M. Determination of the secondary structure content of proteins in aqueous solutions from their amide I and amide II infrared bands. Comparison between classical and partial least-squares methods. Biochemistry 1990;29(37):8771–9.
[113] Lee DC, Haris PI, Chapman D, Mitchell RC. Determination of protein secondary structure using factor analysis of infrared spectra. Biochemistry 1990;29(39):9185–93.
[114] Baumruk V, Pancoska P, Keiderling TA. Predictions of secondary structure using statistical analyses of electronic and vibrational circular dichroism and Fourier transform infrared spectra of proteins in $H_2O$. J Mol Biol 1996;259(4):774–91.
[115] Englander SW, Kallenbach NR. Hydrogen exchange and structural dynamics of proteins and nucleic acids. Q Rev Biophys 1983;16(4):521–655.
[116] Raschke TM, Marqusee S. Hydrogen exchange studies of protein structure. Curr Opin Biotechnol 1998;9(1):80–6.
[117] Alben JO, Caughey WS. Infrared study of bound carbon monoxide in the human red blood cell, isolated hemoglobin, and heme carbonyls. Biochemistry 1968;7(1):175–83.
[118] Riepe ME, Wang JH. Infrared studies on the mechanism of action of carbonic anhydrase. J Biol Chem 1968;243(10):2779–87.
[119] Belasco JG, Knowles JR. Direct observation of substrate distortion by triosephosphate isomerase using Fourier transform infrared spectroscopy. Biochemistry 1980;19(3): 472–7.
[120] Tonge P, Moore G, Wharton C. Fourier-transform infra-red studies of the alkaline isomerization of mitochondrial cytochrome c and the ionization of carboxylic acids. Biochem J 1989;258(2):599–605.
[121] Tonge PJ, Pusztai M, White AJ, Wharton CW, Carey PR. Resonance Raman and Fourier transform infrared spectroscopic studies of the acyl carbonyl group in [3-(5-methyl-2-thienyl) acryloyl] chymotrypsin: evidence for artifacts in the spectra obtained by both techniques. Biochemistry 1991;30(19):4790–5.
[122] Zundel G. Hydrogen bonds with large proton polarizability and proton transfer processes in electrochemistry and biology. Adv Chem Phys 1999;1–217.
[123] Iliadis G, Zundel G, Brzezinski B. Catalytic mechanism of the aspartate proteinase pepsin A: an FTIR study. Biospectroscopy 1997;3(4):291–7.
[124] Bartl F, Palm D, Schinzel R, Zundel G. Proton relay system in the active site of maltodextrinphosphorylase via hydrogen bonds with large proton polarizability: an FT-IR difference spectroscopy study. Eur Biophys J 1999;28(3):200–7.

[125] White AJ, Drabble K, Ward S, Wharton CW. Analysis and elimination of protein perturbation in infrared difference spectra of acyl-chymotrypsin ester carbonyl groups by using 13C isotopic substitution. Biochem J 1992;287(1):317–23.
[126] Thoenges D, Barth A. Direct measurement of enzyme activity with infrared spectroscopy. J Biomol Screen 2002;7(4):353–7.
[127] Pacheco R, Karmali A, Serralheiro M, Haris PI. Application of Fourier transform infrared spectroscopy for monitoring hydrolysis and synthesis reactions catalyzed by a recombinant amidase. Anal Biochem 2005;346(1):49–58.
[128] Luck W. The importance of cooperativity for the properties of liquid water. J Mol Struct 1998;448(2-3):131–42.
[129] Vanderkooi JM, Dashnau JL, Zelent B. Temperature excursion infrared (TEIR) spectroscopy used to study hydrogen bonding between water and biomolecules. Biochim Biophys Acta Proteins Proteom 2005;1749(2):214–33.
[130] Maréchal Y. Observing the water molecule in macromolecules and aqueous media using infrared spectrometry. J Mol Struct 2003;648(1-2):27–47.
[131] Buck U, Huisken F. Infrared spectroscopy of size-selected water and methanol clusters. Chem Rev 2000;100(11):3863–90.
[132] Liltorp K, Maréchal Y. Hydration of lysozyme as observed by infrared spectrometry. Biopolymers 2005;79(4):185–96.
[133] Grdadolnik J, Maréchal Y. Bovine serum albumin observed by infrared spectrometry. II. Hydration mechanisms and interaction configurations of embedded H(2)O molecules. Biopolymers 2001;62(1):54–67.
[134] Barth A, Kreutz W, Mäntele W. Molecular changes in the sarcoplasmic reticulum calcium ATPase during catalytic activity: a Fourier transform infrared (FTIR) study using photolysis of caged ATP to trigger the reaction cycle. FEBS Lett 1990;277(1-2):147–50.
[135] Buchet R, Jona I, Martonosi A. The effect of dicyclohexycarbodiimide and cyclopiazonic acid on the difference FTIR spectra of sarcoplasmic reticulum induced by photolysis of caged-ATP and caged-$Ca^{2+}$. Biochim Biophys Acta Biomembr 1992;1104(1):207–14.
[136] Buchet R, Jona I, Martonosi A. $Ca^{2+}$ release from caged-$Ca^{2+}$ alters the FTIR spectrum of sarcoplasmic reticulum. Biochim Biophys Acta Biomembr 1991;1069(2): 209–17.
[137] Georg H, Barth A, Kreutz W, Siebert F, Mäntele W. Structural changes of sarcoplasmic reticulum $Ca^{2+}$-ATPase upon $Ca^{2+}$ binding studied by simultaneous measurement of infrared absorbance changes and changes of intrinsic protein fluorescence. Biochim Biophys Acta Bioenerg 1994;1188(1-2):139–50.
[138] Barth A, von Germar F, Kreutz W, Mäntele W. Time-resolved infrared spectroscopy of the $Ca^{2+}$-ATPase: the enzyme at work. J Biol Chem 1996;271(48):30637–46.
[139] Tripathi T, Dubey VK. Advances in protein molecular and structural biology methods. 1st ed. Cambridge, MA, USA: Academic Press; 2022. p. 1–714.
[140] Singh DB, Tripathi T. Frontiers in protein structure, function, and dynamics. 1st. Singapore: Springer; 2020. p. 1–458.

CHAPTER 7

# Raman spectroscopy to study biomolecules, their structure, and dynamics

**Mu Su[a], Jiajie Mei[a], Shang Pan[a,b], Junjie Xu[a], Tingting Gu[c], Qiao Li[d], Xiaorong Fan[c], and Zhen Li[a,e]**

[a]College of Resources and Environmental Sciences, Nanjing Agricultural University, Nanjing, Jiangsu, China
[b]College of Agro-grassland Sciences, Nanjing Agricultural University, Nanjing, Jiangsu, China
[c]State Key Laboratory of Crop Genetics and Germplasm Enhancement, Nanjing Agricultural University, Nanjing, Jiangsu, China
[d]College of Veterinary Medicine, Nanjing Agricultural University, Nanjing, Jiangsu, China
[e]Jiangsu Provincial Key Lab for Organic Solid Waste Utilization, Nanjing Agricultural University, Nanjing, China

## 1. Introduction

Raman spectroscopy is one of the most common vibrational spectroscopies. The Raman vibrational bands are characterized by their frequency (energy), intensity (polar character or polarizability), and band shape (chemical environment of vibration). Since the vibrational energy levels are unique to each molecule, the Raman spectrum is able to provide a "fingerprint" of ligands. Therefore the Raman spectra can provide fundamental information on molecular structure, dynamics, and environment. Raman spectroscopy is a powerful technique for biological and medical research due to its nondestructive and noninvasive features (Table 1) [1–3]. It has been widely applied to investigate biological molecules. In particular, the Raman technique has a high spatial resolution, allowing the analysis of biomolecular dynamic variations at micrometer (μm) to nanometer (nm) scale in real time [4,5]. Ongoing improvements in laser, computer, detector, motorized stage, and software have encouraged advanced innovation in Raman instrumentation [6]. With the broadening range of instrumental capabilities (e.g., alternative excitation sources, improved spectral imaging/mapping, and multivariate data analysis) and the increase of specialized applications (e.g., undersea applications, in vivo detection, and deployment on homeland security) [7–9], a variety of technologies have been combined to Raman spectrometer. These techniques include surface-enhanced Raman spectroscopy (SERS), tip-enhanced Raman spectroscopy (TERS), Raman

*Advanced Spectroscopic Methods to Study Biomolecular Structure and Dynamics*
https://doi.org/10.1016/B978-0-323-99127-8.00006-4

**Table 1** A summary of Raman, midinfrared, and near-infrared spectroscopy.

| | Raman | Infrared | Near-IR |
|---|---|---|---|
| Ease of sample preparation | + + | – | + |
| Fingerprinting | + + + | + + + | + + |
| Group frequency | + + + | + + + | – – – |
| Aqueous solution | + + | – – | – – – |
| Low-frequency mode | + + + | – | No |
| Best vibration | Symmetric | Asymmetric | Comb/overtone |

imaging and scanning electron microscopy (RISE), spatially offset Raman spectroscopy (SORS), laser tweezers Raman spectroscopy (LTRS), etc. [3,8,10–15]. In addition, portable Raman spectroscopy has also been designed to achieve the high mobility of the machine. A brief description of their advantages regarding biological sample studies has been summarized in Table 2.

Raman signals of most molecules are usually weak, especially for biological macromolecules. This substantially limited the application of Raman. However, SERS can amplify the intensity of Raman signals by 8–10 orders [10,16,17], thereby showing a series of advantages, e.g., ultrasensitivity, fast speed, multiplexing ability, and portability [18,19], and has been widely applied to biomedical fields [20–24], e.g., investigating pathogens, cells, and tissues [25–28]. TERS [a combination of SERS and scanning probe microscopy (SPM)] improves the spatial resolution down to the nm scale.

**Table 2** Classification of Raman techniques.

| Type | Principles | Advantage |
|---|---|---|
| Fourier transform Raman spectroscopy | Fourier transform technology to collect signals, 1064 nm laser light source | Eliminate fluorescence, high accuracy |
| Confocal Raman spectroscopy | Conjugate focus of light source, sample, and detector to eliminate stray light | High sensitivity, low sample content, informative |
| Surface-enhanced Raman spectroscopy (SERS) | Trace molecules were adsorbed on the surface of Cu, Ag, Au, and other metals and electrodes. The signal enhancement was $10^8$–$10^{10}$ times | High sensitivity and resolution, high molecule specificity, low sample content |

**Table 2** Classification of Raman techniques—cont'd

| Type | Principles | Advantage |
|---|---|---|
| Tip-enhanced Raman spectroscopy (TERS) | Atomic force microscope tip was used to selectively produce SERS. The spatial resolution of the image reaches 20–10 nm | High spatial resolution at nm scale, low sample content, informative |
| Raman imaging and scanning electron microscopy (RISE) | Raman imaging combined with scanning electron microscopy. Chemical information can be acquired with a resolution down to 300 nm | High spatial resolution at nm scale, low sample content, informative |
| Portable Raman spectroscopy | Conjugate focus of light source, sample, and detector to eliminate stray light | Suppresses fluorescence, in situ determination, facilitates analysis of a variety of samples |
| Spatially offset Raman spectroscopy (SORS) | Raman spectra are collected from spatial regions offset from the point of illumination on the sample surface | Eliminate fluorescence, high sensitivity, noncontact inspection, low sample content, remote computing |
| Laser tweezers Raman spectroscopy (LTRS) | Raman microscopy combines with optical tweezers to analyze single, live, moving cells in a medium | Close to the physiological state of the cell, real-time tracking of cell physiological and biochemical processes |

It has been one of the most emerging Raman techniques in the recent two decades [29–31]. TERS has been widely used for organic sample analysis at nm scale, such as RNA, biomembrane, and virus particles [32–34]. RISE microscopy combines the confocal Raman spectrometer and scanning electron microscopy (SEM). The added SEM module allows a resolution of analysis down to 300 nm [35]. Commercial systems of RISE have been available since the 2000s that have been successfully used for the characterization of bacteria, fungi, and human diseases [36–38]. The combination of SEM imaging modes also endows the capability of elemental analysis (EDS), making it possible to obtain elementary information, in addition to the Raman and morphological information. SORS enables the chemical distinction of different layers in stacked systems [39]. The spectra could be collected within 30 s or less without necessitating sample removal or contamination. SORS applications have been successfully applied to investigate seeds [40], bones [41], cells [42], and tissues [12,43]. Recently, LTRS has been increasingly applied to many different biological systems [44]. The combination of Raman microscopy with optical tweezers also made it possible to analyze single, live, and even moving cells [45,46]. The technique is primarily applied to investigate cellular dynamics at the single-cell level or characterize the Raman molecular fingerprint of different cell types. Microbial cells and human peripheral blood cells are two major systems that have been successfully tested by the extensive application of LTRS [14,47–51].

The developments in spectral identification have facilitated the commercial availability of portable Raman spectrometers for in-field applications. Modern portable Raman has been widely applied to forensics, environmental monitoring, and food detection [9,52–55]. Raman microscopy has offered itself a reliable tool in chemistry, physics, biology, archaeology, material, and medicine. Cooperation with other techniques has further developed the potential of Raman spectroscopy with applications in a variety of fields, ranging from microscopic molecular to planetary exploration [56]. This chapter highlights the broad impacts of Raman analysis on biological sciences (Table 3). Many diverse applications are demonstrated to enlighten readers about Raman-based studies and encourage them to apply this technique in both well accepted and new ways. We briefly summarize

**Table 3** Band assignments ($cm^{-1}$) of functional groups in Raman spectra.

| Peak position | Vibrational mode |
|---|---|
| $\nu$(S—S) | 506, 543 |
| S—S disulfide stretching | 524 |
| $\nu$(C—S) stretching | 665 |
| C—S/$CH_2$ rocking | 726 |
| C—N | 717 |
| C—C—$N^+$ | 883 |
| C—N stretching | 1083, 1123, 1150, 1240–1265 |
| $NO_3$ stretching | 1046 |
| $\nu_2$ $PO_4^{3-}$ | 430, 450 |
| $\nu_4$ $PO_4^{3-}$ | 587, 604 |
| O—P—O | 786, 826 |
| P—OH stretching | 918 |
| $\nu_1$ $PO_4^{3-}$ | 960 |
| $\nu_3$ $HPO_4^{2-}$ | 1002–1003 |

**Table 3** Band assignments ($cm^{-1}$) of functional groups in Raman spectra—cont'd

| Peak position | Vibrational mode |
|---|---|
| $\nu_3$ $PO_4^{3-}$ | 1028, 1035,1048, 1073 |
| O—P—O stretching | 1087–1090 |
| O—P—O symmetric stretching, P=O symmetric stretching | 1091 |
| O—P—O antisymmetric stretching | 1238 |
| P=O symmetric stretching | 1080–1158 |
| P=O asymmetric stretching | 1276 |
| (C=O) stretching | 1654, 1667–1680, 1720, 1732, 1746 |
| $\nu$(C—O—C) | 520, 1150, 1155 |
| $CO_3^{2-}$ A type | 1103 |
| $\nu_4$ $CO_3^{2-}$ B type | 756, 1073 |
| $\nu$(C—C—O) | 779–782 |
| C—O—C ring | 883 |
| v(C—O—C) symmetric stretching | 905–918, 1118, 1172–1179, 1204–1208 |
| C—O stretching | 1264 |
| γ(C—O—H) of COOH | 740–747 |
| $\nu$(C—O) + v(C—C) + δ(C—O—H) | 1048 |
| C—O—H bending | 1118 |
| $\nu$(C—O—H) next to aromatic ring + σ(CH) | 1186 |
| C—C twisting | 621, 643 |
| C—C | 883 |
| C—C symmetric stretching | 997 |
| $\nu$(C—C) stretching | 813, 872, 918, 936, 1088, 1117, 1123, 1150 |
| $\nu(C—C)_{trans}$ (phospholipids) | 1056–1064 |
| $\nu$(C—C) (phospholipids) | 1097 |
| $\nu(C—C)_{trans}$ (phospholipids and proteins) | 1127 |
| —C=C— | 1520–1538 |
| C=C in-plane bending | 1603 |
| $\nu$(C—C) aromatic ring + σ(CH) | 1610 |
| $\nu$(C=C) stretching | 1616, 1627–1639, 1652, 1658 |
| δ(C—C—H) aromatic | 847–854 |
| CH=CH bending | 958 |
| $\nu_3$(C—$CH_3$ stretching) | 1004 |
| C—$C_6H_5$ | 1189, 1209 |
| =C—H in-plane bending | 1264 |
| δ(C—C—H) | 1216, 1287 |
| δ(=CH) phospholipids | 1293–1296 |
| C—H | 1031, 1170 |

*Continued*

**Table 3** Band assignments ($cm^{-1}$) of functional groups in Raman spectra—cont'd

| Peak position | Vibrational mode |
|---|---|
| C—H aromatic | 3067 |
| $CH_2$ wag, C—N stretching | 1243 |
| $CH_2$ deformation | 1299–1300 |
| $\nu(CH_2)$ symmetric stretching | 2280, 2845, 2854, 2875 |
| C—$H_2$ antisymmetric stretching | 2888 |
| $\nu(CH_2)$ asymmetric stretching | 2280, 2854, 2875, 2931–2942 |
| $\delta CH_2$ bending | 1320, 1327, 1386 |
| $CH_3$ bend | 1386 |
| $\nu(CH_3)$ symmetric stretching | 2931, 2940, 2971–2984 |
| $CH_3CH_2$ twisting | 1313 |
| $CH_3CH_2$ wagging | 1335–1345 |
| $\delta(CH_2)+\delta(CH_3)$ | 1437–1454, 1465, 1488 |
| $\nu$(HC=CH) stretching | 3010, 3060 |
| $\nu$(C—C—H) aromatic ring | 1585 |

the application and research processes of Raman spectroscopy from aspects of microorganism investigation [36,37,54,57–59], plant classification and diagnostic assessments [60–64], medical science [51,65–70], and food science [1,55,71].

# 2. Applications of Raman spectroscopy

## 2.1 Applications in microbiology

Raman has been successfully applied to investigate the microbial composition and characterization of single-cell organelles (Table 4) [72–78], microbial metabolic products (Table 5) [37,50,56,79], and the classification of microbial species [58,80,81]. It was also used to examine metabolic activities of microorganisms [82], mineralization on/in microbial structure [83], as well as life-detection/biosignature [84,85].

### *2.1.1 In studying viruses*

Several viruses, such as avian influenza virus, avipoxvirus, measles virus, adeno-associated virus, and tobacco mosaic virus (TMV), have been successfully identified by TERS, SERS, or portable Raman spectrometer [86–91]. Raman is able to provide a specific pathway to evaluate virus protein folding in the context of a supramolecular assembly, e.g., the representative S—H band at 2525–2595 $cm^{-1}$ in the Raman spectrum of *Salmonella*

**Table 4** Representative Raman peaks ascribed to microbial excretion.

| Excretion | Origin | Prominent Raman bands |
|---|---|---|
| ***Mycotoxins*** | | |
| Deoxynivalenol | *Fusarium graminearum* *Fusarium culmorum* | 1596, 1553, 1449, 1332, 1364, 1002, 881, 855, 663 |
| Aflatoxin | *Aspergillus flavus* *Aspergillus parasiticus* | 1615, 1550, 1498, 1345, 1272, 1203, 1059, 845 |
| Ochratoxin | *Penicillium verrucosum* various species of *Aspergillus* spp. | 1331, 1076 |
| Fumonisin | *Fusarium verticilloides* *Fusarium moniliforme* | 1366, 1264, 868, 480 |
| Zearalenone | *Fusarium* spp. | 1517, 1448, 880, 762 |
| Patulin | *Aspergillus* spp. | 1354, 1641, 1205 |
| ***Carotenoids*** | | |
| Bacterioruberin | *Halobacterium* sp. | 1505, 1152, 1000 |
| | *Halococcus dombrowskii* | 1507, 1152, 1002 |
| | *Halorubrum sodomense* | 1506, 1152, 1001 |
| | *Haloarcula vallismortis* | 1506, 1151, 1000 |
| | *Rubrobacter radiotolerans* | 1503, 1150, 1000 |
| Bacteriorhodopsin | Pure protein | 1526, 1199, 1168, 1150, 1007 |
| β-Carotene | *Dunaliella tertiolecta* | 1524, 1155, 1001 |
| | *Synechocystis* sp. | 1518, 1155, 1005 |
| | *Cyanobacteria* | 1523, 3053, 1158 |
| Decapreno-carotene | *Natromonas pharaonis* | 1503, 1152, 1000 |
| Dodecapreno-carotene | *Natromonas pharaonis* | 1503, 1447, 1389, 1316, 1285, 1210, 1147 |
| Astaxanthin | *Hematococcus pluvialis* | 1520, 1275, 1192, 1157, 1007 |
| Salinixanthin | *Salinibacter ruber* | 1512, 1155, 1003 |

*Continued*

**Table 4** Representative Raman peaks ascribed to microbial excretion—cont'd

| Excretion | Origin | Prominent Raman bands |
|---|---|---|
| Sarcinaxanthin | *Micrococcus luteus* | 1532, 1157, 1005 |
| Deinoxanthin | *Deinococcus radiodurans* | 1510, 1152, 1003 |
| Xanthomonadin | *Xanthomonas axonopodis* pv. dieffenbachiae | 1529–1531, 1135–1136, 1004 |
| Neurosporene | *Rhodobacter sphaeroides* | 1527, 1288, 1214, 1158, 1006 |
| Spheroidene | *Rhodobacter sphaeroides* | 1524, 1238, 1159, 1170, 1003, 951 |
| Flexirubin | *Flavobacterium johnsoniae* | 1529, 1154, 1133, 1004 |
| | *Flexibacter elegans* | 1515, 1155, 996 |
| ***Chlorophylls*** | | |
| Chlorophyll *a* | *Dunaliella tertiolecta* | 1554, 1437, 1325, 1289, 1233, 1186, 986, 915, 757 |
| Chlorophyll | Microbial crust | 1445, 1327, 915, 745, 508 |
| | *Cyanobacteria* | 3053, 1320 |
| Bacteriochlorophyll *a* | *Rhodobacter sphaeroides* | 1640, 1441, 1371, 1173, 1020, 900, 773, 733 |
| **Phycobiliprotein** | Microbial crust | 1631, 1583, 1370, 1283, 1235, 874, 815, 665 |
| **Scytonemin** | *Chlorogleopsis* | 1590, 1556, 1549, 1172 |
| | *Lyngbya aestuarii* | 1595, 1554, 1173 |
| | Microbial crust | 1596, 1558, 1321, 1172 |
| | *Nostoc* sp. | 1630 |
| | *Cyanobacteria* | 1590, 1549, 1172, 1323 |
| **Mycosporine-like amino acids** | Jarosite superficial crust | 1493, 1414, 1340, 1293, 1215, 1181, 1150, 920, 845, 485 |
| Pulvinic acid | Microbial crust | 1484,1414, 1350 |
| Rhizocarpic acid | *Acarospora chlorophana* | 1031, 1002, 945, 619, 597 |
| Calycin | *Candelariella* sp. | 1605, 1580, 1390, 1365, 1295, 1152, 990, 965, 482 |
| Atranorin | *Cladonia coniocraea* | 1674, 1629, 1605, 1453, 1406, 1379, 1311, 1281, 1115, 1033, 1000, 980, 956, 869, 826, 743, 704, 615, 503, 303 |
| | *Lepraria* spp. | 1674, 1629, 1605, 1406, 1379, 1311, 1281, 1033, 1000, 980, 956, 869, 826, 743, 704, 615, 503, 300 |

**Table 4** Representative Raman peaks ascribed to microbial excretion—cont'd

| Excretion | Origin | Prominent Raman bands |
|---|---|---|
| Gyrophoric acid | *Dirinaria aegialita* | 1663, 1627, 1615, 1453, 1437, 1385, 1302, 1287, 1067, 1046, 855, 782, 587, 567 |
| ***Organic acid*** | | |
| ν(CO) oxalate ion | *Cyanobacteria* | 1493, 1476, 1463, 1085, 901, 711, 527 |
| | *Aspergillus niger* | 1440, 1480, 1590 |
| ***Anthraquinone*** | | |
| Parietin | *Xanthoria elegans* | 1671, 1605, 1380, 1180, 970, 926 |
| | *Caloplaca crosbyae* | 1670, 1608, 1551, 1370, 1275, 1256, 1215, 1198, 1181, 978, 925, 725, 610, 570, 518, 465, 458, 398 |
| | *Cyanobacteria* | 1672, 1613 |
| **Porphyrin** | *Cyanobacteria* | 1453, 1352 |
| **Trehalose** | *Cyanobacteria* | 838 |
| **Erythritol** | *Cyanobacteria* | 699 |

Modified from Wu YL, Shao XQ, Jiao H, Song XW, He K, Li Z. Tracking the fungus-assisted biocorrosion of lead metal by Raman imaging and scanning electron microscopy technique. J Raman Spectrosc 51;2020:508–13. Wynn-Williams DD, Edwards HGM. Proximal analysis of regolith habitats and protective biomolecules in situ by laser Raman spectroscopy: overview of terrestrial Antarctic habitats and Mars analogs. Icarus 2000;144:486–503. Hassan MM, Zareef M, Xu Y, Li HH, Chen QS. SERS based sensor for mycotoxins detection: challenges and improvements. Food Chem 2021;344. Li JJ, Yan H, Tan XC, Lu ZC, Han HY. Cauliflower-inspired 3D SERS substrate for multiple mycotoxins detection. Anal Chem 2019;91:3885–92. Yuan J, Sun CW, Guo XY, Yang TX, Wang H, Fu SY, Li CC, Yang HF. A rapid Raman detection of deoxynivalenol in agricultural products. Food Chem 2017;221:797–802. Jehlicka J, Edwards HGM, Orenc A. Raman spectroscopy of microbial pigments. Appl Environ Microbiol 2014;80:3286–95.

bacteriophage [92]. Different strains of respiratory syncytial virus (RSV) can be distinguished by Raman because of the unique nucleic acid and protein composition with envelope peaks located within 650–1100 $cm^{-1}$ region (at 663, 877, and 1045–1055 $cm^{-1}$) [93]. In addition, Raman has been preliminarily applied to study the infection of Vero cells by herpes simplex virus, varicella-zoster virus, sarcoma virus, HIV-1, and hepatitis B virus (HBV) [94–97]. It was proposed that the Raman peaks at 509, 957, 1002, 1153, 1260, 1512, 1648, and 2305 $cm^{-1}$ can be applied to differentiate patients with/without HBV [97].

**Table 5** Representative Raman peaks of microbial cells.

| Assigned to | Band ($cm^{-1}$) |
|---|---|
| Nucleic acids (A) | 1580, 1340, 1322, 1305, 1268, 734, 725, 713, 647 |
| Nucleic acids (G) | 1580, 1340, 1305, 647 |
| Nucleic acids (T) | 1127, 784, 671 |
| Nucleic acids (C) | 1127, 793, 784 |
| Nucleic acids (U) | 793 |
| DNA (O—P—$O^-$) | 1095, 1085, 1004, 813 |
| CDNA | 936, 899, 859 |
| Lipids | 1450, 1308, 1239 |
| Tryptophan | 1610 |
| Phenylalanine | 1609, 1095, 1032, 1004, 622, 609 |
| Tyrosine | 1607, 1590, 867, 857, 813, 643 |
| Proteins (S—S stretching) | 543 |
| Carbohydrates (COC stretching) | 895 |
| Carbohydrates (C—COO— stretching) | 928 |
| Carbohydrates (glycosidic link) | 1100 |
| Carbohydratescarbohydrates (C—N and C—C stretching) | 1130 |
| Amide I | 1660, 1630 |
| Amide III | 1450, 1305, 1268, 1257 |

Modified from Xie C, Li YQ. Confocal micro-Raman spectroscopy of single biological cells using optical trapping and shifted excitation difference techniques. J Appl Phys 2003;93:2982–86. Xie CG, Li YQ, Tang W, Newton RJ. Study of dynamical process of heat denaturation in optically trapped single microorganisms by near-infrared Raman spectroscopy. J Appl Phys 2003;94:6138–42. Wood BR, Tait B, McNaughton D. Micro-Raman characterisation of the R to T state transition of haemoglobin within a single living erythrocyte. Bba-Mol Cell Res 2001;1539:58–70. Fan C, Hu Z, Mustapha A, Lin M. Rapid detection of food- and waterborne bacteria using surface-enhanced Raman spectroscopy coupled with silver nanosubstrates. Appl Microbiol Biotechnol 2011;92:1053–61. Schuster KC, Urlaub E, Gapes JR. Single-cell analysis of bacteria by Raman microscopy: spectral information on the chemical composition of cells and on the heterogeneity in a culture. J Microbiol Methods 2000;42:29–38.

Virus infection to human cells was successfully detected by Raman spectroscopy after 12h [98,99]. The spectral changes observed after virus infection were probably due to the defense response of the cells. The spectral changes after long virus incubation were also identified, which were due to a combination of virus proliferation and cell defense responses. In particular, $PO_4^{3-}$ (located at ~950 $cm^{-1}$) was proposed as an excellent marker for the proliferation of the virus [98]. Moreover, Raman spectroscopy has the potential to differentiate conditions of human disease. For instance, characteristic Raman bands at 756, 1218, 1672, and 1686 $cm^{-1}$ were observed in the dengue virus-infected sera samples, which can be used to identify typhoid and dengue infection [100]. Therefore, Raman analysis might offer an opportunity to detect unknown human infectious viruses in the absence of human patients.

### 2.1.2 In studying excretion

Raman has been widely applied to detect microbial secretion to monitor their growth conditions, stages, as well as microbiological identification. For example, in recent years, SERS has been used as an efficient tool to detect mycotoxins in cereals or fungi, such as deoxynivalenol, aflatoxin, ochratoxin, fumonisin, zearalenone, and patulin (Table 5) [101–104]. FT-Raman spectroscopy was first used to rapidly screen deoxynivalenol-contaminated wheat and barley meal [105]. A dominated Raman peak at 1449 $cm^{-1}$ (ascribed to the methyl group) was selected to quantitatively analyze deoxynivalenol [106]. However, it is still difficult to accurately identify the peaks of mycotoxins in different plants, as the peaks could be interfered with multiple functional chemical groups in various plant tissues.

Raman is becoming a popular analytical tool for both qualitative and quantitative assessments. Such applications successfully investigated microbial pigments, including carotenoids, chlorophylls, phycobiliprotein, scytonemin, etc. (Table 4) [56,107–109]. The first Raman spectrum of carotenoids was recorded as early as the 1970s [110,111]. Raman spectroscopy confirmed β-carotene accumulations originating from *Dunaliella* in halite inclusions from shallow deposits in Lake Magic (western Australia) [112]. The Raman spectrum of β-carotene is usually characterized by intense peaks at ~1515 and 1157 $cm^{-1}$ [37,50,113,114]. Identification of carotenoids by Raman spectroscopy was then used to confirm the pigment of *Mycoplasma pneumoniae* and successfully distinguish different *M. pneumoniae* strains [115]. Raman successfully demonstrated that yeast could secrete more carotenoids (1526, 1163, and 1009 $cm^{-1}$) when exposed to a high concentration of $Pb^{2+}$ in the liquid medium [37]. In addition to carotenoids, the primary band (1590 $cm^{-1}$) and a set of accompanying bands (1549, 1323, and 1172 $cm^{-1}$) were considered fingerprint peaks for scytonemin (produced by cyanobacteria). Moreover, cyanobacteria also produce other protective pigments based on Raman results [116].

Other microbial excretion (including mycosporine-like amino acid, organic acid, anthraquinone, porphyrin, trehalose, and erythritol) can also be detected by Raman (Table 4). For example, the organic acids secreted by fungi can react with $Pb^{2+}$ to form Pb oxalate (several representative peaks at 1447, 1483, 1593 $cm^{-1}$) [38]. RISE can subsequently be applied to investigate biological mineralization. In addition, Raman spectroscopy can also be used as an astrobiological tool in exploration of Mars [117].

### 2.1.3 In studying bacteria

The components of bacterial cells have been widely studied by SERS, which showed the characteristic spectra of organic acids, carbohydrates, lipids, proteins, and other complex molecular assemblages (Table 5) [74–78]. A deep ultraviolet Raman study on *E. coli* cells showed the vibrational modes of the nucleobases and aromatic amino acids [85]. Specifically, the dominant modes of nucleobases uracil, adenine, guanine, cytosine, and thymine were 1210, 1291, 1440, 1512, and 1647 $cm^{-1}$, respectively. The dominant modes of tyrosine, tryptophan, and phenylalanine were 1591, 1610, and 1595 $cm^{-1}$, respectively.

Gram-negative (G−) and positive (G+) bacterial strains are usually differentiated by their membrane-wall structure. SERS has shown that the spectra of both the negative and positive strains had evident peaks at 1330 and 732 $cm^{-1}$. However, the protein content in the cell wall of the G-strain was higher than that of G+, indicated by the intensity of the 1330 $cm^{-1}$ peak [118]. The dead bacteria can be identified from live ones based on their variational composition of cell walls via SERS [80,119]. In addition, SERS was also used to characterize the complex biomatrix in biofilms [120–122]. A Raman-based microfluidic platform was used to characterize *P. aeruginosa* biofilms at different stages of growth [123]. *P. aeruginosa* biofilms showed Raman peaks for carbohydrates (918, 1123, and 1333 $cm^{-1}$), proteins (746, 918, 1123, 1167, 1223, 1307, 1333, and 1580 $cm^{-1}$), lipids (968, 1123, and 1307 $cm^{-1}$), and nucleic acids (746, 1357, and 1580 $cm^{-1}$). The proteins and carbohydrates accounted for the largest proportion of synthesized substances at the late stage. Single-cell Raman microspectroscopy demonstrated the antibacterial effects of iron oxides in bacterial cells. Goethite particles at the micrometer scale induced severe damage to cell membranes, including the decrease of glycogen (1047 $cm^{-1}$), protein (1448 $cm^{-1}$), and lipopolysaccharide or outer membrane proteins (1662 $cm^{-1}$). In contrast, the toxic effects of nano-sized goethite were mediated by both reactive oxygen species-dependent RNA damage and cell membrane destruction [124]. Furthermore, high-temperature heating decreased the intensity of Raman peaks at 783 and 812 $cm^{-1}$ in single cells, which are assigned to nucleic acids [74].

Identification of live or dead bacteria by employing Raman spectroscopy has also been reported. SERS has been applied to detect the living or dead

*E. coli* in drinking water via the coating of silver nanoparticles on cell walls (Fig. 1) [25,125]. The typical Raman peaks of the cell wall of *E. coli* were located at 624, 652, 735, 955, 1330, and 1456 $cm^{-1}$. Near-IR-Raman was performed to identify 15 causative bacterial pathogens in the bloodstream [126]. In addition, the confocal Raman microscope and NIR-SERS were used to discriminate *Bacillus megaterium*, *Bacillus cereus*, and *Bacillus thuringiensis* spores suspended in deionized water [127–129]. Raman is an appropriate approach to characterize differentiated populations. Moreover, heterogeneities in the microbial population can also be studied via Raman investigation [130].

Recently, Raman techniques have also been used for the safety control of plants in precision agriculture. It can identify plant pathogenic bacteria. *Candidatus Liberibacter* solanacearum infection on tomato leaves was successfully diagnosed by Raman analysis. The combination of Raman spectroscopy and chemometric analysis allowed the accuracy of ~80% while diagnosing *Liberibacter* disease [131]. FT-Raman spectroscopy was used to nondestructively probe whole apple surfaces infected by various microorganisms [132]. The pathogenic and nonpathogenic *E. coli* on apple fruit surfaces could be well differentiated based on the fingerprint region between 1800 and 600 $cm^{-1}$. In addition, SERS was also used to analyze intact bacteria that simulate biological warfare agents, such as *Bacillus globigii*, *B. thuringiensis*, and *Erwinia herbicola* [53].

Raman can be applied to quantify bacterial abundance. For instance, the limit of detection (LOD) values of *E. coli* and *Staphylococcus aureus* detected by SERS were as low as $10^3$ cells/mL [133–135]. Furthermore, *E. coli* in water with a LOD of 8 CFU/mL was also identified via a specific label-based SERS method [136]. Additionally, the LTRS technique combining the advantages of NIR Raman spectroscopy and optical tweezers provides a feasible approach for confirming *B. cereus* spores in a mixed sample containing polystyrene and silica beads [128].

### 2.1.4 In studying fungi

To acquire a complete biomolecule composition of yeast cells, space-resolved Raman spectra were performed to investigate the lipid droplet, cytoplasm, and cell wall [58]. Vegetative cells and ascus walls consisted of both α-glucans (with representative peaks at 550 and 944 $cm^{-1}$) and β-glucans (with a representative peak at 893 $cm^{-1}$). However, the wall of spores was exclusively made of α-glucan, based on its Raman spectrum shown in Fig. 2. In addition, the nucleus, mitochondrion, and septum of

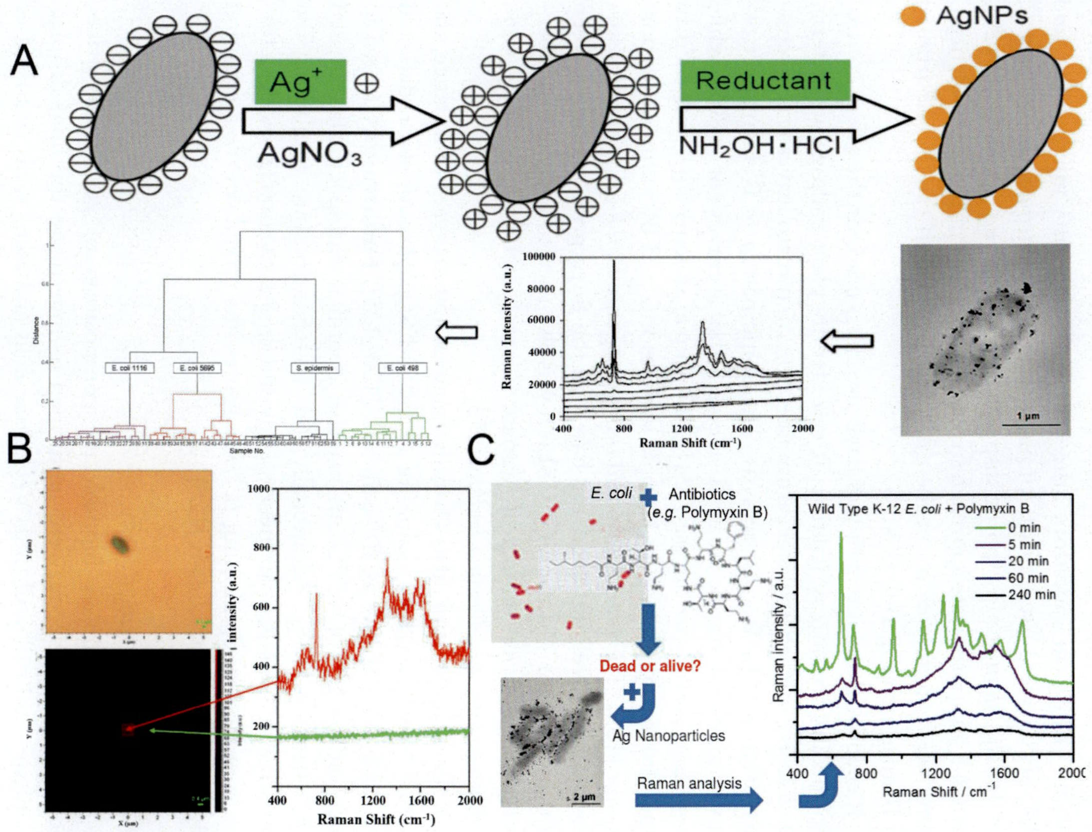


**Fig. 1** (A) Schematic of SERS detection of bacteria in water by in situ coating with Ag NPs. (B) SERS spectrum of a single bacterium. (C) Discrimination of living and dead by SERS. *(Figure reproduced with permission from Liu Y, Zhou H, Hu Z, Yu G, Yang D, Zhao J. Label and label-free based surface-enhanced Raman scattering for pathogen bacteria detection: a review. Biosens Bioelectron 2017;94:131.)*

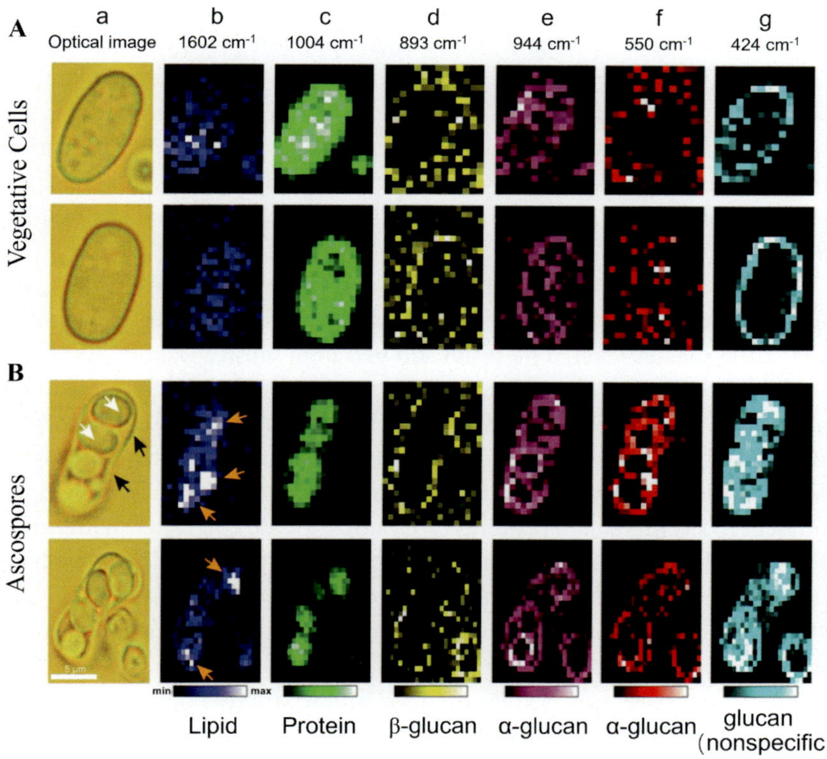


**Fig. 2** Univariate Raman imaging of (A) *Schizosaccharomyces pombe* vegetative cells and (B) spores. (a) Bright-field optical images. Scale bar: 5 μm. White arrows indicate spores; while black arrows indicate ascus walls. (b–g) Raman images constructed for (b) lipids using 1602 cm$^{-1}$ [gold arrows indicate lipid accumulation], (c) proteins using 1004 cm$^{-1}$, (d) β-glucans using 893 cm$^{-1}$, (e) α-glucans using 944 cm$^{-1}$, (f) using 550 cm$^{-1}$, and (g) both glucan anomers using skeletal vibrations at 424 cm$^{-1}$. *(Figure reproduced with permission from Noothalapati H, Sasaki T, Kaino T, Kawamukai M, Ando M, Hamaguchi H, Yamamoto T. Label-free chemical imaging of fungal spore walls by Raman microscopy and multivariate curve resolution analysis. Sci Rep 2016;6.)*

yeast cells, as well as the time-dependent changes of major molecular components during the cell cycle, were identified and elucidated by Raman. The peak at 1602 cm$^{-1}$, together with the 1446 cm$^{-1}$ peak of phospholipids, showed the metabolic activity of mitochondrion in live cells [137,138]. The spatial distribution of various components in yeast cells was also determined by Raman mapping [139]. The C—H stretch (usually between 2850 and 2950 cm$^{-1}$) was a fingerprint peak for most organic matters. The C=O stretch vibration (1731–1765 cm$^{-1}$) and the amide I and the C=C band

(1624–1687 $cm^{-1}$) were assigned to the lipid fraction. The single yeast cell was measured by LTRS, and the Raman spectrum of the yeast cell was determined [74,75]. Furthermore, It was also used for real-time assessment of oxidative injury to single yeast cells via spectral measurement of the C=C bond in lipids [140]. Raman was also successfully performed in mushrooms, which confirmed proteins and polysaccharides as major components in fruiting bodies of *Agaricus bisporus*, based on a series of peaks within the range of 3000–2900, 1258, 1127, 1032, 996, and 885 $cm^{-1}$ [141].

Raman, equipped with the high-dimensional chemometric facility, was used to identify yeasts [59,142]. Three major wine spoilage yeasts were classified at the species level. The result showed high sensitivity, i.e., 98.6% for *S. cerevisiae*, 93.8% for *Z. bailii*, and 92.3% for *B. bruxellensis* [59]. In addition, Raman was also used to differentiate fungal infections of neutrophils [126,143]. The authors established a Raman database of 284 neutrophils infected with different pathogens. The carotenoids (1159 $cm^{-1}$), myeloperoxidases (1519 $cm^{-1}$), and proteins (1657 $cm^{-1}$) were correlated with model infection [143]. Furthermore, the peaks at 751 and 1549 $cm^{-1}$ (tryptophan), 1162 and 1519 $cm^{-1}$ (carotene), 1162 $cm^{-1}$ (tyrosine), and 2761 and 2815 $cm^{-1}$ (lipids) dominated the differentiation between fungal and bacterial infection [143].

Raman has been successfully used for nondestructive detection of plant pathogens directly on the grains of crops [60,144]. The appearance of pathogens in grains, such as *fusarium, Aspergillus niger,* and ergot, can induce the hydrolysis of starch (mainly at 1153, 938, and 864 $cm^{-1}$) and a decrease of lignin (mainly at 1600 and 1633 $cm^{-1}$) in grains. Moreover, the 1547 $cm^{-1}$ peak, which is assigned to carotenoids, can be used to estimate the grain kernel healthiness [144].

## 2.2 Applications in plants

### 2.2.1 In studying pollens

Plant pollen grains consist of complex molecules owing to their biological function. The first Raman spectroscopic work on whole pollen was conducted in the 1990s [145]. Raman was then used to identify their organic molecules, such as sporopollenin (1630, 1605, 1585, 1205, 1170, 855, and 830 $cm^{-1}$), carotenoid pigments (1523, 1155, and 1005 $cm^{-1}$), flavonoid, quinone, and proteins (mainly 1655, 1455, and 1260 $cm^{-1}$), lipids (1745, 1440, 1300, and 1070 $cm^{-1}$), and carbohydrates (around 1450–1300 and 1150–900 $cm^{-1}$) (Table 6) [146–151]. Specifically, FT-Raman spectroscopy is more advantageous in studying the chemical

**Table 6** Representative Raman peaks of plants.

| Assignment | Band ($cm^{-1}$) |
|---|---|
| Carbohydrates | 442, 477, 573, 706–715, 758, 771, 858–864, 900–1150, 1300–1450 |
| Cellulose | 378, 520, 915, 1048, 1118, 1155, 1326, 1375 |
| Proteins | 527, 1007, 1242, 1455, 1655, 1660, 1669 |
| Pectin | 480, 747 |
| Lignin | 646, 915, 1048, 1186, 1264, 1288, 1327, 1601, 1630 |
| Chlorophyll | 744, 915, 985, 1325 |
| Carotenoids | 968, 1000, 1156, 1189, 1279, 1526, 1557 |
| Sporopollenins | 1630, 1605, 1585, 1205, 1170, 855, 830 |
| Phenylalanine | 1003 |
| Cuticular waxes | 1047, 1065, 1081, 1112, 1169, 1371, 1441, 1721, 2853, 2904, 2921 |
| Nitrate | 1046 |
| Piperine | 1153, 1295, 1256 |
| Xylan | 1184, 1218–1226 |
| Lipids | 1745, 1659, 1440, 1300, 1070 |
| Phenolic | 1247, 1605 |

Modified from Schulz H, Baranska M, Baranski R. Potential of NIR-FT-Raman spectroscopy in natural carotenoid analysis. Biopolymers 2005;77:212–21. Schulz H, Baranska M, Quilitzsch R, Schütze W, Losing G. Characterization of peppercorn, pepper oil, and pepper oleoresin by vibrational spectroscopy methods. J Agric Food Chem 2005;53:3358–63. Schulte F, Lingott J, Panne U, Kneipp J. Chemical characterization and classification of pollen. Anal Chem 2008;80:9551–56. Silva C, Vandenabeele P, Edwards H, Oliveira L. NIR-FT-Raman spectroscopic analytical characterization of the fruits, seeds, and phytotherapeutic oils from rosehips. Anal Bioanal Chem 2008;392:1489–96. Bagcioglu M, Zimmermann B, Kohler A. A multiscale vibrational spectroscopic approach for identification and biochemical characterization of pollen. Plos One 2015;10. Trebolazabala J, Maguregui M, Morillas H, de Diego A, Madariaga JM. Portable Raman spectroscopy for an in-situ monitoring the ripening of tomato (*Solanum lycopersicum*) fruits. Spectrochim Acta A 2017;180:138–43. Egging V, Nguyen J, Kurouski D. Detection and identification of fungal infections in intact wheat and sorghum grain using a hand-held Raman spectrometer. Anal Chem 2018;90:8616–21. Farber C, Kurouski D. Detection and identification of plant pathogens on maize kernels with a hand-held Raman spectrometer. Anal Chem 2018;90:3009–12. Farber C, Shires M, Ong K, Byrne D, Kurouski D. Raman spectroscopy as an early detection tool for rose rosette infection. Planta 2019;250:1247–54. Sanchez L, Pant S, Xing Z, Mandadi K, Kurouski D. Rapid and noninvasive diagnostics of Huanglongbing and nutrient deficits on citrus trees with a handheld Raman spectrometer. Anal Bioanal Chem 2019;411:3125–33. Gupta S, Huang CH, Singh GP, Park BS, Chua NH, Ram RJ. Portable Raman leaf-clip sensor for rapid detection of plant stress. Sci Rep 2020;10 (2020). Kendel A, Zimmermann B. Chemical analysis of pollen by FT-Raman and FTIR spectroscopies. Front Plant Sci 2020;11.

composition of pollen wall components (e.g., sporopollenins and pigments). Furthermore, an in situ chemical characterization of rye pollen grains can be achieved (Fig. 3). The Raman 2D mapping showed the abundance and distribution of aromatic amino acids along with other typical molecules in pollens [151].

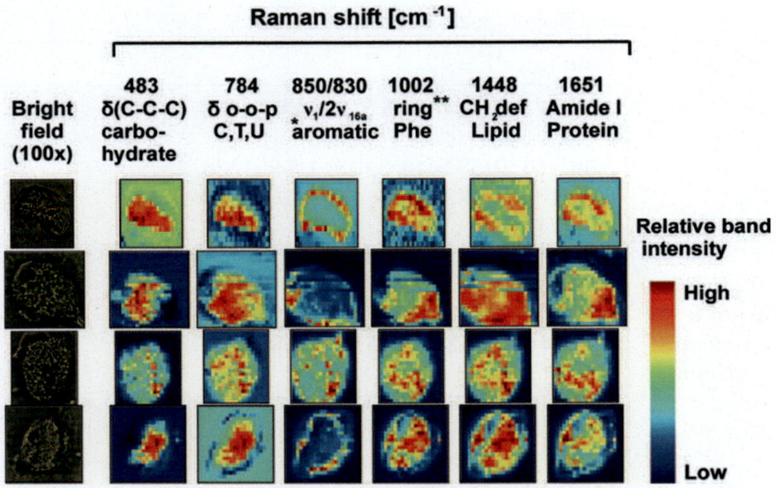


**Fig. 3** Chemical maps of selected molecular groups in sections of rye pollen grains (four examples). For comparison, a photomicrograph (bright field, 100 ×) is displayed (leftmost column). Scale bar: 20 μm. (Abbreviations: $\nu$, stretching; δ, deformation; o-o-p, out-of-plane; Phe, phenylalanine; A, adenine; T, thymine; C, cytosine). *The Fermi doublet at 850/830 cm$^{-1}$ is characteristic of para-disubstituted benzene rings. **The band at 1002 cm$^{-1}$ is due to a breathing vibration of the Phe ring. *(Figure reproduced with permission from Schulte F, Lingott J, Panne U, Kneipp J. Chemical characterization and classification of pollen. Anal Chem 2008;80:9551–6.)*

The chemical composition of pollen varies greatly among families, genera, and even congeneric species. For instance, carotenoid peaks are usually dominant in the spectra of large-leaved linden, sallow, and horse-chestnut pollen [152]. The carotenoid concentration of *Picea* pollen was higher than the rest of the pine family. The sporopollenins (1637, 1610, 1590, 1209, and 1173 cm$^{-1}$) in the pollen grain wall contributed to the differences among the *Pinaceae* family [149].

### 2.2.2 In studying leaves, fruits, and seeds

Raman spectroscopic techniques have been applied to guarantee the economic value of fruits and vegetables in modern agriculture [1,62,63,153–155]. The spectra of *Rosa aff. Rubiginosa* flesh and seeds clarified their critical chemical components, e.g., carotene (1520, 1157, and 1007 cm$^{-1}$) and, most specifically, β-carotene. The fruit seeds usually have characteristic Raman peaks assigned to fatty acids and esters (1745 and 1440 cm$^{-1}$). In contrast, the outer part of the seeds has more complex organic matters, e.g., fatty acids, polysaccharides (1375, 1333, 1152, 1124, 929, and 378 cm$^{-1}$), and polyphenols [155]. This information was used to induce a set of key biomarkers of the main constituents of rosehips. Raman was used to measure the varieties of carotenoids (lycopene, β-carotene, and lutein) and changes in carotenoid content in plants [156–158], as well as in microorganisms. The carotenoids increased from an unripe to a ripe stage of tomato, while the chlorophyll (744, 915, 985, and 1325 $cm^{-1}$) and cuticular waxes (1047, 1065, 1081, 1112, 1169, 1371, 1441, 1721, 2853, 2904, and 2921 $cm^{-1}$) decreased. Moreover, the weakened intensity of the 1553 $cm^{-1}$ peak indicated the decrease of phytofluene during tomato growth, which is a transition compound in the carotenoid biosynthetic pathway [159]. The contents of piperine (1153 for —C—C— stretching, 1256 and 1295 for $CH_2$ vibration) and essential oil in various pepper were analyzed by NIRFT-Raman spectrometry. The piperine is mainly distributed in the whole perisperm of the green fruits and beneath the skin of the black peppercorn [160].

Raman spectroscopy was also used to estimate the fat content in cacao seeds [161,162]. The Raman peaks, at 1659 and 1744 $cm^{-1}$, represent the fatty materials and oleic acid [161]. These two representative peaks were observed as the key indicator of the previously mentioned two chemical species. The spectra of different parts of olive (skin, flesh, and stone) have been evaluated. The intensity of carotenoid peak at 1525 $cm^{-1}$ increased, while the intensity of peaks for phenolic compounds (1605 $cm^{-1}$) decreased during the ripening of olives [163].

Raman was applied to study the conformation of rice globulin under various buffer environments [164]. Globulin is the second major storage protein fraction in rice. It was proposed that hydrophobic interactions and disulfide bonds contributed substantially to stabilizing the conformation and thermal aggregation of rice globulin. The protein in buckwheat showed unfolding and denaturation when suffering extreme pH and high temperature [165]. Raman spectroscopy was also applied to the classification of pure corn and cassava starch samples, together with the quantification of their amylose contents [166]. Furthermore, it is doable to discriminate between transgenic and normal crops during breeding, such as transgenic tobacco and corn, by Raman spectroscopy [167,168].

Raman spectroscopy is sensitive to chemical changes in the healthy and infected tissues in plants. The rose leaf infected by rosette disease can be detected by Raman spectroscopy, based on the elevated peak for lignin (1610 $cm^{-1}$) and decreased peak for aliphatic (1441 $cm^{-1}$) [63]. In addition, the structural characteristics of "Hayward" kiwifruits were also reported [169]. For the pulp spectra of the fruit with elephantiasis, the enhanced Raman peaks at 1530 and 1661 $cm^{-1}$ can be applied as biomarkers of the diseased fruits. Furthermore, a portable Raman probe was used for the early diagnosis of nitrogen deficiency in plants [62]. The nitrate (1045 $cm^{-1}$) was determined as a specific signature of nitrogen status in plant leaves. Apart from that, other metabolites peaks, such as lignin (1604 $cm^{-1}$), carotenoid (1520, 1150, and 1000 $cm^{-1}$), cellulose (1320 $cm^{-1}$), phenylalanine (1003 $cm^{-1}$), and pectin (747 $cm^{-1}$), were also observed (Table 6) [62,63]. The concentrations of anthocyanins (539, 623, and 733 $cm^{-1}$) and photosynthetic carotenoids (1007 and 1157 $cm^{-1}$) in coleus plants were presented with a negative correlation across some abiotic stresses, including salinity, drought, cold, and excess light [61]. Raman spectrometer equipped with 831 nm excitation was used to investigate healthy and Huanglongbing-infected oranges and other grapefruit trees. The technique was also proved efficient in studying trees suffering from nutrient deficiency [153]. A significant increase of lignin (1601–1630 $cm^{-1}$) was detected in the spectra of citrus leaves. Such phenomenon was common for the citrus leaves undergoing biotic and abiotic (nutrient deficiency) stresses. Raman hence has great potential to be utilized in monitoring plant diseases in real time.

## 2.3 Applications in animal science

### *2.3.1 In studying bones*

Bones and teeth are composed of hydroxylapatite (carbonated apatite), water, and collagen. Among the mineral-collagen-water ternary, the mineral content in vertebrate bones varies between 45 and 97 wt% [170,171]. Table 7 shows the spectral bands and the corresponding assignments of bones [65]. The dominant ~960 $cm^{-1}$ peak ($PO_4^{3-}$) is the primary indicator of the appearance of hydroxylapatite [172]. In addition, the $\nu_1$ (960 $cm^{-1}$), $\nu_2$ (430 and 450 $cm^{-1}$), $\nu_3$ (1035, 1048, and 1073 $cm^{-1}$), and $\nu_4$ (587 and 604 $cm^{-1}$) modes of $PO_4^{3-}$ were correlated with characteristic vibrations of P—O vibrations in carbonated apatite [65]. Raman spectroscopy has been used to analyze fluorination reactions that commonly occur in teeth [173]. Once bioapatite particles are partially dissolved in bones and teeth, the mineral can release Ca and P to the F-bearing solution, causing the precipitation of the end product, i.e., fluorapatite. Spectral modeling based on Raman can help to indicate such fluoridation via a kinetically controlled dissolution-reprecipitation process.

Raman spectra of biomineralization structures in bones and teeth can elucidate the pathological tissues and the development of newly mineralized structures. Raman investigation defines the changes in the chemical structure of bone that accompany bone diseases, and characterizes interactions between biocompatible implants and tissues [170,174–177]. The analysis

**Table 7** Band assignments ($cm^{-1}$) in Raman spectra of bone.

| Band attribution | Bone (blood) | Bone |
|---|---|---|
| $\nu_2$ $PO_4^{3-}$ | 430 | 430 |
| $\nu_2$ $PO_4^{3-}$ | 450 | 450 |
| $\nu_4$ $PO_4^{3-}$ | 587 | 587 |
| $\nu_4$ $PO_4^{3-}$ | 604 | 604 |
| $\nu_4$ $CO_3^{2-}$ B type | – | 756 |
| $\nu_1$ $PO_4^{3-}$ | 960 | 960 |
| $\nu_3$ $HPO_4^{2-}$ | 1003 | 1003 |
| $\nu_3$ $PO_4^{3-}$ | 1035 | 1035 |
| $\nu_3$ $PO_4^{3-}$ | 1048 | – |
| $\nu_3$ $PO_4^{3-}$ | 1073 | 1073 |
| $CO_3^{2-}$ A type | 1103 | – |

Bone: bone tissue located at a distance from blood vessels.
Bone (blood): bone tissue located close to (few microns) blood vessels.
Modified from Penel G, Delfosse C, Descamps M, Leroy G. Composition of bone and apatitic biomaterials as revealed by intravital Raman microspectroscopy. Bone 2005;36:893–901.

involving osteoradionecrosis of the mandible demonstrated that bone radionecrosis is associated with changes in bone minerals and organic matters [178]. Irradiation can change the inorganic bone matrix structure and decrease their contents in cells, and then damage the organic bone matrix (mainly type I collagen (1200–1350 $cm^{-1}$)). In poultry science, hens suffer from osteoporosis during their laying period, making their bone an excellent material for investigating bone aging. Raman analysis showed an evident mineralogical change in the femur of laying hen from 11 to 49 weeks. The contents of organic matter (2940 $cm^{-1}$) decreased, and the degree of bone mineralization relatively increased during this period [179]. The intake of low dose Pb by hens can cause a decline in their bone quality, e.g., mineral density and strength. However, the hens might produce more carotenoids (at 1156 and 1515 $cm^{-1}$) in the humerus to resist heavy metal stress [180].

The active research in Raman field is the interaction between bone and coatings (usually calcium phosphate based materials) regarding bone implants. Intravital microscopy has been applied as an appropriate tool to investigate the bone-implant interface. The application is now extended to hemoglobin and tumor tissues [181–183], developing the potential application of Raman spectroscopy to observe cells, bones, or tissues in situ. Intravital Raman microspectroscopy was first applied to study the composition and structure of periosteum [65]. The implanted apatite-based biomaterials remained stable in organisms over time and showed biocompatibility, which was confirmed by the migration of blood vessels into the biomaterials. Meanwhile, the hemoglobin (at 754 $cm^{-1}$) was also observed in the biomaterial regions adjacent to blood vessels [65]. Primary cells and osteoblast-like cells share many similarities regarding their phenotype and biochemical composition. McManus et al. (2012) demonstrated a comprehensive comparison between human primary osteoblasts and osteoblast-like cells to evaluate their osteoinductive potential [184]. They determined the suitability of the osteoblast-like cell line as an alternative to primary osteoblast cells. However, there are also many differences, including the decrease of nucleic acid (at 788, 826, and 857 $cm^{-1}$) concentrations in osteoblast-like cells and the increase of protein (1246–1269 and 1595–1720 $cm^{-1}$) concentrations in primary cells [185].

### 2.3.2 In studying animal cells and tissues

Most cells/tissues comprise nucleic acids, proteins, polysaccharides, lipids, and many other essential components. All diseases, either in humans or other animals, are caused by fundamental changes in cellular and/or tissue biochemistry. The associated delicate changes could be successfully resolved by Raman. The Raman spectroscopic analysis of individual living cells is a powerful tool for investigating complex biological systems. A confocal Raman microscope has been designed to study single human cells and polytene chromosome [186]. Afterward, many biological phenomena in single human cells were studied by Raman, e.g., the cell nucleus and cytoplasm in white blood cells [187,188]. In addition, Raman was used for monitoring the oxygenation process of human erythrocytes [189–191]. Iron, without exception, has been addressed. For example, the enhanced peak at 567 $cm^{-1}$ (attributed to the Fe—$O_2$ stretching mode in the spectra of oxygenated cells) was identified, together with a peak appearing at 419 $cm^{-1}$ that was assigned to the Fe—O—O bending mode [190]. Five peaks at 1396, 1365, 1248, 972, and 662 $cm^{-1}$ were classified as fingerprint peaks for heme aggregation in erythrocytes in oxygenated and deoxygenated states [189]. Furthermore, blood cells (667 and 751 $cm^{-1}$ were assigned to haem) in an optical trap without staining can also be identified by LTRS [192–194]. The reduction in the intensity of several Raman peaks associated with hemoglobin oxygenation and iron atoms indicated the hypoxic capacity of thalassemia red blood cells [191]. In addition, Raman was considered as an excellent tool for monitoring the osteoblast maturity during the differentiation process, especially by the symmetric stretching of phosphate groups [195]. Besides, Raman has been used to identify biochemical changes during cell proliferation. For instance, there was an increase in proteins and nucleic acids, whereas a decrease in lipids and glycogen fractions within tumorigenic cells [70,196].

Raman spectroscopy has great potential in evaluating the quality of meat and fish. The 780 nm laser source is proposed to be suitable for investigating the changes in meat and fish structure due to low fluorescence [197]. The first attempt of Raman spectroscopy to predict free fatty acid composition of unextracted adipose tissue was in 2006 [198]. The effects of aging and cooking of porcine longissimus dorsi were also investigated [199]. Raman spectroscopy has been widely used to evaluate proteins and lipids in fish [200–204]. It was proposed that the temperature, time of freeze-drying, formaldehyde, and plasma powder addition can all influence the structure and composition of fish lipids.

The sequencing of DNA or proteins is procedurally complex. The combination of vibrational spectroscopy and near-field optical techniques gives high contrast in Raman spectroscopy. They allow the direct identification of RNA strands. The TERS was able to map along with a poly(cytosine) and eventually identify and sequence the composition of polymeric biomacromolecules (e.g., DNA, RNA, and peptides) [30,34]. In addition, Raman can be applied to provide information on the molecular composition, secondary structure, and interactions (e.g., chemical microenvironment of molecular subgroups) of DNA-protein complexes [186]. Vo-Dinh et al. (1994) first applied SERS gene probe technology in DNA detection [205]. Later, they also used SERS-active labels as primers to amplify polymerase chain reaction (PCR) specific target DNA sequences [206]. Many studies have proved that SERS is a promising technique for detecting mutated DNA [207]. Using this technique, it is possible to detect even a single molecule of labeled DNA [208], outclassing practically all other analytical methods employed for DNA detection. Based on Raman, it was confirmed that metal addition could induce the structural changes of calf thymus DNA [209]. For example, spectroscopic data showed that $Mn^{2+}$ cations binded to the charged phosphate groups of DNA could stabilize the double-helical structure [210].

Raman has been applied to investigate different types of organs and tissues, including brain, cornea, oral tissue, lung, liver, gastrointestinal tissue, breast, cervical tissue, epithelial tissue, and skin [67,211–213]. The detailed interpretations of their spectra had been introduced [67]. Raman can effectively diagnose human diseases based on their application in cells, tissues, and organs. Raman diagnostics could be particularly helpful for detecting and identifying cancers, e.g., melanoma [214], breast cancer [12,215,216], squamous cell carcinoma [217], uterine cervix cancer [218], basal cell carcinoma

[219], stomach cancer [220], liver cancer [221], colon cancer [222], and skin cancer [223]. The spectra of different tumor tissue showed definite differences. For example, Raman analysis showed significant metabolic differences between healthy tissues and brain tumors, including the intensified Raman signals of nucleic acids (751 $cm^{-1}$) and declined Raman signals of lipids (1437–1444 $cm^{-1}$) in tumor brain samples [70]. Raman spectroscopy was applied to identify different stages of kidney tumors [224]. Additionally, the serum samples from patients with nonsmall cell lung cancer at stages I, II, III, or IV can be effectively discriminated by Raman spectroscopy [225]. For benign colon and prostate tissues, the Raman signals of tyrosine usually have high intensity, while the concentrations of base and lipids increase in neoplastic tissue [222]. For breast tissues, the most significant spectral variations demonstrated that the glucose, lipid, and carotenoid concentrations decreased in malignant tissue, while the hydroxyproline and collagen were elevated [222]. Moreover, the disappeared peaks at ~915, 965, and 1006 $cm^{-1}$ in the NIR-SERS spectra of malignant liver tumors indicated a decrease in the percentage of aromatic amino acids. The vibration modes of the pyrrole ring of oxyhemoglobin in liver cancer patients also changed at the molecular level [221].

Raman spectroscopy has also been proved to be effective in diagnosing many other diseases. The accumulation of $H_2O_2$ in epidermal cells of patients with vitiligo was demonstrated [226]. $H_2O_2$ is well defined at the peak of 875 $cm^{-1}$ assigned to the O—O stretch. It can be successfully removed by a UVB-activated pseudocatalase. NIR-FT Raman spectra were applied to distinguish different skin lesions, including psoriatic skin, chronic dermatitis, eczema, and Kaposi's sarcoma [227]. The normal skin is dominated by type I collagen (with representative peaks at 921, 859, and 816 $cm^{-1}$). However, the increased lipid (2850, 1440, and 1310 $cm^{-1}$) and water (3250 $cm^{-1}$) contents can be detected in the Raman spectra of skin with inflammatory diseases [228]. Raman has been extended to cardiovascular research to effectively distinguish nonatherosclerotic tissues, calcified atherosclerotic plaques, and calcified plaques (Fig. 4) [229]. Beyond that, fatty liver and liver cirrhosis could also be identified by Raman [68,211,230], which might be helpful for the qualification of a liver before transplantation.

## 3. Conclusions

Raman is a powerful tool to evaluate the chemical properties of biosamples due to many advantages such as nondestructive detection, easy sample

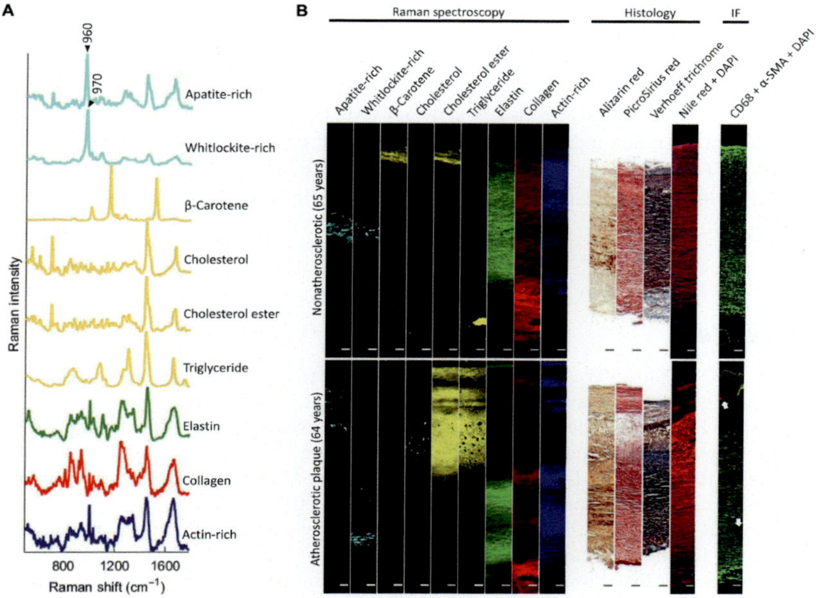


**Fig. 4** Raman imaging of calcified, atherosclerotic aorta. (A) Demixed Raman spectra for nine major components of aortic tissues and (B) representative images from nonatherosclerotic and atherosclerotic (plaque) aorta. *(Figure reproduced with permission from You AYF, Bergholt MS, St-Pierre JP, Kit-Anan W, Pence IJ, Chester AH, Yacoub MH, Bertazzo S, Stevens MM. Raman spectroscopy imaging reveals interplay between atherosclerosis and medial calcification in the human aorta. Sci Adv 2017;3: e1701156.)*

preparation, and rapid measurement. Progresses in Raman instrumentation have provided new perspectives and approaches for their applications to functional analysis of biomolecules in (but not limited to) living organisms, for example, animals, plants, and microbial cells. However, applying the developed methods and platforms for quantitative analysis is still challenging. Aspects of the specific instruments, limitations due to fluorescence, and the specificity of the interpretation of Raman spectra should be resolved in advance. The fluorescence effect from samples, the absorbance of the reaction mixture, and analysis conditions (including laser power, scaning time, and sample orientation) could particularly impede accurate quantification. With the development of electronics and optics, the signal/noise ratio of Raman spectra has been significantly improved, which would make it more widely applicable for biological and medical analysis in the future.

## Acknowledgments

This work was supported by the National Key R&D Program of China (2020YFC1808000), Jiangsu Agricultural Science and Technology Innovation Fund (CX(20)3018), and subproject 4 of Jiangsu middle and late mature garlic industrial cluster construction: Garlic mechanical intelligent operation technology and demonstration and promotion of green production technology. This work was also partially supported by the Program for Student Innovation Through Research and Training (202110307034). Additionally, the authors have no conflict of interest.

## References

[1] Yang DT, Ying YB. Applications of Raman spectroscopy in agricultural products and food analysis: a review. Appl Spectrosc Rev 2011;46:539–60.
[2] Larkin PJ. Infrared and Raman spectroscopy. In: Larkin PJ, editor. IR and Raman spectroscopy principles and spectral interpretation. The USA: Elsevier Inc.; 2011. p. 1–6.
[3] Jones RR, Hooper DC, Zhang LW, Wolverson D, Valev VK. Raman techniques: fundamentals and frontiers. Nanoscale Res Lett 2019;14:231.
[4] Petry R, Schmitt M, Popp J. Raman spectroscopy—a prospective tool in the life sciences. ChemPhysChem 2003;4:14–30.
[5] Matthaus C, Boydston-White S, Miljkovic M, Romeo M, Diem M. Raman and infrared microspectral imaging of mitotic cells. Appl Spectrosc 2006;60:1–8.
[6] Adar F. Evolution and revolution of Raman instrumentation—application of available technologies to spectroscopy and microscopy. In: Lewis IR, Edwards HGM, editors. Handbook of Raman spectroscopy: from the research laboratory to the process line. New York: Marcel Dekker, Inc.; 2001. p. 11–40.
[7] Dubessy J, Caumon M-C, Rull F, Sharma S. Instrumentation in Raman spectroscopy: elementary theory and practice. In: Dubessy J, Caumon M-C, Rull F, editors. Raman spectroscopy applied to earth sciences and cultural heritage. London: European Mineralogical Union and Mineralogical Society of Great Britain & Ireland; 2012. p. 83–172.
[8] Cardell C, Guerra I. An overview of emerging hyphenated SEM-EDX and Raman spectroscopy systems: applications in life, environmental and materials sciences. TrAC Trends Anal Chem 2016;77:156–66.
[9] Izake EL. Forensic and homeland security applications of modern portable Raman spectroscopy. Forensic Sci Int 2010;202:1–8.
[10] Hering K, Cialla D, Ackermann K, Dorfer T, Moller R, Schneidewind H, Mattheis R, Fritzsche W, Rosch P, Popp J. SERS: a versatile tool in chemical and biochemical diagnostics. Anal Bioanal Chem 2008;390:113–24.
[11] Mosca S, Conti C, Stone N, Matousek P. Spatially offset Raman spectroscopy. Nat Rev Methods Primers 2021;1–16.
[12] Keller MD, Wilson RH, Mycek MA, Mahadevan-Jansen A. Monte Carlo model of spatially offset Raman spectroscopy for breast tumor margin analysis. Appl Spectrosc 2010;64:607–14.
[13] Pudney PDA, Hancewicz TM, Cunningham DG. The use of confocal Raman spectroscopy to characterise the microstructure of complex biomaterials: foods. Spectroscopy 2002;16:217–25.
[14] Tang H, Yao H, Wang G, Wang Y, Li YQ, Feng M. NIR Raman spectroscopic investigation of single mitochondria trapped by optical tweezers. Opt Express 2007;15:12708–16.

[15] Doty KC, Muro CK, Bueno J, Halamkova L, Lednev IK. What can Raman spectroscopy do for criminalistics? J Raman Spectrosc 2016;47:39–50.
[16] Rycenga M, Wang ZP, Gordon E, Cobley CM, Schwartz AG, Lo CS, Xia YN. Probing the photothermal effect of gold-based nanocages with surface-enhanced Raman scattering (SERS). Angew Chem Int Ed 2009;48:9924–7.
[17] Smith WE. Practical understanding and use of surface enhanced Raman scattering/surface enhanced resonance Raman scattering in chemical and biological analysis. Chem Soc Rev 2008;37:955–64.
[18] Shi M, Zheng J, Tan Y, Tan G, Li J, Li Y, Li X, Zhou Z, Yang R. Ultrasensitive detection of single nucleotide polymorphism in human mitochondrial DNA utilizing lon-mediated cascade surface enhanced Raman spectroscopy amplification. Anal Chem 2015;87:2734–40.
[19] Jiang X, Yang M, Meng Y, Jiang W, Zhan J. Cysteamine-modified silver nanoparticle aggregates for quantitative SERS sensing of pentachlorophenol with a portable Raman spectrometer. Appl Mater Interfaces 2013;5:6902–8.
[20] Bell SEJ, Sirimuthu NMS. Quantitative surface-enhanced Raman spectroscopy. Chem Soc Rev 2008;37:1012–24.
[21] Graham D, Faulds K. Quantitative SERRS for DNA sequence analysis. Chem Soc Rev 2008;37:1042–51.
[22] Gao F, Du L, Tang D, Yao L, Zhang L. A cascade signal amplification strategy for surface enhanced Raman spectroscopy detection of thrombin based on DNAzyme assistant DNA recycling and rolling circle amplification. Biosens Bioelectron 2015;66:423–30.
[23] Gong T, Zhang N, Kong KV, Goh D, Ying C, Auguste JL, Shum PP, Wei L, Humbert G, Yong KT. Rapid SERS monitoring of lipid-peroxidation-derived protein modifications in cells using photonic crystal fiber sensor. J Biophotonics 2016;9:32–7.
[24] Fan MK, Andrade GFS, Brolo AG. A review on recent advances in the applications of surface-enhanced Raman scattering in analytical chemistry. Anal Chim Acta 2020;1097:1–29.
[25] Liu Y, Zhou H, Hu Z, Yu G, Yang D, Zhao J. Label and label-free based surface-enhanced Raman scattering for pathogen bacteria detection: a review. Biosens Bioelectron 2017;94:131.
[26] Granger JH, Schlotter NE, Crawford AC, Porter MD. Prospects for point-of-care pathogen diagnostics using surface-enhanced Raman scattering (SERS). Chem Soc Rev 2016;45:3865–82.
[27] Laing S, Jamieson LE, Faulds K, Graham D. Surface-enhanced Raman spectroscopy for in vivo biosensing. Nat Rev Chem 2017;1:0060.
[28] Kneipp J. Interrogating cells, tissues, and live animals with new generations of surface-enhanced Raman scattering probes and labels. ACS Nano 2017;11:1136–41.
[29] Deckert V. Tip-enhanced Raman spectroscopy. J Raman Spectrosc 2009;40:1336–7.
[30] Anderson MS. Locally enhanced Raman spectroscopy with an atomic force microscope. Appl Phys Lett 2000;76:3130–2.
[31] Downes A, Mouras R, Elfick A, Mari M. Optimising tip-enhanced optical microscopy. J Raman Spectrosc 2009;40(10):1355–60.
[32] Bohme R, Cialla D, Richter M, Rosch P, Popp J, Deckert V. Biochemical imaging below the diffraction limit—probing cellular membrane related structures by tip-enhanced Raman spectroscopy (TERS). J Biophotonics 2010;3:455–61.
[33] Opilik L, Bauer T, Schmid T, Stadler J, Zenobi R. Nanoscale chemical imaging of segregated lipid domains using tip-enhanced Raman spectroscopy. Phys Chem Chem Phys 2011;13:9978–81.
[34] Bailo E, Deckert V. Tip-enhanced Raman spectroscopy of single RNA strands: towards a novel direct-sequencing method. Angew Chem Int Ed 2008;47:1658–61.

[35] Wille G, Schmidt U, Hollricher O. RISE: correlative confocal Raman and scanning electron microscopy. In: Toporski J, Dieing T, Hollricher O, editors. Confocal Raman microscopy. Springer; 2018. p. 559–80.
[36] Jarvis RM, Brooker A, Goodacre R. Surface-enhanced Raman spectroscopy for bacterial discrimination utilizing a scanning electron microscope with a Raman spectroscopy interface. Anal Chem 2004;76:5198–202.
[37] Wang ZJ, Zhang Y, Jiang L, Qiu JJ, Gao YN, Gu TT, et al. Responses of *Rhodotorula mucilaginosa* under Pb(II) stress: carotenoid production and budding. Environ Microbiol 2022;24:678–88.
[38] Wu YL, Shao XQ, Jiao H, Song XW, He K, Li Z. Tracking the fungus-assisted biocorrosion of lead metal by Raman imaging and scanning electron microscopy technique. J Raman Spectrosc 2020;51:508–13.
[39] Matthiae M, Kristensen A. Hyperspectral spatially offset Raman spectroscopy in a microfluidic channel. Opt Express 2019;27:3782–90.
[40] Conti C, Realini M, Colombo C, Sowoidnich K, Afseth NK, Bertasa M, Botteon A, Matousek P. Noninvasive analysis of thin turbid layers using microscale spatially offset Raman spectroscopy. Anal Chem 2015;87:5810–5.
[41] Di ZY, Hokr BH, Cai H, Wang K, Yakovlev VV, Sokolov AV, Scully MO. Spatially offset Raman microspectroscopy of highly scattering tissue: theory and experiment. J Mod Opt 2014;62:97–101.
[42] Vardaki MZ, Devine DV, Serrano K, Simantiris N, Blades MW, Piret JM, Turner RFB. Defocused spatially offset Raman spectroscopy in media of different optical properties for biomedical applications using a commercial spatially offset Raman spectroscopy device. Appl Spectrosc 2019;74:223–32.
[43] Keller MD, Majumder SK, Mahadevan-Lansen A. Spatially offset Raman spectroscopy of layered soft tissues. Opt Lett 2009;34:926–8.
[44] Li ZH, Zheng ZC, Weng CC, Lin D, Wang QW, Feng SY. The application and progress of laser tweezers Raman spectroscopy in biomedicine. Spectrosc Spectr Anal 2017;37:1123–9.
[45] Snook RD, Harvey TJ, Faria EC, Gardner P. Raman tweezers and their application to the study of singly trapped eukaryotic cells. Integr Biol 2009;1:43–52.
[46] Ai M, Liu JX, Huang SS, Wang GW, Chen XL, Chen ZC, Yao HL. Application and progress of Raman tweezers in single cells. Chin J Anal Chem 2009;37:758–63.
[47] Ojeda JF, Xie CG, Li YQ, Bertrand FE, Wiley J, McConnell TJ. Chromosomal analysis and identification based on optical tweezers and Raman spectroscopy. Opt Express 2006;14:5385–93.
[48] Chen D, Shelenkova L, Li Y, Kempf CR, Sabeinikov A. Laser tweezers Raman spectroscopy potential for studies of complex dynamic cellular processes: single cell bacterial lysis. Anal Chem 2009;81:3227–38.
[49] Huang SS, Chen D, Pelczar PL, Vepachedu VR, Setlow P, Li YQ. Levels of Ca(2+)-dipicolinic acid in individual *Bacillus* spores determined using microfluidic Raman tweezers. J Bacteriol 2007;189:4681–7.
[50] Tao Z, Wang G, Xu X, Yuan Y, Xue W, Li Y. Monitoring and rapid quantification of total carotenoids in *Rhodotorula glutinis* cells using laser tweezers Raman spectroscopy. FEMS Microbiol Lett 2011;42–8.
[51] Ghanashyam C, Shetty S, Bharati S, Chidangil S, Bankapur A. Optical trapping and micro-Raman spectroscopy of functional red blood cells using vortex beam for cell membrane studies. Anal Chem 2021;93:5484–93.
[52] Hargreaves MD, Page K, Munshi T, Tomsett R, Lynch G, Edwards HGM. Analysis of seized drugs using portable Raman spectroscopy in an airport environment—a proof of principle study. J Raman Spectrosc 2008;39:873–80.
[53] Yan F, Vo-Dinh T. Surface-enhanced Raman scattering detection of chemical and biological agents using a portable Raman integrated tunable sensor. Sens Actuators B 2007;121:61–6.

[54] Wood BR, Heraud P, Stojkovic S, Morrison D, Beardall J, McNaughton D. A portable Raman acoustic levitation spectroscopic system for the identification and environmental monitoring of algal cells. Anal Chem 2005;77:4955–61.
[55] Beganovic A, Hawthorne LM, Bach K, Huck CW. Critical review on the utilization of handheld and portable Raman spectrometry in meat science. Foods 2019;8:49.
[56] Wynn-Williams DD, Edwards HGM. Proximal analysis of regolith habitats and protective biomolecules in situ by laser Raman spectroscopy: overview of terrestrial Antarctic habitats and Mars analogs. Icarus 2000;144:486–503.
[57] Jarvis RM, Goodacre R. Discrimination of bacteria using surface-enhanced Raman spectroscopy. Anal Chem 2004;76:40–7.
[58] Noothalapati H, Sasaki T, Kaino T, Kawamukai M, Ando M, Hamaguchi H, et al. Label-free chemical imaging of fungal spore walls by Raman microscopy and multivariate curve resolution analysis. Sci Rep 2016;6:27789.
[59] Rodriguez SB, Thornton MA, Thornton RJ. Raman spectroscopy and chemometrics for identification and strain discrimination of the wine spoilage yeasts *Saccharomyces cerevisiae*, *Zygosaccharomyces bailii*, and *Brettanomyces bruxellensis*. Appl Environ Microbiol 2013;79:6264–70.
[60] Egging V, Nguyen J, Kurouski D. Detection and identification of fungal infections in intact wheat and sorghum grain using a hand-held Raman spectrometer. Anal Chem 2018;90:8616–21.
[61] Altangerel N, Ariunbold GO, Gorman C, Alkahtani MH, Borrego EJ, Bohlmeyer D, Hemmer P, Kolomiets MV, Yuan JS, Scully MO. In vivo diagnostics of early abiotic plant stress response via Raman spectroscopy. Proc Natl Acad Sci U S A 2017;114:3393–6.
[62] Gupta S, Huang CH, Singh GP, Park BS, Chua NH, Ram RJ. Portable Raman leaf-clip sensor for rapid detection of plant stress. Sci Rep 2020;10:20206.
[63] Farber C, Shires M, Ong K, Byrne D, Kurouski D. Raman spectroscopy as an early detection tool for rose rosette infection. Planta 2019;250:1247–54.
[64] Weng S, Hu X, Wang J, Tang L, Li P, Zheng S, Zheng L, Huang L, Xin Z. Advanced application of Raman spectroscopy and surface-enhanced Raman spectroscopy in plant disease diagnostics: a review. J Agric Food Chem 2021;69:2950–64.
[65] Penel G, Delfosse C, Descamps M, Leroy G. Composition of bone and apatitic biomaterials as revealed by intravital Raman microspectroscopy. Bone 2005;36:893–901.
[66] Harada Y, Takamatsu T. Raman molecular imaging of cells and tissues: towards functional diagnostic imaging without labeling. Curr Pharm Biotechnol 2013;14:133–40.
[67] Movasaghi Z, Rehman S, Rehman IU. Raman spectroscopy of biological tissues. Appl Spectrosc Rev 2007;42:493–541.
[68] Gurian E, Giraudi P, Rosso N, Tiribelli C, Bonazza D, Zanconati F, Giuricin M, Palmisano S, de Manzini N, Sergo V, Bonifacio A. Differentiation between stages of non-alcoholic fatty liver diseases using surface-enhanced Raman spectroscopy. Anal Chim Acta 2020;1110:190–8.
[69] Chan JW, Taylor DS, Lane SM, Zwerdling T, Tuscano J, Huser T. Nondestructive identification of individual leukemia cells by laser trapping Raman spectroscopy. Anal Chem 2008;80:2180–7.
[70] Imiela A, Polis B, Polis L, Abramczyk H. Novel strategies of Raman imaging for brain tumor research. Oncotarget 2017;8:85290–310.
[71] Jin HZ, Lu QP, Chen XD, Ding HQ, Gao HZ, Jin SZ. The use of Raman spectroscopy in food processes: a review. Appl Spectrosc Rev 2016;51:12–22.
[72] Ajito K, Torimitsu K. Laser trapping and Raman spectroscopy of single cellular organelles in the nanometer range. Lab Chip 2002;2:11–4.
[73] Rosch P, Harz M, Schmitt M, Peschke KD, Ronneberger O, Burkhardt H, Motzkus HW, Lankers M, Hofer S, Thiele H, Popp J. Chemotaxonomic identification of single bacteria by micro-Raman spectroscopy: application to clean-room-relevant biological contaminations. Appl Environ Microbiol 2005;71:1626–37.

[74] Xie C, Li YQ. Confocal micro-Raman spectroscopy of single biological cells using optical trapping and shifted excitation difference techniques. J Appl Phys 2003;93:2982–6.
[75] Xie CG, Li YQ, Tang W, Newton RJ. Study of dynamical process of heat denaturation in optically trapped single microorganisms by near-infrared Raman spectroscopy. J Appl Phys 2003;94:6138–42.
[76] Wood BR, Tait B, McNaughton D. Micro-Raman characterisation of the R to T state transition of haemoglobin within a single living erythrocyte, Bba-Mol. Cell Res 2001;1539:58–70.
[77] Fan C, Hu Z, Mustapha A, Lin M. Rapid detection of food- and waterborne bacteria using surface-enhanced Raman spectroscopy coupled with silver nanosubstrates. Appl Microbiol Biotechnol 2011;92:1053–61.
[78] Schuster KC, Urlaub E, Gapes JR. Single-cell analysis of bacteria by Raman microscopy: spectral information on the chemical composition of cells and on the heterogeneity in a culture. J Microbiol Methods 2000;42:29–38.
[79] Jehlicka J, Culka A, Mana L, Oren A. Using a portable Raman spectrometer to detect carotenoids of halophilic prokaryotes in synthetic inclusions in NaCl, KCl, and sulfates. Anal Bioanal Chem 2018;410:4437–43.
[80] Wang Y, Lee K, Irudayaraj J. Silver nanosphere SERS probes for sensitive identification of pathogens. J Phys Chem C 2010;114:16122–8.
[81] Kumar V, Kampe B, Rosch P, Popp J. Classification and identification of pigmented cocci bacteria relevant to the soil environment via Raman spectroscopy. Environ Sci Pollut R 2015;22:19317–25.
[82] Sharma SK, Angel SM, Ghosh M, Hubble HW, Lucey PG. Remote pulsed laser Raman spectroscopy system for mineral analysis on planetary surfaces to 66 meters. Appl Spectrosc 2002;56:699–705.
[83] Picard A, Obst M, Schmid G, Zeitvogel F, Kappler A. Limited influence of Si on the preservation of Fe mineral-encrusted microbial cells during experimental diagenesis. Geobiology 2016;14:276–92.
[84] Foucher F. Detection of biosignatures using Raman spectroscopy. Cham, Switzerland: Springer; 2019.
[85] Sapers HM, Hollis JR, Bhartia R, Beegle LW, Orphan VJ, Amend JP. The cell and the sum of its parts: patterns of complexity in biosignatures as revealed by deep UV Raman spectroscopy. Front Microbiol 2019;10:679.
[86] Hermann P, Hermelink A, Lausch V, Holland G, Moller L, Bannert N, Naumann D. Evaluation of tip-enhanced Raman spectroscopy for characterizing different virus strains. Analyst 2011;136:1148–52.
[87] Hoang V, Tripp RA, Rota P, Dluhy RA. Identification of individual genotypes of measles virus using surface enhanced Raman spectroscopy. Analyst 2010; 135:3103–9.
[88] Lim JY, Nam JS, Yang SE, Shin H, Jang YH, Bae GU, Kang T, Lim KI, Choi Y. Identification of newly emerging influenza viruses by surface-enhanced Raman spectroscopy. Anal Chem 2015;87:11652–9.
[89] Hermelink A, Naumann D, Piesker J, Lasch P, Laue M, Hermann P. Towards a correlative approach for characterising single virus particles by transmission electron microscopy and nanoscale Raman spectroscopy. Analyst 2017;142:1342–9.
[90] Song C, Driskell JD, Tripp RA, Cui Y, Zhao Y. The use of a handheld Raman system for virus detection. Proc SPIE Int Soc Opt Eng 2012;8358:83580I.
[91] Cialla D, Deckert-Gaudig T, Budich C, Laue M, Moller R, Naumann D, Deckert V, Popp J. Raman to the limit: tip-enhanced Raman spectroscopic investigations of a single tobacco mosaic virus. J Raman Spectrosc 2009;40:240–3.
[92] Benevides JM, Overman SA, Thomas GJ. Raman spectroscopy of proteins. Curr Protoc Protein Sci 2003;33. 17.18.11–17.18.35.

[93] Shanmukh S, Jones L, Zhao YP, Driskell JD, Tripp RA, Dluhy RA. Identification and classification of respiratory syncytial virus (RSV) strains by surface-enhanced Raman spectroscopy and multivariate statistical techniques. Anal Bioanal Chem 2008;390:1551–5.
[94] Salman A, Shufan E, Zeiri L, Huleihel M. Characterization and detection of Vero cells infected with herpes simplex virus type 1 using Raman spectroscopy and advanced statistical methods. Methods 2014;68:364–70.
[95] Salman A, Shufan E, Zeiri L, Huleihel M. Detection and identification of cancerous murine fibroblasts, transformed by murine sarcoma virus in culture, using Raman spectroscopy and advanced statistical methods. BBA-Gen Subjects 1830;2013:2720–7.
[96] Lee JH, Kim BC, Oh BK, Choi JW. Rapid and sensitive determination of HIV-1 virus based on surface enhanced Raman spectroscopy. J Biomed Nanotechnol 2015;11:2223–30.
[97] Tong DN, Chen C, Zhang JJ, Lv GD, Zheng XX, Zhang ZX, Lv XY. Application of Raman spectroscopy in the detection of hepatitis B virus infection. Photodiagnosis Photodyn Ther 2019;28:248–52.
[98] Moor K, Ohtani K, Myrzakozha D, Zhanserkenova O, Andriana BB, Sato H. Noninvasive and label-free determination of virus infected cells by Raman spectroscopy. J Biomed Opt 2014;19:067003.
[99] Moor K, Kitamura H, Hashimoto K, Sawa M, Andriana BB, Ohtani K, Yagura T, Sato H. Study of virus by Raman spectroscopy. In: Imaging, manipulation, and analysis of biomolecules, cells, and tissues XI, San Francisco CA(US); 2013. 85871X.85871–85871X.85877.
[100] Naseer K, Amin A, Saleem M, Qazi J. Raman spectroscopy based differentiation of typhoid and dengue fever in infected human sera. Spectrochim Acta A 2019;206:197–201.
[101] Hassan MM, Zareef M, Xu Y, Li HH, Chen QS. SERS based sensor for mycotoxins detection: challenges and improvements. Food Chem 2021;344:128652.
[102] Martinez L, He LL. Detection of mycotoxins in food using surface-enhanced Raman spectroscopy: a review. ACS Appl Bio Mater 2021;4:295–310.
[103] Li JJ, Yan H, Tan XC, Lu ZC, Han HY. Cauliflower-inspired 3D SERS substrate for multiple mycotoxins detection. Anal Chem 2019;91:3885–92.
[104] Zhang WJ, Tang SS, Jin YP, Yang CJ, He LD, Wang JY, et al. Multiplex SERS-based lateral flow immunosensor for the detection of major mycotoxins in maize utilizing dual Raman labels and triple test lines. J Hazard Mater 2020;393:122348.
[105] Liu Y, Delwiche SR, Dong Y. Feasibility of FT-Raman spectroscopy for rapid screening for DON toxin in ground wheat and barley. Food Addit Contam Part A 2009;26:1396–401.
[106] Yuan J, Sun CW, Guo XY, Yang TX, Wang H, Fu SY, Li CC, Yang HF. A rapid Raman detection of deoxynivalenol in agricultural products. Food Chem 2017;221:797–802.
[107] Jehlicka J, Edwards HGM, Orenc A. Raman spectroscopy of microbial pigments. Appl Environ Microbiol 2014;80:3286–95.
[108] Kaczor A, Turnau K, Baranska M. In situ Raman imaging of astaxanthin in a single microalgal cell. Analyst 2011;136:1109–12.
[109] Tauber JP, Matthäus C, Lenz C, Hoffmeister D, Popp J. Analysis of basidiomycete pigments in situ by Raman spectroscopy. J Biophotonics 2018;11, e201700369.
[110] Hayashi H, Hamaguchi HO, Tasumi M. Resonance Raman spectra of light-harvesting bacteriochlorophyll a in pigment-protein complexes from purple photosynthetic bacteria. Chem Lett 1983;12:1857–60.
[111] Merlin JC. Resonance Raman-spectroscopy of carotenoids and carotenoid-containing systems. Pure Appl Chem 1985;57:785–92.

[112] Conner AJ, Benison KC. Acidophilic halophilic microorganisms in fluid inclusions in halite from lake Magic, western Australia. Astrobiology 2013;13:850–60.
[113] Addis MF, Tanca A, Landolfo S, Abbondio M, Cutzu R, Biosa G, Pagnozzi D, Uzzau S, Mannazzu I. Proteomic analysis of *Rhodotorula mucilaginosa*: dealing with the issues of a non-conventional yeast. Yeast 2016;33:433–49.
[114] Papaioannou EH, Liakopoulou-Kyriakides M, Christofilos D, Arvanitidis I, Kourouklis G. Raman spectroscopy for intracellular monitoring of carotenoid in *Blakeslea trispora*. Appl Biochem Biotechnol 2009;159:478–87.
[115] Maquelin K, Hoogenboezem T, Jachtenberg J-W, Dumke R, Jacobs E. Raman spectroscopic typing reveals the presence of carotenoids in *Mycoplasma pneumoniae*. Microbiology 2009;155:2068.
[116] Edwards HGM, Holder JM, Wynn-Williams DD. Comparative FT-Raman spectroscopy of Xanthoria lichen-substratum systems from temperate and Antarctic habitats. Soil Biol Biochem 1998;30:1947–53.
[117] Ellery A, Wynn-Williams D, Parnell J, Edwards HGM, Dickensheets D. The role of Raman spectroscopy as an astrobiological tool in the exploration of Mars. J Raman Spectrosc 2004;35:441–57.
[118] Wang P, Pang S, Chen J, McLandsborough L, Nugen SR, Fan M, He L. Label-free mapping of single bacterial cells using surface-enhanced Raman spectroscopy. Analyst 2016;141:1356–62.
[119] Chu H, Huang Y, Zhao Y. Silver nanorod arrays as a surface-enhanced Raman scattering substrate for foodborne pathogenic bacteria detection. Appl Spectrosc 2008;62:922–31.
[120] Chao Y, Tong Z. Surface-enhanced Raman scattering (SERS) revealing chemical variation during biofilm formation: from initial attachment to mature biofilm. Anal Bioanal Chem 2012;404:1465–75.
[121] Ivleva NP, Wagner M, Horn H, Niessner R, Haisch C. In situ surface-enhanced Raman scattering analysis of biofilm. Anal Chem 2008;80:8538–44.
[122] Ivleva NP, Wagner M, Szkola A, Horn H, Niessner R, Haisch C. Label-free in situ SERS imaging of biofilms. J Phys Chem B 2010;114:10184–94.
[123] Feng JS, de la Fuente-Núñez C, Trimble MJ, Xu J, Hancock REW, Lu XN. An in situ Raman spectroscopy-based microfluidic "lab-on-a-chip" platform for non-destructive and continuous characterization of *Pseudomonas aeruginosa* biofilms. Chem Commun 2015;51:8966.
[124] Liu ZR, Mukherjee M, Wu YC, Huang QY, Cai P. Increased particle size of goethite enhances the antibacterial effect on human pathogen Escherichia coli O157:H7: a Raman spectroscopic study. J Hazard Mater 2021;405:124174.
[125] Zhou HB, Yang DT, Ivleva NP, Mircescu NE, Niessner R, Haisch C. SERS detection of bacteria in water by in situ coating with Ag nanoparticles. Anal Chem 2014;86:1525–33.
[126] Maquelin K, Kirschner C, Choo-Smith LP, Ngo-Thi NA, van Vreeswijk T, Stammler M, Endtz HP, Bruining HA, Naumann D, Puppels GJ. Prospective study of the performance of vibrational spectroscopies for rapid identification of bacterial and fungal pathogens recovered from blood cultures. J Clin Microbiol 2003;41:324–9.
[127] Alexander TA, Gillespie JB, Pellegrino P, Fell N, Wood GL, Salamo G. Near-infrared surface-enhanced-Raman-scattering (SERS) mediated identification of single, optically trapped, bacterial spores. In: Proceedings of SPIE; 2003. p. 1121–9.
[128] Chan JW, Esposito AP, Talley CE, Hollars CW, Lane SM, Huser T. Reagentless identification of single bacterial spores in aqueous solution by confocal laser tweezers Raman spectroscopy. Anal Chem 2004;76:599–603.
[129] Esposito AP, Talley CE, Huser T, Hollars CW, Schaldach CM, Lane SM. Analysis of single bacterial spores by micro-Raman spectroscopy. Appl Spectrosc 2003;57:868–71.

[130] Schuster KC, Reese I, Urlaub E, Gapes JR, Lendl B. Multidimensional information on the chemical composition of single bacterial cells by confocal Raman microspectroscopy. Anal Chem 2000;72:5529–34.
[131] Sanchez L, Ermolenkov A, Tang XT, Tamborindeguy C, Kurouski D. Non-invasive diagnostics of *Liberibacter* disease on tomatoes using a hand-held Raman spectrometer. Planta 2020;251:64.
[132] Yang H, Irudayaraj J. Rapid detection of foodborne microorganisms on food surface using Fourier transform Raman spectroscopy. J Mol Struct 2003;646:35–43.
[133] Wang C, Wang J, Min L, Qu X, Zhang K, Zhen R, Xiao R, Wang S. A rapid SERS method for label-free bacteria detection using polyethylenimine-modified Au-coated magnetic microspheres and Au@Ag nanoparticles. Analyst 2016;141:6226–38.
[134] Efrima S, Zeiri L. Understanding SERS of bacteria. J Raman Spectrosc 2009;40:277–88.
[135] Zeiri L, Bronk BV, Shabtai Y, Eichler J, Efrima S. Surface-enhanced Raman spectroscopy as a tool for probing specific biochemical components in bacteria. Appl Spectrosc 2004;58:33–40.
[136] Guven B, Basaran-Akgul N, Temur E, Tamer U, Boyaci IH. SERS-based sandwich immunoassay using antibody coated magnetic nanoparticles for *Escherichia coli* enumeration. Analyst 2011;136:740–8.
[137] Huang YS, Karashima T, Yamamoto M, Ogura T, Hamaguchi H. Raman spectroscopic signature of life in a living yeast cell. J Raman Spectrosc 2004;35:525–6.
[138] Huang YS, Karashima T, Yamamoto M, Hamaguchii H. Molecular-level pursuit of yeast mitosis by time- and space-resolved Raman spectroscopy. J Raman Spectrosc 2003;34:1–3.
[139] Rosch P, Harz M, Schmitt M, Popp J. Raman spectroscopic identification of single yeast cells. J Raman Spectrosc 2005;36:377–9.
[140] Chang WT, Lin HL, Chen HC, Wu YM, Chen WJ, Lee YT, Liau I. Real-time molecular assessment on oxidative injury of single cells using Raman spectroscopy. J Raman Spectrosc 2009;40:1194–9.
[141] Edwards HGM, Russell NC, Weinstein R, Wynnwilliams DD. Fourier-transform Raman-spectroscopic study of fungi. J Raman Spectrosc 1995;26:911–6.
[142] Iversen JA, Berg RW, Ahring BK. Quantitative monitoring of yeast fermentation using Raman spectroscopy. Anal Bioanal Chem 2014;406:4911–9.
[143] Arend N, Pittner A, Ramoji A, Mondol AS, Dahms M, Ruger J, Kurzai O, Schie IW, Bauer M, Popp J, Neugebauer U. Detection and differentiation of bacterial and fungal infection of neutrophils from peripheral blood using Raman spectroscopy. Anal Chem 2020;92:10560–8.
[144] Farber C, Kurouski D. Detection and identification of plant pathogens on maize kernels with a hand-held Raman spectrometer. Anal Chem 2018;90:3009–12.
[145] Manoharan R, Ghiamati E, Britton KA, Nelson WH, Sperry JF. Resonance Raman-spectra of aqueous pollen suspensions with 222.5–242.4-Nm pulsed laser excitation. Appl Spectrosc 1991;45:307–11.
[146] Schulte F, Joseph V, Panne U, Kneipp J. Applications of Raman and surface-enhanced Raman scattering to the analysis of eukaryotic samples. In: Matousek P, Morris MD, editors. Biological and medical physics, biomedical engineering. Springer Heidelberg Dordrecht London New York; 2010. p. 71–95.
[147] Bako E, Deli J, Toth G. HPLC study on the carotenoid composition of Calendula products. J Biochem Biophys Methods 2002;53:241–50.
[148] Schulz H, Baranska M, Baranski R. Potential of NIR-FT-Raman spectroscopy in natural carotenoid analysis. Biopolymers 2005;77:212–21.
[149] Bagcioglu M, Zimmermann B, Kohler A. A multiscale vibrational spectroscopic approach for identification and biochemical characterization of pollen. Plos One 2015;10:e0137899.

[150] Kendel A, Zimmermann B. Chemical analysis of pollen by FT-Raman and FTIR spectroscopies. Front Plant Sci 2020;11:352.
[151] Schulte F, Lingott J, Panne U, Kneipp J. Chemical characterization and classification of pollen. Anal Chem 2008;80:9551–6.
[152] Weesie RJ, Merlin JC, Lugtenburg J, Britton G, Jansen FJHM, Cornard JP. Semiempirical and Raman spectroscopic studies of carotenoids. Biospectroscopy 1999;5:19–33.
[153] Sanchez L, Pant S, Xing Z, Mandadi K, Kurouski D. Rapid and noninvasive diagnostics of Huanglongbing and nutrient deficits on citrus trees with a handheld Raman spectrometer. Anal Bioanal Chem 2019;411:3125–33.
[154] Synytsya A, Čopı´Ková J, Matějka P, Machovič V. Fourier transform Raman and infrared spectroscopy of pectins. Carbohydr Polym 2003;54:97–106.
[155] Silva C, Vandenabeele P, Edwards H, Oliveira L. NIR-FT-Raman spectroscopic analytical characterization of the fruits, seeds, and phytotherapeutic oils from rosehips. Anal Bioanal Chem 2008;392:1489–96.
[156] Pudney PDA, Gambelli L, Gidley MJ. Confocal Raman microspectroscopic study of the molecular status of carotenoids in tomato fruits and foods. Appl Spectrosc 2011;65:127–35.
[157] Baranska M, Schütze W, Schulz H. Determination of lycopene and β-carotene content in tomato fruits and related products: comparison of FT-Raman, ATR-IR, and NIR spectroscopy. Anal Chem 2006;78:8456–61.
[158] Bicanic D, Dimitrovski D, Luterotti S, Markovi K, Twisk CV, Buijnsters JG, Dóka O. Correlation of trans-lycopene measurements by the HPLC method with the optothermal and photoacoustic signals and the color readings of fresh tomato homogenates. Food Biophys 2010;5:24–33.
[159] Trebolazabala J, Maguregui M, Morillas H, de Diego A, Madariaga JM. Portable Raman spectroscopy for an in-situ monitoring the ripening of tomato (*Solanum lycopersicum*) fruits. Spectrochim Acta A 2017;180:138–43.
[160] Schulz H, Baranska M, Quilitzsch R, Schütze W, Losing G. Characterization of peppercorn, pepper oil, and pepper oleoresin by vibrational spectroscopy methods. J Agric Food Chem 2005;53:3358–63.
[161] Edwards H, Villar S, Oliveira L, Hyaric ML. Analytical Raman spectroscopic study of cacao seeds and their chemical extracts. Anal Chim Acta 2005;538:175–80.
[162] Celedón A, Aguilera JM. Applications of microprobe Raman spectroscopy in food science. Food Sci Technol Int 2002;8:101–8.
[163] Lopez-Sanchez M, Ayora-Canada MJ, Molina-Diaz A. Olive fruit growth and ripening as seen by vibrational spectroscopy. J Agric Food Chem 2010;58:82–7.
[164] Ellepola SW, Choi SM, Phillips DL, Ma CY. Raman spectroscopic study of rice globulin. J Cereal Sci 2006;43:85–93.
[165] Choi SM, Ma CY. Structural characterization of globulin from common buckwheat (*Fagopyrum esculentum* Moench) using circular dichroism and Raman spectroscopy. Food Chem 2007;102:150–60.
[166] Almei Da MR, Alves RS, Nascimbem L, Stephani R, Oliveira LC. Determination of amylose content in starch using Raman spectroscopy and multivariate calibration analysis. Anal Bioanal Chem 2010;397:2693–701.
[167] Dib SR, Silva TV, Neto J, Guimares L, Ferreira EC. Raman spectroscopy for discriminating transgenic corns. Vib Spectrosc 2021;112, 103183.
[168] Stewart D, Yahiaoui N, Mcdougall GJ, Myton K, Marque C, Haigh BJ. Fourier-transform infrared and Raman spectroscopic evidence for the incorporation of cinnamaldehydes into the lignin of transgenic tobacco (*Nicotiana tabacum* L.) plants with reduced expression of cinnamyl alcohol dehydrogenase. Planta 1997;201:311–8.
[169] Bonora S, Francioso O, Tugnoli V, Prodi A, Foggia M, Righi V, Nipoti P, Filippini G, Pisi A. Structural characteristics of 'Hayward' kiwifruits from elephantiasis-affected

plants studied by DRIFT, FT-Raman, NMR, and SEM techniques. J Agric Food Chem 2009;57:4827–32.
[170] Carden A, Morris MD. Application of vibrational spectroscopy to the study of mineralized tissues (review). J Biomed Opt 2000;5:259–68.
[171] Steele A, Fries MD, Pasteris JD. Geoscience meets biology: Raman spectroscopy in geobiology and biomineralization. Elements 2020;16:111–6.
[172] Wang SJ, Zhang PH, Kong XF, Xie SD, Li Q, Li Z, Zhou ZL. Delicate changes of bioapatite mineral in pig femur with addition of dietary xylooligosaccharide: evidences from Raman spectroscopy and ICP. Anim Sci J 2017;88:1820–6.
[173] Pasteris JD, Ding DY. Experimental fluoridation of nanocrystalline apatite. Am Mineral 2009;94:53–63.
[174] Penel G, Leroy G, Rey C, Bres E. MicroRaman spectral study of the $PO_4$ and $CO_3$ vibrational modes in synthetic and biological apatites. Calcif Tissue Int 1998;63:475–81.
[175] Tarnowski CP, Ignelzi MA, Morris MD. Mineralization of developing mouse calvaria as revealed by Raman microspectroscopy. J Bone Miner Res 2002;17:1118–26.
[176] Freeman JJ, Wopenka B, Silva MJ, Pasteris JD. Raman spectroscopic detection of changes in bioapatite in mouse femora as a function of age and in vitro fluoride treatment. Calcif Tissue Int 2001;68:156–62.
[177] Freeman JJ, Silva MJ. Separation of the Raman spectral signatures of bioapatite and collagen in compact mouse bone bleached with hydrogen peroxide. Appl Spectrosc 2002;56:770–5.
[178] Lakshmi RJ, Alexander M, Kurien J, Mahato KK, Kartha VB. Osteoradionecrosis (ORN) of the mandible: a laser Raman spectroscopic study. Appl Spectrosc 2003;57:1100–16.
[179] Wang J, Hu YX, Wu YL, Liu YW, Liu GQ, Yan ZJ, Li Q, Zhou ZL, Li Z. Influences of bioapatite mineral and fibril structure on the mechanical properties of chicken bone during the laying period. Poult Sci 2019;98:6393–9.
[180] Wang SJ, Li Q, Gao YN, Zhou ZL, Li Z. Influences of lead exposure on its accumulation in organs, meat, eggs and bone during laying period of hens. Poult Sci 2021;100:101249.
[181] Torres IP, Terner J, Pittman RN, Somera LG, Ward KR. Hemoglobin oxygen saturation measurements using resonance Raman intravital microscopy. Am J Physiol Heart Circ Physiol 2005;289:H488–95.
[182] McVeigh PZ, Wilson BC. Intravital confocal Raman microscopy with multiplexed SERS contrast agents. Proc SPIE 2012;8234:82340D–2.
[183] Yildirim T, Matthaus C, Press AT, Schubert S, Bauer M, Popp J, et al. Uptake of retinoic acid-modified PMMA nanoparticles in LX-2 and liver tissue by Raman imaging and intravital microscopy. Macromol Biosci 2017;17:1700064.
[184] McManus LL, Bonnier F, Burke GA, Meenan BJ, Boyd AR, Byrne HJ. Assessment of an osteoblast-like cell line as a model for human primary osteoblasts using Raman spectroscopy. Analyst 2012;137:1559–69.
[185] Notingher I, Jell G, Lohbauer U, Salih V, Hench LL. In situ non-invasive spectral discrimination between bone cell phenotypes used in tissue engineering. J Cell Biochem 2004;92:1180–92.
[186] Puppels GJ, Demul FFM, Otto C, Greve J, Robertnicoud M, Arndtjovin DJ, Jovin TM. Studying single living cells and chromosomes by confocal Raman microspectroscopy. Nature 1990;347:301–3.
[187] Puppels GJ, Garritsen HSP, Kummer JA, Greve J. Carotenoids located in human lymphocyte subpopulations and natural-killer-cells by Raman microspectroscopy. Cytometry 1993;14:251–6.
[188] Puppels GJ, Schut TCB, Sijtsema NM, Grond M, Maraboeuf F, Degrauw CG, Figdor CG, Greve J. Development and application of Raman microspectroscopic and Raman imaging techniques for cell biological studies. J Mol Struct 1995;347:477–83.

[189] Wood BR, Hammer L, Davis L, McNaughton D. Raman microspectroscopy and imaging provides insights into heme aggregation and denaturation within human erythrocytes. J Biomed Opt 2005;10:014005.
[190] Wood BR, Caspers P, Puppels GJ, Pandiancherri S, McNaughton D. Resonance Raman spectroscopy of red blood cells using near-infrared laser excitation. Anal Bioanal Chem 2007;387:1691–703.
[191] Rao S, Balint S, Cossins B, Guallar V, Petrov D. Raman study of mechanically induced oxygenation state transition of red blood cells using optical tweezers. Biophys J 2009;96:209–16.
[192] Xie CG, Dinno MA, Li YQ. Near-infrared Raman spectroscopy of single optically trapped biological cells. Opt Lett 2002;27:249–51.
[193] Liu R, Zheng LN, Matthews DL, Satake N, Chan JW. Power dependent oxygenation state transition of red blood cells in a single beam optical trap. Appl Phys Lett 2011;99, 043702.
[194] De Luca AC, Rusciano G, Ciancia R, Martinelli V, Pesce G, Rotoli B, Selvaggi L, Sasso A. Spectroscopical and mechanical characterization of normal and thalassemic red blood cells by Raman tweezers. Opt Express 2008;16:7943–57.
[195] Chiang HK, Peng FY, Hung SC, Feng YC. In situ Raman spectroscopic monitoring of hydroxyapatite as human mesenchymal stem cells differentiate into osteoblasts. J Raman Spectrosc 2009;40:546–9.
[196] Mourant JR, Short KW, Carpenter S, Kunapareddy N, Coburn L, Powers TM, et al. Biochemical differences in tumorigenic and nontumorigenic cells measured by Raman and infrared spectroscopy. J Biomed Opt 2005;10:031106.
[197] Marquardt BJ, Wold JP. Raman analysis of fish: a potential method for rapid quality screening. LWT Food Sci Technol 2004;37:1–8.
[198] Beattie JR, Bell S, Borgaard C, Fearon A, Moss BW. Prediction of adipose tissue composition using Raman spectroscopy: average properties and individual fatty acids. Lipids 2006;41:287–94.
[199] Beattie JR, Bell S, Borggaard C, Moss BW. Preliminary investigations on the effects of ageing and cooking on the Raman spectra of porcine longissimus dorsi. Meat Sci 2008;80:1205–11.
[200] Sarkardei S, Howell NK. The effects of freeze-drying and storage on the FT-Raman spectra of Atlantic mackerel (*Scomber scombrus*) and horse mackerel (*Trachurus trachurus*). Food Chem 2007;103:62–70.
[201] Herrero AM, Cambero MI, Ordonez JA, Hoz L, Carmona P. Plasma powder as cold-set binding agent for meat system: rheological and Raman spectroscopy study. Food Chem 2009;113:493–9.
[202] Herrero AM, Carmona P, Careche M. Raman spectroscopic study of structural changes in hake (*Merluccius merluccius* L.) muscle proteins during frozen storage. J Agric Food Chem 2004;52:2147–53.
[203] Thawornchinsombut S, Park JW, Meng G, Li-Chan E. Raman spectroscopy determines structural changes associated with gelation properties of fish proteins recovered at alkaline pH. J Agric Food Chem 2006;54:2178–87.
[204] Fu X, Xue C, Jiang L, Miao B, Yong X. Structural changes in squid (*Loligo japonica*) collagen after modification by formaldehyde. J Sci Food Agric 2008;88:2663–8.
[205] Vo-Dinh T, Houck K, Stokes DL. Surface-enhanced Raman gene probes. Anal Chem 1994;66:3379–83.
[206] Vo-Dinh T, Wang HN, Scaffidi J. Plasmonic nanoprobes for SERS biosensing and bioimaging. J Biophotonics 2010;3:89–102.
[207] Pyrak E, Krajczewski J, Kowalik A, Kudelski A, Jaworska A. Surface enhanced Raman spectroscopy for DNA biosensors—How far are we? Molecules 2019;24:4423.

[208] Lim DK, Jeon KS, Kim HM, Nam JM, Suh YD. Nanogap-engineerable Raman-active nanodumbbells for single-molecule detection. Nat Mater 2010;9:60–7.
[209] Tajmirriahi HA, Langlais M, Savoie R. A laser Raman spectroscopic study of the interaction of calf-thymus DNA with Cu(II) and Pb(II) ions: metal ion binding and DNA conformational changes. Nucleic Acids Res 1988;16:751–62.
[210] Muntean CM, Misselwitz R, Dostal L, Welfle H. $Mn^{2+}$-DNA interactions in aqueous systems: a Raman spectroscopic study. Spectrosc Int J 2006;20:29–35.
[211] Yan S, Cui SS, Ke K, Zhao BX, Liu XL, Yue SH, Wang P. Hyperspectral stimulated Raman scattering microscopy unravels aberrant accumulation of saturated fat in human liver cancer. Anal Chem 2018;90:6362–6.
[212] Dybas J, Marzec KM, Pacia MZ, Kochan K, Czamara K, Chrabaszcz K, Staniszewska-Slezak E, Malek K, Baranska M, Kaczor A. Raman spectroscopy as a sensitive probe of soft tissue composition—imaging of cross-sections of various organs vs. single spectra of tissue homogenates. TrAC Trends Anal Chem 2016;85:117–27.
[213] Ember KJI, Hoeve MA, McAughtrie SL, Bergholt MS, Dwyer BJ, Stevens MM, et al. Raman spectroscopy and regenerative medicine: a review. NPJ Regen Med 2017;2:12.
[214] Gniadecka M, Philipsen PA, Sigurdsson S, Wessel S, Nielsen OF, Christensen DH, Hercogova J, Rossen K, Thomsen HK, Gniadecki R, Hansen LK, Wulf HC. Melanoma diagnosis by Raman spectroscopy and neural networks: structure alterations in proteins and lipids in intact cancer tissue. J Invest Dermatol 2004;122:443–9.
[215] Shafer-Peltier KE, Haka AS, Fitzmaurice M, Crowe J, Myles J, Dasari RR, Feld MS. Raman microspectroscopic model of human breast tissue: implications for breast cancer diagnosis in vivo. J Raman Spectrosc 2002;33:552–63.
[216] Haka AS, Volynskaya Z, Gardecki JA, Nazemi J, Shenk R, Wang N, Dasari RR, Fitzmaurice M, Feld MS. Diagnosing breast cancer using Raman spectroscopy: prospective analysis. J Biomed Opt 2009;14, 054023.
[217] Krishna CM, Sockalingum GD, Kurien J, Rao L, Venteo L, Pluot M, Manfait M, Kartha VB. Micro-Raman spectroscopy for optical pathology of oral squamous cell carcinoma. Appl Spectrosc 2004;58:1128–35.
[218] Kamemoto LE, Misra AK, Sharma SK, Goodman MT, Luk H, Dykes AC, Acosta T. -Near-infrared micro-Raman spectroscopy for in vitro detection of cervical cancer. Appl Spectrosc 2010;64:255–61.
[219] Bodanese B, Silveira L, Albertini R, Zangaro RA, Pacheco MTT. Differentiating normal and basal cell carcinoma human skin tissues in vitro using dispersive Raman spectroscopy: a comparison between principal components analysis and simplified biochemical models. Photomed Laser Surg 2010;28:S119–27.
[220] Ling XF, Xu YZ, Weng SF, Li WH, Zhi X, Hammaker RM, Fateley WG, Wang F, Zhou XS, Soloway RD, Ferraro JR, Wu JG. Investigation of normal and malignant tissue samples from the human stomach using Fourier transform Raman spectroscopy. Appl Spectrosc 2002;56:570–3.
[221] Liu RM, Xiong Y, Tang WY, Guo Y, Yan XH, Si MZ. Near-infrared surface-enhanced Raman spectroscopy (NIR-SERS) studies on oxyheamoglobin (OxyHb) of liver cancer based on PVA-Ag nanofilm. J Raman Spectrosc 2013;44:362–9.
[222] Stone N, Kendall C, Smith J, Crow P, Barr H. Raman spectroscopy for identification of epithelial cancers. Faraday Discuss 2004;126:141–57.
[223] Lieber CA, Majumder SK, Ellis DL, Billheimer DD, Mahadevan-Jansen A. In vivo nonmelanoma skin cancer diagnosis using Raman microspectroscopy. Lasers Surg Med 2008;40:461–7.
[224] Mert S, Ozbek E, Otunctemur A, Culha M. Kidney tumor staging using surface-enhanced Raman scattering. J Biomed Opt 2015;20, 047002.

[225] Wang H, Zhang SH, Wan LM, Sun H, Tan J, Su QC. Screening and staging for non-small cell lung cancer by serum laser Raman spectroscopy. Spectrochim Acta A 2018;201:34–8.
[226] Schallreuter KU, Moore J, Wood JM, Beazley WD, Hibberts NA. In vivo and in vitro evidence for hydrogen peroxide ($H_2O_2$) accumulation in the epidermis of patients with vitiligo and its successful removal by a UVB-activated pseudocatalase. J Investig Dermatol Symp Proc 1999;4:91–6.
[227] Schrader B, Dippel B, Erb I, Keller S, Lochte T, Schulz H, Tatsch E, Wessel S. NIR Raman spectroscopy in medicine and biology: results and aspects. J Mol Struct 1999;481:21–32.
[228] Fendel S, Schrader B. Investigation of skin and skin lesions by NIR-FT-Raman spectroscopy. Fresen J Anal Chem 1998;360:609–13.
[229] You AYF, Bergholt MS, St-Pierre JP, Kit-Anan W, Pence IJ, Chester AH, Yacoub MH, Bertazzo S, Stevens MM. Raman spectroscopy imaging reveals interplay between atherosclerosis and medial calcification in the human aorta. Sci Adv 2017;3, e1701156.
[230] Pacia MZ, Czamara K, Zebala M, Kus E, Chlopicki S, Kaczor A. Rapid diagnostics of liver steatosis by Raman spectroscopy via fiber optic probe: a pilot study. Analyst 2018;143:4723–31.

CHAPTER 8

# Spectroscopic investigation of biomolecular dynamics using light scattering methods

**Eva Rose M. Balog**
School of Mathematical and Physical Sciences, University of New England, Biddeford, ME, United States

## 1. Introduction

There are many distinct techniques generally referred to as light scattering methods. The term does not typically encompass primary experimental techniques such as X-ray crystallography, small-angle X-ray scattering (SAXS), and Raman spectroscopy, despite their basis in the scattering of electromagnetic waves. Instead, it refers to methods of determining particle parameters, most fundamentally size, by illuminating a suspension of dispersed particles with a laser beam. The angle and intensity of the scattered light provide information about the size of the particles. Different commercially available techniques may vary in wavelength of the light, angle of incidence, number of light sources, and number and position of detectors. The most suitable technique for a given situation depends on the nature of the sample, the desired information, and practical considerations like cost, availability, and speed. The two principal methods discussed herein are dynamic light scattering (DLS), also sometimes referred to as quasi-elastic light scattering (QELS) or photon correlation spectroscopy (PCS), and static light scattering (SLS), also sometimes referred to as the classical light scattering or multiangle light scattering (MALS).

An enjoyable aspect of light scattering is that studying the underlying theory leads to more profound questions, such as the definition of a particle and the dual nature of light. However, for practical purposes, a particle characterized by light scattering techniques is a discrete region with a complex refractive index different from the surrounding medium [1]. This definition clarifies that light scattering techniques are theoretically agnostic with respect to the phase (solid, liquid, and gas) and composition (organic, inorganic) of the particle. Typical industrial uses include quality management of substances such as powders, pigments, aerosols such as asthma inhalers and

*Advanced Spectroscopic Methods to Study Biomolecular Structure and Dynamics*
https://doi.org/10.1016/B978-0-323-99127-8.00011-8

agricultural chemicals, milk and other foods and beverages, vaccines, and other pharmaceuticals, ensuring uniformity of products and detecting unwanted aggregates and contaminants such as microplastics. While there are numerous fascinating examples of the application of certain light scattering techniques to biomolecular dynamics, of many which are discussed herein, light scattering is perhaps more common to applications in polymer chemistry, physical colloid chemistry, materials science, nanotechnology, and earth sciences. Light scattering methods would not be considered among the standard tools introduced, e.g., in an undergraduate biochemistry curriculum, save for perhaps certain quantitative analysis or instrumental analysis courses. Limited access within academic chemistry and biology departments, combined with the expense of instrumentation, creates barriers for biochemists and biophysicists interested in using light scattering techniques. However, the ease and speed with which most light scattering experiments can be performed makes visiting a colleague's laboratory, even at another institution, an appropriate undertaking. As interdisciplinary research at the intersection of nanotechnology and biomedicine continues to emerge and advance, we predict that light scattering techniques will receive broader dissemination and interest.

Similar to biomolecular nuclear magnetic resonance (NMR) spectroscopy and atomic force microscopy (AFM), a complete mathematical understanding of the theory behind light scattering techniques, including the physics of optics and diffusion, is not strictly necessary to obtain useful data to answer a particular question. Those interested in the general history, theory, applications, and limitations of light scattering will find many fine treatments elsewhere [2–9]. Additionally, the major instrumentation manufacturers offer detailed manuals, tutorials, and support. Malvern Panalytical provides a series of freely accessible white papers and technical notes [10], Horiba provides a "Particle Education" resource [11], and Wyatt developed "Light Scattering University," an intensive 3-day training course [12]. Ample high-quality theoretical and practical resources, including lectures and webinars from both industrial and academic sources, are also available on YouTube. An accessible and practical introduction to DLS analysis of proteins for undergraduate and graduate students can be found in Kern et al., wherein it is acknowledged that while light scattering is generally introduced in chemistry and physics courses, there are few resources to be found in the biology or biochemical education literature [13,14]. Here, we need only to be able to refer to certain parameters and understand how they are calculated.

In principle, DLS makes use of the random movement of particles in the solution. The motion of the particles creates fluctuations in the intensity of scattered light. The measured data in a DLS experiment is the correlation curve, which mathematically describes the time dependence of the scattering intensity. The correlation function provides information about the distribution of diffusion coefficients in solution—the smaller the particle, the faster the intensity will change. Practically, for most biomolecules, only the translational motion will produce sufficiently large fluctuations in scattering patterns. The distributions of diffusion coefficients are converted into distributions of particle size using the Stokes-Einstein relationship $D = \frac{k_b T}{6\pi\eta R_h}$, in which $D$ is the translational diffusion coefficient, $\eta$ is viscosity, and $R_h$ is the hydrodynamic radius. Typically, the dynamics accessible by DLS measurements occur on a timescale >100 ns for particles ~1 to ~6 μm. DLS is widely used in monitoring the formation of biomolecular aggregates and assemblies through their diffusional properties [15]. For a broad overview of modern applications of DLS techniques in chemistry and biology, an excellent review by Stetefeld et al. includes a timeline of major events in the development of light scattering theory and analysis as well as a practical guide to the study of proteins and nucleic acids, their homogeneity, hydrodynamic radius of globular proteins (Table 1) and their interactions, using modern instrumentation [16].

DLS is frequently used as a gold standard to compare complementary or novel particle sizing techniques [17]. Diffusion-ordered NMR spectroscopy (DOSY-NMR) and DLS measure diffusion coefficient and $R_h$. Several interesting side-by-side comparisons demonstrate the benefits and limitations of each technique for different samples and experimental goals [18,19]. DOSY-NMR has the advantage of being combined with other

**Table 1** Hydrodynamic radius of common globular proteins.

| Protein | $M_w$ (kDa) | Cong. (g/L) | $R_h$ (nm) |
|---|---|---|---|
| Thyroglobulin | 670 | 2 | 8.71 ± 0.21 |
| Ferritin | 440 | 3 | 7.17 ± 0.24 |
| Aldolase | 158 | 4 | 4.98 ± 0.08 |
| Conalbumin | 75 | 5 | 3.72 ± 0.08 |
| Ovalbumin | 44 | 5 | 2.98 ± 0.04 |
| Carbonic anhydrase | 29 | 6 | 2.37 ± 0.05 |
| Aprotinin | 6.5 | 8 | 1.82 ± 0.05 |

Table adapted with permission from Stetefeld J, McKenna SA, Patel TR. Dynamic light scattering: a practical guide and applications in biomedical sciences. Biophys Rev 2016;8:409–27.

NMR experiments, e.g., to combine diffusion measurements with residue-specific information [20]. On the other hand, DLS is more sensitive to the presence of higher molecular weight species, making it an excellent choice to monitor aggregation or oligomerization [18,21]. Commercial benchtop DLS instruments can also measure zeta ($\zeta$) potential, the electric potential at the particle–fluid interface, providing the opportunity for additional insight into molecular organization and stabilization [22–24]. Nonelectrophoretic DLS measurements performed in the presence of an applied electric field allow investigation of the influence of voltage on biomolecular dynamics [25,26].

SLS makes use of the fact that the light scattering properties of a particle are strongly dependent on particle size [27]. The data obtained are the intensity of scattered light as a function of scattering angle. A MALS instrument can collect such data at multiple angles simultaneously. It can be used to determine the radius of gyration ($R_g$) of a particle, molecular mass, and concentration. The technique of size-exclusion chromatography coupled to MALS (SEC-MALS) is particularly useful for characterizing proteins and their complexes and aggregates [28]. I will focus on select recent applications of DLS and SLS most closely related to protein dynamics per se rather than protein interactions.

## 2. Basics of light scattering

Light scattering can be a powerful tool when combined with other structural, thermodynamic, and/or kinetic methods to achieve a comprehensive picture of any biomolecular system. Notably, there are discoveries made using light scattering as a primary method. For example, Baussay et al. show how the combination of SLS and DLS can be used to its fullest potential in their study of the effects of ionic strength, concentration, and heat-induced denaturation on the self-association of β-lactoglobulin [29]. They observed an increase in the apparent translational diffusion coefficient $D$ at lower ionic strength due to repulsive electrostatic interactions and a decrease in $D$ as a function of β-lactoglobulin concentration due to increased friction between proteins and screening by counterions [29]. Most interestingly, in addition to measuring $R_h$, they used DLS to probe the degree of flexibility of β-lactoglobulin aggregates by analyzing the distribution of relaxation times from intensity autocorrelation functions [30]. By determining how $D$ changed with the scattering wave vector $q$, they concluded that aggregates were more flexible at low ionic strength versus high ionic strength [29]. This kind of insight into aggregate dynamics has apparent applications to functional biomolecular assemblies such as self-assembling protein nanomaterials.

Another example of extracting maximum value from modern light scattering experiments can be found in the work of Jachimska et al. [31] Using a Malvern Zetasizer Nano ZS instrument and samples of bovine serum albumin (BSA) and human serum albumin (HSA), they measured $D$, $R_h$, and sample polydispersity (Fig. 1). Using known values for the densities of BSA and HSA, they also determined that the shape of both proteins deviated from spherical. Further calculations allowed them to estimate the overall

| Protein / Property | BSA | HSA |
|---|---|---|
| **Molecular weight [Da]** | 67000 | 69000 |
| **Specific density [g·cm$^{-3}$]** | 1.35 | 1.35 |
| **Specific volume [nm$^3$]** | 82.4 | 85 |
| **Equivalent sphere radius [nm]** | 2.70 | 2.72 |
| **Hydrodynamic radius $R_h$ [nm]** | 3.3 – 4.3 | 3.3 – 4.1 |
| **Geometrical dimensions, spheroid [nm]** | 9.5 x 5 x 5 | 9.5 x 5 x 5 |
| **Geometrical volume for spheroid [nm$^3$]** | 124 | 124 |
| **Porosity** | 0.33 | 0.33 |
| **Molecular shapes** | **HSA, BSA**<br>Monomer Dimer<br>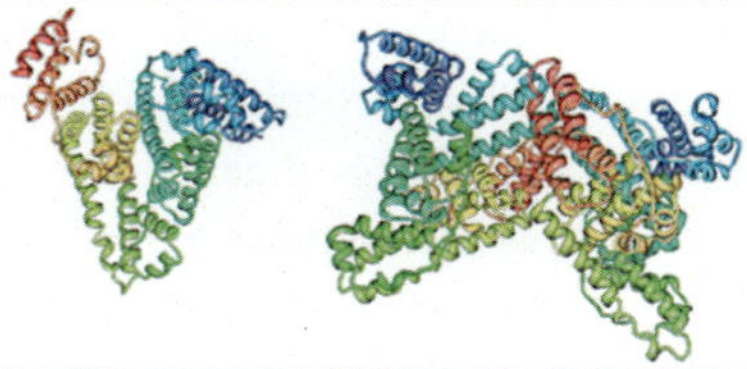 | |

**Fig. 1** Molecular shapes and physicochemical properties of HSA and BSA proteins. *(Figure adapted with permission from Jachimska B, Wasilewska M, Adamczyk Z. Characterization of globular protein solutions by dynamic light scattering, electrophoretic mobility, and viscosity measurements. Langmuir 2008;24:6866–72.)*

dimensions and void volume of these proteins. Since the crystal structures of BSA and HSA are available, the authors could compare the estimated values to those in the literature. They studied the effects of concentration, pH, and temperature on $R_h$, allowing the determination of the protein melting temperature. Further, using electrophoretic zeta potential measurements, they determined the isoelectric points and the number of uncompensated charges on the proteins under various conditions. Finally, they calculated the intrinsic viscosity of the proteins and compared their values to theoretical values, providing insight into the form and composition of oligomers and aggregates.

Conformational dynamics and oligomerization are often linked, and it is desirable to study them concurrently to determine the order of molecular events. O'Donnell et al. described and provided a detailed approach to combining SEC-MALS and FRET to study the oligomerization and conformational changes in the dynamin-related protein atlastin [32]. Because MALS data provides absolute molecular weights to distinguish monomer and dimer states, the kinetics and orientation data obtained from FRET experiments are more meaningful. The determination of binding stoichiometries via DLS particle sizing differences is only reliable for highly reproducible and well-controlled experimental systems. For example, FG-nucleoporins contain multiple intrinsically disordered Phe-Gly (FG) repeat regions that bind specifically to other proteins and facilitate their rapid transport through nuclear pore complexes [33,34]. Hayama et al. used DLS titration experiments to measure the stoichiometry of linear FG motifs binding to transport factors, providing a model of how multivalency contributes to transport [35].

## 3. Applications of light scattering methods

### 3.1 Protein folding and unfolding

Light scattering techniques have well-demonstrated applications in mechanistic investigations of protein folding [36,37]. Light scattering measurements are commonly used to measure conformation-dependent properties of polymers, such as the hydrodynamic radius ($R_h$) and the radius of gyration ($R_g$), both of which are expected to change as a polypeptide chain undergoes collapse and adopts a secondary structure [38,39]. The invention of laser light sources and the development of new instrumentations in the 1960s and 1970s enabled modern applications of light scattering to biomacromolecules, such as monitoring the pH-dependent helix-coil transitions of poly-L-lysine [40,41]. Time-resolved instruments and stopped-flow methods were soon developed to extend kinetic applications to the subsecond region [42–44].

A 2005 chapter in *Protein Folding Handbook* by Gast and Modler provides a thorough survey of applications of SLS and DLS to protein folding and unfolding, including a useful table displaying the hydrodynamic dimensions of selected proteins under native and various denaturing conditions [37]. The studies represented in this table demonstrate that the method of denaturation has implications for the ensemble-average dimensions of the denatured state. For example, yeast phosphoglycerate kinase (PGK) has $R_h = 2.97$ nm in its native state, while guanidine hydrochloride unfolded PGK has $R_h = 5.66$ nm, and acid-induced unfolding (pH = 2) of PGK results in $R_h = 7.42$ nm [45]. Such studies contribute to the elucidation of environmentally dependent folding and unfolding trajectories [46]. Demonstrating how light scattering methods can complement other methods, Mitra et al. used DLS in combination with CD and fluorescence spectroscopy to study the compressibility and hydration of HSA during unfolding and refolding [47]. They found that the $R_h$ of HSA increased from ~10 nm at room temperature to ~40 nm upon thermal unfolding [47]. Comparing this $R_h$ to the diameter of the X-ray structure, they approximated a hydration layer thickness of 6 Å [48]. Fluorescence anisotropy was used to determine that the larger size at higher temperatures was due to unfolding rather than aggregation. The temperature-dependent CD was used to correlate secondary structure content with folding and unfolding processes. Ultrasound and densimetric techniques were used to obtain apparent specific volume and adiabatic compressibility, providing further insights into the hydration layer and extent of the collapse of the protein. Since few proteins have folding and unfolding pathways well characterized as those of PGK and HSA, there is much territory left to explore, including challenges to long-standing orthodoxies in protein folding [49,50]. It must be acknowledged that small-angle scattering techniques (SAXS and SANS) are expected to cover much of the same territory more directly and at higher resolution [51]. Some specialized setups exist that allow for simultaneous DLS and small-angle scattering measurements [52,53]. Sonje et al. evaluated the stability of lactate dehydrogenase (LDH) under different conditions using a combination of SANS to measure conformational changes and DLS to provide quantitative size information and monitor aggregation [54]. In any case, it is perhaps important to mention that the 2005 edition of the protein folding methods handbook contained a chapter on light scattering methods, while a similar volume in 2022 did not and instead favored NMR, fluorescence spectroscopy, and single-molecule techniques [37,55].

## 3.2 Conformational fluctuations, disorder, and transitions

The conformational dynamics that are central to a protein's function may be considered separately from folding and unfolding processes [56]. In soft materials (e.g., synthetic polymers), stochastic motions that occur on characteristic timescales and between different structural units of matter are referred to as relaxation processes. In comparison to synthetic polymers, the relaxation processes of biological macromolecules are incompletely described [57]. Specialized approaches to light scattering measurements and data analysis can offer insights into segmental dynamics, the main structural relaxation in proteins [57,58].

A highly active area of inquiry for applications of light scattering methods in biomolecular dynamics is the study of intrinsically disordered proteins (IDPs), intrinsically disordered protein regions (IDRs), and stimuli-responsive, phase-separating protein polymers. Traditionally, the signature methods for characterizing ensembles of conformational states of IDPs are NMR and SAXS, to be complemented with single-molecule fluorescence techniques (e.g., FRET, FCS) to discriminate structural transitions and subpopulations [59–61]. However, these methods become complicated if insolubility and aggregation arise. Furthermore, fluorescent labeling can result in the formation of a complex with very different dynamics than unlabeled protein. On the other hand, light scattering is well suited to continue to provide useful information about unlabeled proteins and polydisperse solutions. Indeed, with respect to the dynamics of IDPs and unfolded proteins, single-molecule FRET is recognized to be complementary to light scattering techniques [62].

The principles and methods of polymer physics (coil-globule transition, Flory-Huggins theory, etc.) have historically been obvious starting points for the investigation of IDPs [63,64]. However, the hydrodynamic behavior of disordered polypeptide chains may or may not conform to ideal polymer behavior [65,66]. DLS methods have the potential to make a substantive contribution to the knowledge of biomolecular dynamics to bridge the gap between what is known about polymer hydrodynamics and how biomolecules deviate from ideality or theory, either intrinsically or due to environmental conditions in the biological milieu. In discussing the challenges of applying polymer theory to IDPs, Hofmann et al. reported that we are limited by a lack of knowledge about the Θ-point for polypeptide chains, the

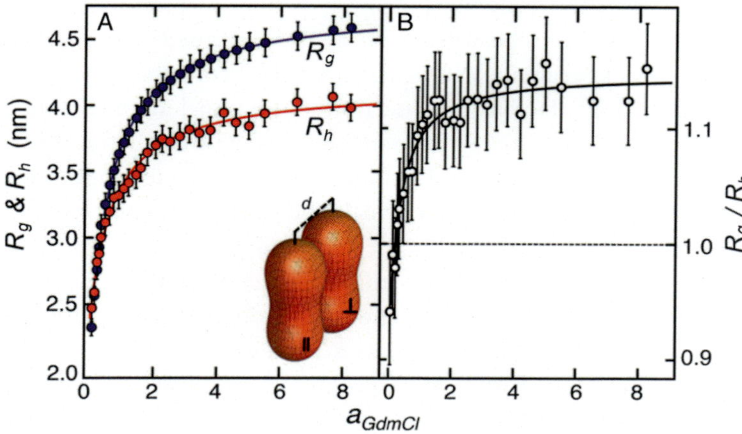


**Fig. 2** Comparison of radii of gyration ($R_g$) and hydrodynamic radius ($R_h$) of hCyp at various guanidium chloride activity. (A) $R_g$ for Cyp16 (blue circles) and $R_h$ determined by 2fFCS (red circles). (B) $R_g/R_h$ as a function of guanidium chloride activity. *(Figure adapted with permission from Hofmann H, Soranno A, Borgia A, Gast K, Nettels D, Schuler B. Polymer scaling laws of unfolded and intrinsically disordered proteins quantified with single-molecule spectroscopy. Proc Natl Acad Sci U S A 2012;109:16155–60.)*

critical point of the transition between a swollen disordered chain that would exist in good solvent conditions, and a collapsed disordered chain that would occur in poor solvent conditions (Fig. 2) [67]. The density of a polymer, and therefore information about its **Θ** conditions, can be represented by the ratio of $R_g/R_h$, with very dense states corresponding to higher values and reduced conformational entropy [68]. Again demonstrating its use as a validation tool, Hofmann et al. used DLS as a label-free confirmation of the accuracy of $R_h$ values obtained by dual-focus FCS [67].

A greater understanding of the applications of polymer theory to biopolymers should facilitate their use in place of synthetic polymers in biomaterials, biosensors, etc. Thus we ought to see more studies like Fluegel et al., in which the Kuhn statistical segment length of ELPs was determined by DLS analysis [69]. Briefly, the Kuhn length is a measure of the stiffness or flexibility of a polymer chain and is related to the persistence length, the length over which the chain "persists" in the same direction as the first bond [70]. In the case of ELPs, Kuhn length values were sought to help distinguish between models of ELP behavior with similar structures but different dynamics. To determine the Kuhn length, the authors measured $R_h$ by DLS using the usual application of Stokes' law. Then, they fit their measured $R_h$ values to calculated $R_h$ values obtained by the application of a worm-like chain model to their polymers, making certain assumptions about the dimensions of an ELP repeat based on peptide geometry constraints. The best fits of the experimental and calculated data correspond to the more accurate values for effective cross-section and Kuhn length. Kuhn length values determined using DLS represent the average overall conformations, making it a complementary and orthogonal approach to single-molecule force spectroscopy techniques, in which the application of external force can itself be a variable [71].

Beyond IDP dynamics, chain stiffness is an important parameter in understanding protein dynamics in globular protein folding pathways, the mechanics of structural and filamentous proteins, and biopolymer-based nanoengineered materials, in which programmable distances and orientations may be desirable [72–74]. With respect to structural proteins and their networks, Janmey et al. determined the bending stiffness of individual actin filaments by performing DLS on intact actin meshworks [75]. While their approach relied on relatively sophisticated theoretical analysis and measurements taken at multiple scattering angles, it nevertheless provided a nonintrusive, label-free method of investigating semiflexible chain dynamics [76]. An updated application of DLS to dynamic polymer networks such as protein hydrogels is found in the work of Cai et al., in which the authors describe the applications of dynamic light scattering microrheology (DLSμR) to dilute polymer solutions, covalently and dynamically cross-linked polymer gels, and active biological fluids [77]. Performed on a commercial benchtop instrument, DLSμR involves measuring the scattering from "tracer particles" embedded in the sample. The movement of tracer particles varies with the physical properties of their microenvironment, generating different correlation functions from which viscoelastic moduli can be determined. Interesting future applications of DLSμR include not only the study of the dynamics of various biological polymer networks and fluids but also the dynamics of biomolecules that are encapsulated within soft materials.

With respect to bionanomaterials, Fazelinia et al. used DLS to demonstrate the temperature-dependent contraction and elongation and the programmed assembly of two engineered protein polymers consisting of helical blocks separated by ELP repeats [78]. Independently, each polymer acts as a monomer in solution. The ELP repeats are fully hydrated and extended at low temperatures. With increasing temperature, the polymers contract, displaying a substantial and reversible reduction in $R_h$ from ~12 to ~6 nm. When mixed together, the two polymers assemble into larger structures, appearing as >1 μm particles via DLS via specific coiled-coil interactions.

TEM revealed these structures to be a 2D meshwork with tunable pore size. This provides a patent example of how DLS particle sizing data must not be assumed to be accurate at face value.

It is now broadly appreciated that the sequence correlates of disorder in proteins are interrelated with phase separation behavior [79]. Stimuli-responsive protein polymers such as ELPs, resilin-like polymers (RLPs), and FG-nucleoporins all undergo reversible phase separation. Peran et al. provide a detailed procedure for using SLS to determine coexistence points—points along a curve representing the conditions at which two phases are in equilibrium—of IDPs that undergo liquid-liquid phase separation (LLPS), toward comparing the experimental phase diagram of an IDP onto a theoretical model from polymer physics [80]. Their method applies to proteins that have either a lower critical solution temperature (LCST) or an upper critical solution temperature (UCST) and relies on the increase in scattering intensity that accompanies the formation of polymer-rich droplets. While phase separation may be characterized by turbidity measurements, more information is accessible through DLS for particle sizing. Weitzhandler et al. use SLS and DLS to describe the morphology and phase behavior of ELP-RLP block copolypeptides [81]. This combination allows the determination of the aforementioned shape factor $R_g/R_h$ as well as the number of polymer chains per self-assembled nanostructure.

## 4. Conclusion and future perspectives

Much remains to explore using light scattering methods to investigate biomolecular dynamics. Future applications will be bolstered by advances in computational methods and instrumentation development, such as powerful combinations like SEC-MALS, DLS-SAXS, DLS-SANS, and DLS-microfluidics [53,82–84]. The contributions of artificial intelligence to the knowledge of the folding and dynamics of IDPs and IDRs will require experimental validation. Perhaps an interest in light-activatable materials will lead to simultaneous manipulation and measurement of biomolecular structures and assemblies using light scattering techniques, expanding the optobiochemical toolkit [85,86].

## References

[1] Turzhitsky V, Qiu L, Itzkan I, Novikov AA, Kotelev MS, Getmanskiy M, Vinokurov VA, Muradov AV, Perelman LT. Spectroscopy of scattered light for the characterization of micro and nanoscale objects in biology and medicine. Appl Spectrosc 2014;68:133–54.

[2] Pecora R. Quasi-elastic light scattering from macromolecules. Annu Rev Biophys Bioeng 1972;1:257–76.
[3] Schurr JM, Bloomfield V. Dynamic light scattering of biopolymers and biocolloid. CRC Crit Rev Biochem 1977;4:371–431.
[4] Bloomfield VA. Quasi-elastic light scattering applications in biochemistry and biology. Annu Rev Biophys Bioeng 1981;10:421–50.
[5] Harding SE, Jumel K. Light scattering. Curr Protoc Protein Sci 1998;11:7.8.1–7.8.14.
[6] Berne BJ, Pecora R. Dynamic light scattering: with applications to chemistry, biology, and physics. Courier Corporation; 2000.
[7] Gun'ko VM, Klyueva AV, Levchuk YN, Leboda R. Photon correlation spectroscopy investigations of proteins. Adv Colloid Interface Sci 2003;105:201–328.
[8] Zakharov P, Scheffold F. Advances in dynamic light scattering techniques. In: Light scattering reviews, vol. 4. Springer; 2009. p. 433–67.
[9] Fischer K, Schmidt M. Pitfalls and novel applications of particle sizing by dynamic light scattering. Biomaterials 2016;98:79–91.
[10] Anonymous Malvern Panalytical's Knowledge Center | Malvern Panalytical, https://www.malvernpanalytical.com/en/learn/knowledge-center. [Accessed 14 November 2021].
[11] Anonymous Particle Education, https://www.horiba.com/en_en/products/scientific/particle-characterization/particle-education/. [Accessed 14 November 2021].
[12] Anonymous Light Scattering University, https://www.wyatt.com/events/training/light-scattering-university.html. [Accessed 14 November 2021].
[13] Lorber B, Fischer F, Bailly M, Roy H, Kern D. Protein analysis by dynamic light scattering: methods and techniques for students. Biochem Mol Biol Educ 2012;40:372–82.
[14] Santos NC, Castanho MA. Teaching light scattering spectroscopy: the dimension and shape of tobacco mosaic virus. Biophys J 1996;71:1641–50.
[15] Grimaldo M, Roosen-Runge F, Zhang F, Schreiber F, Seydel T. Dynamics of proteins in solution. Q Rev Biophys 2019;52.
[16] Stetefeld J, McKenna SA, Patel TR. Dynamic light scattering: a practical guide and applications in biomedical sciences. Biophys Rev 2016;8:409–27.
[17] Marvin L, Paiva W, Gill N, Morales MA, Halpern JM, Vesenka J, Balog ERM. Flow imaging microscopy as a novel tool for high-throughput evaluation of elastin-like polymer coacervates. PLoS One 2019;14, e0216406.
[18] Patil SM, Keire DA, Chen K. Comparison of NMR and dynamic light scattering for measuring diffusion coefficients of formulated insulin: implications for particle size distribution measurements in drug products. AAPS J 2017;19:1760–6.
[19] Zhang C, Jin Z, Zeng B, Wang W, Palui G, Mattoussi H. Characterizing the Brownian diffusion of nanocolloids and molecular solutions: diffusion-ordered NMR spectroscopy vs dynamic light scattering. J Phys Chem B 2020;124:4631–50.
[20] Lee J, Park SH, Cavagnero S, Lee JH. High-resolution diffusion measurements of proteins by NMR under near-physiological conditions. Anal Chem 2020;92:5073–81.
[21] Joshi S, Khatri LR, Kumar A, Rathore AS. Monitoring size and oligomeric-state distribution of therapeutic mAbs by NMR and DLS: trastuzumab as a case study. J Pharm Biomed Anal 2021;195, 113841.
[22] Bhattacharjee S. DLS and zeta potential—what they are and what they are not? J Control Release 2016;235:337–51.
[23] Grisham DR, Nanda V. Zeta potential prediction from protein structure in general aqueous electrolyte solutions. Langmuir 2020;36:13799–803.
[24] Welsh TJ, Krainer G, Espinosa JR, Joseph JA, Sridhar A, Jahnel M, Arter WE, Saar KL, Alberti S, Collepardo-Guevara R, Knowles TPJ. Surface electrostatics govern the emulsion stability of biomolecular condensates. Nano Lett 2022;22:612–21.
[25] Ren T, Roberge EJ, Csoros JR, Seitz WR, Balog ERM, Halpern JM. Application of voltage in dynamic light scattering particle size analysis. J Vis Exp 2020;, e60257.

[26] LaFreniere J, Roberge E, Ren T, Seitz WR, Balog ERM, Halpern JM. Insights on the lower critical solution temperature behavior of pNIPAM in an applied electric field. ECS Trans 2020;97:709–15.
[27] Wyatt PJ. Light scattering and the absolute characterization of macromolecules. Anal Chim Acta 1993;272:1–40.
[28] Sarkar P, Akhavantabib N, D'Arcy S. Comprehensive analysis of histone-binding proteins with multi-angle light scattering. Methods 2020;184:93–101.
[29] Baussay K, Bon CL, Nicolai T, Durand D, Busnel J. Influence of the ionic strength on the heat-induced aggregation of the globular protein beta-lactoglobulin at pH 7. Int J Biol Macromol 2004;34:21–8.
[30] Le Bon C, Nicolai T, Durand D. Growth and structure of aggregates of heat-denatured β-lactoglobulin. Int J Food Sci Technol 1999;34:451–65.
[31] Jachimska B, Wasilewska M, Adamczyk Z. Characterization of globular protein solutions by dynamic light scattering, electrophoretic mobility, and viscosity measurements. Langmuir 2008;24:6866–72.
[32] O'Donnell JP, Kelly CM, Sondermann H. Nucleotide-dependent dimerization and conformational switching of atlastin. Methods Mol Biol 2020;2159:93–113.
[33] Lyngdoh DL, Nag N, Uversky VN, Tripathi T. Prevalence and functionality of intrinsic disorder in human FG-nucleoporins. Int J Biol Macromol 2021;175:156–70.
[34] Nag N, Sasidharan S, Uversky VN, Saudagar P, Tripathi T. Phase separation of FG-nucleoporins in nuclear pore complexes. Biochim Biophys Acta Mol Cell Res 2022;1869(4);119205.
[35] Hayama R, Sparks S, Hecht LM, Dutta K, Karp JM, Cabana CM, Rout MP, Cowburn D. Thermodynamic characterization of the multivalent interactions underlying rapid and selective translocation through the nuclear pore complex. J Biol Chem 2018;293:4555–63.
[36] Gast K, Damaschun G, Misselwitz R, Zirwer D. Application of dynamic light scattering to studies of protein folding kinetics. Eur Biophys J 1992;21:357–62.
[37] Gast K, Modler AJ. Studying protein folding and aggregation by laser light scattering. In: Protein folding handbook. John Wiley & Sons, Ltd; 2005. p. 673–709.
[38] Nicoli DF, Benedek GB. Study of thermal denaturation of lysozyme and other globular proteins by light-scattering spectroscopy. Biopolymers 1976;15:2421–37.
[39] Haran G. How, when and why proteins collapse: the relation to folding. Curr Opin Struct Biol 2012;22:14–20.
[40] Jamieson AM, Mack L, Walton AG. Quasielastic light scattering investigation of the isothermal "helix to extended-coil" transition of poly-L-lysine HBr. Biopolymers 1972;11:2267–79.
[41] Lee WI, Schurr JM. Dynamic light scattering studies of poly-L-lysine HBr in the presence of added salt. Biopolymers 1974;13:903–8.
[42] Feng HP, Scherl DS, Widom J. Lifetime of the histone octamer studied by continuous-flow quasielastic light scattering: test of a model for nucleosome transcription. Biochemistry 1993;32:7824–31.
[43] Gast K, Zirwer D, Damaschun G. Time-resolved dynamic light scattering as a method to monitor compaction during protein folding. Macromol Symp 2000;162:205–20.
[44] Gast K, Nppert A, Mller-Frohne M, Zirwer D, Damaschun G. Stopped-flow dynamic light scattering as a method to monitor compaction during protein folding. Eur Biophys J 1997;3:211–9.
[45] Damaschun G, Damaschun H, Gast K, Zirwer D. Proteins can adopt totally different folded conformations. J Mol Biol 1999;291:715–25.
[46] Li Q, Scholl ZN, Marszalek PE. Unraveling the mechanical unfolding pathways of a multidomain protein: phosphoglycerate kinase. Biophys J 2018;115:46–58.

[47] Mitra RK, Sinha SS, Pal SK. Hydration in protein folding: thermal unfolding/refolding of human serum albumin. Langmuir 2007;23:10224–9.
[48] He XM, Carter DC. Atomic structure and chemistry of human serum albumin. Nature 1992;358:209–15.
[49] Rose GD. Protein folding—seeing is deceiving. Protein Sci 2021;30:1606–16.
[50] Sorokina I, Mushegian AR, Koonin EV. Is protein folding a thermodynamically unfavorable, active, energy-dependent process? Int J Mol Sci 2022;23.
[51] Smilgies D, Folta-Stogniew E. Molecular weight-gyration radius relation of globular proteins: a comparison of light scattering, small-angle X-ray scattering and structure-based data. J Appl Cryst 2015;48:1604–6.
[52] Falke S, Dierks K, Blanchet C, Graewert M, Cipriani F, Meijers R, Svergun D, Betzel C. Multi-channel in situ dynamic light scattering instrumentation enhancing biological small-angle X-ray scattering experiments at the PETRA III beamline P12. J Synchrotron Radiat 2018;25:361–72.
[53] Nigro V, Angelini R, King S, Franco S, Buratti E, Bomboi F, Mahmoudi N, Corvasce F, Scaccia R, Church A, Charleston T, Ruzicka B. Apparatus for simultaneous dynamic light scattering–small angle neutron scattering investigations of dynamics and structure in soft matter. Rev Sci Instrum 2021;92, 023907.
[54] Sonje J, Thakral S, Krueger S, Suryanarayanan R. Reversible self-association in lactate dehydrogenase during freeze-thaw in buffered solutions using neutron scattering. Mol Pharm 2021;18:4459–74.
[55] Ahmed IA, Alonso-Caballero A, Baxa MC, Best RB, Campos LA, Cerminara M, Chu X, Chung HS, Contessoto VG, Czaplewski C, D'Amelio N, de Alba E, de Oliveira VM, De Sancho D, Ferreiro DU, Gai F, Garcia AE, Gołaś E, Gopich IV, Guzovsky AB, Holla A, Ibarra-Molero B, Karczyńska AS, Krupa P, Kubelka GS, Kubelka J, Lapidus LJ, Leite VBP, Lipska AG, Liwo A, Lubecka EA, Makowski M, Malhotra P, Mothi N, Mozolewska MA, Mukherjee D, Muñoz V, Naganathan AN, Nagpal S, Ołdziej S, Perez-Jimenez R, Sanchez-Ruiz JM, Schafer NP, Schönfelder J, Schuler B, Sieradzan AK, Sosnick TR, Udgaonkar JB, Veeramuthu Natarajan S, Wang Z, Wirecki T, Witalka D, Wolynes PG, Ye X, Zosel F. Protein folding: methods and protocols. In: Methods in molecular biology, vol. 2376. New York, NY: Springer US; 2022.
[56] Tripathi T. Calculation of thermodynamic parameters of protein unfolding using far-ultraviolet circular dichroism. J Proteins Proteomics 2013;4(2):85–91.
[57] Khodadadi S, Sokolov AP. Protein dynamics: from rattling in a cage to structural relaxation. Soft Matter 2015;11:4984–98.
[58] Melillo JH, Gabriel JP, Pabst F, Blochowicz T, Cerveny S. Dynamics of aqueous peptide solutions in folded and disordered states examined by dynamic light scattering and dielectric spectroscopy. Phys Chem Chem Phys 2021;23:15020–9.
[59] Drake JA, Pettitt BM. Physical chemistry of the protein backbone: enabling the mechanisms of intrinsic protein disorder. J Phys Chem B 2020;124:4379–90.
[60] Nag N, Chetri PB, Uversky VN, Giri R, Tripathi T. Experimental methods to study intrinsically disordered proteins. In: Tripathi T, Dubey VK, editors. Advances in protein molecular and structural biology methods. USA: Academic Press; 2022. p. 505–33, ISBN:978-032-39-0264-9. https://doi.org/10.1016/B978-0-323-90264-9.00031-3.
[61] Kumar P, Bhardwaj A, Uversky VN, Tripathi T, Giri R. Computational methods to study intrinsically disordered proteins. In: Tripathi T, Dubey VK, editors. Advances in protein molecular and structural biology methods. USA: Academic Press; 2022. p. 489–504, ISBN:978-032-39-0264-9. https://doi.org/10.1016/B978-0-323-90264-9.00030-1.
[62] Schuler B, Soranno A, Hofmann H, Nettels D. Single-molecule FRET spectroscopy and the polymer physics of unfolded and intrinsically disordered proteins. Annu Rev Biophys 2016;45:207–31.

[63] Dignon GL, Zheng W, Best RB, Kim YC, Mittal J. Relation between single-molecule properties and phase behavior of intrinsically disordered proteins. Proc Natl Acad Sci 2018;115:9929–34.
[64] Chan HS, Dill KA. Polymer principles in protein structure and stability. Annu Rev Biophys Biophys Chem 1991;20:447–90.
[65] Zhou H. Dimensions of denatured protein chains from hydrodynamic data. J Phys Chem B 2002;106:5769–75.
[66] Kohn JE, Millett IS, Jacob J, Zagrovic B, Dillon TM, Cingel N, Dothager RS, Seifert S, Thiyagarajan P, Sosnick TR, Hasan MZ, Pande VS, Ruczinski I, Doniach S, Plaxco KW. Random-coil behavior and the dimensions of chemically unfolded proteins. Proc Natl Acad Sci U S A 2004;101:12491–6.
[67] Hofmann H, Soranno A, Borgia A, Gast K, Nettels D, Schuler B. Polymer scaling laws of unfolded and intrinsically disordered proteins quantified with single-molecule spectroscopy. Proc Natl Acad Sci U S A 2012;109:16155–60.
[68] Nygaard M, Kragelund BB, Papaleo E, Lindorff-Larsen K. An efficient method for estimating the hydrodynamic radius of disordered protein conformations. Biophys J 2017;113:550–7.
[69] Fluegel S, Fischer K, McDaniel JR, Chilkoti A, Schmidt M. Chain stiffness of elastin-like polypeptides. Biomacromolecules 2010;11:3216–8.
[70] Cantor CR, Schimmel PR. Biophysical chemistry. New York: Freeman; 1980.
[71] Janshoff A, Neitzert M, Oberdörfer Y, Fuchs H. Force spectroscopy of molecular systems—single molecule spectroscopy of polymers and biomolecules. Angew Chem Int Ed 2000;39:3212–37.
[72] Collu G, Bierig T, Krebs A, Engilberge S, Varma N, Guixà-González R, Sharpe T, Deupi X, Olieric V, Poghosyan E, Benoit RM. Chimeric single α-helical domains as rigid fusion protein connections for protein nanotechnology and structural biology. Structure 2022;30:95–106.e7.
[73] Chubynsky M, Hespenheide B, Jacobs DJ, Kuhn LA, Lei M, Menor S, et al. Constraint theory applied to proteins. Nanotechnol Res J 2008;2(1):61–72.
[74] Wen Q, Janmey PA. Polymer physics of the cytoskeleton. Curr Opin Solid State Mater Sci 2011;15:177–82.
[75] Janmey PA, Hvidt S, Käs J, Lerche D, Maggs A, Sackmann E, Schliwa M, Stossel TP. The mechanical properties of actin gels. Elastic modulus and filament motions. J Biol Chem 1994;269:32503–13.
[76] Farge E, Maggs A. Dynamic scattering from semiflexible polymers. Macromolecules 1993;26:5041–4.
[77] Cai PC, Krajina BA, Kratochvil MJ, Zou L, Zhu A, Burgener EB, Bollyky PL, Milla CE, Webber MJ, Spakowitz AJ, Heilshorn SC. Dynamic light scattering microrheology for soft and living materials. Soft Matter 2021;17:1929–39.
[78] Fazelinia H, Balog ERM, Desireddy A, Chakraborty S, Sheehan CJ, Strauss CEM, Martinez JS. Genetically engineered elastomeric polymer network through protein zipper assembly. ChemistrySelect 2017;2:5008–12.
[79] Quiroz FG, Chilkoti A. Sequence heuristics to encode phase behaviour in intrinsically disordered protein polymers. Nat Mater 2015;14:1164–71.
[80] Peran I, Martin EW, Mittag T. Walking along a protein phase diagram to determine coexistence points by static light scattering. Methods Mol Biol 2020;2141:715–30.
[81] Weitzhandler I, Dzuricky M, Hoffmann I, Quiroz FG, Gradzielski M, Chilkoti A. Micellar self-assembly of perfectly sequence-defined recombinant resilin-like/elastin-like block copolypeptides. Biomacromolecules 2017;18:2419–26.
[82] Some D, Amartely H, Tsadok A, Lebendiker M. Characterization of proteins by size-exclusion chromatography coupled to multi-angle light scattering (SEC-MALS). JoVE 2019; e59615.

[83] Falke S, Dierks K, Blanchet C, Graewert M, Cipriani F, Meijers R, Svergun D, Betzel C. Multi-channel in situ dynamic light scattering instrumentation enhancing biological small-angle X-ray scattering experiments at the PETRA III beamline P12. J Synchrotron Radiat 2018;25:361–72.
[84] Destremaut F, Salmon J, Qi L, Chapel J. Microfluidics with on-line dynamic light scattering for size measurements. Lab Chip 2009;9:3289–96.
[85] Ross TD, Lee HJ, Qu Z, Banks RA, Phillips R, Thomson M. Controlling organization and forces in active matter through optically-defined boundaries. Nature 2019; 572:224–9.
[86] Seong J, Lin MZ. Optobiochemistry: genetically encoded control of protein activity by light. Annu Rev Biochem 2021;90:475–501.

CHAPTER 9

# Protein footprinting by mass spectrometry: H/D exchange, specific amino acid labeling, and fast photochemical oxidation of proteins

**Ravi Kant, Austin B. Moyle, Prashant N. Jethva, and Michael L. Gross**
Department of Chemistry, Washington University in Saint Louis, St. Louis, MO, United States

## 1. Introduction

Spectroscopic methods play a vital role in following protein high-order structural changes, mapping interactions, and determining binding affinity. Mass spectrometry (MS) is evolving to contribute to solving structural problems in protein science. Unlike many spectroscopic approaches, MS has a significant advantage in that it affords regional and even amino acid residue information by taking advantage of bottom-up proteomics, an approach that determines the primary structure for proteins in mixtures as complex as protein lysates. Hence, the approach reviewed here has been termed structural proteomics. Furthermore, the output is fast, sensitive, and informative. For disclosure, MS methods do not yet give structural information in the form of atomic coordinates as is the case with X-ray crystallography, nuclear magnetic resonance (NMR) spectroscopy, or cryo-electron microscopy (cryo-EM); instead, it maps changes in the structure such as those caused by mutation, ligand binding, aggregation, and protein folding/unfolding to name a few instances. This information is certainly structural, although not of the highest spatial resolution.

To work, the protein must be modified in some way to provide its "footprint." Moreover, the modifications must be amenable for detection and location by MS and tandem MS coupled to high-performance liquid chromatography (HPLC). This capability has been evolving for over 50 years as mass spectrometrists have been challenged to develop sensitive ionization methods that work with nonvolatile biomolecules, to design tandem

*Advanced Spectroscopic Methods to Study Biomolecular Structure and Dynamics*
https://doi.org/10.1016/B978-0-323-99127-8.00017-9

spectrometers that provide sensitive fragmentation of biomolecules and that are sensitive and fast, yielding product-ion mass spectra on the time scale of fractions of a second, and to interface these tandem spectrometers to high-performance chromatography working at flow rates of microliters/min and fast computers to process the massive data sets (thousands of spectra per second) that this approach produces. This field is growing rapidly now that instrumentation is highly evolved and is no longer an impediment. An encyclopedic review is not possible; indeed, a recent review is 100 printed pages and contains over 1000 references [1]. Further, there are other approaches besides footprinting that include native MS, crosslinking, ion mobility, and new approaches to induce fragmentation; however, these are not covered here. We covered a discussion of three approaches to footprinting: hydrogen/deuterium exchange (HDX), specific amino acid footprinting, and fast, free-radical (or other reactive species) footprinting, with the latter carried out on our platform called "fast photochemical oxidation of proteins" or FPOP. We do not wish to offend our colleagues in this area by omitting their work, but for efficiency, we have chosen examples of our work to illustrate our capabilities. The reader is informed that there are so many examples that justify our approach of presenting a tutorial review that emphasizes our work. We begin with the most heavily used approach, HDX mass spectrometry (HDX MS).

## 2. Hydrogen-deuterium exchange mass spectrometry

### 2.1 Introduction to HDX-MS

HDX-MS is a reversible footprinting approach and an important method for studying protein structure, dynamics, and function. HDX-MS provides peptide-level information and complements existing high-resolution techniques (i.e., X-ray crystallography, NMR, and cryo-EM). Historically, the application of HDX is not new, and it began with the first reported HDX experiment on dried pork insulin by Kaj Linderstrøm-Lang in 1954 [2]. HDX was later coupled with NMR to investigate small proteins [3–5]. HDX-NMR is still utilized and provides high spatial resolution information at the residue level, albeit only for small proteins. Because NMR is limited by the protein quantity and size, HDX coupled with MS provides an opportunity to overcome the shortcomings of NMR. Conventionally, the HDX-MS setup includes reversed-phase high-performance liquid chromatography (RP-HPLC) connected to a mass spectrometer. Practically, there is no protein size limit as proteins are cleaved by acid-stable proteases, separated by RP-HPLC, and measured at high mass resolving power on many instruments available today. Another advantage of HDX is that small quantities of protein (in picomoles) are adequate. Besides the usual HDX-MS experiment that occurs in seconds to hours, there are microfluidics-based and other fast HDX-MS experiments that reduce the labeling from milliseconds to seconds [6–8]. Moreover, the recent robotic advances in HDX-MS (i.e., automatic sample preparation, injection, and data collection) have further increased its application in structural MS [9].

## 2.2 HDX-MS workflow and mechanism

The conventional bottom-up HDX-MS workflow includes (1) equilibration in $D_2O$ (labeling), (2) acidified denaturant quench of the exchange, (3) digestion with acid-stable proteases, (4) chromatographic separation and mass spectrometric measurements of the molecular mass of the peptide, and (5) semiautomatic data analysis (Fig. 1). In the first step, proteins in solution (typically in protonated buffer) are diluted into a $D_2O$-based buffer that allows the labile hydrogens to undergo exchange with deuterium from $D_2O$. All hydrogens that are attached to heteroatoms (including those on the amide backbone (N—H) and side chains) can exchange with $D_2O$. The HDX of the protein side chains is not measured owing to their

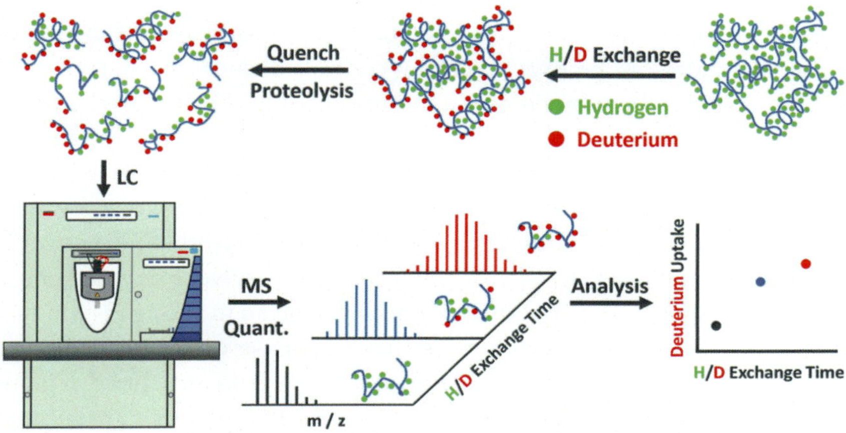


**Fig. 1** Schematic illustration of a bottom-up HDX workflow. *Green and red dots* in the protein structure indicate hydrogen and deuterium atoms in the peptide bonds, respectively. *(Adapted with permission from Liu XR, Zhang MM, Gross ML. Mass spectrometry-based protein footprinting for higher-order structure analysis: fundamentals and applications. Chem Rev 2020;120:4355–454. https://doi.org/10.1021/acs.chemrev.9b00815. Copyright 2020, American Chemical Society.)*

fast back exchange, whereas Hs bonded to backbone amides exchange in minutes to hours and can be measured by a conventional HDX-MS approach [10–12]. The exchange mechanism is both acid- and base-catalyzed, leading to a minimum exchange rate at a low pH of ~2.5 [13,14]. Therefore, to minimize back exchange, deuterated samples at different time points are prepared at pH 7 and are quenched with solutions containing denaturants (i.e., urea or guanidine-hydrochloride) at a pH of ~2.5. After quenching, the samples are digested with acid-stable proteases (i.e., pepsin, fungal XIII, and/or nepenthesin) [15,16]. A combination of multiple proteases ensures that small- to medium-size peptides are generated to provide good spatial resolution. The proteolytic step can be performed on ice, at room temperature, or at higher temperatures if the protein is resistant to proteases. The peptides are separated by RP-HPLC at low temperatures and detected using a mass spectrometer. The shift in the mass owing to deuterium incorporation can be calculated using several semiautomatic data analysis software packages [1]. This software calculates the centroid from the isotopic distribution of either constituent peptides or the intact protein in an undeuterated state and compares it with the centroid shift (increase in mass) of the same peptide or protein at longer times of HDX when the mass increases. A plot of HDX extent (*Y*-axis) versus time (*X*-axis) shows the kinetics of HDX. The rate at which amide hydrogens are replaced with deuterium is dependent upon several factors (i.e., pH, temperature, and biophysical properties of the protein (structure and dynamics)). Of these, pH and temperature can be experimentally controlled; thus, the shift in mass (Da) because of D incorporation can be attributed to the biophysical properties of the protein.

In a compact protein (native-like structure), the kinetics of the exchange vary depending upon the solvent accessibility and H-bonding [1]. For example, those parts of the protein involved in stable H-bonding networks (α-helices and/or β-sheets) exchange slower than the solvent-exposed, more disordered regions. If parts of the protein are involved in strong H-bonding or exist in solvent-inaccessible regions, they may never exchange deuterium under physiological conditions. Many of the N-Hs in structured regions exchange owing to thermally driven structural perturbations (breathing) in which H-bonds are in a dynamic equilibrium involving closed and open states. The hydrogen bond in the open state is transiently broken and eventually replaced with D from the solvent, and N-D goes back to the closed state. The scheme following this paragraph summarizes the mechanism of HDX of amide hydrogens [1,14,17].

$$NH_{closed} \underset{k_{closed}}{\overset{k_{open}}{\rightleftarrows}} NH_{open} \xrightarrow{k_{ch}} ND_{open} \underset{k_{open}}{\overset{k_{closed}}{\rightleftarrows}} ND_{closed}$$

In theory, the rate constants ($k_{HDX}$) for the exchange of every amino acid except proline (that lacks N–H) depend on $k_{open}$, $k_{closed}$, and $k_{ch}$, as shown in the above equation. The observed rate constant of HDX ($k_{HDX}$) is determined by two factors, namely, solvent accessibility and hydrogen bond stability of amide hydrogens. For loop and flexible regions, these two factors have a minimal impact on exchange, and overall, $k_{HDX}$ depends on $k_{ch}$. For the less flexible and/or ordered regions, the HDX can be categorized into EX1 and EX2 [11,18]. The majority of proteins exhibit the EX2 limit of HDX under solution conditions. In this scenario, $k_{cl}$ is greater than $k_{ch}$, and $k_{HDX}$ is determined by both $k_{open}$ and $k_{ch}$, in which $k_{open} = k_{open}/k_{closed}$. The EX1 regime is less common, $k_{ch}$ is greater than $k_{cl}$, and $k_{HDX}$ depends upon $k_{open}$ [18,19]. The EX1 regime provides kinetics, and the EX2 regime provides the thermodynamic aspect of the protein unfolding [20,21]. Proteins also undergo HDX, which could be a mixture of both EX1- and EX2-based mechanisms [18,22]. The majority of the HDX experiments are differential, where two states (free vs ligand-bound state, wild-type vs mutant) are compared, and the deuterium differences are calculated by taking the difference in HDX for one state and that of the other. These experiments reveal both local and allosteric or remote conformational effects.

The HDX for one state is conducted when a structural property of a protein needs to be investigated. In this kind of experiment, back exchange correction is critical before assigning the dynamic properties in distinct regions of a protein. Temperature-dependent HDX can also differentiate between local and global dynamics at the peptide level [23]. To investigate the global dynamics at the intact protein level, HDX at the protein level is an option, where all the experimental steps are the same as in conventional HDX-MS except that the protein is not cleaved by acid-stable proteases [24,25]. The global HDX provides an overall picture of the conformational states of the protein, whereas protease-based HDX-MS provides high-resolution information at both secondary and tertiary structure levels (high-order structure determination).

## 2.3 Recent applications

Owing to advances in technology, HDX-MS applications are now widespread and used to solve several problems. In this section, we will discuss epitope mapping of Duffy binding protein (DBP) [26], small-molecule binding to apoE3 [27], metal (calcium ions) binding to troponin [28], and calprotectin [29].

### 2.3.1 Epitope mapping, DBP, and inhibitory antibodies

In the area of biotherapeutics, HDX-MS has become an important tool for identifying the therapeutic antibody binding site on an antigen (epitope mapping). In a study performed in our lab, HDX-MS was implemented to identify epitopes for three inhibitory antibodies (mAbs), 2D10, 2H2, and 2C6, and one noninhibitory antibody, 3D10, against *Plasmodium vivax* Duffy binding protein (PvDBP) from mosquitos implicated in malaria. PvDBP binds to the DBP-II domain of the Duffy antigen receptor for chemokines (DARC) that is located on the host erythrocytes. This interaction is crucial for the infection of *P. vivax* into red blood cells [30–35]. A crystal structure of the 2D10-bound DBP complex is available [26].

Owing to the difficulties in crystallizing DBP complexes with mAbs 2C6 and 3D10, HDX-MS was implemented for epitope mapping. Differential HDX experiments show a significant reduction in HDX of the peptides spanning residues 408–442 in the 2D10-bound DBP (Fig. 2A). The HDX findings are consistent with the crystal structure of the 2D10-bound DBP complex. In the crystal structure, the epitope region encompasses the tail end of the helical bundle on SD3 (Fig. 2B). The HDX protection profile for 2H2 is similar to that of 2D10, indicating similar epitopes for both antibodies (Fig. 2A and B). To corroborate the binding site of 2D10 and 2H2, the binding residues were further validated by mutagenesis and competitive enzyme-linked immunosorbent assay (ELISA). The binding sites of 2D10 and 2H2 are similar but not identical (overlapping binding sites), as confirmed by mutagenesis. For 2C6, there is HDX protection for peptides from a different location (residues 471–484), indicating a distinct helical face epitope on SD3 rather than on the other two antibodies (Fig. 2A and B). This finding is supported by complementary data from small-angle X-ray scattering (SAXS), sequence analysis, and ELISA. The noninhibitory antibody 3D10 shows the N-terminal end of SD1 as a binding site. Overall, this study reveals that the tail end of the helix on SD3 is a hotspot where inhibitory antibodies bind and inhibit the interaction between DBP and DARC. This motif should be included in the future DBP-II vaccine design to afford treatment against malaria. The study is a typical illustration of HDX for delineating an epitope.

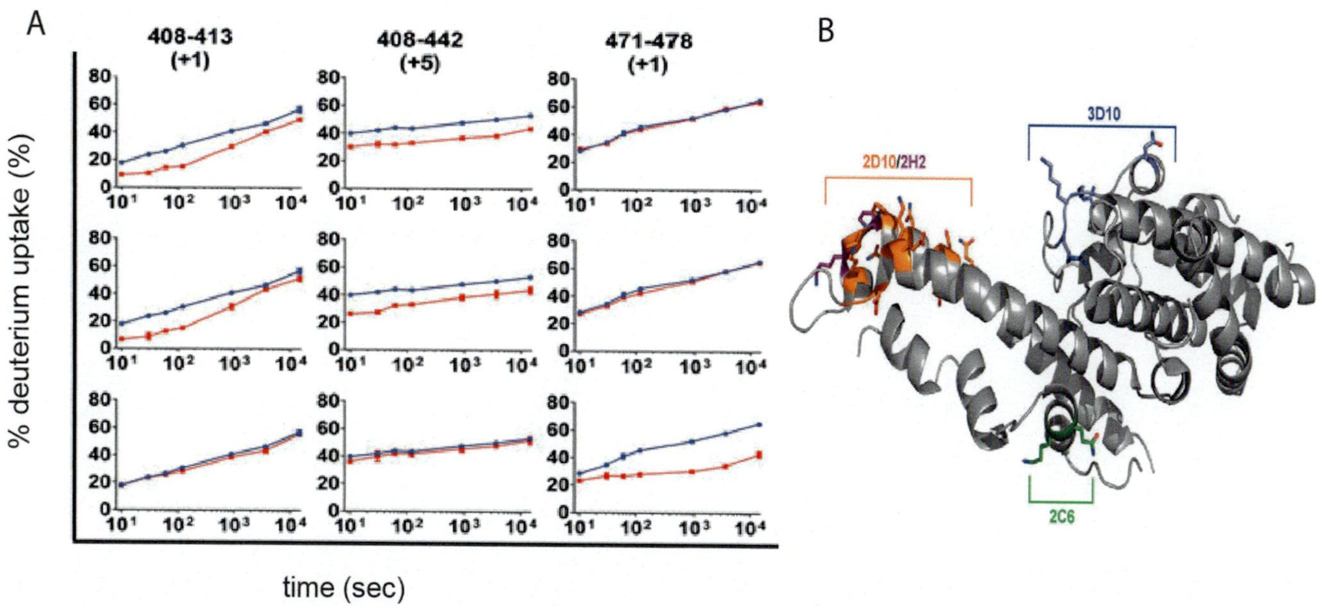


**Fig. 2** (A) Kinetic plots show the comparison of three peptides from DBP in both apo *(blue)* and holo states *(red)*. Each row shows a different monoclonal antibody: top 2D10, middle 2H2, and bottom 2C6. The peptides with reduced HDX in the bound state are putative binding sites/epitopes on DBP. The peptides with no significant differences can be considered as controls. (B) The epitopes of 2D10, 2H2, 2C6, and 3D10 mapped on the PvDBP crystal structure; 2D10 in *orange*, 2H2 specific residues in *purple*, 3D10 in *blue*, and 2C6 in *green*. *(Adapted with permission from Chen E, Salinas ND, Huang Y, Ntumngia F, Plasencia MD, Gross ML, Adams JH, Tolia NH. Broadly neutralizing epitopes in the Plasmodium vivax vaccine candidate Duffy Binding Protein. Proc Natl Acad Sci 2016;113:6277. https://doi.org/10.1073/pnas.1600488113. Copyright 2016, National Academy of Sciences.)*

### 2.3.2 Small-molecule binding study, ApoE3 and EZ482

A second study shows an unusual application of HDX to determine the affinity of binding in addition to locating the binding site. The example utilizes the apoE family, which comprises three isoforms that vary by single amino acid substitution at two positions out of 299 residues. These positions are Cys112/Cys158, Cys112/Arg158, and Arg112/Arg158 in E2, E3, and E4 isoforms, respectively. ApoE2 and E3 seem to have little or no role in the etiology of Alzheimer's disease (AD); however, apoE4 is primarily associated with late-onset AD [36–38]. The underlying mechanism behind this difference remains elusive. We speculated that late-onset AD might be a consequence of the alteration in structural properties of apoE4 relative to apoE3. In this study, we used HDX to examine the effect of a small molecule, a potential drug candidate EZ482, on the structural dynamics of apoE3. After assessing the location of the ligand-induced changes, the binding affinity of apoE3 was determined by fitting a titration curve (PLIMSTEX) to obtain affinity. In the first step of PLIMSTEX (Protein-Ligand Interaction by MS, Titration, and H/D EXchange), curves are obtained by titrating a fixed concentration of apoE3 with increasing concentrations of the ligand (Fig. 3). Three different trends are observed: for example, peptides from the C-terminus show a decrease in uptake (a putative binding region), peptides from the N-terminus show an increase in uptake (remote conformational change), and some peptides from N- and C-termini show no change (unaffected region). Specifically, the PLIMSTEX curves of the C-terminus, 229–235, 234–243, and 258–265, show binding. The binding constants for the first two peptides (229–235, 234–243) from the C-terminus are similar, whereas the binding constant is slightly larger for 258–265. This difference in binding constants indicates distinct protein conformations at the C-terminus. One of the characteristics of apoE3 is the tendency to self-oligomerize even at low-micromolar concentrations. The C-terminus, which seems to be a putative binding region based on our HDX data, can also participate in the self-association of apoE monomers. This raises the question of whether the measured binding affinity by PLIMSTEX taken from processing HDX readouts is accurate. The PLIMSTEX-based measured binding constants are supported by previously reported binding affinities of apoE determined with fluorophores and fluorescence-based assays. Moreover, previous analytical ultracentrifugation shows that the status of apoE oligomers is not influenced by EZ482. Therefore, we are confident that the mathematical fit of the PLIMSTEX curve (assuming a 1:1 binding system) is independent of the self-association of apoE.

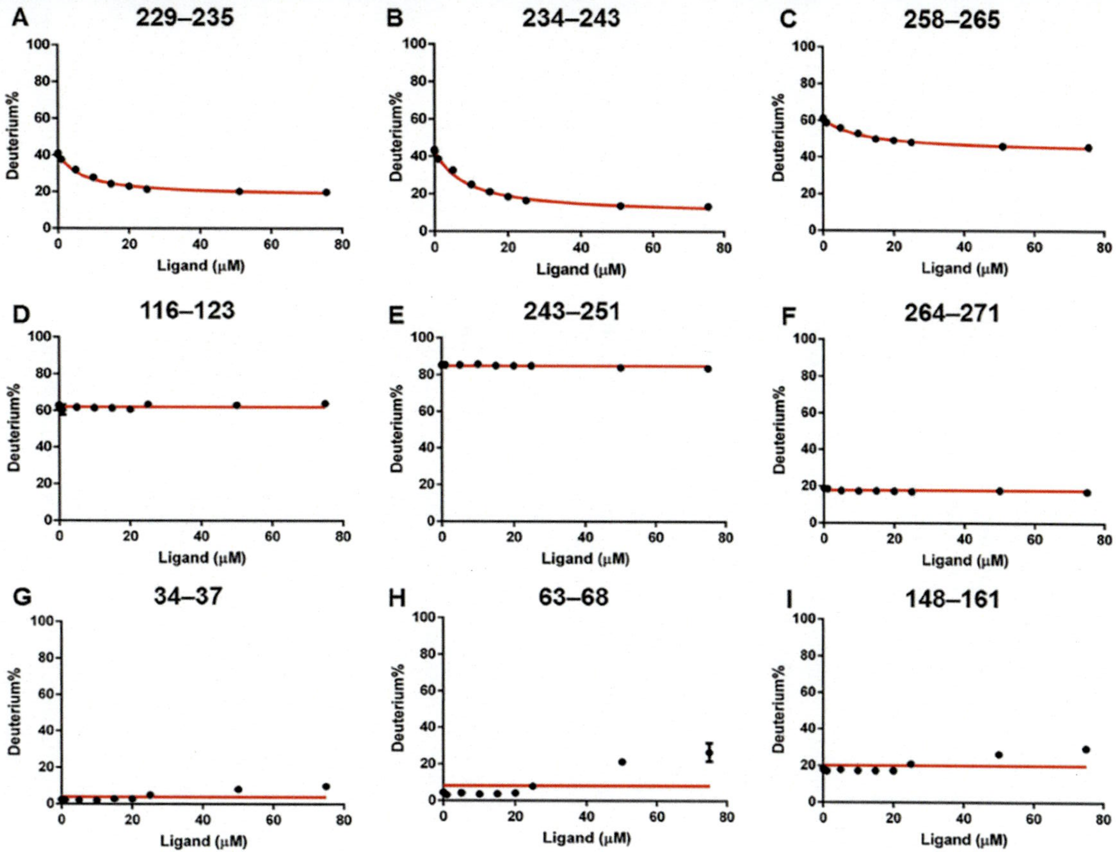


**Fig. 3** The peptides represent the PLIMSTEX curve obtained for ApoE at several concentrations of ligands. A decrease in HDX shows the putative binding site (A–C), an increase in HDX (G–I) shows remote conformational changes, and no differences in HDX refer to unaffected regions (D–F). *Red lines* in the curve are 1:1 fitting curves. *(Adapted with permission from Wang H, Rempel DL, Giblin D, Frieden C, Gross ML. Peptide-level interactions between proteins and small-molecule drug candidates by two hydrogen–deuterium exchange MS-based methods: the example of apolipoprotein E3. Anal Chem 2017:89(20):10687–95. https://doi.org/10.1021/acs.analchem.7b01121.)*

Lipids induce a large-scale conformational change in apoE, and investigating EZ482 on apoE3 should provide insight into regions of distal changes. To follow this, a modified SUPREX (Stability of Unpurified Proteins from Rates of H/D EXchange) was implemented. SUPREX is an HDX variant that measures both protein-folding energy changes and binding constants [39]. Traditional SUPREX involves HDX measurements at different time intervals in varying denaturant concentrations. In this approach, HDX is fixed for 2 min, and the urea concentration is varied. The EZ482 bound state, especially in the putative binding regions (229–235, 234–243, and 258–265), undergoes less HDX than the unbound state, in agreement with PLIMSTEX. A closer look shows that the SUPREX curve of the EZ482 bound state shifts further to the right compared to the unbound state, indicating stability from ligand binding. For most parts of the C-terminus, no differences occur for PLIMSTEX and SUPREX, indicating a flexible or solvent-accessible nature. This study also points to a correlation between relative unfolding upon ligand binding ($\Delta C_{1/2}\%$) and amplitude of the PLIMSTEX curve. For example, a peptide from 234 to 243 retains the largest $\Delta C_{1/2}\%$ and $\Delta$HDX%, indicating this region to be most sensitive to ligand binding among the putative binding regions. Overall, this study introduces modified approaches of HDX to determine the binding interfaces, affinities, and thermodynamics of a protein/ligand system.

### 2.3.3 Metal-binding study: Troponin C and $Ca^{2+}$ binding

In another example of obtaining structural information, we turn to the $Ca^{2+}$ binding protein, troponin C (TnC). Troponin plays a vital role in muscle contraction in mammals, birds, and some invertebrates [40–42]. In the skeletal muscle, TnC can bind four $Ca^{2+}$ ions with four EF-hands (protein binding sites). The two C-terminal EF-hands (EF-III and EF-IV) have a higher affinity for $Ca^{2+}$ than the other two EF-hands (EF-III and EF-IV) [43,44]. Despite several biophysical studies involving NMR, fluorescence, circular dichroism, calorimetry, and $Ca^{2+}$-selective electrode measurements, the binding order of $Ca^{2+}$ on TnC remained unclear. To fill this gap, we used PLIMSTEX not only to assess the binding affinity but also to determine the binding order of $Ca^{2+}$ to the EF-hands of TnC. In the first step, a differential global HDX measurement comparing $Ca^{2+}$ bound and unbound showed a decrease in uptake that indicates a rigid structure of TnC upon $Ca^{2+}$ binding. After confirming $Ca^{2+}$-induced conformational change, the equilibrium constant for $Ca^{2+}$ binding to TnC was determined by titrating a fixed TnC concentration (PLIMSTEX) with $Ca^{2+}$. The kinetics of

D uptake was also measured for two fixed protein concentrations, 0.3 μM (~in $K_d$ range) and 15 μM (~100 times $K_d$). The titration at the latter concentration resulted in a sharp break curve with a stoichiometry ratio of 4, which is consistent with previously reported four $Ca^{2+}$ binding to TnC. The four binding constants ($K_a$'s, in the low μM range) of TnC, determined by PLIMSTEX, agree within a factor of 5 with previous findings.

After confirming the stoichiometry and affinity, the binding order of $Ca^{2+}$ was revealed by examining the variation of the deuterium uptake of the peptides representing EF-hands as a function of $Ca^{2+}$-bound TnC states. The first HDX protection occurs only in the peptide (100–109) from the EF-III hand (transition TnC-0 $Ca^{2+}$ to TnC-1 $Ca^{2+}$), whereas peptides spanning the other three EF-hands remain unchanged. This indicates that the first $Ca^{2+}$ binds to EFIII. Similarly, peptides from EF-IV show HDX protection (a change of 6-4 deuteriums) in transitioning from TnC-0 $Ca^{2+}$ to TnC-1 $Ca^{2+}$, indicating that EF-IV is the second $Ca^{2+}$ binding site. In both transitions, TnC-1 $Ca^{2+}$ and TnC-2 $Ca^{2+}$, peptides from EF-I and EF-II are not HDX-protected, but in the TnC-3 $Ca^{2+}$ transition, peptides representing EFI show HDX protection. Finally, in the TnC-4 $Ca^{2+}$ transition, EF-II shows HDX protection (Fig. 4). Overall, the $Ca^{2+}$ binding can be assigned as follows: EF-III > EF-IV > EF-I > EF-II. This binding order of $Ca^{2+}$ to TnC is consistent with the higher binding affinity of EF-III, EF-IV, and relatively lower affinities of EF-I and EF-II hands.

### 2.3.4 Metal-binding study: Calprotectin and $Ca^{2+}$ binding

Calprotectin (CP) is a $Ca^{2+}$ and a transition-metal-ion binding protein and an essential component of the innate immune system [43]. In response to microbial infection, neutrophils and epithelial cells secrete CP in the extracellular space (ECS) to sequester transition metals $Mn^{2+}$, $Fe^{2+}$, $Ni^{2+}$, and $Zn^{2+}$, which are necessary for microbial survival. In the extracellular environment, $Ca^{2+}$ binds to CP and enhances not only its binding affinity to transition metal ions but also its antimicrobial activity and resistance to the proteolytic activity of serine proteases [43,45–49]. Upon $Ca^{2+}$ binding, the heterodimer transitions into a heterotetramer [50–52], raising the question of the mechanism behind this oligomerization of human calprotectin (hCP).

To address this critical question, we implemented a novel integrated MS-based structural proteomics approach that includes HDX-MS, PLIMSTEX, and native MS. We chose an hCP, CP-Ser, and an S100A8(C42S)/S100A9(C3S) variant of hCP in an oligomeric state. Each subunit has two $Ca^{2+}$ binding sites, one at the N-terminus, a noncanonical EF-hand, and the

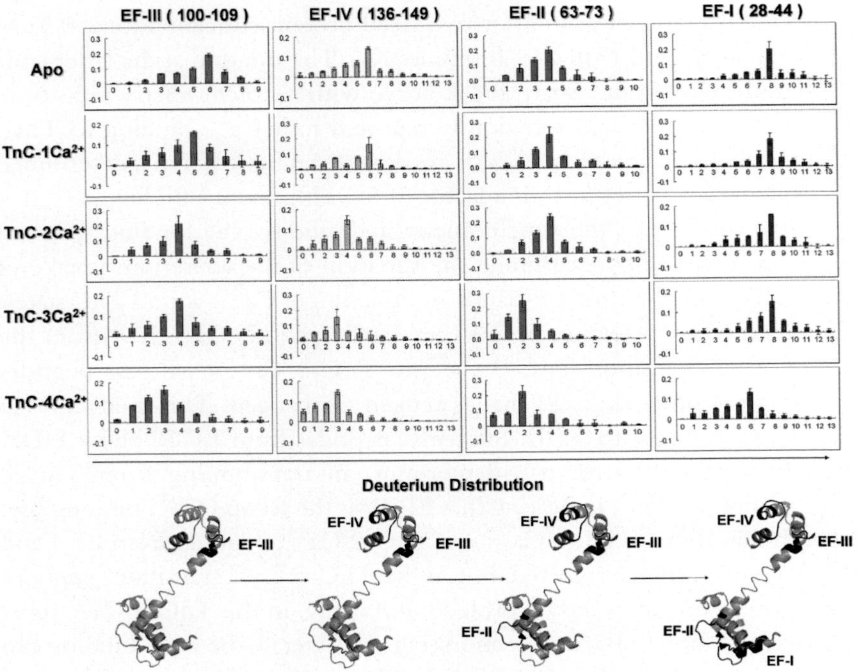


**Fig. 4** Top: Each column represents the deuterium distributions of the peptides representing EF-hands in the left to right order, III, IV, II, and I. Each row denotes the distinct TnC states, first row (apo), second row (apo + 1$Ca^{2+}$), third row (apo + 2$Ca^{2+}$), fourth row (apo + 3$Ca^{2+}$), and final row (apo + 4$Ca^{2+}$). Bottom: The crystal structure of TnC (PDB: 1TCF) shows the four EF-hands (in *black*) in their binding order. *(Adapted with permission from Huang RYC, Rempel DL, Gross ML. HD exchange and PLIMSTEX determine the affinities and order of binding of $Ca^{2+}$ with troponin C. Biochemistry 2011;50:5426–35. https://doi.org/10.1021/bi200377c. Copyright 2011, American Chemical Society.)*

other at the C-terminus, a canonical EF-hand. A differential HDX experiment comparing $Ca^{2+}$-bound and unbound hCP shows protection for $Ca^{2+}$ binding not only at the N- and C-terminal EF-hands but also at the heterotetrameric interface and other remote regions in both S100A8 and S100A9 subunits. In comparison to the N-terminal EF-hand, HDX protection in the C-terminus is greater, indicating relatively stronger binding of $Ca^{2+}$ with the EF-hands of the C-terminus (Fig. 5A). Interestingly, $Ca^{2+}$ binding has a minimal effect on the C-terminal tail, which is known to play an important role in transition metal sequestration. Using HDX data as a foundation, we implemented PLIMSTEX, where CP-Ser is titrated with $Ca^{2+}$ to determine the binding affinity and binding order.

PLIMSTEX-based sharp-break curves reveal that the CP-Ser heterodimer binds four $Ca^{2+}$ ions (Fig. 5B). To find the number of $Ca^{2+}$ ions required for conversion from the dimer to a tetramer, native mass spectra of CP-Ser (15 μM) in the presence of 0–500 μM $Ca^{2+}$ ions show that the stoichiometry is 8, in agreement with the PLIMSTEX shark break curve for the dimer. Additionally, native MS also showed that heterodimers with 4 $Ca^{2+}$ prefer a tetrameric state.

Unlike troponin C, the PLIMSTEX trend of hCP is similar and does not readily reveal the binding order of $Ca^{2+}$ to CP-Ser. To unravel the binding order, mathematical modeling shows that the first two $Ca^{2+}$ bind at the

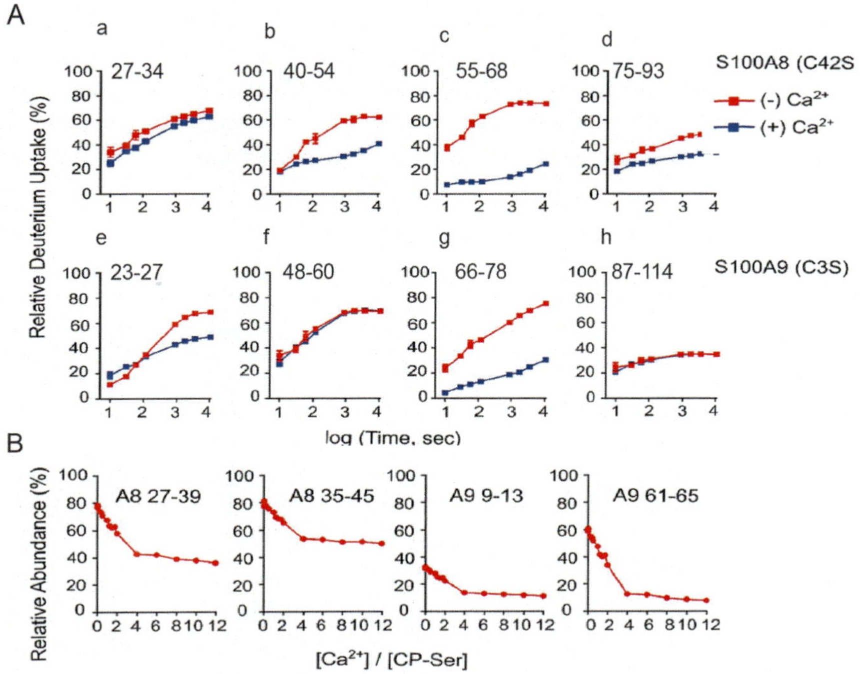


**Fig. 5** (A) Represents the PLIMSTEX data for the titration of CP-ser with $Ca^{2+}$ ions. a–d and e–h data are from S100A8/C42S AND S100A9/C3S, respectively. Peptides representing EF-hands (a, c, e, g) and other regions undergoing remote conformational change (b, d, f, h) are displayed. (B) The plots represent sharp-break PLIMSTEX curves obtained for both subunits and affording stoichiometry. *(Adapted with permission from Adhikari J, Stephan JR, Rempel DL, Nolan EM, Gross ML. Calcium binding to the innate immune protein human calprotectin revealed by integrated mass spectrometry. J Am Chem Soc 2020;142:13372–83. https://doi.org/10.1021/jacs.9b11950. Copyright 2020, American Chemical Society.)*

C-terminal canonical EF-hands with high affinity (the same disassociation constant $K_1$, and $K_2$, in the low micromolar range), whereas the other two $Ca^{2+}$ bind at the N-terminal, noncanonical EF-hands with low affinity ($K_3$, twofold larger than $K_1$ and $K_2$ and $K_4$, sevenfold larger than $K_1$ and $K_2$). Further, we speculate that the S100A9 noncanonical EF-hand that binds the third $Ca^{2+}$ ion undergoes slow conformational changes, whereas S100A9 binds to the fourth $Ca^{2+}$ ion, giving a sharp change in conformation. Interestingly, the best fit gives a much lower tetramerization $K_d$ that indicates a strong binding interface between heterodimers.

Overall, this study supports a model in which hCP exists in the dimeric form in the cytoplasm (low nanomolar concentration) and is half bound to $Ca^{2+}$ in that medium. When hCP is released in ECS, where $Ca^{2+}$ is at a micromolar concentration, a third $Ca^{2+}$ ion binds and sets the stage for binding a fourth $Ca^{2+}$. Our modeling further suggests that the driving force for tetramerization with four $Ca^{2+}$ ions is so favorable that it compensates for the very weak binding affinity of the fourth $Ca^{2+}$ ion and shifts the equilibrium toward a functionally relevant heterotetrameric state.

## 3. Specific amino acid labeling

### 3.1 Introduction

Complementary to broadly reactive HDX-MS data, nonlaser-initiated, specific amino acid labeling affords an approach to modify, or "footprint," specific amino acid side chain(s), usually with high specificity [53]. Such reagents are small molecules that react more slowly than free radicals, and their activation does not require a plasma, laser, or synchrotron [54–56]. The reagents are readily available from commercial sources, making protein footprinting by this approach accessible to any laboratory with high-performance mass spectrometry. Although this category of footprinting includes reagents such as *N*-ethyl maleimide (NEM) [57,58] (Cys), ethyl acetimidate hydrochloride (ETAT) [59,60] (Lys), and methylglyoxal (MG) [59,61] (Arg), it generally affords high specificity and also offers broad applicability when using reagents reactive to the relevant functional groups. This is the case for glycine ethyl ester (GEE) [62,63] and benzhydrazide (BHD) [64] (acid footprinters) and diethyl pyrocarbonate (DEPC) [65,66] (nucleophiles). Generally, high specificity allows the user to target only residues of interest, and this approach divides the modified LC peptide signal by fewer species to simplify data analysis and increase sensitivity. If greater protein coverage is desired, multiple reagents can be applied in parallel experiments to afford complementary data sets combined for targeted analysis at the residue level across the protein sequence [59]. Typically, reaction yields are higher for specific amino acid labeling than for radical-based chemistries because the reagent can be in large excess, and the reaction time can be minutes (not milliseconds as for radicals), giving a greater dynamic range, easier detection, and easier relative quantification for differential measurements. In general, specific amino acid labeling elucidates protein side-chain dynamics, providing an efficacious complement to the HDX-MS reporting of backbone dynamics.

The typical specific amino acid labeling workflow involves adding the reagent from a concentrated stock in excess (generally 5–2000 M equivalents) to the protein in buffer solution (Fig. 6). After an appropriate reaction time (minutes to hours at room temperature (RT)), a quenching reagent with reactive functional groups is introduced in excess of the footprinter (commonly $\geq$50 equivalents) to consume the reagent and halt the reaction. Alternatively, rapid centrifugation spin/SEC desalting columns can also remove the labeling reagent. Overall, protein footprinting with this technique can be completed in less than an hour. Unlike labile HDX reactions, the usual high stability of irreversibly attached footprints (exception is DEPC [66,67]) permits an extensive bottom-up proteomics workflow, including overnight digestion using one or more specific proteases (i.e., trypsin, chymotrypsin, and GluC). Following isolation and digestion, multihour gradients for high-resolution LC are used to separate the peptides (Fig. 6) and MS and MS/MS are used to identify the peptides and locate the modification sites. The modified peptides often have altered hydrophobicity compared to the unmodified, and the LC retention time shifts accordingly. The ability to separate singly modified peptides with different modification sites is especially desirable as it allows for subresidue level resolution for peptides containing several reactive sites. The extracted ion chromatograms for modified peptides are used to quantify the modification, allowing for comparative studies of multistate systems (Fig. 6) to identify changes in protein solvent accessible surface area (SASA) and dynamics.

Although some reagents are well established for protein footprinting (GEE and DEPC), the field continues to grow as reagents with more desirable attributes are repurposed for footprinting. For example, BHD, as a footprinter of Glu and Asp, affords the user a one-step synthesis combined with isotopic encoding, improved reactivity for Glu and Asp, and no hydrolysis product compared to that in the well-established GEE footprinting [64]. Similarly, recently introduced benzoyl fluoride (BF), where fast acyl transfer

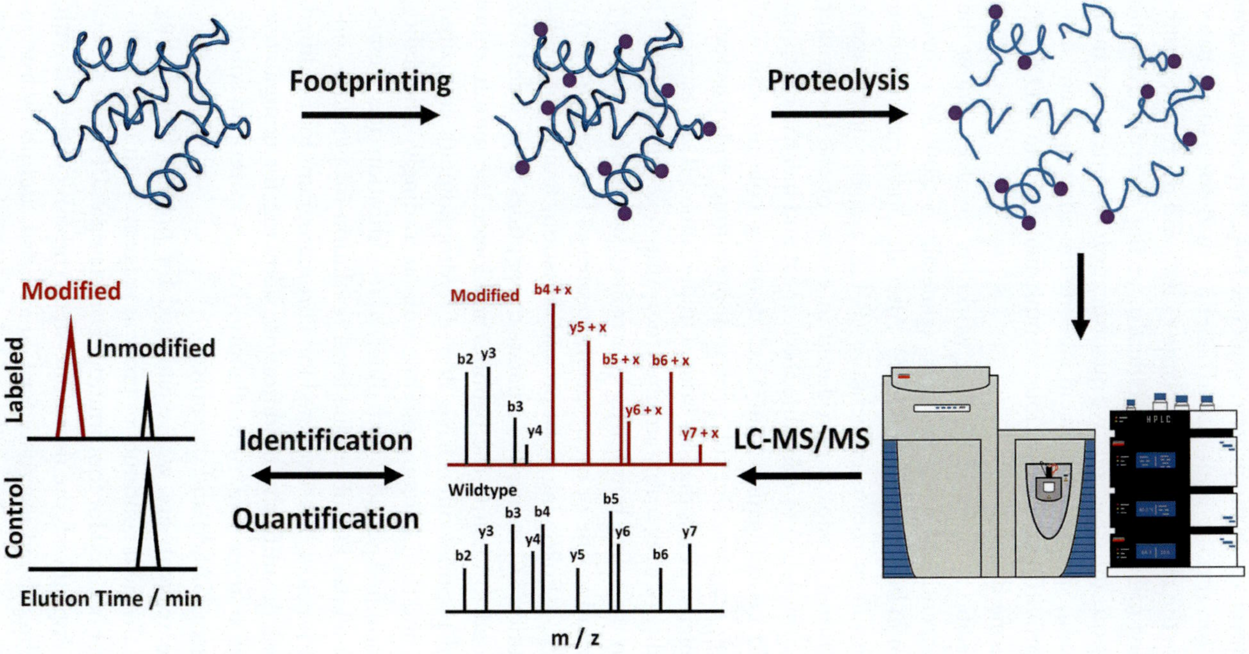


**Fig. 6** Workflow for specific amino acid labeling. A buffered protein is subjected to modification by footprinting, and the reaction is quenched. The labeled, intact protein is subjected to in-solution digestion, and the modified peptides are submitted to high-resolution LC-MS/MS. Analysis of the fragment b- and y-ions containing the modified residue *(red)* determines the modification site by shifts in mass corresponding to the reagent, and quantification can be performed by integrating extracted ion chromatograms (EICs) corresponding to the modified and unmodified peptides. The resulting EIC peak areas for unmodified *(black)* and modified *(red)* peptides are ratioed to quantify the modification and monitor changes in protein dynamics. *(Adapted with permission from Liu XR, Zhang MM, Gross ML. Mass spectrometry-based protein footprinting for higher-order structure analysis: fundamentals and applications. Chem Rev 2020;120:4355–454. https://doi.org/10.1021/acs.chemrev.9b00815. Copyright 2020, American Chemical Society.)*

chemistry is modulated by the electronegative leaving group fluorine, affords 10× greater reactivity with weakly nucleophilic, OH-containing Tyr compared to DEPC and increased hydrophobicity for the modified peptides for improved chromatographic separation [68].

## 3.2 Recent applications

### 3.2.1 FMO orientation between the membrane and chlorosome, elucidated by GEE footprinting

GEE footprinting uses a two-step labeling process pioneered by Hoare and Koshland [69], requiring 1-ethyl-3-(3-(dimethylamino)propyl)-carbodiimide (EDC) coupling chemistry to modify acidic residues Asp and Glu. Using GEE to modify solvent-accessible Asp/Glu residues, Blankenship and Gross groups elucidated key interactions for the Fenna-Matthews-Olsen (FMO) protein, a large photosynthesis complex [70]. The antenna protein is a critical energy-transfer vehicle in green-sulfur-bacteria photosynthesis. The chlorosome, a protein antenna complex, absorbs light and transfers it through the FMO protein to the reaction center (RC) in the cytoplasmic membrane (CM) (Fig. 7A). FMO is a disk-shaped trimer of three identical 40 kDa subunits, with each subunit containing seven bacterial chlorophyll *a* (Bchl a) antenna molecules. The X-ray crystal structure for FMO is known, but its orientations and interactions with the chlorosome and RC have not been elucidated. Such a structure-function relationship is critical for understanding a protein that could yield clues for light-harvesting mechanisms.

To probe the orientation of FMO protein between the membrane and chlorosome, we designed a three-state system: (1) free, soluble FMO protein, (2) FMO and membrane-bound RC (no chlorosome), and (3) chlorosome, FMO, and membrane-bound RC (Fig. 7B). Each state was labeled with Asp/Glu-reactive GEE and submitted to LC-MS/MS to measure changes in protein SASA and dynamics. The modified peptides resulting from digestion are grouped into three categories. The first (Fig. 7C) are those that showed decreased GEE modification for FMO in the complete system of the native membrane and chlorosome *(blue)* and a system that is chlorosome-depleted *(green)* relative to free, soluble FMO control *(red)*. These results indicate that such peptides, mapped to structure (Fig. 7D), are involved in interactions with the membrane. The second group of peptides shows increased GEE modification for a system that is 'chlorosome-depleted (Fig. 7E, *green*). This is hypothesized to occur because the concentration of the GEE reagent had to be increased to compensate for

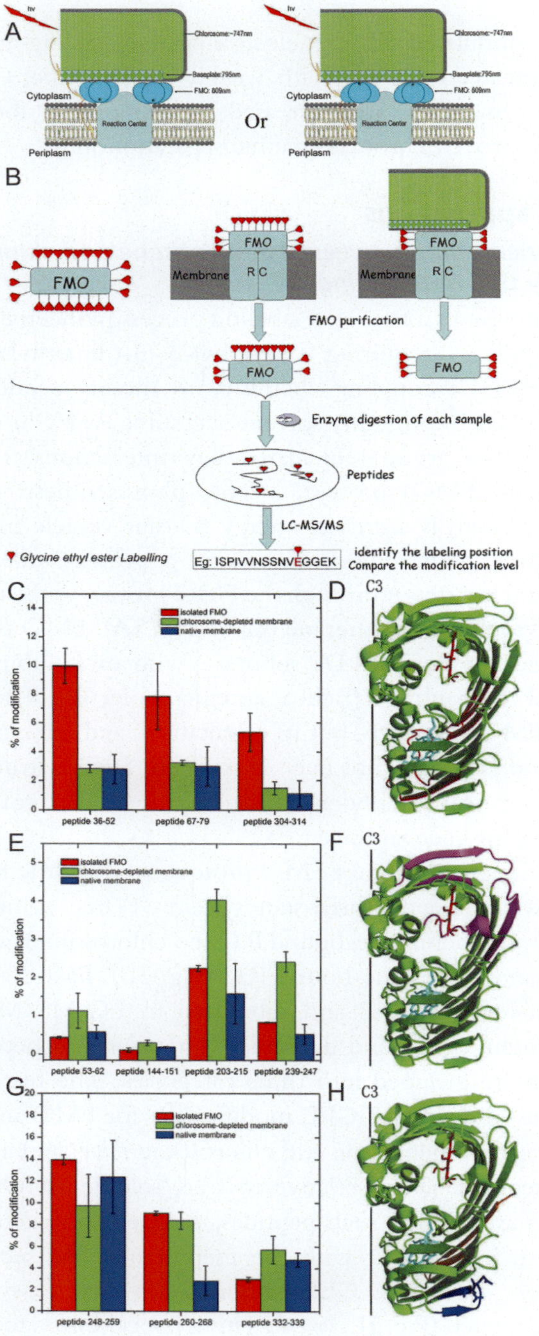

A
hv
Chlorosome:~747nm
Baseplate:795nm
FMO: 809nm
Cytoplasm
Reaction Center
Periplasm
Or
B
FMO
Membrane
R C
FMO purification
Enzyme digestion of each sample
Peptides
LC-MS/MS
Glycine ethyl ester Labelling
Eg: ISPIVVNSSNVEGGEK
identify the labeling position
Compare the modification level
C
isolated FMO
chlorosome-depleted membrane
native membrane
% of modification
peptide 36-52
peptide 67-79
peptide 304-314
D
C3
E
peptide 53-62
peptide 144-151
peptide 203-215
peptide 239-247
F
G
peptide 248-259
peptide 260-268
peptide 332-339
H


*(Continued)*

**Fig. 7** (A) Model of the photosystem, showing energy transfer from the chlorosome to the reaction center. Two possible models of FMO orientation between the chlorosome and membrane, including the site of interest Bchl a #3 denoted with star. (B) Overview of GEE labeling to probe SASA of FMO regions, including unbound, chlorosome-depleted membrane-bound, and native membrane-bound. The FMO protein is enzymatically digested and submitted to LC-MS/MS, where GEE modifications are quantified and compared across conditions. (C) GEE modification for peptides 36–53, 67–79, and 304–314 showing decreases for the chlorosome-depleted membrane and native membrane, with regions mapped onto the structure. (D, E) GEE modification for peptides 53–62, 144–151, 203–215, and 239–247, showing increased modification for the chlorosome-depleted membrane relative to other states, with regions mapped onto the structure. (F, G) Peptides 248–259, 260–268, and 332–339, showing mixed character across membrane-bound conditions, mapped onto the structure. (H) Bchl a #3 and #1 are shown as *cyan* and *red*, respectively. *(Adapted with permission from Wen J, Zhang H., Gross ML, Blankenship RE. Membrane orientation of the FMO antenna protein from* Chlorobaculum tepidum *as determined by mass spectrometry-based footprinting. Proc Natl Acad Sci 2009;106:6134–9. https://doi.org/10.1073/pnas.0901691106. Copyright 2009, Proceedings of the National Academy of Sciences.)*

higher relative levels of modifiable sites in the membrane. When situated in the native membrane *(blue)*, there is meaningful protection relative to the chlorosome-depleted system *(green)*, indicating that the regions represented by those peptides (Fig. 7F) are sites of interaction with the chlorosome. The third category shows mixed character, with peptides 248–259 and 260–268 showing similar or higher GEE modification for the free FMO relative to bound states (Fig. 7G). The protein regions represented by these peptides are on the middle and right of the protein (Fig. 7H, *orange*). Peptide 332–339 is in a loop-like region of the protein predicted to have greater solvent accessibility (Fig. 7H, *blue*), so the modification is expected to correlate with relative GEE concentration.

Taken together, GEE labeling of FMO protein in complex with chlorosome-depleted and native membranes elucidates the orientation of FMO. The Bchl *a* #3 side consists of peptides involved in the first group (Fig. 7D), whereas the Bchl *a* #1 side affords peptides involved in the second group (Fig. 7F). These data resolve a controversial hypothesis from theoretical predictions that the Bchl *a* #3 side interacts with membrane RC, with Bchl *a* #3 being a key trap to transfer energy to the CM; conversely, the Bchl *a* #1 side interacts with the chlorosome (Fig. 7A, left). These data also support the presence of an eighth antenna pigment that transfers energy between FMO and CM, indicating that FMO protein is not tightly packed against CM owing to high modification in the bound states in group A and peptide 332–339 on the Bchl *a* #3 side. FMO is also likely not tightly packed against chlorosomes as free FMO and native protein-bound FMO experience comparable modification for peptides on the Bchl *a* #1 side. Indeed, both interfaces allow for solvent permeability. In summary, a single GEE labeling experiment provides insight into both binding interfaces, protein orientation, and protein dynamics for a challenging and important system.

### 3.2.2 Siderocalin: Footprint Arg and Lys

Highly specific amino acid labeling reagents used in parallel can provide complementary, targeted insights. ETAT and MG are footprinting reagents for Lys and Arg, respectively (Fig. 8A and B) [60,61]. Using native MS, HDX-MS, ETAT labeling, and MG labeling while working with the Henderson group at Washington University, the Gross group characterized interactions between an iron sequestering protein (siderocalin, Scn) and an iron sequestering siderophore, linear enterobactin (lin-Ent), associated with pathogenic *E. coli*, under ferric and aferric conditions [59]. Both humans and bacteria produce agents to rob the other of energy-rich iron. Although bacteria employ small-molecule chelators, specifically siderophores, humans secrete Scn, a 20.8 kDa protein, to sequester siderophores. Pathogenic *E. coli* hydrolyzes enterobactin to form lin-Ent, implicated in virulent urinary tract infections. Because the structure of Scn in complex with lin-Ent is not known, footprinting MS was employed. Before footprinting, native MS confirmed lin-Ent and Fe-lin-Ent binding with Scn. This was a critical quality control step in covalent labeling MS to ensure ligand binding, sample purity, and the expected oligomerization state before devoting necessary instruments and human resources for footprinting MS. Then, HDX-MS was employed to provide insight into peptide level backbone dynamics upon lin-Ent binding. Specifically, peptides from the β-barrel core of Scn that are hypothesized to hold siderophores generally showed protection, including peptides containing R83, K127, and K136, which provide electrostatic/cation-pi interactions with the siderophore ligand. These differential results are particularly striking owing to the already rigid, ordered structure of the protein core, which limits the dynamic range of HDX.

To improve the spatial resolution while elucidating side-chain specific interactions with lin-Ent and Fe-lin-Ent, ETAT and MG labeling were performed in a three-state experiment. Most significantly, peptides containing R83, K127, and K136 demonstrated significant decreases in modification for the ligand-bound states relative to unbound (Fig. 8C and D). Additionally,

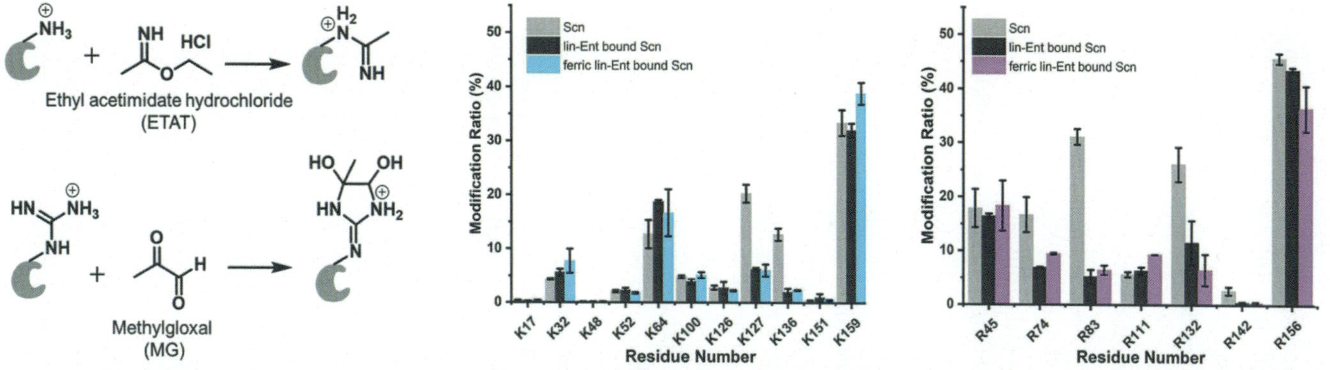


**Fig. 8** (A) Modification of amine-containing Lys with ETAT. (B) Modification of guanidino-containing Arg with MG. (C) Differential labeling of Lys by ETAT of three states of Scn: free control Scn, lin-Ent bound Scn, and ferric lin-Ent bound Scn. (D) Differential labeling of Arg by MG of three state Scn: free control Scn, lin-Ent bound Scn, and ferric lin-Ent bound Scn. *(Adapted with permission from Guo C, Steinberg LK, Cheng M, Song JH, Henderson JP, Gross ML. Site-specific siderocalin binding to ferric and ferric-free enterobactin as revealed by mass spectrometry. ACS Chem Biol 2020;15:1154–60. https://doi.org/10.1021/acschembio.9b00741. Copyright 2020, American Chemical Society.)*

R74 and R132 showed reductions in MG modification, corresponding well to HDX-MS protection in nearby regions. Thus, a reorientation may occur upon ligand binding. Overall, the protection observed for Scn with both ferric and aferric Ent suggests that Scn may be proactively combating bacterial siderophores even before they chelate iron.

### *3.2.3 Protocol for the development of amino acid specific footprinting*

Similar to HDX- and FPOP-MS, the use of specific amino acid footprinting reagents requires method validation because each reagent affords unique reactivity specific to a functional group and the protein. There is currently no standardized and accepted reagent validation workflow. Because labeling times range from seconds to hours, high footprinting reagent concentrations and extended labeling times can perturb protein higher-order structures (HOSs), giving mixed footprints of both the native structure and a new structure that may have undergone changes in HOS owing to the presence of the reagent or modifications, rendering covalent labeling (CL) measurements inaccurate for monitoring native structure. Although the usual expectation is that the user performs some biochemical or MS-based measurements to ensure that the labeling reagent or modification does not perturb HOS (e.g., circular dichroism [71], "sharp-break" in labeling kinetic curve [66], Poisson distribution [72]), this quality control is not consistently employed, and there is currently no standardized method to monitor CL-induced HOS perturbation.

The field of HDX has experienced decades of careful evaluation (e.g., pH, T, ionic strength, and AA specificity dependence; data analysis) [13,73–75]. Similarly, radical-initiated oxidation of amino acid side chains includes detailed kinetic measurements and exploration of mechanisms [76,77]. These subfields of protein footprinting deal with only one probe (e.g., $D_2O$, •OH), allowing a deep understanding from the collective contributions of many investigators. In contrast, irreversible, specific amino acid footprinting includes a breadth of probes, with many only announced and under development or only used by a handful of specialized research groups. Indeed, without the widespread adoption of amino acid-specific footprinting, a standardized protocol for developing and evaluating footprinting reagents is not guaranteed.

No specific amino acid labeling reagent has been better studied for covalent labeling than DEPC. This is largely because of the pioneering efforts of the Richard Vachet group. Their efforts include protein level kinetics applied to a model protein and two-state aggregation systems with a second-order kinetic model [66], detection of reversible DEPC modification on weak nucleophiles Ser/Thr [67], covalent label transfer between residues [78,79], measurement of local microenvironment effects on reactivity with weak nucleophilic residues Ser/Thr [80], and most recently the use of a microenvironment effect as a means to score Rosetta structures for protein structure prediction [81], among many others. Indeed, such thorough evaluations and applications of DEPC provide examples for evaluating a footprinter; however, this requires many years of focused effort, which is not realistic during the initial introduction of new reagents to the scientific community.

Traditionally, covalent labeling reagents are validated through three key steps: (1) modification of peptides or intact protein, (2) measurement of kinetics and evaluation of modification-induced HOS perturbation, and (3) quantification of differential footprinting by MS, demonstrating statistically significant changes in protein dynamics for a model protein system. Using BHD and DEPC as examples of covalent labeling reagents to modify the acidic residues Glu/Asp and nucleophiles, respectively, to give broad coverage complementary to •OH footprinting, the Gross and Vachet groups have validated these reagents for protein footprinting [64,66]. Their efforts will be used as examples of protocols for reagent development. Generally, reagent validation should be applied to increasingly complex systems. Before a footprinter is used to measure protein dynamics, a simple system containing one or two possible modification sites should be tested. Five-membered cyclic peptides with no termini and little HOS may be used to measure modification reactivity and specificity of BHD and $d_5$-BHD for Asp and Glu on a cyclic peptide c(RGDFE) (Fig. 9A–C). We monitored changes in mass corresponding to the addition of BHD and $d_5$-BHD to the cyclic peptide and observed a shift in LC elution time (Fig. 9A). The modified cyclic peptides were submitted to collision-induced dissociation (CID) to determine the footprinting site (Fig. 9B and C) (Ngoka and Gross proposed the fragmentation mechanism and nomenclature for sequencing cyclic peptides, allowing the user to quickly identify modified fragments for otherwise complex spectra [82]). Having demonstrated specificity with side-chain functional groups, global protein modification can be monitored. When the modification is plotted as a function of reagent concentration, footprinting reaction kinetics can be determined by fitting the modification extent to second-order rate kinetics, as is shown for myoglobin modified by DEPC (Fig. 9D) [66]. The method also allows for measuring reagent-induced perturbation of protein HOS; it is hypothesized that local protein HOS

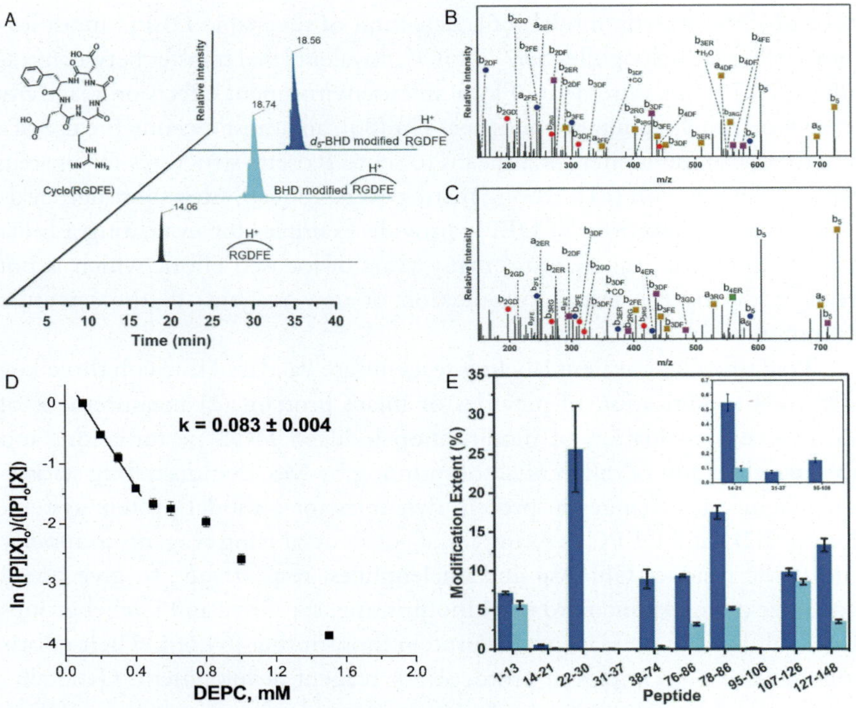


**Fig. 9** (A) Extracted ion chromatograms for unmodified, BHD-modified, and $d_5$-BHD-modified c(RGDFE). MS2 spectra for BHD-modified (B) and $d_5$-BHD-modified (C) c(RGDFE). Color codings indicate the following: *orange circle*, loss of $NH_3$; *blue circle*, loss of $H_2O$; *yellow square*, modification by BHD or $d_5$-BHD; *green square*, modification by BHD or $d_5$-BHD and loss of $H_2O$; *pink square*, the addition of modification by BHD or $d_5$-BHD and loss of $NH_3$. (D) Modification of myoglobin as a function of DEPC concentration fitted to a second-order kinetic model. (E) BHD modification of calmodulin tryptic peptides. *(Adapted with permission from (A–C) and (E) Guo C, Cheng M, Gross ML. Protein-metal-ion interactions studied by mass spectrometry-based footprinting with isotope-encoded benzhydrazide. Anal Chem 2019;91:1416–23. https://doi.org/10.1021/acs.analchem.8b04088; (D) Mendoza VL, Vachet RW. Protein surface mapping using diethylpyrocarbonate with mass spectrometric detection. Anal Chem 2008;80:2895–904. https://doi.org/10.1021/ac701999b. Copyright 2008, 2019, American Chemical Society.)*

becomes more exposed to concentrations at which the modification extent breaks from linearity. Experiments such as this are critical quality-control steps to ensure that the data are efficacious for drawing conclusions about native protein dynamics and HOS. Alternatively, other biophysical assays may be employed to test perturbation, such as circular dichroism [71].

The reagent should not be used at concentrations for which there is evidence of HOS perturbation for the evaluated protein system.

Finally, the reagent may be applied to a well-characterized protein system for its intended purpose of measuring protein HOS and dynamics. Peptide-level quantification of BHD footprinting for apo and holo CaM was monitored to demonstrate the efficacy of BHD in measuring changes in HOS and dynamics upon calcium binding to peptides containing acidic residues (Fig. 9E). The results indicate meaningful decreases in BHD modifications for each region where the protein binds calcium at an EF-hand or becomes more rigid, as is the case for the linker region: EF-1, represented by peptides 14–21, 22–30, and 31–37; EF-2, represented by peptide 38–74; the linker represented by peptides 76–86 and 78–86, which changes from a flexible loop to a rigid α-helix; EF-3, represented by peptide 95–106; and EF-4, represented by peptide 127–148 (Fig. 9E). Taken together, these data indicate that BHD is a reliable footprinter for Asp and Glu.

Other important attributes that are not often measured at reagent conception are temperature dependence, pH dependence, quench efficacy, and reversibility. Indeed, the introduction of a novel covalent labeling reagent requires significant validation such that the method is useful and efficacious for other groups, including those that do not specialize in covalent labeling.

## 4. FPOP for protein structural studies

### 4.1 Introduction

FPOP was first introduced by Hambly and Gross in 2005 [55], and since then, it has been utilized in several studies to probe changes in protein structure induced by external or internal stimuli (e.g., binding to a ligand or another protein, mutation, protein stress) [53]. The elegant yet straightforward FPOP methodology provides a convenient and practical way to label irreversibly solvent-exposed regions of a protein. This allows a protein to be rigorously examined by bottom-up proteomics at the domain level, peptide level, and residue level. The commonly used protein footprinting approach HDX-MS yields reversible labeling, and, hence, peptide separation must be completed in 30 min, or preferably 15 min or less, to keep back exchange manageable. Over the past 15 years, FPOP has flourished in a variety of settings to study protein higher-order structures [1], epitope mapping [83], protein-ligand interactions [84], protein hidden conformations [85], and protein folding [86] to name a few.

A typical experimental setup for FPOP uses a 248 nm KrF laser and a protein-solution flow system (Fig. 10) [87]. The protein is mixed with ~15 mM $H_2O_2$ (0.04%) and low mM of a radical scavenger (typically a free amino acid such as histidine, glutamine, or methionine). The protein solution containing hydrogen peroxide flows through a 150 μm inner diameter capillary (fused silica tubing) that is placed perpendicular to the laser beam (Fig. 10B). The flow of the protein and laser pulses are started simultaneously, and when the protein solution reaches the transparent window, the laser photolyzes $H_2O_2$ within nanoseconds. The resulting hydroxyl radicals (•OH) quickly label the protein's solvent-exposed side chains. The lifetime of •OH radicals is controlled by the self-association rate constant of •OH and the reactivity of the scavenger amino acids with •OH [76]. The addition of 20 mM glutamine effectively shortens the lifetime of •OH radicals to ~1 μs, indicating that FPOP affords microsecond time resolution. The outlet of the FPOP flow system is placed in a tube containing catalase enzyme and methionine to break down quickly any remaining $H_2O_2$ from the system, effectively stopping the reaction (for details, please consult [53,88]). Reactions of hydroxyl radicals cause protein oxidation by abstracting a hydrogen from the side chain, and sequential reactions ultimately lead to oxidative labeling of the protein with mass increases of +16, +32, and so on, which are separated by 15.9949 Da in mass (readily recognized by high-resolving-power MS). This outcome can be viewed as a replacement of a H with a OH.

The FPOP-labeled protein is usually analyzed at three different levels (see below). If the protein is sufficiently small (<100 kDa), then the global level can be readily interrogated by MS of the intact protein to monitor differences in FPOP labeling, so the two states (e.g., unbound vs bound) can be compared. Although protein level interrogation is the least informative, it can confirm, for example, that the FPOP experiment is working properly and that the protein binds to a ligand. Most studies, however, utilize bottom-up proteomics approaches to investigate peptide and residue level comparisons in FPOP labeling. The protein is typically digested by specific proteases such as trypsin, chymotrypsin, or Glu-C. Digested peptides are loaded onto LC-MS/MS instrumentation, where they are separated and monitored in the data-dependent analysis (DDA) mode. Differential modification by FPOP is determined, and differences between the two states are confirmed to be significant by the statistics test. MS/MS can often pinpoint the change at the residue level. In the following, we will discuss a few FPOP studies to exemplify its versatility and adaptability to a variety of biophysical studies.

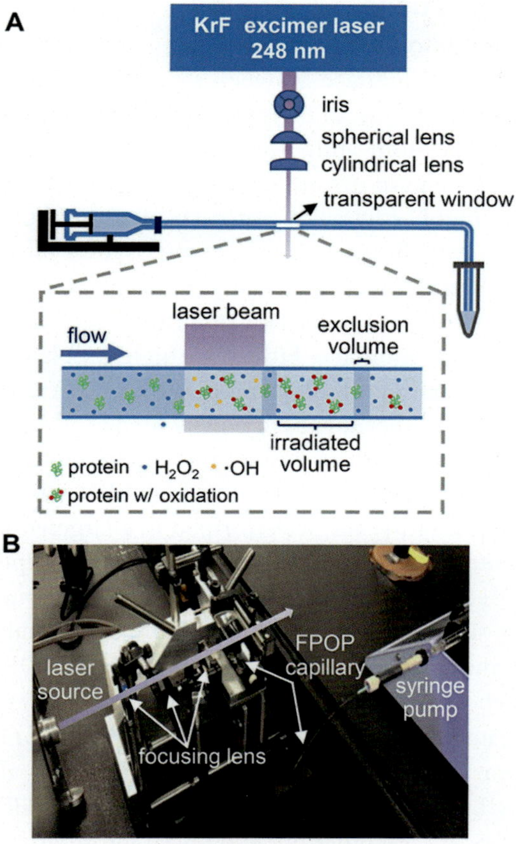


**Fig. 10** FPOP setup. (A) Schematic of the platform. The laser beam *(violet)* generated from an excimer laser *(blue square)* is focused through an iris and then by two convex lenses. The FPOP capillary *(blue line)* made of silica tubing is connected to a syringe pump. The transparent window without polyimide coating indicates the location of laser irradiation (a blow-up of the transparent window is shown in the *dashed box*). A tube containing catalase and Met is placed at the end of the FPOP capillary to collect the modified sample. (B) Photo of the apparatus. Primary components are labeled, and the *violet arrow* represents a visual pathway of the laser beam perpendicular to the FPOP capillary. *(Adapted with permission from Li KS, Shi L, Gross ML. Mass spectrometry-based fast photochemical oxidation of proteins (FPOP) for higher order structure characterization. Acc Chem Res 2018;51:736–44. https://doi.org/10.1021/acs.accounts.7b00593. Copyright 2018, American Chemical Society.)*

## 4.2 Recent applications

### *4.2.1 Epitope mapping using FPOP*

Localizing the binding interface between the antigen (epitope) and therapeutic antibody (paratope) is an important milestone in understanding the mechanism of action of an antibody. The nature of an epitope targeted by an antibody also helps in streamlining its development. An antibody targeting a novel yet highly conserved epitope is often attractive for its broadly neutralizing ability [89], limited off-target effects, and ability to patent. Various biophysical methods such as X-ray crystallography [90,91], cryo-EM [92–94], and NMR [95] are important reporters with a high structural resolution for the epitope interface. However, high sample quantities, special equipment, a high turnaround time, and limited human resources are common obstacles [96]. Another approach widely used for this analysis is alanine scanning mutagenesis. If performed in a nontargeted manner, this method may be labor-intensive and often requires multiple cycles of mutations to confirm assignments. However, if performed in a targeted manner, alanine scanning mutagenesis can validate observations from complementary techniques [97,98]. Because MS has small sample size requirements, is not limited by protein size, and has an intermediate structural resolution, its associated footprinting approaches, such as HDX and FPOP, have emerged as prime approaches for epitope mapping.

To demonstrate the capabilities of HDX and FPOP, our lab collaborated with investigators at Bristol-Myers Squibb to map the epitope of Interleukin-23 (IL-23) [98]. This study indicates that HDX and FPOP not only complement each other but also reveal the ability of FPOP to provide residue-level information to pinpoint key interacting residues. Those data are invaluable for assisting the design of targeted alanine scanning mutagenesis for validation. IL-23 is a proinflammatory cytokine consisting of two subunits (p19 and p40) that are linked via two disulfide bonds [99]. Antibodies targeting IL-23 have the potential to treat a variety of autoimmune disorders, motivating their development [100–102]. In this study [98], epitope mapping of a commercial therapeutic antibody, 7B7 Fab (fragment antigen binding), targeting the p19 subunit of IL-23, was carried out. In the first set of experiments, HDX and FPOP were performed for unbound and Fab7B7-bound IL-23p19. Both identified five different regions showing a binding-induced reduction in footprints (labeling). Among the five, three regions are common, indicating complementarity for both approaches and showing applicability to study conformational epitopes (Fig. 11). LC-MS/MS scans were analyzed to assign modifications to residues in each

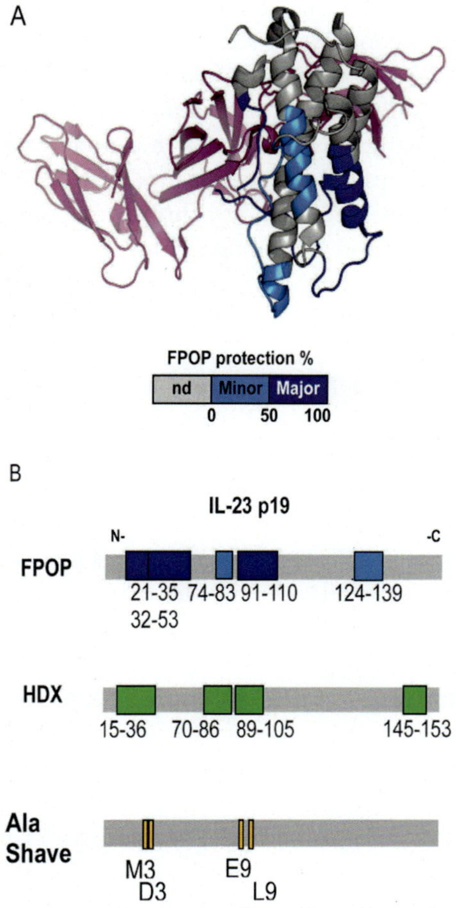


**Fig. 11** (A) Epitope regions determined by FPOP mapped on the crystal structure of IL-23. Color code: no significant difference *(gray)*, minor epitope region *(cyan)*, and major epitope region *(blue)*. The p40 subunit is colored in *purple*. (B) Epitope regions determined by FPOP, HDX, and Ala Shave Energetics as mapped on the linear sequence of the IL-23 p19. *(Adapted with permission from Li J, Wei H, Krystek Jr SR, Bond D, Brender TM, Cohen D, Feiner J, Hamacher N, Harshman J, Huang RY, Julien SH, Lin Z, Moore K, Mueller L, Noriega C, Sejwal P, Sheppard P, Stevens B, Chen G, Tymiak AA, Gross ML, Schneeweis LA. Mapping the energetic epitope of an antibody/ interleukin-23 interaction with hydrogen/deuterium exchange, fast photochemical oxidation of proteins mass spectrometry, and alanine shave mutagenesis. Anal Chem 2017;89:2250–8. https://doi.org/10.1021/acs.analchem.6b03058. Copyright 2017, American Chemical Society.)*

peptide. The data point to nine residues showing a statistically significant reduction in FPOP labeling. Using stringent criteria of accepting only those residues that showed more than a twofold reduction in FPOP labeling, we identified residues W26, M35, and L96/L97/P98 as critical binders. Extracted ion chromatograms (EICs) for the modified peptides extracted from the LC-MS/MS runs for three residues, L96/L97/P98, on the same peptide could not be resolved owing to chromatographic overlap and non-confirmatory fragmentation of peptide. Nevertheless, the FPOP results restrict the oxidative labeling to just three out of 20 residues for the 91–110 peptide. This outcome drastically reduces the complexity of locating key interacting residues.

To validate HDX and FPOP results, alanine-scanning mutants were designed based on SASA values and optimum docking area (ODA) scores. By performing surface plasmon resonance binding using Biacore assays for mutants, residues M35, D36, E93, and L97 were identified as the primary energetic contributors to the epitope binding. Considering all the results, we have shown that MS-based footprinting is less sample- and time-consuming and can accurately identify the epitope with spatial resolution at some residues.

### 4.2.2 FPOP for studying the protein aggregation

Nearly 50 proteins/peptides are known to form pathologic amyloids [103]. However, our fundamental understanding of their aggregation behavior is still lacking; therefore, no therapeutic strategy to prevent amyloid formation has been developed. Perhaps the best-known amyloid-forming protein is amyloid-beta (Aβ) owing to its involvement in Alzheimer's disease (AD). Among the different isoforms present, $A\beta_{1-42}$ is the most pathogenic [104]. The "holy grail" in aggregation studies is to characterize early aggregation events and gain an understanding of the structures and morphology of toxic oligomers that cause disease. High-resolution probes such as X-ray crystallography, cryo-EM, and NMR, as mentioned earlier, are time-consuming and are useful for characterizing the start and final states of aggregation [105]. However, they are not suitable for studying dynamic, soluble oligomers, especially when they exist in a highly heterogeneous state. MS-based footprinting may contribute to our understanding of the aggregation and to embrace the complexity encountered from in vitro to in vivo experiments.

In a second representative FPOP study [106], 10 μM $A\beta_{1-42}$ was allowed to aggregate without any agitation at 25°C for 48 h. At different time points, aliquots of the reaction were withdrawn, and FPOP footprinting was carried out in its inherently "pulsed" manner. FPOP-induced oxidative labeling decreased as a function of aggregation time, indicating that a greater number of monomers are converting into higher-order, solvent-excluded species (Fig. 12A–D). At the peptide level, a fraction species labeled versus time plot showed three sigmoidal curves indicating that protein undergoes several nucleation steps en route to rapid exponential growth (Fig. 12E). Each brief pause corresponds to a conformational rearrangement of the oligomer (nucleation), and each exponential phase may be an expansion of a nucleated oligomer to a higher-order structure.

Based on these and other literature data, we proposed a scheme for in vitro aggregation. The first transition is attributed to the formation of dimer D via monomer association (transition A → B and solid M curve). At nearly 30% of conversion of the monomer to the dimer, the dimer undergoes structural rearrangement (transition B → C and dashed black curve D) to become small oligomers. Once the latter form, they act as seeds to promote the formation of high-molecular-weight oligomers (protofibrils; transition C → D and gray dashed line curve D*). When ~80% of monomers are consumed, structural rearrangement of protofibrils occurs (D → E transition, Fig. 12E and F). Such rearrangements lead to further expansion of protofibrils by their likely lateral association and formation of mature fibrils (transition E → F and solid gray D** curve). This further shifts the equilibrium, with an additional 10% monomer consumed before the aggregation reaction reaches a steady state.

To localize the structural changes at the peptide level and further at the residue level, Lys-N digestion of an FPOP-treated sample was performed. Lys-N digestion of $A\beta_{1-42}$ provides three peptides that conveniently cover the N-terminus, middle, and C-terminal regions. Time-course footprinting shows that the N-terminal region undergoes small, probably statistically insignificant changes, supporting the hypothesis that it is not involved in aggregation. Conversely, the middle (Fig. 12G) and the C-terminal regions (Fig. 12H) show FPOP modification like those of the intact protein. This includes a rapid drop in FPOP modification for the middle region, suggesting that it is involved in forming low molecular species and hence is critical for aggregation. Furthermore, MS/MS analysis pinpoints Phe19/Phe20 residues to be critical for aggregation. Given that Phe is hydrophobic, this residue may initiate hydrophobicity-driven oligomerization of $A\beta_{1-42}$ although in cell or in vivo footprinting are needed and should be future goals.

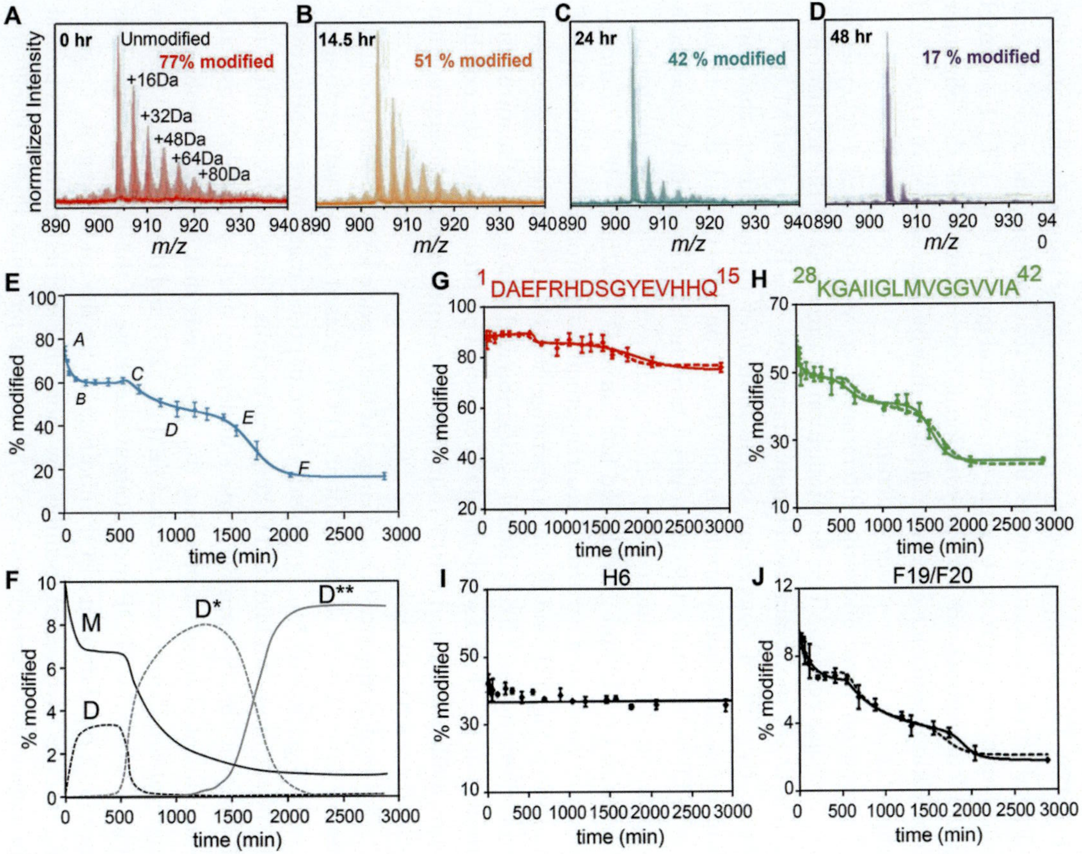


**Fig. 12** See figure legend on opposite page.

*(Continued)*

Overall, FPOP illuminates the heterogeneity in $A\beta_{1-42}$ aggregation. A comparison of FPOP and HDX results reveals a final aggregation step only seen by FPOP. That step may be a structural reorganization (e.g., twisting of protofibrils to accommodate their likely lateral association and formation of mature fibrils). The absence of such a transition in HDX suggests that protofibril association is driven by the side-chain interactions, and the protein backbone is not undergoing major change during these transitions.

### *4.2.3 Submillisecond folding probed by FPOP*

Understanding protein folding is an important biophysical challenge [107,108]. Success not only assists the design of new proteins for tailored functions but also may offer explanations for why some intrinsically disordered proteins misfold to form toxic amyloids. Two models for protein folding are currently pursued [109]. Model 1, hydrophobic collapse, involves a compression in protein structure starting from the unfolded polypeptide, followed by the emergence of secondary and tertiary structures. Model 2 is a nucleation condensation mechanism that invokes the formation of critical secondary structural elements during early folding around which the rest of the polypeptide folds. Because these structural changes happen very fast during folding, they are difficult to study. Nevertheless, attempts have been made by using rapid mixing [110,111], temperature-jump [112], or

---

**Fig. 12** FPOP and kinetic modeling characterize the time-dependent aggregation of $A\beta_{1-42}$ at protein, peptide, and amino-acid residue levels. (A–D) Mass spectra showing extents of FPOP modification for intact $A\beta_{1-42}$ (5+ charge) as a function of incubation time. The modification percentages are shown in each panel. (E) Characterization of $A\beta_{1-42}$ aggregation on the global (full-polypeptide) level by a kinetic simulation. Points represent experimental data, and the solid curve is a model fit based on two autocatalytic reactions. (F) Concentration (in monomeric equivalents) change of representative $A\beta_{1-42}$ species (M, monomer; D, paranuclei; D*, protofibrils; and D**, fibrils) from kinetic simulation. (G, H) Aggregation curves of N-terminal region 1–15 and C-terminal region 28–42. (I, J) Aggregation curves of representative $A\beta_{1-42}$ residues H6 and F19/F20. In panels G–J, the *solid and dashed curves* are model fits independent of or constrained by the global rates, respectively. *(Adapted with permission from Li KS, Shi L, Gross ML. Mass spectrometry-based fast photochemical oxidation of proteins (FPOP) for higher order structure characterization. Acc Chem Res 2018;51:736–44. https://doi.org/10.1021/acs.accounts.7b00593; Li KS, Rempel DL, Gross ML. Conformational-sensitive fast photochemical oxidation of proteins and mass spectrometry characterize amyloid beta 1–42 aggregation. J Am Chem Soc 2016;138:12090–8. https://doi.org/10.1021/jacs.6b07543. Copyright 2016, 2018 American Chemical Society.)*

pH-jump experiments [113] coupled with fluorescence, isotopically labeled carbonyl group monitoring by infrared (IR) spectroscopy [114], NMR [115,116], HDX-MS [20,117], and others [118,119]. Most of these probes afford little structural resolution or need substantial isotopic labeling of protein, which may affect protein stability and dynamics.

In 2012, our lab introduced the temperature jump coupled with FPOP to study submillisecond folding at the residue level with spatial structural resolution [86]. The barstar protein was selected because it undergoes reversible cold denaturation at 0°C in the presence of 1.2 M GdnCl, and a rapid rise in temperature causes it to fold back to its native conformation [120–122]. This reversible cold denaturation allows for interrogating the folding transition with a two-laser experiment. In the beginning, cold denatured barstar flows through the FPOP capillary (Fig. 13A) to the transparent window, where it is irradiated by an Nd:YAG laser to raise the solution temperature ~20°C, driving its folding. At the same time, a carefully timed, KrF laser focused on the same transparent window photolyzes hydrogen peroxide mixed with protein [123]. The resultant pool of OH radicals rapidly labels the exposed

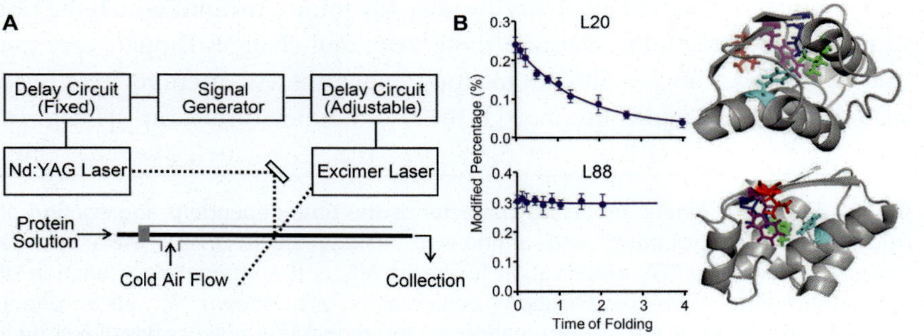


**Fig. 13** Folding of barstar characterized by a T-jump FPOP. (A) Schematic representation of the two-laser FPOP platform. The transparent window is located at the intersection of the two laser beams *(dashed lines)*. (B) Left, the extent of footprinting of two representative residues as a function of the protein (barstar) folding time. *Solid lines* in the plots are obtained from kinetic fitting. Right, five critical residues identified by LC-MS/MS mapped onto the native barstar structure. Two views are provided to show the side chains of the amino acids L88 *(red)*, F74 *(cyan)*, I5 *(blue)*, L20 *(green)*, and W53 *(purple)*. *(Adapted with permission from Chen J, Rempel DL, Gau BC, Gross ML. Fast photochemical oxidation of proteins and mass spectrometry follow submillisecond protein folding at the amino-acid level. J Am Chem Soc 2012;134:18724–31. https://doi.org/10.1021/ja307606f; Li KS, Shi L, Gross ML. Mass spectrometry-based fast photochemical oxidation of proteins (FPOP) for higher order structure characterization. Acc Chem Res 2018;51:736–44. https://doi.org/10.1021/acs.accounts.7b00593. Copyright 2012, 2018 American Chemical Society.)*

side chains. By carefully adjusting the time separation for firing the two lasers, the time course of a folding reaction can be followed (a "pump-probe" approach in physical chemistry). As folding proceeds, side chains become increasingly buried, and the overall extent of labeling decreases with time [86]. Using trypsin and GluC digestion enables a peptide level time-course measurement of change in FPOP labeling to be performed, followed by residue level analysis by LC-MS/MS. Broadly, 10 residues are significantly modified and classified into four categories: (1) residues showing little or no change in the modification (e.g., residues E57, L62, and R75). These residues are part of helix3 and helix4; (2) residues showing no change in modification during the first stage of folding but have reduced labeling in the second stage of folding; examples are residues W53 and L88 (Fig. 13B). Both residues are buried in the hydrophobic core, indicating consolidation of a hydrophobic core after critical nucleus formation; (3) residues that complete a decreased extent of FPOP labeling during the first stage; examples are H17, L20, and L24, which are part of helix1, indicating that helix1 is completely folded in 2 ms in the first stage of folding (Fig. 13B); (4) residues I5 (β sheet1) and F74 (helix3) situated remote to the hydrophobic structural core, which show partial folding in the early folding intermediate state and achieve complete folding during the second phase where consolidation of the folding nucleus and expansion of secondary and tertiary structure occur. Together, FPOP coupled to temperature jump strongly suggests that Barstar follows a nucleation condensation mechanism where helix1 folds completely along with the partial folding of β sheet1 and helix3. Once this nucleus condenses, it allows the structure of the rest of the protein to fold in a second slower phase of folding.

## 5. Conclusions

MS now plays a vital role alongside optical spectroscopy in studying the biophysical properties of proteins. Although MS cannot provide high-resolution structural details with atomic coordinates, such as NMR, X-ray, and cryo-EM, it can elucidate structural changes induced by mutation, ligand binding (including epitope mapping), time of folding, and aggregation. It can uncover the orientation of a protein associated with a membrane by examining a footprint obtained at different levels of complexity. In the absence of a differential experiment, footprinting can distinguish dynamic, disordered, structured, and buried regions. Perhaps with the development of appropriate modeling, footprinting results can be rationalized in terms of coarse-grained structure by using solvent accessibility as a constraint in structure assignment, much like NMR and crosslinking using the distance between atoms or groups of atoms as a constraint.

To illustrate MS-based capabilities, we described three areas of footprinting and showed examples taken from our work on how they can be used in structural biology. Those methods include hydrogen/deuterium exchange, specific amino acid footprinting, and OH radical footprinting performed on an FPOP platform. Each method has advantages and disadvantages. HDX is a general footprinting method (except for Pro), but it "labels" the protein in a reversible manner. HDX is usually not fast, and its kinetics is followed for seconds, minutes, and hours. Specific amino acid labeling is also not fast, but it provides highly specific information, and combination reagents (footprinters) can be chosen to increase its breadth of coverage. Its disadvantage is that there is uncertainty whether the footprinting is "benign"; that is, whether early stages of labeling perturb the protein so that later stages have footprint regions that are unfolded from the wild-type structure (to deal with this question, we included a brief discussion on validation). FPOP has the advantages of broad coverage and fast chemistry, faster than protein unfolding. Its disadvantages are that it needs specialized equipment (laser, although commercially available alternatives to the laser are available) and its coverage is limited by the relatively low concentrations of footprinters (low mM compared to 50 M for $D_2O$ in HDX and tens of mM for specific amino acid footprinting). As usual, compromise should be sought, and a good choice is to use a combination of footprinting as was illustrated for epitope mapping.

## Acknowledgments

The chapter was prepared with the support of the National Institutes of Health (Grants P41GM103422, R24GM136766, and R01131008).

## References

[1] Liu XR, Zhang MM, Gross ML. Mass spectrometry-based protein footprinting for higher-order structure analysis: fundamentals and applications. Chem Rev 2020;120:4355–454. https://doi.org/10.1021/acs.chemrev.9b00815.

[2] Hvidt A, Linderstrøm-Lang K. Exchange of hydrogen atoms in insulin with deuterium atoms in aqueous solutions. Biochim Biophys Acta 1954;14:574–5. https://doi.org/10.1016/0006-3002(54)90241-3.

[3] Kossiakoff AA. Protein dynamics investigated by the neutron diffraction–hydrogen exchange technique. Nature 1982;296:713–21. https://doi.org/10.1038/296713a0.

[4] Saunders M, Wishnia A. Nuclear magnetic resonance spectra of proteins. Ann N Y Acad Sci 1958;70:870–4. https://doi.org/10.1111/j.1749-6632.1958.tb35437.x.

[5] Wishnia A, Saunders M. The nature of the slowly exchanging protons of ribonuclease. J Am Chem Soc 1962;84:4235–9. https://doi.org/10.1021/ja00881a008.
[6] Liuni P, Rob T, Wilson DJ. A microfluidic reactor for rapid, low-pressure proteolysis with on-chip electrospray ionization. Rapid Commun Mass Spectrom 2010;24:315–20. https://doi.org/10.1002/rcm.4391.
[7] Liuni P, Olkhov-Mitsel E, Orellana A, Wilson DJ. Measuring kinetic isotope effects in enzyme reactions using time-resolved electrospray mass spectrometry. Anal Chem 2013;85:3758–64. https://doi.org/10.1021/ac400191t.
[8] Rob T, Liuni P, Gill PK, Zhu S, Balachandran N, Berti PJ, Wilson DJ. Measuring dynamics in weakly structured regions of proteins using microfluidics-enabled subsecond H/D exchange mass spectrometry. Anal Chem 2012;84:3771–9. https://doi.org/10.1021/ac300365u.
[9] Watson MJ, Harkewicz R, Hodge EA, Vorauer C, Palmer J, Lee KK, Guttman M. Simple platform for automating decoupled LC-MS analysis of hydrogen/deuterium exchange samples. J Am Soc Mass Spectrom 2021;32:597–600. https://doi.org/10.1021/jasms.0c00341.
[10] Konermann L, Pan J, Liu Y-H. Hydrogen exchange mass spectrometry for studying protein structure and dynamics. Chem Soc Rev 2011;40:1224–34. https://doi.org/10.1039/C0CS00113A.
[11] Englander SW, Downer NW, Teitelbaum H. Hydrogen exchange. Annu Rev Biochem 1972;41:903–24. https://doi.org/10.1146/annurev.bi.41.070172.004351.
[12] Englander SW, Kallenbach NR. Hydrogen exchange and structural dynamics of proteins and nucleic acids. Q Rev Biophys 1983;16:521–655. https://doi.org/10.1017/S0033583500005217.
[13] Bai Y, Milne JS, Mayne L, Englander SW. Primary structure effects on peptide group hydrogen exchange. Proteins 1993;17:75–86. https://doi.org/10.1002/prot.340170110.
[14] Smith DL, Deng Y, Zhang Z. Probing the non-covalent structure of proteins by amide hydrogen exchange and mass spectrometry. J Mass Spectrom 1997;32:135–46. https://doi.org/10.1002/(SICI)1096-9888(199702)32:2<135::AID-JMS486>3.0.CO;2-M.
[15] Hamuro Y, Zhang T. High-resolution HDX-MS of cytochrome c using pepsin/fungal protease type XIII mixed bed column. J Am Soc Mass Spectrom 2019;30:227–34. https://doi.org/10.1007/s13361-018-2087-7.
[16] Rey M, Yang M, Burns KM, Yu Y, Lees-Miller SP, Schriemer DC. Nepenthesin from monkey cups for hydrogen/deuterium exchange mass spectrometry. Mol Cell Proteomics 2013;12:464–72. https://doi.org/10.1074/mcp.M112.025221.
[17] Hvidt A, Nielsen SO. Hydrogen exchange in proteins. Adv Protein Chem 1966;21:287–386. https://doi.org/10.1016/s0065-3233(08)60129-1.
[18] Konermann L, Tong X, Pan Y. Protein structure and dynamics studied by mass spectrometry: H/D exchange, hydroxyl radical labeling, and related approaches. J Mass Spectrom 2008;43:1021–36. https://doi.org/10.1002/jms.1435.
[19] Englander SW, Sosnick TR, Englander JJ, Mayne L. Mechanisms and uses of hydrogen exchange. Curr Opin Struct Biol 1996;6:18–23. https://doi.org/10.1016/S0959-440X(96)80090-X.
[20] Jethva PN, Udgaonkar JB. Modulation of the extent of cooperative structural change during protein folding by chemical denaturant. J Phys Chem B 2017;121:8263–75. https://doi.org/10.1021/acs.jpcb.7b04473.
[21] Malhotra P, Udgaonkar JB. Tuning cooperativity on the free energy landscape of protein folding. Biochemistry 2015;54:3431–41. https://doi.org/10.1021/acs.biochem.5b00247.

[22] Konermann L, Rodriguez AD, Sowole MA. Type 1 and type 2 scenarios in hydrogen exchange mass spectrometry studies on protein–ligand complexes. Analyst 2014;139:6078–87. https://doi.org/10.1039/C4AN01307G.
[23] Tajoddin NN, Konermann L. Analysis of temperature-dependent H/D exchange mass spectrometry experiments. Anal Chem 2020;92:10058–67. https://doi.org/10.1021/acs.analchem.0c01828.
[24] Kant R, Llauró A, Rayaprolu V, Qazi S, de Pablo PJ, Douglas T, Bothner B. Changes in the stability and biomechanics of P22 bacteriophage capsid during maturation. Biochim Biophys Acta Gen Subj 1862;2018:1492–504. https://doi.org/10.1016/j.bbagen.2018.03.006.
[25] Jethva PN, Udgaonkar JB. The osmolyte TMAO modulates protein folding cooperativity by altering global protein stability. Biochemistry 2018;57:5851–63. https://doi.org/10.1021/acs.biochem.8b00698.
[26] Chen E, Salinas ND, Huang Y, Ntumngia F, Plasencia MD, Gross ML, Adams JH, Tolia NH. Broadly neutralizing epitopes in the Plasmodium vivax vaccine candidate Duffy Binding Protein. Proc Natl Acad Sci U S A 2016;113:6277. https://doi.org/10.1073/pnas.1600488113.
[27] Mondal T, Wang H, DeKoster GT, Baban B, Gross ML, Frieden C. ApoE: in vitro studies of a small molecule effector. Biochemistry 2016;55:2613–21. https://doi.org/10.1021/acs.biochem.6b00324.
[28] Huang RYC, Rempel DL, Gross ML. HD exchange and PLIMSTEX determine the affinities and order of binding of $Ca^{2+}$ with troponin C. Biochemistry 2011;50:5426–35. https://doi.org/10.1021/bi200377c.
[29] Adhikari J, Stephan JR, Rempel DL, Nolan EM, Gross ML. Calcium binding to the innate immune protein human calprotectin revealed by integrated mass spectrometry. J Am Chem Soc 2020;142:13372–83. https://doi.org/10.1021/jacs.9b11950.
[30] Batchelor JD, Malpede BM, Omattage NS, DeKoster GT, Henzler-Wildman KA, Tolia NH. Red blood cell invasion by Plasmodium vivax: structural basis for DBP engagement of DARC. PLoS Pathog 2014;10, e1003869. https://doi.org/10.1371/journal.ppat.1003869.
[31] VanBuskirk KM, Sevova E, Adams JH. Conserved residues in the Plasmodium vivax Duffy-binding protein ligand domain are critical for erythrocyte receptor recognition. Proc Natl Acad Sci U S A 2004;101:15754–9. https://doi.org/10.1073/pnas.0405421101.
[32] Chitnis CE, Chaudhuri A, Horuk R, Pogo AO, Miller LH. The domain on the Duffy blood group antigen for binding Plasmodium vivax and P. knowlesi malarial parasites to erythrocytes. J Exp Med 1996;184:1531–6. https://doi.org/10.1084/jem.184.4.1531.
[33] Wertheimer SP, Barnwell JW. Plasmodium vivax interaction with the human Duffy blood group glycoprotein: identification of a parasite receptor-like protein. Exp Parasitol 1989;69:340–50. https://doi.org/10.1016/0014-4894(89)90083-0.
[34] Miller LH, Mason SJ, Clyde DF, McGinniss MH. The resistance factor to Plasmodium vivax in blacks. The Duffy-blood-group genotype, FyFy. N Engl J Med 1976;295:302–4. https://doi.org/10.1056/nejm197608052950602.
[35] Miller LH, Mason SJ, Dvorak JA, McGinniss MH, Rothman IK. Erythrocyte receptors for (Plasmodium knowlesi) malaria: Duffy blood group determinants. Science 1975;189:561–3. https://doi.org/10.1126/science.1145213.
[36] Liu C-C, Kanekiyo T, Xu H, Bu G. Apolipoprotein E and Alzheimer disease: risk, mechanisms and therapy. Nat Rev Neurol 2013;9:106–18. https://doi.org/10.1038/nrneurol.2012.263.
[37] Corder EH, Saunders AM, Strittmatter WJ, Schmechel DE, Gaskell PC, Small GW, Roses AD, Haines JL, Pericak-Vance MA. Gene dose of apolipoprotein E type 4 allele

and the risk of Alzheimer's disease in late onset families. Science 1993;261:921–3. https://doi.org/10.1126/science.8346443.
[38] Wang M, Chen J, Turko IV. 15N-labeled full-length apolipoprotein E4 as an internal standard for mass spectrometry quantification of apolipoprotein E isoforms. Anal Chem 2012;84:8340–4. https://doi.org/10.1021/ac3018873.
[39] Powell KD, Ghaemmaghami S, Wang MZ, Ma L, Oas TG, Fitzgerald MC. A general mass spectrometry-based assay for the quantitation of protein-ligand binding interactions in solution. J Am Chem Soc 2002;124:10256–7. https://doi.org/10.1021/ja026574g.
[40] Zot AS, Potter JD. Structural aspects of troponin-tropomyosin regulation of skeletal muscle contraction. Annu Rev Biophys Biophys Chem 1987;16:535–59. https://doi.org/10.1146/annurev.bb.16.060187.002535.
[41] Gagné SM, Tsuda S, Li MX, Smillie LB, Sykes BD. Structures of the troponin C regulatory domains in the apo and calcium-saturated states. Nat Struct Biol 1995;2:784–9. https://doi.org/10.1038/nsb0995-784.
[42] Filatov VL, Katrukha AG, Bulargina TV, Gusev NB. Troponin: structure, properties, and mechanism of functioning. Biochemistry (Mosc) 1999;64:969–85.
[43] Zygiel EM, Nolan EM. Transition metal sequestration by the host-defense protein calprotectin. Annu Rev Biochem 2018;87:621–43. https://doi.org/10.1146/annurev-biochem-062917-012312.
[44] Li Z, Gergely J, Tao T. Proximity relationships between residue 117 of rabbit skeletal troponin-I and residues in troponin-C and actin. Biophys J 2001;81:321–33. https://doi.org/10.1016/s0006-3495(01)75702-5.
[45] Hood MI, Mortensen BL, Moore JL, Zhang Y, Kehl-Fie TE, Sugitani N, Chazin WJ, Caprioli RM, Skaar EP. Identification of an Acinetobacter baumannii zinc acquisition system that facilitates resistance to calprotectin-mediated zinc sequestration. PLoS Pathog 2012;8, e1003068. https://doi.org/10.1371/journal.ppat.1003068.
[46] Liu JZ, Jellbauer S, Poe AJ, Ton V, Pesciaroli M, Kehl-Fie TE, Restrepo NA, Hosking MP, Edwards RA, Battistoni A, Pasquali P, Lane TE, Chazin WJ, Vogl T, Roth J, Skaar EP, Raffatellu M. Zinc sequestration by the neutrophil protein calprotectin enhances Salmonella growth in the inflamed gut. Cell Host Microbe 2012;11:227–39. https://doi.org/10.1016/j.chom.2012.01.017.
[47] Nakashige TG, Zygiel EM, Drennan CL, Nolan EM. Nickel sequestration by the host-defense protein human calprotectin. J Am Chem Soc 2017;139:8828–36. https://doi.org/10.1021/jacs.7b01212.
[48] Nakashige TG, Zhang B, Krebs C, Nolan EM. Human calprotectin is an iron-sequestering host-defense protein. Nat Chem Biol 2015;11:765–71. https://doi.org/10.1038/nchembio.1891.
[49] Corbin BD, Seeley EH, Raab A, Feldmann J, Miller MR, Torres VJ, Anderson KL, Dattilo BM, Dunman PM, Gerads R, Caprioli RM, Nacken W, Chazin WJ, Skaar EP. Metal chelation and inhibition of bacterial growth in tissue abscesses. Science 2008;319:962–5. https://doi.org/10.1126/science.1152449.
[50] Strupat K, Rogniaux H, Van Dorsselaer A, Roth J, Vogl T. Calcium-induced noncovalently linked tetramers of MRP8 and MRP14 are confirmed by electrospray ionization-mass analysis. J Am Soc Mass Spectrom 2000;11:780–8. https://doi.org/10.1016/s1044-0305(00)00150-1.
[51] Vogl T, Roth J, Sorg C, Hillenkamp F, Strupat K. Calcium-induced noncovalently linked tetramers of MRP8 and MRP14 detected by ultraviolet matrix-assisted laser desorption/ionization mass spectrometry. J Am Soc Mass Spectrom 1999;10:1124–30. https://doi.org/10.1016/s1044-0305(99)00085-9.
[52] Brophy MB, Hayden JA, Nolan EM. Calcium ion gradients modulate the zinc affinity and antibacterial activity of human calprotectin. J Am Chem Soc 2012;134:18089–100. https://doi.org/10.1021/ja307974e.

[53] Liu XR, Rempel DL, Gross ML. Protein higher-order-structure determination by fast photochemical oxidation of proteins and mass spectrometry analysis. Nat Protoc 2020;15:3942–70. https://doi.org/10.1038/s41596-020-0396-3.
[54] Minkoff BB, Blatz JM, Choudhury FA, Benjamin D, Shohet JL, Sussman MR. Plasma-generated OH radical production for analyzing three-dimensional structure in protein therapeutics. Sci Rep 2017;7:12946. https://doi.org/10.1038/s41598-017-13371-7.
[55] Hambly DM, Gross ML. Laser flash photolysis of hydrogen peroxide to oxidize protein solvent-accessible residues on the microsecond timescale. J Am Soc Mass Spectrom 2005;16:2057–63. https://doi.org/10.1016/j.jasms.2005.09.008.
[56] Maleknia SD, Brenowitz M, Chance MR. Millisecond radiolytic modification of peptides by synchrotron X-rays identified by mass spectrometry. Anal Chem 1999;71:3965–73. https://doi.org/10.1021/ac990500e.
[57] Everett EA, Falick AM, Reich NO. Identification of a critical cysteine in EcoRI DNA methyltransferase by mass spectrometry. J Biol Chem 1990;265:17713–9. https://doi.org/10.1016/S0021-9258(18)38222-X.
[58] Apuy JL, Chen X, Russell DH, Baldwin TO, Giedroc DP. Ratiometric pulsed alkylation/mass spectrometry of the cysteine pairs in individual zinc fingers of MRE-binding transcription factor-1 (MTF-1) as a probe of zinc chelate stability. Biochemistry 2001;40:15164–75. https://doi.org/10.1021/bi0112208.
[59] Guo C, Steinberg LK, Cheng M, Song JH, Henderson JP, Gross ML. Site-specific siderocalin binding to ferric and ferric-free enterobactin as revealed by mass spectrometry. ACS Chem Biol 2020;15:1154–60. https://doi.org/10.1021/acschembio.9b00741.
[60] Sekiguchi T, Oshiro S, Goingo EM, Nosoh Y. Chemical modification of ε-amino groups in glutamine synthetase from Bacillus stearothermophilus with ethyl acetimidate. J Biochem 1979;85:75–8. https://doi.org/10.1093/oxfordjournals.jbchem.a132333.
[61] Gao Y, Wang Y. Site-selective modifications of arginine residues in human hemoglobin induced by methylglyoxal. Biochemistry 2006;45:15654–60. https://doi.org/10.1021/bi061410o.
[62] Sanderson RJ, Mosbaugh DW. Identification of specific carboxyl groups on uracil-DNA glycosylase inhibitor protein that are required for activity. J Biol Chem 1996;271:29170–81. https://doi.org/10.1074/jbc.271.46.29170.
[63] Zhang H, Wen J, Huang RYC, Blankenship RE, Gross ML. Mass spectrometry-based carboxyl footprinting of proteins: method evaluation. Int J Mass Spectrom 2012;312:78–86. https://doi.org/10.1016/j.ijms.2011.07.015.
[64] Guo C, Cheng M, Gross ML. Protein-metal-ion interactions studied by mass spectrometry-based footprinting with isotope-encoded benzhydrazide. Anal Chem 2019;91:1416–23. https://doi.org/10.1021/acs.analchem.8b04088.
[65] Rosén CG, Fedorcsák I. Studies on the action of diethyl pyrocarbonate on proteins. Biochim Biophys Acta Gen Subj 1966;130:401–5. https://doi.org/10.1016/0304-4165(66)90236-4.
[66] Mendoza VL, Vachet RW. Protein surface mapping using diethylpyrocarbonate with mass spectrometric detection. Anal Chem 2008;80:2895–904. https://doi.org/10.1021/ac701999b.
[67] Zhou Y, Vachet RW. Increased protein structural resolution from diethylpyrocarbonate-based covalent labeling and mass spectrometric detection. J Am Soc Mass Spectrom 2012;23:708–17. https://doi.org/10.1007/s13361-011-0332-4.
[68] Moyle AB, Cheng M, Wagner ND, Gross ML. Benzoyl transfer for footprinting alcohol-containing residues in higher order structural applications of mass-

spectrometry-based proteomics. Anal Chem 2022;94:1520–4. https://doi.org/10.1021/acs.analchem.1c04659.
[69] Hoare DG, Koshland DE. A procedure for the selective modification of carboxyl groups in proteins. J Am Chem Soc 1966;88:2057–8. https://doi.org/10.1021/ja00961a045.
[70] Wen J, Zhang H, Gross ML, Blankenship RE. Membrane orientation of the FMO antenna protein from Chlorobaculum tepidum as determined by mass spectrometry-based footprinting. Proc Natl Acad Sci U S A 2009;106:6134–9. https://doi.org/10.1073/pnas.0901691106.
[71] Cheng M, Asuru A, Kiselar J, Mathai G, Chance MR, Gross ML. Fast protein footprinting by X-ray mediated radical trifluoromethylation. J Am Soc Mass Spectrom 2020;31:1019–24. https://doi.org/10.1021/jasms.0c00085.
[72] Gau BC, Sharp JS, Rempel DL, Gross ML. Fast photochemical oxidation of protein footprints faster than protein unfolding. Anal Chem 2009;81:6563–71. https://doi.org/10.1021/ac901054w.
[73] Masson GR, Burke JE, Ahn NG, Anand GS, Borchers C, Brier S, Bou-Assaf GM, Engen JR, Englander SW, Faber J, Garlish R, Griffin PR, Gross ML, Guttman M, Hamuro Y, Heck AJR, Houde D, Iacob RE, Jørgensen TJD, Kaltashov IA, Klinman JP, Konermann L, Man P, Mayne L, Pascal BD, Reichmann D, Skehel M, Snijder J, Strutzenberg TS, Underbakke ES, Wagner C, Wales TE, Walters BT, Weis DD, Wilson DJ, Wintrode PL, Zhang Z, Zheng J, Schriemer DC, Rand KD. Recommendations for performing, interpreting and reporting hydrogen deuterium exchange mass spectrometry (HDX-MS) experiments. Nat Methods 2019;16:595–602. https://doi.org/10.1038/s41592-019-0459-y.
[74] Englander SW, Mayne L, Bai Y, Sosnick TR. Hydrogen exchange: the modern legacy of Linderstrøm-Lang. Protein Sci 1997;6:1101–9. https://doi.org/10.1002/pro.5560060517.
[75] Kim PS, Baldwin RL. Influence of charge on the rate of amide proton exchange. Biochemistry 1982;21:1–5. https://doi.org/10.1021/bi00530a001.
[76] Xu G, Chance MR. Hydroxyl radical-mediated modification of proteins as probes for structural proteomics. Chem Rev 2007;107:3514–43. https://doi.org/10.1021/cr0682047.
[77] Liu XR, Zhang MM, Zhang B, Rempel DL, Gross ML. Hydroxyl-radical reaction pathways for the fast photochemical oxidation of proteins platform as revealed by 18O isotopic labeling. Anal Chem 2019;91:9238–45. https://doi.org/10.1021/acs.analchem.9b02134.
[78] Borotto NB, Degraan-Weber N, Zhou Y, Vachet RW. Label scrambling during CID of covalently labeled peptide ions. J Am Soc Mass Spectrom 2014;25:1739–46. https://doi.org/10.1007/s13361-014-0962-4.
[79] Zhou Y, Vachet RW. Diethylpyrocarbonate labeling for the structural analysis of proteins: label scrambling in solution and how to avoid it. J Am Soc Mass Spectrom 2012;23:899–907. https://doi.org/10.1007/s13361-012-0349-3.
[80] Limpikirati P, Pan X, Vachet RW. Covalent labeling with diethylpyrocarbonate: sensitive to the residue microenvironment, providing improved analysis of protein higher order structure by mass spectrometry. Anal Chem 2019;91:8516–23. https://doi.org/10.1021/acs.analchem.9b01732.
[81] Biehn SE, Limpikirati P, Vachet RW, Lindert S. Utilization of hydrophobic microenvironment sensitivity in diethylpyrocarbonate labeling for protein structure prediction. Anal Chem 2021;93:8188–95. https://doi.org/10.1021/acs.analchem.1c00395.
[82] Ngoka LCM, Gross ML. Multistep tandem mass spectrometry for sequencing cyclic peptides in an ion-trap mass spectrometer. J Am Soc Mass Spectrom 1999;10:732–46. https://doi.org/10.1016/S1044-0305(99)00049-5.

[83] Yan Y, Chen G, Wei H, Huang RY, Mo J, Rempel DL, Tymiak AA, Gross ML. Fast photochemical oxidation of proteins (FPOP) maps the epitope of EGFR binding to adnectin. J Am Soc Mass Spectrom 2014;25:2084–92. https://doi.org/10.1007/s13361-014-0993-x.
[84] Liu XR, Zhang MM, Rempel DL, Gross ML. Protein-ligand interaction by ligand titration, fast photochemical oxidation of proteins and mass spectrometry: LITPOMS. J Am Soc Mass Spectrom 2019;30:213–7. https://doi.org/10.1007/s13361-018-2076-x.
[85] Hart KM, Ho CM, Dutta S, Gross ML, Bowman GR. Modelling proteins' hidden conformations to predict antibiotic resistance. Nat Commun 2016;7:12965. https://doi.org/10.1038/ncomms12965.
[86] Chen J, Rempel DL, Gau BC, Gross ML. Fast photochemical oxidation of proteins and mass spectrometry follow submillisecond protein folding at the amino-acid level. J Am Chem Soc 2012;134:18724–31. https://doi.org/10.1021/ja307606f.
[87] Li KS, Shi L, Gross ML. Mass spectrometry-based fast photochemical oxidation of proteins (FPOP) for higher order structure characterization. Acc Chem Res 2018;51:736–44. https://doi.org/10.1021/acs.accounts.7b00593.
[88] Zhang B, Cheng M, Rempel D, Gross ML. Implementing fast photochemical oxidation of proteins (FPOP) as a footprinting approach to solve diverse problems in structural biology. Methods 2018;144:94–103. https://doi.org/10.1016/j.ymeth.2018.05.016.
[89] Kim AS, Kafai NM, Winkler ES, Gilliland Jr TC, Cottle EL, Earnest JT, Jethva PN, Kaplonek P, Shah AP, Fong RH, Davidson E, Malonis RJ, Quiroz JA, Williamson LE, Vang L, Mack M, Crowe Jr JE, Doranz BJ, Lai JR, Alter G, Gross ML, Klimstra WB, Fremont DH, Diamond MS. Pan-protective anti-alphavirus human antibodies target a conserved E1 protein epitope. Cell 2021;184:4414–29. e4419 https://doi.org/10.1016/j.cell.2021.07.006.
[90] Saul FA, Alzari PM. Crystallographic studies of antigen-antibody interactions. Methods Mol Biol 1996;66:11–23. https://doi.org/10.1385/0-89603-375-9:11.
[91] VanBlargan LA, Errico JM, Kafai NM, Burgomaster KE, Jethva PN, Broeckel RM, Meade-White K, Nelson CA, Himansu S, Wang D, Handley SA, Gross ML, Best SM, Pierson TC, Fremont DH, Diamond MS. Broadly neutralizing monoclonal antibodies protect against multiple tick-borne flaviviruses. J Exp Med 2021;218. https://doi.org/10.1084/jem.20210174.
[92] Li N, Li Z, Fu Y, Cao S. Cryo-EM studies of virus-antibody immune complexes. Virol Sin 2020;35:1–13. https://doi.org/10.1007/s12250-019-00190-5.
[93] Bianchi M, Turner HL, Nogal B, Cottrell CA, Oyen D, Pauthner M, Bastidas R, Nedellec R, McCoy LE, Wilson IA, Burton DR, Ward AB, Hangartner L. - Electron-microscopy-based epitope mapping defines specificities of polyclonal antibodies elicited during HIV-1 BG505 envelope trimer immunization. Immunity 2018;49:288–300. e288 https://doi.org/10.1016/j.immuni.2018.07.009.
[94] Basore K, Kim AS, Nelson CA, Zhang R, Smith BK, Uranga C, Vang L, Cheng M, Gross ML, Smith J, Diamond MS, Fremont DH. Cryo-EM structure of chikungunya virus in complex with the Mxra8 receptor. Cell 2019;177:1725–37. e1716 https://doi.org/10.1016/j.cell.2019.04.006.
[95] Rosen O, Anglister J. Epitope mapping of antibody-antigen complexes by nuclear magnetic resonance spectroscopy. Methods Mol Biol 2009;524:37–57. https://doi.org/10.1007/978-1-59745-450-6_3.
[96] Abbott WM, Damschroder MM, Lowe DC. Current approaches to fine mapping of antigen-antibody interactions. Immunology 2014;142:526–35. https://doi.org/10.1111/imm.12284.

[97] Chen Y, Wiesmann C, Fuh G, Li B, Christinger HW, McKay P, de Vos AM, Lowman HB. Selection and analysis of an optimized anti-VEGF antibody: crystal structure of an affinity-matured Fab in complex with antigen. J Mol Biol 1999;293:865–81. https://doi.org/10.1006/jmbi.1999.3192.
[98] Li J, Wei H, Krystek Jr SR, Bond D, Brender TM, Cohen D, Feiner J, Hamacher N, Harshman J, Huang RY, Julien SH, Lin Z, Moore K, Mueller L, Noriega C, Sejwal P, Sheppard P, Stevens B, Chen G, Tymiak AA, Gross ML, Schneeweis LA. Mapping the energetic epitope of an antibody/interleukin-23 interaction with hydrogen/deuterium exchange, fast photochemical oxidation of proteins mass spectrometry, and alanine shave mutagenesis. Anal Chem 2017;89:2250–8. https://doi.org/10.1021/acs.analchem.6b03058.
[99] Lupardus PJ, Garcia KC. The structure of interleukin-23 reveals the molecular basis of p40 subunit sharing with interleukin-12. J Mol Biol 2008;382:931–41. https://doi.org/10.1016/j.jmb.2008.07.051.
[100] Griffiths CE, Strober BE, van de Kerkhof P, Ho V, Fidelus-Gort R, Yeilding N, Guzzo C, Xia Y, Zhou B, Li S, Dooley LT, Goldstein NH, Menter A, A.S. Group. Comparison of ustekinumab and etanercept for moderate-to-severe psoriasis. N Engl J Med 2010;362:118–28. https://doi.org/10.1056/NEJMoa0810652.
[101] Sandborn WJ, Gasink C, Gao LL, Blank MA, Johanns J, Guzzo C, Sands BE, Hanauer SB, Targan S, Rutgeerts P, Ghosh S, de Villiers WJ, Panaccione R, Greenberg G, Schreiber S, Lichtiger S, Feagan BG, C.S. Group. Ustekinumab induction and maintenance therapy in refractory Crohn's disease. N Engl J Med 2012;367:1519–28. https://doi.org/10.1056/NEJMoa1203572.
[102] Gaffen SL, Jain R, Garg AV, Cua DJ. The IL-23-IL-17 immune axis: from mechanisms to therapeutic testing. Nat Rev Immunol 2014;14:585–600. https://doi.org/10.1038/nri3707.
[103] Chiti F, Dobson CM. Protein misfolding, amyloid formation, and human disease: a summary of progress over the last decade. Annu Rev Biochem 2017;86:27–68. https://doi.org/10.1146/annurev-biochem-061516-045115.
[104] Klein WL, Stine Jr WB, Teplow DB. Small assemblies of unmodified amyloid beta-protein are the proximate neurotoxin in Alzheimer's disease. Neurobiol Aging 2004;25:569–80. https://doi.org/10.1016/j.neurobiolaging.2004.02.010.
[105] Tripathi T, Dubey VK. Advances in protein molecular and structural biology methods. 1st ed. Cambridge, MA, USA: Academic Press; 2022. p. 1–714.
[106] Li KS, Rempel DL, Gross ML. Conformational-sensitive fast photochemical oxidation of proteins and mass spectrometry characterize amyloid beta 1–42 aggregation. J Am Chem Soc 2016;138:12090–8. https://doi.org/10.1021/jacs.6b07543.
[107] Dill KA, Ozkan SB, Shell MS, Weikl TR. The protein folding problem. Annu Rev Biophys 2008;37:289–316. https://doi.org/10.1146/annurev.biophys.37.092707.153558.
[108] Singh DB, Tripathi T. Frontiers in protein structure, function, and dynamics. 1st. Singapore: Springer; 2020. p. 1–458.
[109] Udgaonkar JB. Multiple routes and structural heterogeneity in protein folding. Annu Rev Biophys 2008;37:489–510. https://doi.org/10.1146/annurev.biophys.37.032807.125920.
[110] Xu M, Beresneva O, Rosario R, Roder H. Microsecond folding dynamics of apomyoglobin at acidic pH. J Phys Chem B 2012;116:7014–25. https://doi.org/10.1021/jp3012365.
[111] Goluguri RR, Udgaonkar JB. Microsecond rearrangements of hydrophobic clusters in an initially collapsed globule prime structure formation during the folding of a small protein. J Mol Biol 2016;428:3102–17. https://doi.org/10.1016/j.jmb.2016.06.015.

[112] Eigen M, DeMaeyer L. Technique of organic chemistry. vol. VIII. New York: John Wiley and Sons, Inc.; 1963. 2.
[113] Mallik R, Udgaonkar JB, Krishnamoorthy G. Kinetics of proton transfer in a green fluorescent protein: a laser-induced pH jump study. J Chem Sci 2003;115:307–17.
[114] Brewer SH, Song B, Raleigh DP, Dyer RB. Residue specific resolution of protein folding dynamics using isotope-edited infrared temperature jump spectroscopy. Biochemistry 2007;46:3279–85.
[115] Udgaonkar JB, Baldwin RL. NMR evidence for an early framework intermediate on the folding pathway of ribonuclease A. Nature 1988;335:694–9.
[116] Neudecker P, Robustelli P, Cavalli A, Walsh P, Lundstrom P, Zarrine-Afsar A, Sharpe S, Vendruscolo M, Kay LE. Structure of an intermediate state in protein folding and aggregation. Science 2012;336:362–6. https://doi.org/10.1126/science.1214203.
[117] Englander SW, Mayne L, Kan ZY, Hu W. Protein folding-how and why: by hydrogen exchange, fragment separation, and mass spectrometry. Annu Rev Biophys 2016;45:135–52. https://doi.org/10.1146/annurev-biophys-062215-011121.
[118] Dobson CM. Experimental investigation of protein folding and misfolding. Methods 2004;34:4–14. https://doi.org/10.1016/j.ymeth.2004.03.002.
[119] Zhang H, Gong W, Wu S, Perrett S. Studying protein folding in health and disease using biophysical approaches. Emerg Top Life Sci 2021;5:29–38. https://doi.org/10.1042/ETLS20200317.
[120] Khurana R, Hate AT, Nath U, Udgaonkar JB. pH dependence of the stability of barstar to chemical and thermal denaturation. Protein Sci 1995;4:1133–44. https://doi.org/10.1002/pro.5560040612.
[121] Schreiber G, Fersht AR. The refolding of cis- and trans-peptidylprolyl isomers of barstar. Biochemistry 1993;32:11195–203. https://doi.org/10.1021/bi00092a032.
[122] Shastry MC, Udgaonkar JB. The folding mechanism of barstar: evidence for multiple pathways and multiple intermediates. J Mol Biol 1995;247:1013–27. https://doi.org/10.1006/jmbi.1994.0196.
[123] Chen J, Rempel DL, Gross ML. Temperature jump and fast photochemical oxidation probe submillisecond protein folding. J Am Chem Soc 2010;132:15502–4. https://doi.org/10.1021/ja106518d.

# CHAPTER 10

# Small-angle scattering techniques for biomolecular structure and dynamics

**Andrea Mathilde Mebert**[a,b,*], **María Emilia Villanueva**[c,d,*], **Gabriel Ibrahin Tovar**[a,c,*], **Jonás José Perez Bravo**[c,e,f,*], **and Guillermo Javier Copello**[a,c,*]

[a]Universidad de Buenos Aires, Facultad de Farmacia y Bioquímica, Departamento de Ciencias Químicas, Buenos Aires, Argentina
[b]CONICET—Centro de Investigación y Desarrollo en Criotecnología de Alimentos (CIDCA), La Plata, Argentina
[c]CONICET—Universidad de Buenos Aires, Instituto de Química y Metabolismo del Fármaco (IQUIMEFA), Buenos Aires, Argentina
[d]Departamento de Ciencias Básicas, Universidad Nacional de Luján (UNLu), Buenos Aires, Argentina
[e]Grupo de Aplicaciones de Materiales Biocompatibles, Departamento de Química, Facultad de Ingeniería, Universidad de Buenos Aires (UBA), CABA, Argentina
[f]CONICET-Universidad de Buenos Aires, Instituto de Tecnología de Polímeros y Nanotecnología (ITPN-UBA-CONICET), Buenos Aires, Argentina

## 1. Introduction to small-angle scattering experiments

Small-angle scattering (SAS) techniques provide useful information, typically around the mesoscale (1–100 nm), but they could bring the information up to 1000 nm for certain instrumental arrangements. Contrary to microscopies, a direct image is not obtained in SAS techniques. The scattered light, X-ray, or neutron radiation is analyzed to deduce the structure. Light scattering from homogeneous media (gas, solid, or liquid) is because of electron density fluctuations. The contrast in X-ray scattering results from the differences in the electron densities and, for neutron beams, arises from the differences in the nature of the atomic nucleus. The X-rays and neutrons can sample similar structural dimensions because of their comparable wavelengths. From another point of view, although electrons surrounding the atomic nuclei are the target of light and X-ray interaction, the neutrons are scattered by the nucleus itself. Although experiments and data treatment of SAS techniques are different, the fundamental theories and models are quite similar. The main difference is the mechanism by which the incident radiation interacts with the matter. In all approaches, a

* Equally contributed to this work.

*Advanced Spectroscopic Methods to Study Biomolecular Structure and Dynamics*
https://doi.org/10.1016/B978-0-323-99127-8.00015-5

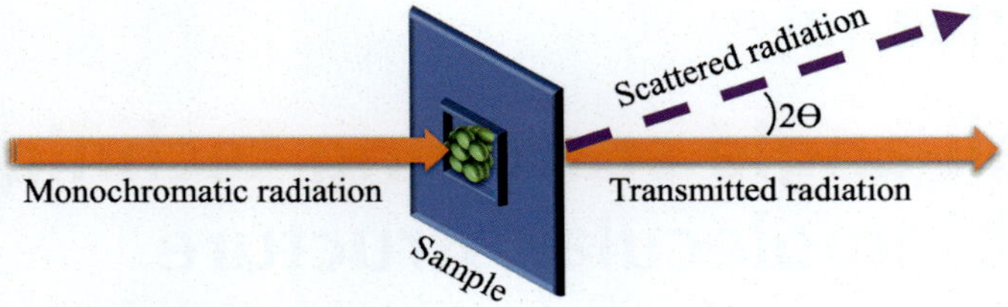


**Fig. 1** Representation of SAS techniques.

collimated radiation interacts with a sample, and the scattered pattern is analyzed. A perpendicular monochromatic beam interacts with the sample, and the scattered radiation is recorded on a detector, as represented in Fig. 1. Time-averaged intensity is measured as a function of the scattering angle and then, for isotropic systems, radially averaged to produce a chart of intensity versus the scattering vector ($q$) [1,2].

## 1.1 Instrumental layouts

### *1.1.1 Small-angle X-ray scattering*

In an elementary small-angle X-ray scattering (SAXS) experiment, the sample is placed in the path of an X-ray beam so that the transmitted and scattered radiations reach a detector. As the scattered radiation intensity given by the sample needs to be measured as a function of $q$ at small angles, which are very close to the nonscattered radiation from the incident beam, it turns the instrumentation design into a real challenge (Fig. 2). At this configuration, a highly collimated beam is needed to allow the low-angle scattered radiation to be recorded at a very short distance from the primary transmitted beam without interference. Also, the detector should be placed far from the sample, and the path needs to be evacuated for many applications.

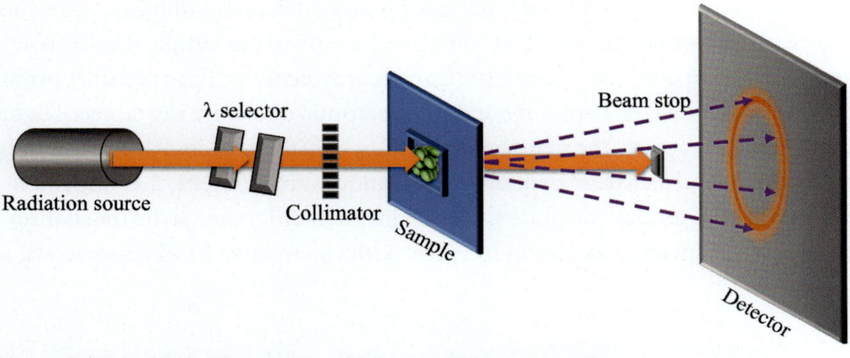


**Fig. 2** Schematic representation of SAXS and SANS experimental arrangement.

To obtain shape data of macromolecules from biological origin, one can expect that the signal appears at angles ranging from 0.15 to 3 degrees. These signals can be detected when a detector is placed 1.5 m from the sample. This range comprises spatial dimensions around 3–60 nm (considering Guinier approximation), which includes most protein structural dimensions. In many cases, proteins have sizes in the low-nm range, and, to avoid detection artifacts, it is preferable to select shorter sample-to-detector distances. For example, if the desired protein has a radius of gyration ($R_g$) around 10 nm, a distance between the sample and the detector of 1 m would allow monitoring spatial dimensions between 2 and 45 nm (assuming $\lambda = 1.54$ Å). From the same perspective, if shorter sample-to-detector distances (typically below 0.5 m) are used, the angles that the assay will monitor will be larger than 5 degrees, and smaller spatial distances will be probed. In this case, we will be referring to wide-angle X-ray scattering (WAXS), and crystallographic data of the sample will be acquired. When it is possible to acquire data at the SAXS and WAXS ranges, it is possible to overlap the scattering profiles, and full-range information can be obtained. WAXS can bring similar information to X-ray diffraction (XRD) because it probes a similar angle range, but it has the advantage that hydrated and liquid samples can be analyzed. On the other hand, when larger structures or aggregates are studied, higher sample-to-detector distances should be used to record the intensity at lower scattering vectors, $q$. When the sample-to-detector distances are higher than 3 m, this technique is named UltraSAXS (USAXS). Dimensions near 1 μm can be explored if distances of 20 m are used.

To increase the sensitivity (e.g., for weakly scattering samples), background (parasitic) scattering should be minimized. Here, the sample holders or quartz capillaries are inserted into a vacuum level evacuated beam path to avoid parasitic air scattering [2,3]. Fig. 2 shows the commonly experimental X-ray source, an X-ray focusing system, a sample holder, and a photon counting detector experimental setup used for the SAXS technique. A beam stop in front of the detector is crucial, especially in weak scatterers, as in the case of biological macromolecules. The primary role of the beam stop is to block the nonscattered radiation intensity of the transmitted beam. Thus, it avoids distortions during data collection and protects the detector from the flux of the transmitted intensity [4].

### *1.1.2 Small-angle neutron scattering*

The fundamentals of a small-angle neutron scattering (SANS) experiment are identical to those of SAXS, but here, the primary beam consists of neutrons that are elastically scattered by the sample [5]. SANS experimental arrangement is quite similar to that of SAXS, and most of the information obtained from both techniques may overlap. Also, sample-to-detector distances that set the scattering vector, $q$, ranges, and therefore the dimensions are studied. As SANS techniques require a neutron reactor or a spallation source, measurements can only be performed in a few large facilities around the world. Thus, in SANS, the available beam time is very limited, whereas SAXS has the advantage that in addition to synchrotron beamlines, in-house laboratory instruments may be readily available when needed [6].

Neutron beams, generated at neutron reactors or spallation sources, are moderated and collimated in the direction of the instrument. Then, the desired wavelength and its direction are selected by optical elements located in the beam path before the beam reaches the sample. The desired velocity and monochromaticity of the beam are acquired by a velocity selector. A monochromator is employed to select the neutron's wavelength. The collimator is used to limit the neutron beam divergence. After the radiation strikes the sample, scattered neutrons are collected on a detector. Usually, a position-sensitive detector with a variable distance from the sample is typically placed in an evacuated tank and used to determine the neutron position. As in USAXS, higher distances in the sample-to-detector position give rise to USANS, which assesses higher spatial dimensions, which can be up to 20 μm. In the middle of SANS and USANS, the neutron community locates very small-angle neutron scattering (VSANS), which probes structures between the formers [7,8]. A simplified schematic of a SANS instrument is shown in Fig. 2. SANS is considered a nondestructive technique because of the type of interactions between neutrons and the sample.

### 1.1.3 Small-angle light scattering

Small-angle light scattering (SALS) instruments have the advantage of rapid detection and simplicity of the equipment. It was initially implemented using an arc lamp as a radiation source and a photographic film as a detector. To control the polarization of the incident and scattered radiation beams, a polarizer was placed before and an analyzer after the sample. As a detector, a photomultiplier tube placed on a small-angle goniometer was employed. By scanning one angle at a time, the scattering profile (Intensity, $I$ vs scattering vector, $q$) could be constructed [9–11]. Later on, a helium-neon (He-Ne) laser was set as the light source. Also, a photodiode array was implemented to measure the angular profile at once, leading to time-resolved measurements. Additionally, for the detection of two-dimensional (2D) data, a charge-coupled device would be the detector of choice (Fig. 3). The spatial

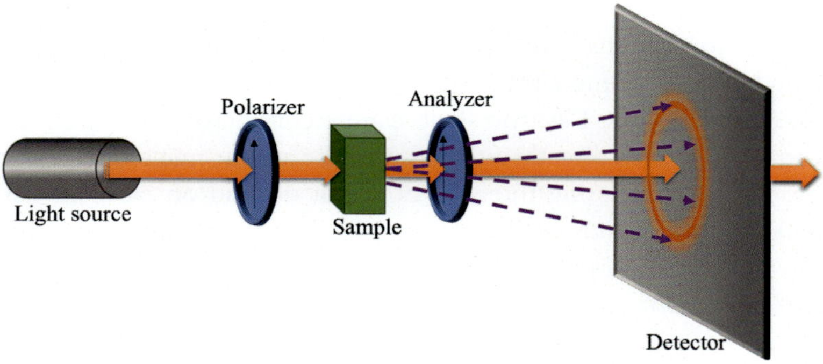


**Fig. 3** The experimental arrangement for photographic light scattering from films.

dimensions sampled by SALS are larger than those for SAXS and SANS [12–14]. The reason for this is that the scattering vector decreases with the increase of the $\lambda$ of the incident beam, which is typically in the visible range, according to $q=(4\pi/\lambda)\ \sin(\Theta)$, with $2\Theta$ being the radial angle of scattering.

## 1.2 Beam sources and their interaction with the sample

### *1.2.1 SAXS*

SAXS experiments are performed with X-ray radiation sources. The X-ray radiation interacts with the sample electrons. The electrons oscillate at the same frequency as that of the wave because of the interaction of the electric field of the incident X-ray radiation and the electrons. In a homogeneous medium, the scattered intensity depends on the electron density fluctuations in the beam path. In "home" laboratory experiments, X-ray tubes are used as radiation sources. These tubes contain a heated cathode, often copper, in vacuum. Synchrotron X-ray sources provide a higher X-ray flux but require large facilities. The X-rays are produced by accelerated charged particles (electrons) at almost the speed of light that emit electromagnetic energy upon curving their path. As a result, a brilliance $10^{12}$ times superior to the bench-top instruments can be generated. Higher fluxes increase the signal-to-noise ratio, reduce exposure times, and, thus, allow more rapid dynamic measurements. Moreover, a vast increase in the distance between the sample and the detector is feasible. High fluxes would also be related to a higher probability of radiation damage. For an unknown sample, it is necessary to establish the exposure time at which the signal-to-noise ratio is optimum and radiation damage is negligible [15,16].

### 1.2.2 SANS

SANS uses sources that provide neutron radiation of varying intensities. Neutron beams are penetrating nonionizing radiations with variable wavelengths (and energies). Electron shells can be considered transparent to the neutron beam, which directly interacts with the atomic nucleus. Contrary to X-rays, neutron scattering intensity does not depend on electron density fluctuations but depends on the characteristics of the neutron-nucleus system interactions. In this sense, the relative X-ray scattering intensity of a particular component in the sample is somehow predictable because the scattering length density (SLD) increases with the atomic number of the elements. Therefore, the heavier the element, the higher the SLD. On the contrary, the scattering length varies irregularly from one nucleus to another. Thus, for the same element, the scattering length is characteristic of a particular isotope. For example, the scattering length density for $H_2O$ is $-5.6 \times 10^{-7}$ Å$^{-2}$ when it is $6.4 \times 10^{-6}$ Å$^{-2}$ for $D_2O$. This will settle the basis for the scattering contrast variation technique that is often used in SANS and mentioned later.

Neutrons are produced in large research infrastructures by fission, fusion reactions, spallation, and so on. The neutrons generated must often be moderated (liquid hydrogen/water/deuterium, solid methane) to lower their energy before using them in scattering experiments. Neutrons that are not cooled are thermal neutrons and have $\lambda$ between 0.1 and 0.4 nm. Cold or "slow" neutrons are obtained after cooling in the cold source moderator liquids and may have wavelengths from 0.5 to 1.5 nm. The neutron cooling process reduces the flux, which decreases sensitivity. On the other hand, cold neutrons allow higher spatial resolution for scattering techniques [17,18]. Independent of the source, because of the nature of the interaction of the neutron with the sample, the radiation damage because of neutron exposure is negligible. Unfortunately, this also causes that even at large facilities, the signal-to-noise ratio is typically low compared to SAXS, even when comparing spallation sources to SAXS bench-top instruments. A SAXS assay that may take 1 min in a synchrotron beamline or 10–20 min in a laboratory instrumental setup could take from 10 min to more than 1 h in a nuclear reactor beamline. Thus, many dynamic measurements possible in SAXS cannot be performed in SANS instruments.

### 1.2.3 SALS

In contrast to X-rays, when light radiation strikes an atom, it only interacts with the electrons on the outer shell because it has the longest wavelength

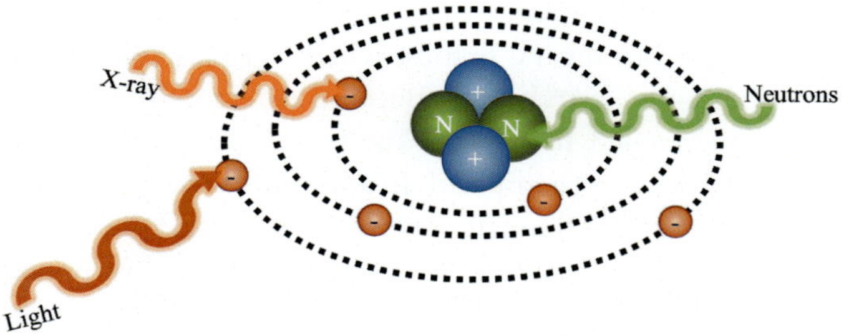


**Fig. 4** Scheme of light, X-ray, and neutron radiation interaction with the electrons and neutrons present in an atom.

and thus the lowest energy out of the three radiation sources mentioned (Fig. 4). Consequently, the fluctuations of polarization predominantly determine the scattered intensity of light radiation [1]. Most commercially available He-Ne lasers are based on the 632.8 nm line, although these lasers can emit several lines in the visible and infrared ranges.

## 1.3 Large facilities and bench-top instruments

Bench-top instruments are powerful tools, and good quality results can be obtained, even considering that the X-ray flux from the tubes is lower than that of the synchrotrons. These instruments have a low background that allows precise measurements even at low angles. Nevertheless, SAXS assays for the analysis of biological molecules are preferably performed on synchrotron sources, whose brilliance greatly exceeds the ones of the X-ray tubes. Data can be collected in a very short period of time (that can be less than a second up to a few minutes), whereas with in-home X-ray tubes, the acquisition time usually is between 10 min and the hour range. On the other hand, the sample amount required is almost the same. Possible radiation damage to the samples has to be considered in synchrotron experiments [19]. Specific gadgets can be implemented in synchrotron SAXS beamlines. For example, 96-well plates can be interposed in the beam path for automatized multiple detections 30–60 min, which allows screening of multiple experimental parameters and obtaining structure information. The development of novel detectors enables gathering time-resolved information from gel permeation chromatography eluents in large facilities such as synchrotrons [3].

As mentioned earlier, decent information can be obtained from bench-top instruments. Nevertheless, one relevant use is screening samples and conditions to choose and reproduce at synchrotron radiation facilities to efficiently use the beamline allocated time. Bench-top instruments are commercially available from several brands, which are mainly dedicated to the development of X-ray instruments. Because detector-to-sample distances of up to 3 m are needed (or even 10 m for USAXS), these instruments require big laboratories to be installed. In addition to their cost, this requirement makes in-home SAXS instruments less widespread than other routine analysis instruments. Nevertheless, access to bench-top instruments allows the study of fresh samples, which is not always possible at synchrotron facilities [6].

## 2. Structural studies

When proteins are analyzed in real systems, their dynamic folding transitions in aqueous systems are important to their biological roles. SAXS and SANS are powerful techniques to analyze the size and shape of proteins, typically in solution. Even though SAS has become a vital technique to evaluate protein structure, other complementary instruments are needed for studying the dynamical structures of proteins, in combination with structural studies employing solution nuclear magnetic resonance (solution NMR) and other complementary methods such as cryo-electron microscopy (cryo-EM) and circular dichroism (CD) spectroscopy. SAS experiments are performed mostly at diluted protein concentrations ($<10$ mg/mL), although information about the protein intermolecular interaction effect and protein aggregates can be obtained by working at higher concentrations.

In the case of the scattering for diluted protein solutions, different parameters can be obtained, including relevant information such as molecular size, the total number of molecules for oligomers, and the overall shape. Combined with theoretical simulations, the three-dimensional (3D) structure of the proteins can be obtained. In a more concentrated liquid solution, the behavior of the protein may be different because of the interaction in the short- and long-range interactions. In comparison, the structural analysis of proteins as a solid state by SAS is poorly studied. The experimental study of SAXS and SANS of proteins with a low solvent content (also in a solid state) is associated with several practical issues. In particular, the scattering on the particle surface can lay under the beam stop and, consequently, hamper the collection of structural data from the scattering at the smaller angles (lowest $q$ range).

One of the main features of SAS techniques is the association of the scattering profiles with structural and intermolecular interaction information.

For this type of analysis, many mathematical models and interpretations of the intensity curves have been developed over the years. These theoretical models will be further discussed later in another section.

## 2.1 Single-protein analysis

In this section, we discuss the analysis of a single-protein structure (shape and size) in dilute solutions. Hence, some relevant limits need to be considered in order to evaluate a single-protein solution analysis:

**(i)** *Proteins are identical:* This means that they will be considered monodisperse, and there should be a narrow standard deviation when analyzing their radius of gyration ($R_g$) or other spatial dimensions (if the protein shape is significantly different from a sphere). This deviation will arise from protein flexibility (folding and unfolding transitions spectral contributions) and instrumental smearing.

**(ii)** *The system is isotropic:* The molecular distribution in the solution is randomized by Brownian motion, and all positions of a single macromolecule will be averaged.

**(iii)** *Macromolecular instantaneous positions have no spatial correlation:* The solution should be diluted enough that intermolecular interaction can be neglected.

**(iv)** *The electron density of the protein is homogeneous along with its structure.*

**(v)** *Proteins are dissolved in an aqueous system with a particular electron density* [20].

Experimentally diluted systems guarantee, without interference effects, that each molecule presents an independent contribution to the overall intensity curve. The SAS methods allow investigating the structure, including for disordered proteins in partially or completely disordered systems, and do not require complex preparation procedures. The main features to consider are the lower limit for protein dilution, obtaining a good signal-to-noise ratio, and the upper limit for protein concentration to avoid protein-protein interaction effects. The experimental SAXS and SANS intensity associated with a consortium of proteins of the same nature in a dilute system can be obtained by subtracting the parasitic scattering intensity produced by aqueous dispersion (e.g., buffer) and the sample holder contribution [21]. Under these conditions, the intensity profile will account for the average scattering of each molecule assuming all positions. When orientation directors are absent and the orientation is random, the scattered pattern is isotropic. From another point of view, the stability of the biomolecule in a particular buffer solution should be considered. Some conditions can induce protein aggregations even at high dilutions and promote specific conformations or even particular structural folding/unfolding or denaturation, for example, because of reduction or shuffling of disulfide bridges.

SANS presents a particular case among SAS techniques because the scattering length depends on the isotopic composition of the sample. This property gives rise to the contrast variation technique. In a very simple procedure, by diluting the sample in different $H_2O$:$D_2O$ ratios, the scattering length contrast between the macromolecules can be maximized or annulled (contrast matching), allowing the exaltation or the smoothing of specific moieties of the sample, such as glycosylated residues, in one solvent condition with respect to the other [7]. The typical contrast matching condition for proteins is near 40% $D_2O$, whereas for lipids and DNA, the typical contrast matching conditions are around 10% and 60% $D_2O$, respectively. The contrast variation technique has a less powerful analog in SAXS, which varies salts or sucrose contents. Unfortunately, the electron density variation ranges that can be achieved using this technique are much narrower than the $H_2O$:$D_2O$ contrast variation technique, highlighting the relevance of the SANS studies.

Single-protein analysis has been performed in different reports to evaluate the structure of a single protein employing SAXS and SANS. Yeast switch protein serves as a practical example of 3D reconstruction structure models applied to proteins [22]. In this study, the sequence information indicated that these proteins contain a coiled-coil domain hanging from a DNA binding domain. A 3D reconstruction revealed a boom shape and demonstrated that the coiled-coil sequence protrudes from the DNA binding domain by about 11 nm. Although this type of model is speculative, it is useful for understanding the role of a molecule. For this case, it turned out to be a useful tool because it allowed the researchers to know that the coiled-coil handle domain bears the actin-binding domain. Another interesting example of SAXS 3D structure modeling was focused on β-mannosidase crystallography studies and studies of lipoxygenase [23,24], oligomerization, and ligand binding-related conformations. Also, the conformational changes associated with two retinoic acid receptors, RXR and RAR, on the oligomerization and ligand binding, respectively, were studied [25]. The authors discussed the applications in RNA structural reconstruction. In another field, by SANS, De Kruif studied casein-isolated micelles associated with colloids found in mammalian milk [26]. The researchers showed that the casein micelles of cow's milk were colloidal and homogeneous particles.

They described that the casein micelles are supported in the structure of a protein matrix, wherein clusters of calcium phosphate are dispersed. Furthermore, they verified that the protein matrix has a similar length scale on the density variations.

## 2.2 Aggregates and protein material

Most protein research is performed at dilute concentrations (usually, proteins are in the micromolar range, and nucleic acids are in the nanomolar range). Under dilute conditions, proteins cannot interact with each other. Thus, SAXS and SANS scattering profiles are ruled by intramolecular interference and provide information on the protein's overall structure and shape. In this section, we will focus on analyzing the concentrated protein solutions. These are presented with a concentration of up to 400 mg/mL in the case of cell compartments and up to 100 mg/mL in therapeutic formulations, such as for monoclonal antibodies [27,28]. High-protein concentrations generate the possibility for intermolecular encounters. Undoubtedly, the concentrated proteins would have significant conformational changes in their conformational ensembles and dynamic complexity in real scenarios. Therefore, quantifying these interactions will take us one step closer to elucidating biochemical processes as they occur in reality. Accordingly, important information on intermolecular interactions of proteins in concentrated solutions can be obtained using SAXS and SANS. The SAS signals from a concentrated protein solution can provide information related to the interferences of scattered waves of the different molecules present in the solvent. This contains information on the distribution of interatomic distances, and typically, they are Angstroms scale and can probe intermolecular interactions quantitatively.

Researchers have employed SAXS and SANS to evaluate different concentrated protein systems with myoglobin and bovine pancreatic trypsin inhibitor (BPTI) [29,30]. They used SANS to explore an inherently messy protein in myoglobin and BPTI solutions, and SAXS measurements were used to measure the intermolecular interactions. In addition, for this purpose, they used SANS to assess the assembly of the N protein of bacteriophage λ (λ N protein) at myoglobin and BPTI-concentrated systems. Their results reported that the analysis by SANS showed that the cumulus of proteins affected the conformation of the λ N protein because of their interaction. Zhang et al. employed SANS and SAXS to analyze protein-protein interactions in high-ionic strength solutions [31]. In comparison to SAXS, SANS results were less sensitive to variations in the hydration shell of the protein because of the presence of ions. This makes the SAXS-SANS complementarity very useful. The authors find that the interaction among proteins was dependent on the nature of the added salt. They observed that the Hofmeister series was inverted when considering the tendency of the second virial coefficient.

## 3. Dynamics analysis

Global information on the size and shape of biomolecules in a solution can be obtained by SAS techniques. Hence, these techniques are instrumental in characterizing biomolecules under different conditions. To understand the function and mechanisms of biological macromolecules, collecting information on the structure and dynamics is of fundamental importance and may also provide important information about the inner structure of biomaterials [32–34]. Information about molecular aggregation and assembly states, folding, unfolding, and shape conformation in a solution can be obtained with these techniques [35]. In analogy to static assays, real-time recollection of data can be performed. Stimulus-triggered conformational changes that occur in the nanosecond to millisecond range can thus be monitored [36,37]. As a consequence, nonequilibrium processes such as protein folding/unfolding can be studied. Additionally, SAS techniques are quite versatile. They can be applied to monodisperse solutions of relatively rigid and well-folded macromolecules and also to complex mixtures of different types of biomolecular systems that have conformational flexibility and polydispersity, such as disordered (macro)molecules, oligomeric mixtures, RNA folding, multidomain proteins, or intrinsically disordered proteins [35].

### 3.1 Sample environment

The sample environment can be modified to characterize the properties and morphology of the material in a steady state. Also, it can be manipulated to bring the material out of equilibrium and monitor the structural transformation on the pathway to equilibrium during the experiment. Hence, several parameters such as pH and ionic strength can be varied, or solvents or other reagents can be added [38–41]. In addition, when the experiments involve the interaction between two reactants, it is possible to choose between continuous-flow mixing and stopped-flow mixing [42–44]. Stopped-flow experiments rapidly mix two solutions by injecting them into an observation chamber. The injection flow is stopped, and data are collected as the system evolves in time [42]. The advantages of stopped-flow mixers are described in the literature. To increase the signal intensity, a large beam area can be used because the beam cross-section has the same dimensions as the sample cell. Multiple syringes can be accommodated, and their content can be mixed in different proportions. Data can also be obtained at different times, even after mixing is complete. However, disadvantages have also been reported; for example, measuring the most rapid events can be difficult, and concentrations may be too low because together with mixing, a dilution is generated. In addition, the beam section should be smaller than the entire cell; thus, not all the sample within the cell interacts with the beam. When the sample is immobile, radiation damage can occur, and it may be severe for long beam exposure times [45–47]. Stopped-flow experiments have been implemented to find out the structural changes in proteins as a function of time. For example, Josts et al. [48] employed time-resolved SAXS, initiated by stopped-flow mixing, to elucidate the structural aspects of transporter MsbA during binding and hydrolysis of adenosine triphosphate to fully describe the protein conformational changes of nucleotide-binding domain dimerization and the following dissociation.

SAS assays can be hyphenated to continuous-flow mixers [46,47]. In this case, mixing is accomplished by rapidly diffusing reagents from a lateral to a central channel, and the data are obtained at different focus sites across the outlet line. Each location is associated with an established time after mixing. Microfluidics platforms can also be used [49,50]. For example, Komorowski et al. [51] used microfluidic sample environments for SAXS to study the agglomeration of lipid vesicles in the presence of divalent cations because of electrostatic interaction. Different conditions have been assayed to study the bilayer structure and the interlayer distance between adhered vesicles. In order to vary the surface charge density of the membrane, different 1,2-dioleoyl-*sn*-glycero-3-phosphocholine (DOPC): 1,2-dioleoyl-*sn*-glycero-3-phospho-L-serine (DOPS) were employed, as well as several cations, such as $Ca^{2+}$, $Sr^{2+}$, and $Zn^{2+}$. In this technology, every flowing molecule scatters and sums up to the overall scattering profile; thus, samples are efficiently used. In addition, because the sample flows, radiation damage is not an issue to worry about. The sample can be subjected to a wide range of environments by changing the composition of the fluids in the lateral channels. As a result, the sample concentration is always the same despite reaction conditions because mixing follows from diffusion and not dilution. When using the best versions of the mixer, mixing and data collection can be accomplished at submillisecond periods. On the other hand, diffusive broadening makes it difficult to obtain data at time scales higher than 1 s after the mixing [45,47].

## 3.2 Changes in temperature

Temperature is one of the basic parameters that can be varied during an experiment. Temperature control is important for biomolecular research because it can change many of its properties. Both SAXS and SANS have been performed by controlling the temperature with Peltier hot stage systems with an operating temperature range from −10°C to +60°C. Also, if more extreme conditions are desired to be studied, using a single-cell calorimetry system with an operating temperature range from −195°C to 600°C, assays can be carried out at rates up to 130 °C/min.

Multiple examples of temperature-controlled SANS and SAXS studies are available in the literature. For instance, Sreij et al. [52] studied the phase and aggregation behavior of 1,2-dimyristoyl-*sn*-glycero-3-phosphocholine (DMPC) bilayers at different concentrations of saponin aescin. When the aescin concentration is low, pure unilamellar vesicles can be found, and at higher concentrations, aescin provokes the complete decomposition of the DMPC membrane. Temperatures below and above the main phase-transition temperature of DMPC ($Tm = 23.6$°C) were used in this study. Thus, saponin-lipid interaction was investigated at different phase states of DMPC. The SANS technique was chosen to study the relation of aescin content with conserved stable small unilamellar vesicle structures below and above the *Tm* of DMPC. With the help of model analysis, it was seen that both the membrane size and thickness increase the rise in aescin amount in both phases, which would be an effect of a successful aescin incorporation. SAXS was employed to elucidate the effect of temperature on the structures for aescin levels above the critical micelle concentration of aescin. When working at such aescin concentrations, the DMPC membrane became soluble, forming smaller bilayer fragments below and around *Tm*. They observed that the lamellar phases presented a recombination behavior that was related to the temperature. Also, structures depicting lamellar-lamellar phase separation near *Tm* were shown.

SALS experiments of varying temperatures can also be performed; the temperature control is usually obtained using a Peltier plate. For example, Bey et al. [53] investigated the effect of temperature on the structure of the lysing enzyme—polyallylamine hydrochloride coacervate droplets, using SALS varying temperatures from 25°C to 65°C. The diameter of the coacervates was established as a function of temperature.

## 3.3 Changes in pressure

The pressure is a relatively easy parameter to vary in a controlled way. This allows the reduction/elimination of large pressure gradients over the sample, producing more robust results. Most high-pressure experiments use hydrostatic pressure. Depending on the cell design, pressures up to 500 MPa can be applied [54]. The pressure cells may consist of a stainless-steel compartment where the beam path goes through perpendicularly placed windows that can bear the applied pressure, for example, made from diamond.

SAXS coupled with a high-pressure system was employed to study the effect of high-pressure processes on the structure and dynamics of milk components [55]. The work proved that SAXS is useful for decoding the inner structure of materials while varying the pressure. The results confirmed that at high pressures, the solubilization of colloidal calcium phosphate and the dissociation of casein micelles into smaller-sized casein clusters occurred. The results also showed a modification of nanoscale protein inhomogeneities. The pressure effect on conformational behavior of biomolecules is not routinely assessed by SANS. This is because sample compartments need to be robust and generate the minimum neutron dispersion to minimize the loss of signal-to-noise ratio [7]. Nevertheless, a few studies are available in the literature. For example, Banachowicz et al. used a high-pressure cell using a sapphire window to study the structure of glucose/xylose isomerase as a function of pressure. It was shown that the enzyme structure, when dissolved in a liquid phase and the crystal state, is very similar and is not prone to vary when subjected to pressures up to 150 MPa [56].

## 3.4 Shear and rheology

In polymer processing, the combination of shear stress with SAXS and SANS is a powerful tool. Rheo-SAS techniques allow the assessment of structural features such as interchain correlations, segmental orientations [57,58], and real-time monitoring of biomolecular structure development and their association with the media viscosity [59]. The selection of the cell material introduces a difference in the simplicity of implementing rheological measurements hyphenated with SANS or SAXS. The wall thickness should be sufficiently small at the beam position to minimize the absorption and background scattering by the cell. Although in rheo-SANS, it is easy to find an appropriate flow cell material, in SAXS, the selection of the cell material is often not as simple. The most widely used material is polycarbonate because of its noncrystalline structure, transparency, low density, and good dimensional stability [60]. In addition, the amount of material required in SANS is small, keeping deuteration costs to a minimum. On the other hand, in rheo-SAXS, larger quantities are needed, which also limits the versatility of this technique.

SALS systems can also be hyphenated with rheometers. In these experiments, the light source is a laser diode, and the SALS profiles are obtained using a detector placed below a screen. Examples of SALS coupled with rheology can also be found in the literature [59,61,62]. For instance, in an interesting study, Nguyen et al. [63] induced solution crystallization of liquid poly (3-hexylthiophene) (P3HT) and monitored crystallization kinetics by in situ rheo-SANS and rheo-SALS.

## 3.5 Magnetic fields

SAXS experiments can be performed under magnetic fields. For magnetic fields below 1.5 T, a simple system based on two permanent magnets is usually used. The magnet can be rotated around a vertical axis, so the X-rays can pass through the sample perpendicular or parallel to the magnetic field. The magnetic field is adjusted by changing the gap between the permanent magnets. In addition, a superconducting magnet of 1.5 T with three field orientations can be used. Magnetic fields can be applied both along and transverse to the incident beam directions. SANS experiments can also be performed under a magnetic field. It can be a static field at 90 degrees with respect to the neutron beam along the neutron path or a radio frequency field produced by a solenoidal coil [64]. However, biomolecules are not often characterized under a magnetic field. Nevertheless, some examples can be found in the literature. For instance, Meddeb et al. [65] investigated the cellulose nanocrystal periodic organization in *n*-methylformamide when subjected to a 0.7 T magnetic field and compared it to the crystal ordering observed in water. It was shown that the presence of a magnetic field could induce the order in the chiral nematic phase of cellulose nanocrystals in a nonaqueous polar medium, leading to a certain degree of homogeneous ordering.

## 3.6 Electric fields

Biomolecules exposed to an electric field can be characterized by employing SAXS using a high-brilliance synchrotron radiation. For example, Rajnak et al. [66] investigated a transformer oil-based magnetic nanofluid when exposed to an electric field. For this, two sample holders with electrodes were prepared. In one of them, the electrodes were electrically insulated from the sample, and in the other, the sample was in direct electrical contact with the electrodes. To power the electrodes, a direct current voltage was applied up to 6 kV for the contact mode and up to 4 kV for the noncontact mode. The SAXS measurements started a few seconds after the application of the voltage. No scattering changes were observed for the noncontact mode, reflecting the unchanged initial structure of the nanofluid in the electric field. On the other hand, the stochastic variation in the scattering intensity under various voltages was detected in the direct contact mode.

In the case of SANS, electric fields have not been widely studied as sample environments. The main problem with using static electric fields is that the voltage required as a function of sample thickness is considerable. Using too high a field risks electric breakdown, damaging the sample [36].

## 4. Models

Mathematical modeling can significantly enhance the impact of SAS data. Mathematical models are used to fit SAS data in a way that the scattering profile takes a physical sense. The modeling sample analysis process for SAS is described in Fig. 5. The reciprocal-space fit should be considered cautiously because different models often give a plausible fit. For this reason, modeling must be aided and complemented with prior knowledge from other techniques, such as NMR.

### 4.1 General models for the SAS pattern

In general, SAS scattering profiles present three zones (Guinier, fractal, and Porod) (Fig. 6). The scattering vector from the Guinier to the fractal region is defined as $q_g$, and that from the fractal to the Porod region is defined as $q_p$. The dimensions of the elementary particles can be obtained from $2\pi/q_p$. The values of the power-law exponent $\alpha$ can be determined from a log-log plot of intensity versus $q$, *considering* $I(q) \sim q^{-\alpha}$. The mass fractal dimension ($D_m$) is equal to the slope of the $-\alpha$ line when $1 < \alpha < 3$. From the same plot, if $3 < \alpha < 4$, the surface fractal dimension ($D_s$) is equal to $6 - \alpha$ [67].

#### *4.1.1 Guinier region*

The Guinier fit allows obtaining information about the total size of the molecule and the quality of the acquired data. Nonlinearities in a Guinier fit are indicative of problems in the sample. Guinier's approximation states that at low $q$, the scattering profile can be approximated as follows [68]:

$$I(q) \approx NI(0)e^{-q^2\left(\frac{R_g^2}{3}\right)} \tag{1}$$

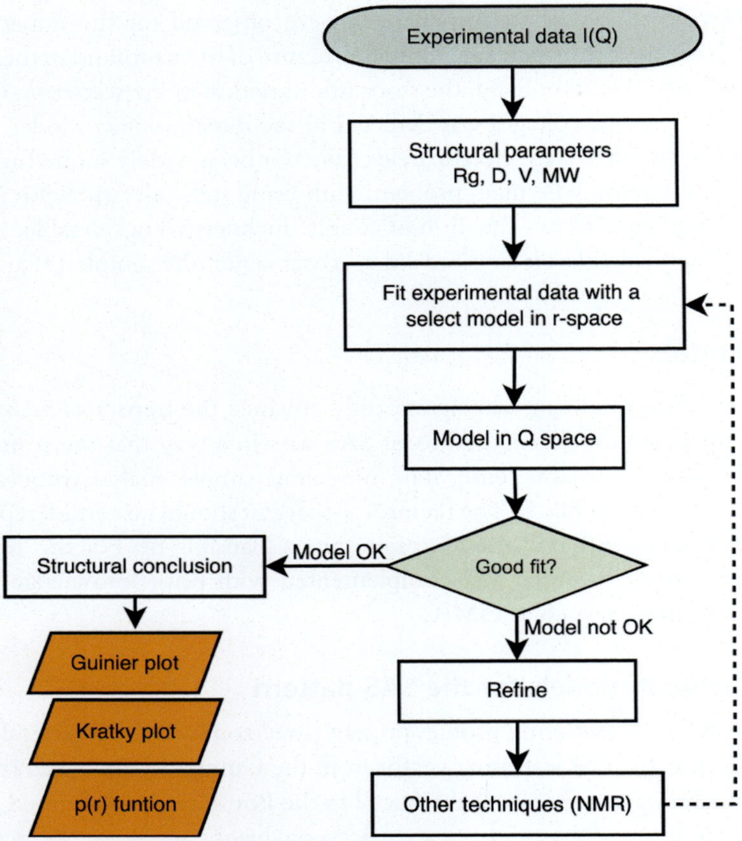


**Fig. 5** Scheme for the analysis of SAS data.

where $I(0)$ is the intensity at $q = 0$ and $R_g$ is the radius of gyration. The $R_g$ describes the mass distribution of the particle around its center of gravity, defined by

$$R_g = \left(\frac{1}{n_e}\int_{V_1} r^2 \rho(\vec{r})\,d\vec{r}\right) \qquad (2)$$

where $n_e$ is the total number of electrons in the particle and $V_1$ is the volume of the particle. The Guinier approximation enables a calculation of the $R_g$, which is obtained by a linear fit to a plot of $lnI(q)$ versus $q^2$, called the Guinier plot. The linear fit is carried out in the well-known "Guinier region."

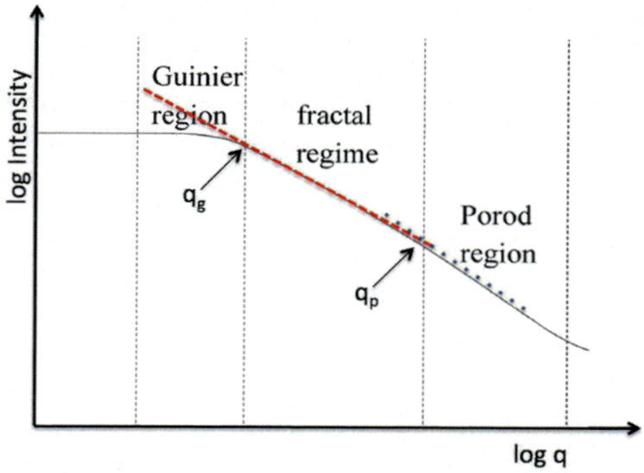


**Fig. 6** General SAS pattern by Guinier, fractal, and Porod regions. *(Reproduced with permission from Seftel EM, Niarchos M, Vordos N, Nolan JW, Mertens M, Mitropoulos ACh, Vansant EF, Cool P. LDH and $TiO_2$/LDH-type nanocomposite systems: a systematic study on structural characteristics. Microporous Mesoporous Mater 2015;203:208–15. https://doi.org/10.1016/j.micromeso.2014.10.029.)*

Consider a spherically shaped monodisperse particle of radius $R$; according to Eqs. (1) and (2), $R_g = \sqrt{(3/5)}R$, the Guinier plot would present at low $q$ a straight line. A negative slope is expected according to $a = (R_g)^{2/3}$. Also, $R_g$ can be determined by $R_g = \sqrt{(-3a)}$. These assumptions are acceptable for low $q$ values. For example, for homogeneous particles, this should be correct up to $1.3/R_g$.

The Guinier plot can indicate the quality of the samples. A linear profile at low $q$ also implies monodispersity without protein-protein interactions or aggregates. On the contrary, if the samples deviate from this behavior, the plot will not show a straight line, and data should be discarded and reassayed under optimal conditions or interpreted by a specific model that considers heterogeneity, aggregation, and/or intermolecular repulsion. When heterogeneous or aggregated particles are present in the sample, the Guinier plot does not follow a straight line at low angles. This is also true when there is repulsion between particles. To perform the measurements so that the Guinier law can be applied, typically, the protein level should be from 1 to 10 mg/mL [69].

### *4.1.2 Fractal regime*

Fractal information is very useful for understanding the structure and dynamic behavior of polymers. From SAS profiles, one can obtain information such as the fractal dimensions, chain density, scaling factor, the number of fractal iterations, and the total number of structural units from the fractal [70].

The description and understanding of fractals have also been simplified using an analytical expression for the scattering intensity and the fractal $R_g$. More recently, the "deterministic approach" for the analysis of $I(q)$ has also been extended to surface fractals, obtaining that the scale factor is different. However, the range of parameters is the same [71,72].

Intensity curves typically present a simple power-law decay on a double logarithmic scale within a given $q$-range, which borders the outer and inner fractal cutoffs $q_{oc}$ and $q_{ic}$, respectively, within which the fractal regime lies. From $I(q) \propto q^{-\alpha}$, the fractal dimension can be obtained from the slope, $\alpha$. The value of $\alpha$ is related to the Euclidean dimensionality of the object: $\alpha = 4$ for 3D objects (spheres, cubes), $\alpha = 2$ for 2D structures (disks), and $\alpha = 1$ for 1D (rods) [73]. According to this, the fractal geometry needs to be considered for these objects. For these cases, the profile could be modeled with the following equation [73,74]:

$$I(q) \propto q^{D_s - 2(D_m + D_p) + 2d} \tag{3}$$

where $D_m$, $D_s$, and $D_p$ are the mass, surface, and pore fractal dimensions, respectively, and $d$ is the topological dimension of the space into which the fractal is immersed. For a mass fractal, $D_s = D_m < d$ and $D_p = d$, and Eq. (3) becomes

$$I(q) \propto q^{-D_m} \tag{4}$$

For surface fractals, $D_m = D_p = d$ and $d - 1 < D_s < d$. Therefore, Eq. (4) can be represented as

$$I(q) \propto q^{-(2d - D_s)} \tag{5}$$

The SAS technique has the advantage over other techniques of providing the dimensions $D_m$ and $D_s$ experimentally.

### *4.1.3 Porod region*

Porod's law assumes that for a folded particle, $I(q)$ decays with a slope as $q^{-4}$, with a linear relationship to the interphase surface area ($S_1$) [75]. In this sense, Porod's law suggests that the profile should reach constant values as $q$ tends to infinity [76]. This asymptotic behavior is infrequently detected because of the contribution of shape information from the macromolecules at the high $q$ boundary. In a diluted solution of N particles (monodisperse and isotropic), Porod's law can be presented as

$$I(q) = \frac{2\pi(\Delta\rho)^2 N \cdot S_1}{q^4} (q \to \infty) \tag{6}$$

As a general rule, this asymptote should be proportional to $S$ (total interfacial area) when $q$ tends to infinity. Consider that the integral in reciprocal space associated with a particle, Q, is given by

$$Q = \int_0^{\infty} 4\pi q^2 I(q)\,dq \tag{7}$$

Furthermore, assuming $Q = N \cdot Q^1$, from Eqs. (6) and (7), the surface-to-volume ratio can be obtained according to

$$\frac{S_1}{V_1} = 4\frac{\pi^2}{Q}\left[q^4\, I(q)\right] \tag{8}$$

The particles and/or the matrix present fluctuations in electron density that constantly contribute to the intensity of the scattering. $q^4 I(q)$ versus $q^4$ plots present a linear relationship for high $q$, and these fluctuations are minimized in this representation. In these plots, $I(q)q^4$ is equal to $A + Bq^4$, where $A$ and $B$ are the linear and angular coefficients, respectively. When $q$ tends to infinity, Porod's law is satisfied for $[I(q) - B] = A/q^4$, and $S$ or $S_1/V_1$ can be calculated using $A$.

### *4.1.4 Unified model*

The scattering profile contains the information provided by all the features and regions mentioned earlier. Thus, the analysis of hierarchically organized systems may present some deviations for each region when analyzed separately. For a global understanding of complex systems with many dimensional levels, the implementation of analytical models, such as the "unified model" developed by Beaucage [77–79] and later adapted by Hammouda [80,81], is of great relevance. These models were proposed for simultaneously describing the regions mentioned above, obtaining the $R_g$ and scattering exponents and scaling factors for every structural level. The Beaucage model can describe systems with more than one structural level and the correspondent fractal regime according to

$$I(q) \simeq \sum\nolimits_n^{i=1} \left( G_i \exp\left(\frac{-q^2 R_{g,i}^2}{3}\right) + B_i \exp\left(\frac{-q^2 R_{g,i+1}^2}{3}\right) \left(\frac{erf\left(qkR_{g,i}/\sqrt{6}\right)}{q}\right)^{\alpha_1} \right) \tag{9}$$

where $i$ corresponds to the larger structure, $G_i$ and $B_i$ are the Guinier and Power-law prefactors, respectively, *erf* is the error function, and $k$ is an empirical constant.

Another way to treat data is the indirect Fourier transform (IFT), which is usually very useful when some sample data, such as the shape and size of the dispersion objects, are unknown. Implementations of this method can be performed with the GNOM program [82] for the reading and interpretation of one-dimensional (1D) dispersion curves and to evaluate the particle distance distribution function $p(r)$ for monodisperse systems or the size distribution function $D(R)$ for polydisperse systems.

The determination of size distribution $n(r)$, the form factor $f(qr)$, and the structure factor $S(qr)$ can be performed by strict methods. They can be determined according to

$$I(q) = N(\Delta\rho)^2 \int n(r)[V(r)f(qr)]^2 S(qr)dr \tag{10}$$

where $N$ is the sum of scatterers, $\Delta\rho$ is the electron density contrast between the particle and matrix, and $V(r)$ is the volume of the particle. Pedersen has presented many equations for $f(qr)$ (or $P(qr) = [f(qr)]^2$) to be implemented for several shapes [83]. $S(qr)$ describes the interparticle scattering, that is, $S(qr) \equiv 1$ for a noninteracting system or in the dilute limit. In SAXS analyses, it is common to express the structure factor as a hard sphere for interacting objects.

## 4.2 Biomolecular models for the SAS pattern

SAS is a fundamental tool in the study of biological macromolecules. In the past 3 decades, SAS has emerged as an essential method in biomolecular dynamics because of significant advances in efficient hardware, methods, and software for data modeling and analysis. The major advantage of the method lies in its ability to provide structural information about the general shape of a macromolecule or assembly in solution in partially or completely disordered systems. Several models describing atomic motions within proteins have been proposed to interpret experimental data for SAS. They consist of the description of the potentials that govern the movements of protein atoms and/or solvent/hydration atoms. Especially on SANS [84,85], the contrast variation technique and deuteration of the sample can reveal the position of each subunit within a complex sample, as has been described for the *Escherichia coli* ribosome [86,87]. Also, the allosteric conformational changes can be studied in detail [88]. The increasing availability of SAXS in synchrotron beamlines as well as bench-top instrument sources has made SAS data acquisition more frequent, which has helped in the continuous improvement of the analysis software.

### 4.2.1 Protein solutions

When polydisperse and nonspherical samples are analyzed, the scattering intensity can be modeled with methods such as "decoupling" and "average structure factor" estimation [83,89]. These models imply that the spatial location and orientation of the macromolecule are not associated. As expected, these approximations are not accurate for concentrated systems and/or anisotropic ones. Many proteins do not present perfectly spherical shapes but are monodisperse, and the two approximations give similar outcomes. Therefore, the expression that uses the average structure factor approximation is [90]

$$I(q) = N_P(\Delta\rho)^2 V_P^2 P(q) S^-(q) \tag{11}$$

where $N_P$ is the number of proteins per unit volume, $V_P$ is the volume of a single protein, and $P(q)$ is the form factor of a given protein. For example, an ellipsoid form factor can be used to model a protein as ovalbumin [91].

$$P(q) \equiv \langle |F(q)|^2 \rangle = \int_0^1 dx \left| \frac{3(\sin u - u \cos u)}{u^3} \right|^2 \tag{12}$$

$$u = qb\left[\left(\frac{a}{b}\right)^2 x^2 + (1-x)^2\right]^{\frac{1}{2}}$$

When the solution is diluted enough, the solvent electron density, $\rho_0$, governs one of the systems. For a system with two scatterers, with the second being a protein with $\rho_1$, the contrast of electron density can be described as

$$\Delta\rho = (\rho_1 - \rho_0) \tag{13}$$

Protein-protein interaction should be negligible in dilute solutions. Thus, this condition ensures the independent contribution of each protein particle, without interfering effects, to the total scattering intensity. Furthermore, the protein can have a negative or positive charge depending on the ionic strength (<100 mM). The load-induced interaction can be described using the Coulombic potential screening developed by Hayter and Penfold [92,93]. The model is based on the interaction potential $U_{SC}(r)$ between charged colloidal particles consisting of a hard-sphere plus a screened Coulomb potential.

$$Usc(r) = \begin{cases} \dfrac{z^2 e^2}{\varepsilon(1+k_D R)^2} \dfrac{\exp[-k_D(r-2R)]}{r} & \text{for } r > 2R \\ \infty & \text{for } r \leq 2R \end{cases} \tag{14}$$

where $z$ is the charge of the protein, $e$ is the electronic charge, and $\varepsilon$ is the dielectric constant of the solvent. The ionic strength, $I_{\text{ion}}$, determines the Debye screening length, $k_D$, which is defined as

$$K_D^{-1} = \frac{\epsilon\epsilon_0 RT}{2\rho F^2 I_{\text{ion}}} \tag{15}$$

From the characteristic function $\lambda 0(r)$ associated with a protein, which for a dilute monodisperse particle system depends exclusively on the geometry (size and shape) of the particle, the "distance distribution function" $p(r)$ is defined as $p(r) = 4\pi r^2 \lambda 0(r)$. The average orientation of the scattering intensity by a protein $I_1(q)$ can be written as a function of $p(r)$:

$$I_1(q) = (\Delta\rho)^2 V_1 \int_0^{D_{\max}} p(r) \frac{\sin qr}{qr} dr \tag{16}$$

Consequently, the distance distribution function $p(r)$ is determined by

$$p(r) = \frac{1}{8\pi^3 (\Delta\rho)^2 V_1} \int I_1(q) \frac{\sin qr}{qr} dr \tag{17}$$

A simple inspection of $p(r)$ allows one to approximate the shape and size of macromolecules in solution. $p(r)$ can be calculated by an inverse Fourier transform of $I(q)$. The largest particle dimension, $D_{\max}$, can be extrapolated as the distance at which the $p(r)$ drops to zero. Fig. 7 shows the expected profiles and $p(r)$ functions for particles of different but similar $D_{\max}$. When particles present two subunits, the $p(r)$ functions will present two peaks accounting for intrasubunit and intersubunit distances [94].

### 4.2.2 Membranes

Just as visible light leads to scattering because of differences in refractive index, for example, because of contrast between air and the material, contrasts in scattering length density result in the detection of the components of the material in SAXS and SANS. For X-rays, the contrast originates in differences in the electron density of two components of the sample. In the case of neutrons, the contrast originates in the particular structure of each nucleus in the sample, which in turn would lead to a specific interaction with neutrons. This contrast difference obtained by SAXS and SANS is of great interest for the study of complex systems such as membranes (see Fig. 8A) [95].

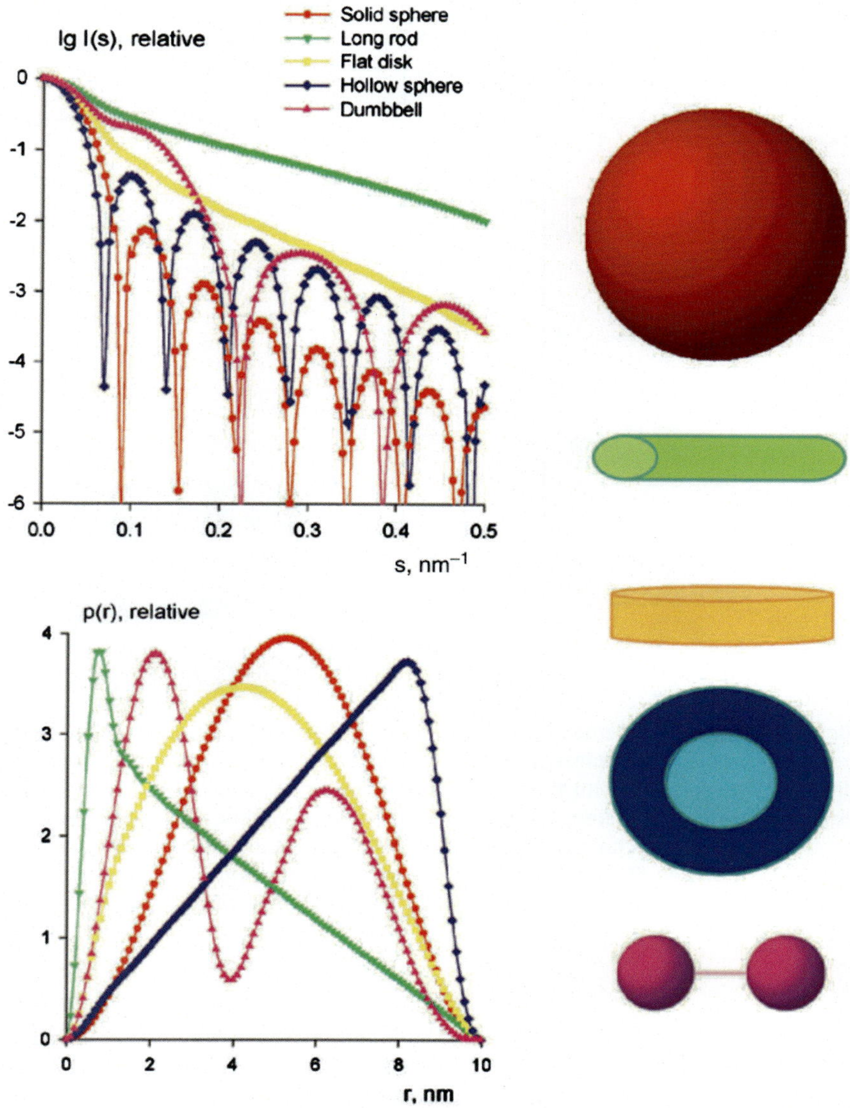


**Fig. 7** Log of scattering intensities and distance distribution functions of geometrical bodies. *(Reproduced with permission from Svergun DI, Koch MHJ. Small-angle scattering studies of biological macromolecules in solution. Rep Prog Phys 2003;66:1735–82. https://doi.org/10.1088/0034-4885/66/10/R05.)*

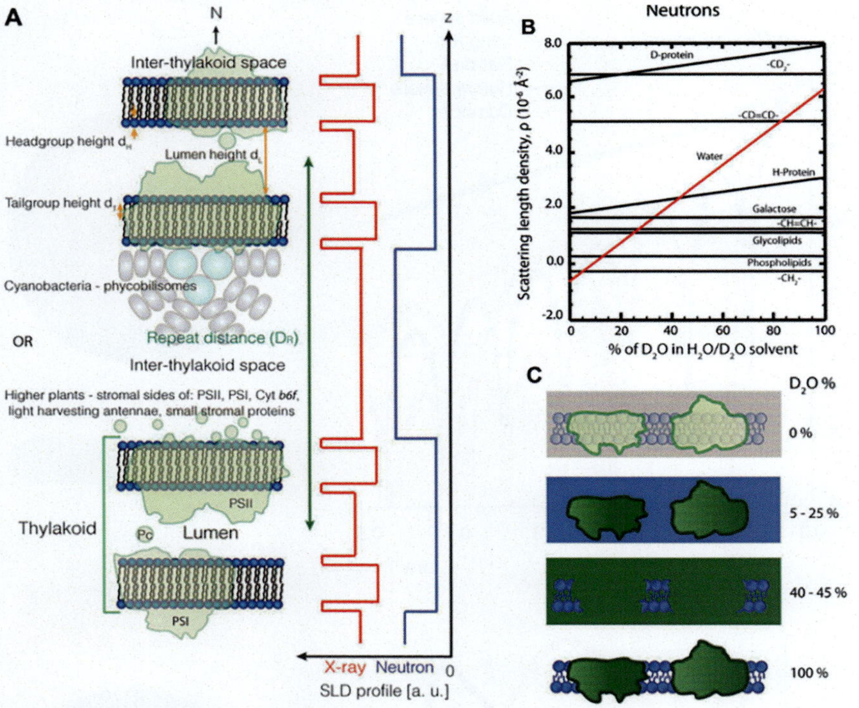


**Fig. 8** (A) Structural scheme of a photosynthetic membrane. Neutron *(blue)* and X-ray *(red)* SLD profiles are represented. (B) Comparison of the SLD of the solvent with different $D_2O$ contents with the neutron SLD of several biomolecules. (C) Schematic visualization of the different components of a photosynthetic membrane with the contrast variation technique. *(Reproduced with permission from Jakubauskas D, Mortensen K, Jensen PE, Kirkensgaard JJK. Small-angle X-ray and neutron scattering on photosynthetic membranes. Front Chem 2021;9:631370. https://doi.org/10.3389/fchem.2021.631370.)*

In SANS, the deuteration of biomolecules and/or solvents offers the possibility of matching the intensity of specific components within multicomponent systems, which allows determining the composition of particles and low-resolution structures of complex multicomponent particles, such as most biological samples [96].

According to the model expression for the stack of Nallet et al. [97] for lyotropic liquid crystalline lamellar phases, Fig. 8B illustrates the scattering length density profile for the cyanobacterial thylakoid membrane stack, with a basic building block of a double-layered thylakoid [98]. Using the contrast variation technique, the structure could be efficiently described. This enables the extraction of detailed information that is very difficult to obtain in other ways, allowing the independent analysis of proteins or lipids in an ensembled membrane (Fig. 8C). In this case, considering the lamellar distance $D$, the model can be expressed as

$$I(q) = 2\pi \frac{P(q) \cdot S(q)}{D \cdot q^2} \tag{18}$$

where $P(q)$ accounts for the unit cell of the thylakoids (e.g., the double bilayer) and $S(q)$ describes their stacking. The lamellar $S(q)$ from Nallet et al. [97] is given by

$$S(q) = 1 + 2\sum_{n=1}^{N-1} \left(1 - \frac{n}{N}\right) \cos\left(nq \cdot D\right) \exp\left(-q^2 D^2 \alpha(n)\right) \tag{19}$$

with

$$\alpha(n) = \frac{\eta_{cp}}{4\pi^2} \left( \ln\left(\pi n\right) + \gamma_E \right) \tag{20}$$

where $N$ is the stacked number of lamellae, $\gamma_E$ is Euler's constant, and $\eta_{cp}$ is the Caillé parameter. Low values depict a rigid membrane with a high bending modulus.

## 4.3 Deep learning methods for the reconstruction of 3D models

With the arrival of machine learning (ML) and artificial intelligence (AI) in labs, processing SAXS data to reconstruct high-quality 3D models is a challenge still under study. When working with ML/AI, it is necessary to test the developed algorithm against sufficient data. Using sufficient SAXS data, 3D density maps can be built for biomolecules, including folding of polymer chains, assembly of virtual atoms, approximation of the shape of the envelope, iterative stages, and database search methods.

Using ML methods, Franke et al. have developed a classification algorithm to categorize SAXS profiles associated with specific molecular shapes [99]. Using large amounts of SAXS data, the algorithm was fed and then used to predict molecular shapes of unknown structures (see Fig. 9). Also, there are successful attempts to integrate SAXS data to obtain 3D structures using the prediction/modeling/simulation of molecules [100,101]. Large databases are used for these models; the shapes are extracted from real protein complexes and represented by 3D Zernike polynomials to retrieve 3D models that match the experimental SAXS profiles. A commonly used

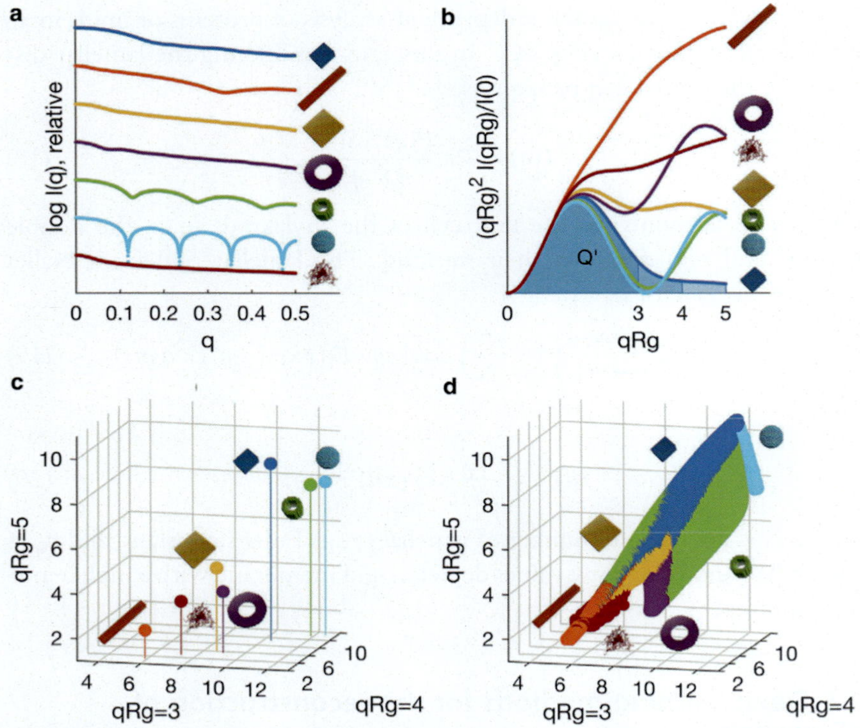


**Fig. 9** SAXS data for 488,000 scattering patterns. (A) Scattering patterns of geometric objects on a logarithmic scale, (B) normalized Kratky plot, and (C and D) in space $V'$. Colors by shape: compact *(dark blue)*, extended *(orange)*, flat *(yellow)*, ring *(purple)*, compact-hollow *(green)*, hollow sphere *(light blue)*, and random chain *(dark red)*, also indicated by the corresponding pictograms. *(Reproduced with permission from Franke D, Jeffries CM, Svergun DI. Machine learning methods for X-ray scattering data analysis from biomacromolecular solutions. Biophys J 2018;114:2485–92. https://doi.org/10.1016/j.bpj.2018.04.018.)*

piece of software is ToolBox SASTBX with the option sastbx.shapeup [102]. Inspired by AI, other work used deep learning methods, where an automatic encoder for 3D protein models was programmed to compile information from 3D shapes into vectors of a 200-dimensional latent space. This method (Fig. 10) can optimize vectors using genetic algorithms to build 3D models (both shape and radius) that are reconstructed from 1D SAXS data [103].

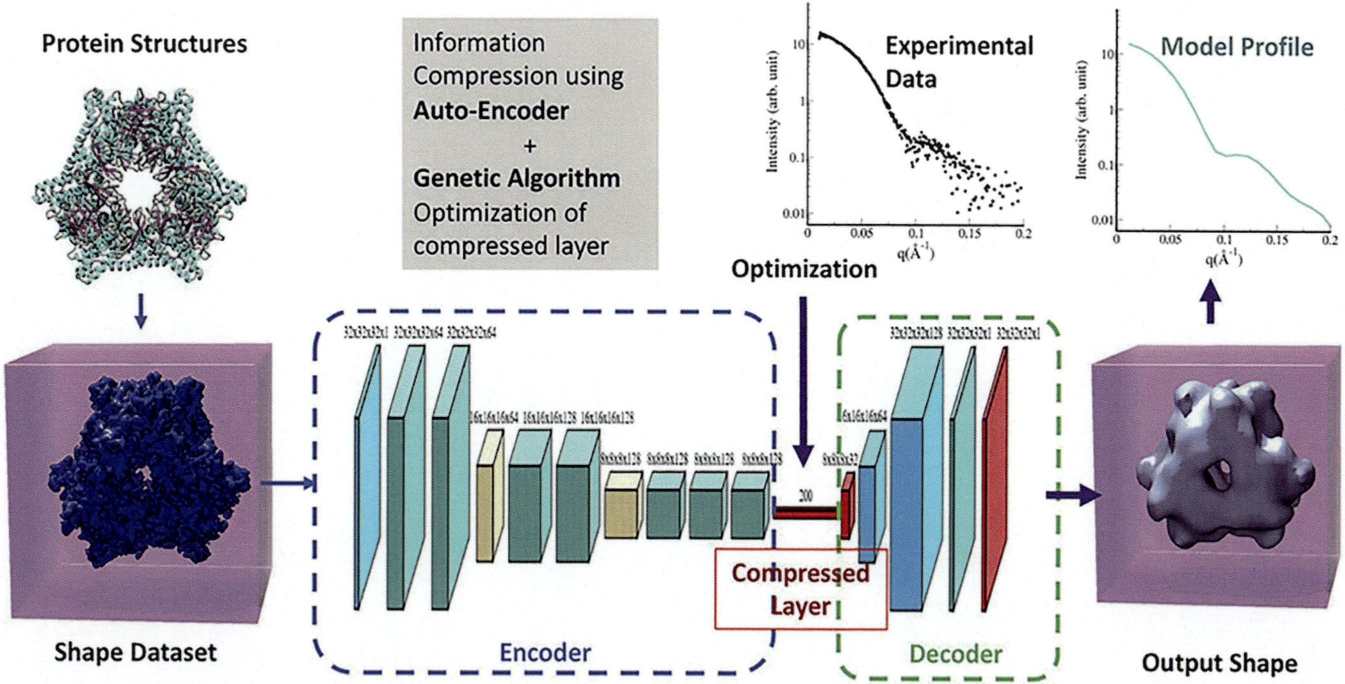


**Fig. 10** Deep learning autoencoder flow diagram to represent 3D protein models. *(Reproduced with permission from He H, Liu C, Liu H. Model reconstruction from small-angle X-ray scattering data using deep learning methods. iScience 2020;23:100906. https://doi.org/10.1016/j.isci.2020.100906.)*

# 5. Conclusions

The advances in the field of SAS techniques have demonstrated how useful they can be for biomolecular analysis, not just for pure systems but also for complex ones and even under dynamic conditions. Nevertheless, the physical background required to approach these techniques, principally regarding the interpretation of the results, is not negligible. This has been a significant disadvantage that has discouraged their use for many years by researchers, who supported their findings with the classical techniques of their field, such as crystallography or circular dichroism spectroscopy. Since their beginnings, SAS techniques have been dominated by physicists, which has also hampered the communication between them and biomolecular researchers. However, as discussed above, the unquestionable relevance of SAS in their field has encouraged researchers from different areas, such as molecular biology, to get involved with a technique unfamiliar to their typical academic formation. Given the advantages of incorporating SAS techniques or even modern 3D NMR studies, many research groups have been including physicists, software engineers, and bioinformatics researchers in their staff. The tendency to create interdisciplinary groups has become common across many fields and will lead to the growth of implementation of SAS techniques in biomolecular research in the near future.

Another disadvantage of SAS techniques that has been slowly being overcome over the years was the scarcity of bench-top instruments, which allow obtaining routine information or performing preliminary assays before requesting the beam time from large facilities. This is still true for SANS-related investigations, but the growing availability of SAXS bench-top instruments and the similarity of the interpretation of the results obtained by both techniques help reduce the gap from the laboratory to the large facility.

From another point of view, one of the most attractive capabilities of SAS techniques is that they can be applied to analyze many sample environments and even perform dynamic analysis in a versatile way that many other instruments cannot. Also, with novel bench-top instruments, large facilities, powerful modeling software, and a growing scientific community, SAS has been able to gain an indisputable place in the field of biomolecular analysis together with traditional techniques, such as crystallography or circular dichroism, or even more modern ones, such as NMR spectroscopy. Because of the versatility and potentiality of SAS techniques in dynamic studies, they may likely emerge as the gold standard techniques for monitoring structural biochemical and biological events in real time.

## Acknowledgments

J.J.P.B. and G.I.T.J. are grateful for his postdoctoral and doctoral fellowships granted by Consejo Nacional de Investigaciones Científicas y Técnicas (CONICET). This work was supported with grants from Universidad de Buenos Aires (UBACyT 20020170100125BA, UBACyT 20020190200022BA) and Agencia Nacional de Promoción Científica y Tecnológica (PICT 2018-01731, PICT 2019-03757, PICT 2018-01225).

## References

[1] Hsiao BS, Zuo F, Mao Y, Schick C. Experimental techniques. In: Piorkowska E, Rutledge GC, editors. Handbook of polymer crystallization. Hoboken, NJ, USA: John Wiley & Sons, Inc.; 2013. p. 1–30. https://doi.org/10.1002/9781118541838.ch1.

[2] Sanjeeva Murthy N. Scattering techniques for structural analysis of biomaterials. In: Characterization of biomaterials. Elsevier; 2013. p. 34–72. https://doi.org/10.1533/9780857093684.34.

[3] Brosey CA, Tainer JA. Evolving SAXS versatility: solution X-ray scattering for macromolecular architecture, functional landscapes, and integrative structural biology. Curr Opin Struct Biol 2019;58:197–213. https://doi.org/10.1016/j.sbi.2019.04.004.

[4] Blanchet CE, Hermes C, Svergun DI, Fiedler S. A small and robust active beamstop for scattering experiments on high-brilliance undulator beamlines. J Synchrotron Rad 2015;22:461–4. https://doi.org/10.1107/S160057751402829X.

[5] Strobl M, Harti R, Gruenzweig C, Woracek R, Plomp J. Small angle scattering in neutron imaging—a review. J Imaging 2017;3:64. https://doi.org/10.3390/jimaging3040064.

[6] Bolze J, Kogan V, Beckers D, Fransen M. High-performance small- and wide-angle X-ray scattering (SAXS/WAXS) experiments on a multi-functional laboratory goniometer platform with easily exchangeable X-ray modules. Rev Sci Instrum 2018;89, 085115. https://doi.org/10.1063/1.5041949.

[7] Mahieu E, Gabel F. Biological small-angle neutron scattering: recent results and development. Acta Crystallogr D Struct Biol 2018;74:715–26. https://doi.org/10.1107/S2059798318005016.

[8] Neylon C. Small angle neutron and X-ray scattering in structural biology: recent examples from the literature. Eur Biophys J 2008;37:531–41. https://doi.org/10.1007/s00249-008-0259-2.

[9] Nishida K, Ogawa H, Matsuba G, Konishi T, Kanaya T. A high-resolution small-angle light scattering instrument for soft matter studies. J Appl Crystallogr 2008;41:723–8. https://doi.org/10.1107/S002188980801265X.

[10] Stein RS, Keane JJ, Norris FH, Bettelheim FA, Wilson PR. Some light-scattering studies of the texture of crystalline polymers*. Ann N Y Acad Sci 2006;83:37–59. https://doi.org/10.1111/j.1749-6632.1960.tb40882.x.

[11] Stein RS, Rhodes MB. Photographic light scattering by polyethylene films. J Appl Phys 1960;31:1873–84. https://doi.org/10.1063/1.1735468.

[12] Kyu T. Time-resolved characterization of polymer phase transitions. In: Comprehensive polymer science and supplements. Elsevier; 1989. p. 307–46. https://doi.org/10.1016/B978-0-08-096701-1.00247-0.

[13] Okada T, Saito H, Inoue T. Time-resolved light scattering studies on the early stage of crystallization in isotactic polypropylene. Macromolecules 1992;25:1908–11. https://doi.org/10.1021/ma00033a011.

[14] Samuels RJ, Pulver MI. Structured polymer properties. Phys Today 1975;28:56–7. https://doi.org/10.1063/1.3068969.

[15] Jones KW. Synchrotron-radiation induced X-ray emission (SRIXE). In: Handbook of X-ray spectrometry. New York, NY: Marcel Dekker, Inc; 1999.
[16] Polizzi S, Spinozzi F. Small angle X-ray scattering (SAXS) with synchrotron radiation sources. In: Mobilio S, Boscherini F, Meneghini C, editors. Synchrotron radiation. Berlin Heidelberg, Berlin, Heidelberg: Springer; 2015. p. 337–59. https://doi.org/10.1007/978-3-642-55315-8_11.
[17] Carpenter JM, Yelon WB. 2. Neutron sources. In: Methods in experimental physics. Elsevier; 1986. p. 99–196. https://doi.org/10.1016/S0076-695X(08)60555-4.
[18] Jacques DA, Trewhella J. Small-angle scattering for structural biology-expanding the frontier while avoiding the pitfalls: small-angle scattering for structural biology. Protein Sci 2010;19:642–57. https://doi.org/10.1002/pro.351.
[19] Kikhney AG, Svergun DI. A practical guide to small angle X-ray scattering (SAXS) of flexible and intrinsically disordered proteins. FEBS Lett 2015;589:2570–7. https://doi.org/10.1016/j.febslet.2015.08.027.
[20] Lamas DG, de Oliveira Neto M, Kellermann G, Craievich AF. X-ray diffraction and scattering by nanomaterials. In: Nanocharacterization techniques. Elsevier; 2017. p. 111–82. https://doi.org/10.1016/B978-0-323-49778-7.00005-9.
[21] Piiadov V, Ares de Araújo E, Oliveira Neto M, Craievich AF, Polikarpov I. SAXSMoW 2.0: online calculator of the molecular weight of proteins in dilute solution from experimental SAXS data measured on a relative scale. Protein Sci 2019;28:454–63. https://doi.org/10.1002/pro.3528.
[22] Bada M, Walther D, Arcangioli B, Doniach S, Delarue M. Solution structural studies and low-resolution model of the Schizosaccharomyces pombe sap1 protein. J Mol Biol 2000;300:563–74. https://doi.org/10.1006/jmbi.2000.3854.
[23] Costenaro L, Grossmann JG, Ebel C, Maxwell A. Small-angle X-ray scattering reveals the solution structure of the full-length DNA gyrase A subunit. Structure 2005;13:287–96. https://doi.org/10.1016/j.str.2004.12.011.
[24] Hammel M, Walther M, Prassl R, Kuhn H. Structural flexibility of the N-terminal β-barrel domain of 15-lipoxygenase-1 probed by small angle X-ray scattering. Functional consequences for activity regulation and membrane binding. J Mol Biol 2004;343:917–29. https://doi.org/10.1016/j.jmb.2004.08.076.
[25] Fujisawa T, Kostyukova A, Maéda Y. The shapes and sizes of two domains of tropomodulin, the P-end-capping protein of actin-tropomyosin. FEBS Lett 2001;498:67–71. https://doi.org/10.1016/S0014-5793(01)02498-X.
[26] De Kruif CG. The structure of casein micelles: a review of small-angle scattering data. J Appl Crystallogr 2014;47:1479–89. https://doi.org/10.1107/S1600576714014563.
[27] Luh LM, Hänsel R, Löhr F, Kirchner DK, Krauskopf K, Pitzius S, Schäfer B, Tufar P, Corbeski I, Güntert P, Dötsch V. Molecular crowding drives active Pin1 into nonspecific complexes with endogenous proteins prior to substrate recognition. J Am Chem Soc 2013;135:13796–803. https://doi.org/10.1021/ja405244v.
[28] Miklos AC, Sumpter M, Zhou H-X. Competitive interactions of ligands and macromolecular crowders with maltose binding protein. PLoS One 2013;8, e74969. https://doi.org/10.1371/journal.pone.0074969.
[29] Goldenberg DP, Argyle B. Self crowding of globular proteins studied by small-angle X-ray scattering. Biophys J 2014;106:895–904. https://doi.org/10.1016/j.bpj.2013.12.004.
[30] Goldenberg DP, Argyle B. Minimal effects of macromolecular crowding on an intrinsically disordered protein: a small-angle neutron scattering study. Biophys J 2014;106:905–14. https://doi.org/10.1016/j.bpj.2013.12.003.
[31] Zhang F, Roosen-Runge F, Skoda MWA, Jacobs RMJ, Wolf M, Callow P, Frielinghaus H, Pipich V, Prévost S, Schreiber F. Hydration and interactions in protein

solutions containing concentrated electrolytes studied by small-angle scattering. Phys Chem Chem Phys 2012;14:2483. https://doi.org/10.1039/c2cp23460b.
[32] Ibrahim Z, Martel A, Moulin M, Kim HS, Härtlein M, Franzetti B, Gabel F. Time-resolved neutron scattering provides new insight into protein substrate processing by a AAA+ unfoldase. Sci Rep 2017;7:40948. https://doi.org/10.1038/srep40948.
[33] Sato D, Ikeguchi M. Mechanisms of ferritin assembly studied by time-resolved small-angle X-ray scattering. Biophys Rev 2019;11:449–55. https://doi.org/10.1007/s12551-019-00538-x.
[34] Zamora RA, Gutiérrez-Cerón C, Fernandes JA, Abarca G. Advanced surface characterization techniques in nano- and biomaterials. In: Alarcon EI, Ahumada M, editors. Nanoengineering materials for biomedical uses. Cham: Springer International Publishing; 2019. p. 35–55. https://doi.org/10.1007/978-3-030-31261-9_3.
[35] Lombardo D, Calandra P, Kiselev MA. Structural characterization of biomaterials by means of small angle X-rays and neutron scattering (SAXS and SANS), and light scattering experiments. Molecules 2020;25:5624. https://doi.org/10.3390/molecules25235624.
[36] Urban VS, Heller WT, Katsaras J, Bras W. Soft matter sample environments for time-resolved small angle neutron scattering experiments: a review. Appl Sci 2021;11:5566. https://doi.org/10.3390/app11125566.
[37] Van Vleet MJ, Weng T, Li X, Schmidt JR. In situ, time-resolved, and mechanistic studies of metal–organic framework nucleation and growth. Chem Rev 2018;118:3681–721. https://doi.org/10.1021/acs.chemrev.7b00582.
[38] Keidel R, Ghavami A, Lugo DM, Lotze G, Virtanen O, Beumers P, Pedersen JS, Bardow A, Winkler RG, Richtering W. Time-resolved structural evolution during the collapse of responsive hydrogels: the microgel-to-particle transition. Sci Adv 2018;4:eaao7086. https://doi.org/10.1126/sciadv.aao7086.
[39] Martin EW, Harmon TS, Hopkins JB, Chakravarthy S, Incicco JJ, Schuck P, Soranno A, Mittag T. A multi-step nucleation process determines the kinetics of prion-like domain phase separation. Nat Commun 2021;12:4513. https://doi.org/10.1038/s41467-021-24727-z.
[40] Tuukkanen AT, Spilotros A, Svergun DI. Progress in small-angle scattering from biological solutions at high-brilliance synchrotrons. IUCrJ 2017;4:518–28. https://doi.org/10.1107/S2052252517008740.
[41] Wang W, Li L, Henzler K, Lu Y, Wang J, Han H, Tian Y, Wang Y, Zhou Z, Lotze G, Narayanan T, Ballauff M, Guo X. Protein immobilization onto cationic spherical polyelectrolyte brushes studied by small angle X-ray scattering. Biomacromolecules 2017;18:1574–81. https://doi.org/10.1021/acs.biomac.7b00164.
[42] Kelley EG, Nguyen MHL, Marquardt D, Maranville BB, Murphy RP. Measuring the time-evolution of nanoscale materials with stopped-flow and small-angle neutron scattering. JoVE 2021;62873. https://doi.org/10.3791/62873.
[43] Takahashi R, Narayanan T, Sato T. Growth kinetics of polyelectrolyte complexes formed from oppositely-charged homopolymers studied by time-resolved ultra-small-angle X-ray scattering. J Phys Chem Lett 2017;8:737–41. https://doi.org/10.1021/acs.jpclett.6b02957.
[44] Wrede O, Reimann Y, Lülsdorf S, Emmrich D, Schneider K, Schmid AJ, Zauser D, Hannappel Y, Beyer A, Schweins R, Gölzhäuser A, Hellweg T, Sottmann T. Volume phase transition kinetics of smart N-n-propylacrylamide microgels studied by time-resolved pressure jump small angle neutron scattering. Sci Rep 2018;8:13781. https://doi.org/10.1038/s41598-018-31976-4.
[45] Feng J, Kriechbaum M, Liu L(E). In situ capabilities of small angle X-ray scattering. Nanotechnol Rev 2019;8:352–69. https://doi.org/10.1515/ntrev-2019-0032.

[46] Ilhan-Ayisigi E, Yaldiz B, Bor G, Yaghmur A, Yesil-Celiktas O. Advances in microfluidic synthesis and coupling with synchrotron SAXS for continuous production and real-time structural characterization of nano-self-assemblies. Colloids Surf B: Biointerfaces 2021;201, 111633. https://doi.org/10.1016/j.colsurfb.2021.111633.
[47] Pollack L. Time resolved SAXS and RNA folding. Biopolymers 2011;95:543–9. https://doi.org/10.1002/bip.21604.
[48] Josts I, Gao Y, Monteiro DCF, Niebling S, Nitsche J, Veith K, Gräwert TW, Blanchet CE, Schroer MA, Huse N, Pearson AR, Svergun DI, Tidow H. Structural kinetics of MsbA investigated by stopped-flow time-resolved small-angle X-ray scattering. Structure 2020;28:348–54. e3 https://doi.org/10.1016/j.str.2019.12.001.
[49] Adamo M, Poulos AS, Miller RM, Lopez CG, Martel A, Porcar L, Cabral JT. Rapid contrast matching by microfluidic SANS. Lab Chip 2017;17:1559–69. https://doi.org/10.1039/C7LC00179G.
[50] Silva BFB. SAXS on a chip: from dynamics of phase transitions to alignment phenomena at interfaces studied with microfluidic devices. Phys Chem Chem Phys 2017;19:23690–703. https://doi.org/10.1039/C7CP02736B.
[51] Komorowski K, Schaeper J, Sztucki M, Sharpnack L, Brehm G, Köster S, Salditt T. Vesicle adhesion in the electrostatic strong-coupling regime studied by time-resolved small-angle X-ray scattering. Soft Matter 2020;16:4142–54. https://doi.org/10.1039/D0SM00259C.
[52] Sreij R, Dargel C, Hannappel Y, Jestin J, Prévost S, Dattani R, Wrede O, Hellweg T. Temperature dependent self-organization of DMPC membranes promoted by intermediate amounts of the saponin aescin. Biochim Biophys Acta Biomembr 2019;1861:897–906.
[53] Bey H, Gtari W, Aschi A. Study of the complex coacervation mechanism between the lysing enzyme from T. harzianum and polyallylamine hydrochloride. Int J Biol Macromol 2019;124:780–7. https://doi.org/10.1016/j.ijbiomac.2018.11.266.
[54] Brooks NJ, Gauthe BLLE, Terrill NJ, Rogers SE, Templer RH, Ces O, Seddon JM. Automated high pressure cell for pressure jump x-ray diffraction. Rev Sci Instrum 2010;81, 064103. https://doi.org/10.1063/1.3449332.
[55] Yang S, Tyler AII, Ahrné L, Kirkensgaard JJK. Skimmed milk structural dynamics during high hydrostatic pressure processing from in situ SAXS. Food Res Int 2021;147, 110527. https://doi.org/10.1016/j.foodres.2021.110527.
[56] Banachowicz E, Kozak M, Patkowski A, Meier G, Kohlbrecher J. High-pressure small-angle neutron scattering studies of glucose isomerase conformation in solution. J Appl Crystallogr 2009;42:461–8. https://doi.org/10.1107/S0021889809007456.
[57] Papagiannopoulos A, Pispas S. Protein- and nanoparticle-loaded hydrogels studied by small-angle scattering and rheology techniques. In: Thakur VK, Thakur MK, editors. Hydrogels, gels horizons: from science to smart materials. Singapore: Springer Singapore; 2018. p. 113–43. https://doi.org/10.1007/978-981-10-6077-9_5.
[58] Shui Y, Huang L, Wei C, Chen J, Song L, Sun G, Lu A, Liu D. Intrinsic properties of the matrix and interface of filler reinforced silicone rubber: an in situ Rheo-SANS and constitutive model study. Compos Commun 2021;23, 100547. https://doi.org/10.1016/j.coco.2020.100547.
[59] Xu H-N, Tang Y-Y, Ouyang X-K. Shear-induced breakup of cellulose nanocrystal aggregates. Langmuir 2017;33:235–42. https://doi.org/10.1021/acs.langmuir.6b03807.
[60] Narayanan T, Dattani R, Möller J, Kwaśniewski P. A microvolume shear cell for combined rheology and x-ray scattering experiments. Rev Sci Instrum 2020;91, 085102. https://doi.org/10.1063/5.0012905.
[61] Pignon F, Challamel M, De Geyer A, Elchamaa M, Semeraro EF, Hengl N, Jean B, Putaux J-L, Gicquel E, Bras J, Prevost S, Sztucki M, Narayanan T, Djeridi

H. Breakdown and buildup mechanisms of cellulose nanocrystal suspensions under shear and upon relaxation probed by SAXS and SALS. Carbohydr Polym 2021;260, 117751. https://doi.org/10.1016/j.carbpol.2021.117751.
[62] Qi W, Yu J, Zhang Z, Xu H-N. Effect of pH on the aggregation behavior of cellulose nanocrystals in aqueous medium. Mater Res Express 2019;6, 125078. https://doi.org/10.1088/2053-1591/ab5974.
[63] Nguyen NA, Shen H, Liu Y, Mackay ME. Kinetics and mechanism of poly(3-hexylthiophene) crystallization in solution under shear flow. Macromolecules 2020;53:5795–804. https://doi.org/10.1021/acs.macromol.0c00717.
[64] Dewhurst CD, Grillo I, Honecker D, Bonnaud M, Jacques M, Amrouni C, Perillo-Marcone A, Manzin G, Cubitt R. The small-angle neutron scattering instrument D33 at the Institut Laue–Langevin. J Appl Crystallogr 2016;49:1–14. https://doi.org/10.1107/S1600576715021792.
[65] Barhoumi Meddeb A, Chae I, Han A, Kim SH, Ounaies Z. Magnetic field effects on cellulose nanocrystal ordering in a non-aqueous solvent. Cellulose 2020;27:7901–10. https://doi.org/10.1007/s10570-020-03320-5.
[66] Rajnak M, Garamus VM, Timko M, Kopcansky P, Paulovicova K, Kurimsky J, Dolnik B, Cimbala R. Small angle X-ray scattering study of magnetic nanofluid exposed to an electric field. Acta Phys Pol A 2020;137:942–4. https://doi.org/10.12693/APhysPolA.137.942.
[67] Bale HD, Schmidt PW. Small-angle X-ray-scattering investigation of submicroscopic porosity with fractal properties. Phys Rev Lett 1984;53:596–9. https://doi.org/10.1103/PhysRevLett.53.596.
[68] Guinier A, Fournet G, Walker CB, Vineyard GH. Small-angle scattering of X-rays. Phys Today 1956;9:38–9. https://doi.org/10.1063/1.3060069.
[69] Choi KH, Morais M. Use of small-angle X-ray scattering to investigate the structure and function of dengue virus NS3 and NS5. In: Padmanabhan R, Vasudevan SG, editors. Dengue, methods in molecular biology. New York, New York, NY: Springer; 2014. p. 241–52. https://doi.org/10.1007/978-1-4939-0348-1_15.
[70] Cherny AY, Anitas EM, Osipov VA, Kuklin AI. Deterministic fractals: extracting additional information from small-angle scattering data. Phys Rev E 2011;84, 036203. https://doi.org/10.1103/PhysRevE.84.036203.
[71] Cherny AY, Anitas EM, Osipov VA, Kuklin AI. Small-angle scattering from the Cantor surface fractal on the plane and the Koch snowflake. Phys Chem Chem Phys 2017;19:2261–8. https://doi.org/10.1039/C6CP07496K.
[72] Cherny AY, Anitas EM, Osipov VA, Kuklin AI. Scattering from surface fractals in terms of composing mass fractals. J Appl Crystallogr 2017;50:919–31. https://doi.org/10.1107/S1600576717005696.
[73] Schmidt PW. Interpretation of small-angle scattering curves proportional to a negative power of the scattering vector. J Appl Crystallogr 1982;15:567–9. https://doi.org/10.1107/S002188988201259X.
[74] Pfeifer P, Ehrburger-Dolle F, Rieker TP, González MT, Hoffman WP, Molina-Sabio M, Rodríguez-Reinoso F, Schmidt PW, Voss DJ. Nearly space-filling fractal networks of carbon nanopores. Phys Rev Lett 2002;88, 115502. https://doi.org/10.1103/PhysRevLett.88.115502.
[75] Porod G. Die Röntgenkleinwinkelstreuung von dichtgepackten kolloiden Systemen: I. Teil. Kolloid-Zeitschrift 1951;124:83–114. https://doi.org/10.1007/BF01512792.
[76] Glatter O, Kratky O, editors. Small angle x-ray scattering. New York: Academic Press, London; 1982.
[77] Beaucage G. Small-angle scattering from polymeric mass fractals of arbitrary mass-fractal dimension. J Appl Crystallogr 1996;29:134–46. https://doi.org/10.1107/S0021889895011605.

[78] Beaucage G. Approximations leading to a unified exponential/power-law approach to small-angle scattering. J Appl Crystallogr 1995;28:717–28. https://doi.org/10.1107/S0021889895005292.
[79] Beaucage G, Schaefer DW. Structural studies of complex systems using small-angle scattering: a unified Guinier/power-law approach. J Non-Cryst Solids 1994;172–174:797–805. https://doi.org/10.1016/0022-3093(94)90581-9.
[80] Hammouda B. A new Guinier–Porod model. J Appl Crystallogr 2010;43:716–9. https://doi.org/10.1107/S0021889810015773.
[81] Hammouda B. Analysis of the Beaucage model. J Appl Crystallogr 2010;43:1474–8. https://doi.org/10.1107/S0021889810033856.
[82] Svergun DI. Determination of the regularization parameter in indirect-transform methods using perceptual criteria. J Appl Crystallogr 1992;25:495–503. https://doi.org/10.1107/S0021889892001663.
[83] Pedersen JS. Analysis of small-angle scattering data from colloids and polymer solutions: modeling and least-squares fitting. Adv Colloid Interf Sci 1997;70:171–210. https://doi.org/10.1016/S0001-8686(97)00312-6.
[84] Cusack S, Jacrot B, Leberman R, May R, Timmins P, Zaccai G. Neutron scattering. Nature 1989;339:330. https://doi.org/10.1038/339330a0.
[85] Jacrot B. The study of biological structures by neutron scattering from solution. Rep Prog Phys 1976;39:911–53. https://doi.org/10.1088/0034-4885/39/10/001.
[86] Capel MS, Engelman DM, Freeborn BR, Kjeldgaard M, Langer JA, Ramakrishnan V, Schindler DG, Schneider DK, Schoenborn BP, Sillers I-Y, Yabuki S, Moore PB. A complete mapping of the proteins in the small ribosomal subunit of *Escherichia coli*. Science 1987;238:1403–6. https://doi.org/10.1126/science.3317832.
[87] Ramakrishnan V. Distribution of protein and RNA in the 30 *S* ribosomal subunit. Science 1986;231:1562–4. https://doi.org/10.1126/science.3513310.
[88] Putnam CD, Hammel M, Hura GL, Tainer JA. X-ray solution scattering (SAXS) combined with crystallography and computation: defining accurate macromolecular structures, conformations and assemblies in solution. Quart Rev Biophys 2007;40:191–285. https://doi.org/10.1017/S0033583507004635.
[89] Chen S-H, Lin T-L. 16. Colloidal solutions. In: Methods in experimental physics. Elsevier; 1987. p. 489–543. https://doi.org/10.1016/S0076-695X(08)60576-1.
[90] Chen SH. Small angle neutron scattering studies of the structure and interaction in micellar and microemulsion systems. Annu Rev Phys Chem 1986;37:351–99. https://doi.org/10.1146/annurev.pc.37.100186.002031.
[91] Ianeselli L, Zhang F, Skoda MWA, Jacobs RMJ, Martin RA, Callow S, Prévost S, Schreiber F. Protein–protein interactions in ovalbumin solutions studied by small-angle scattering: effect of ionic strength and the chemical nature of cations. J Phys Chem B 2010;114:3776–83. https://doi.org/10.1021/jp9112156.
[92] Hansen J-P, Hayter JB. A rescaled MSA structure factor for dilute charged colloidal dispersions. Mol Phys 1982;46:651–6. https://doi.org/10.1080/00268978200101471.
[93] Hayter JB, Penfold J. An analytic structure factor for macroion solutions. Mol Phys 1981;42:109–18. https://doi.org/10.1080/00268978100100091.
[94] Svergun DI, Koch MHJ. Small-angle scattering studies of biological macromolecules in solution. Rep Prog Phys 2003;66:1735–82. https://doi.org/10.1088/0034-4885/66/10/R05.
[95] Jakubauskas D, Mortensen K, Jensen PE, Kirkensgaard JJK. Small-angle X-ray and neutron scattering on photosynthetic membranes. Front Chem 2021;9, 631370. https://doi.org/10.3389/fchem.2021.631370.
[96] Breyton C, Gabel F, Lethier M, Flayhan A, Durand G, Jault J-M, Juillan-Binard C, Imbert L, Moulin M, Ravaud S, Härtlein M, Ebel C. Small angle neutron scattering

for the study of solubilised membrane proteins. Eur Phys J E 2013;36:71. https://doi.org/10.1140/epje/i2013-13071-6.
[97] Nallet F, Laversanne R, Roux D. Modelling X-ray or neutron scattering spectra of lyotropic lamellar phases: interplay between form and structure factors. J Phys II France 1993;3:487–502. https://doi.org/10.1051/jp2:1993146.
[98] Jakubauskas D, Kowalewska Ł, Sokolova AV, Garvey CJ, Mortensen K, Jensen PE, Kirkensgaard JJK. Ultrastructural modeling of small angle scattering from photosynthetic membranes. Sci Rep 2019;9:19405. https://doi.org/10.1038/s41598-019-55423-0.
[99] Franke D, Jeffries CM, Svergun DI. Machine learning methods for X-ray scattering data analysis from biomacromolecular solutions. Biophys J 2018;114:2485–92. https://doi.org/10.1016/j.bpj.2018.04.018.
[100] Hura GL, Hodge CD, Rosenberg D, Guzenko D, Duarte JM, Monastyrskyy B, Grudinin S, Kryshtafovych A, Tainer JA, Fidelis K, Tsutakawa SE. Small angle X-ray scattering-assisted protein structure prediction in CASP13 and emergence of solution structure differences. Proteins 2019;87:1298–314. https://doi.org/10.1002/prot.25827.
[101] Karczyńska AS, Mozolewska MA, Krupa P, Giełdoń A, Liwo A, Czaplewski C. Prediction of protein structure with the coarse-grained UNRES force field assisted by small X-ray scattering data and knowledge-based information. Proteins 2018;86:228–39. https://doi.org/10.1002/prot.25421.
[102] Liu H, Hexemer A, Zwart PH. The *Small Angle Scattering ToolBox* (*SASTBX*): an open-source software for biomolecular small-angle scattering. J Appl Crystallogr 2012;45:587–93. https://doi.org/10.1107/S0021889812015786.
[103] He H, Liu C, Liu H. Model reconstruction from small-angle X-ray scattering data using deep learning methods. iScience 2020;23, 100906. https://doi.org/10.1016/j.isci.2020.100906.

CHAPTER 11

# Advances in X-ray crystallography methods to study structural dynamics of macromolecules

Ali A. Kermani[a,*], Swati Aggarwal[b], and Alireza Ghanbarpour[c]
[a]Department of Molecular, Cellular, and Developmental Biology, University of Michigan, Ann Arbor, MI, United States
[b]BioMAX, MAXIV Laboratory, Lund, Sweden
[c]Department of Biology, Massachusetts Institute of Technology, Cambridge, MA, United States

## 1. Introduction

X-ray crystallography has been one of the greatest scientific tools developed in the 20th century. This technique has enabled molecular biologists to understand and explain some of the most critical fundamental processes of biological systems. Structure determination of proteins using X-ray crystallography was initiated when John Kendrew and Max Perutz unraveled the structure of myoglobin [1] and hemoglobin [2], respectively, in the late 1950s. Other significant milestones in the field of protein X-ray crystallography are the structure elucidation of nucleic acid-protein complexes by Aaron Klug [3]; the structure of the photosynthetic reaction center by Johann Deisenhofer, Hartmut Michel, and Robert Huber [4]; the structure of F1 ATPase by John Walker [5]; the structure of potassium channels by Roderick Mackinnon [6]; the structure of RNA polymerase by Roger Kornberg [7]; and the first structure of G-protein coupled receptor (GPCR) by Brian Kobilka [8].

According to the Protein Data Bank (PDB) statistics, X-ray crystallography (with over 167,000 entries as of July 2022) is the primary technique for structure determination of both cytosolic and membrane proteins (https://www.rcsb.org/stats/growth/growth-xray). Membrane proteins account for 20%–30% of the proteome in most organisms [9] and play critical roles in a wide range of biological processes such as signal transduction, respiration, molecular transport, and cell-cell communication in eukaryotes

* Current address: Department of Structural Biology, St. Jude Children's Research Hospital, Memphis, TN, United States.

*Advanced Spectroscopic Methods to Study Biomolecular Structure and Dynamics*
https://doi.org/10.1016/B978-0-323-99127-8.00020-9

and conferring multidrug resistance in pathogenic bacteria. Membrane proteins, mainly GPCRs, ion channels, and receptor tyrosine kinases, are the target of more than 60% of the current drugs on the market [10,11]. The structural information of membrane proteins has facilitated our understanding of the molecular details of signal transduction [12–15] and selective ion conduction [16,17] across the cell membrane and to design novel structure-based drugs [18].

However, elucidating the structure of proteins using X-ray crystallography is associated with multiple obstacles that impede the rate of solving new crystal structures. Some of these challenges include difficulty in obtaining large amounts of the target protein; extracting and solubilizing membrane proteins; stabilizing purified proteins; generating single, large, and well-diffracting crystals; phasing novel protein structures; and finally dealing with crystallographic defects such as high levels of anisotropy and mosaicity [19]. Recent technological advances have aided crystallographers in overcoming some of these challenges [20,21]. Advances in sample preparation, automation, and computational programs minimize the sample volume, lower the cost, and accelerate the process of generating crystals, screening crystals, data collection, and data processing. They enable high-throughput crystal screening, processing data collected from twinned or multicrystal samples, and generating single crystals from liquids by applying pressure [22]. In this chapter, we summarize some of the recent developments in the field of protein X-ray crystallography that have enabled the high-resolution structure determination of challenging proteins, particularly membrane proteins. These include (i) novel detergents and solubilization reagents, (ii) strategies to improve stabilization and increase crystallization likelihood, (iii) methods to assess the homogeneity and purity of protein samples, (iv) new crystallization methods, (v) new crystallization additives, and (vi) advances in instrument and data-processing software.

## 2. Protein extraction and purification

The first bottleneck in macromolecular X-ray crystallography is to generate sufficient amounts of target protein in stable, monodispersed, and aggregation-free states [23]. This can be achieved by optimizations and strategical modulations of several conditions [20].

### 2.1 Detergents and surfactants

Compared to soluble proteins, membrane proteins mostly express in low quantities and are more prone to denaturation and aggregation [19]. This is because membrane proteins are embedded in the phospholipid bilayer, and thus, their extraction requires utilizing amphiphilic molecules, such as detergents. The amphiphilic detergents with a hydrophobic tail and a hydrophilic head group enable them to solubilize membrane proteins by enclosing the hydrophobic core of membrane proteins, whereas the loops and hydrophilic regions remain exposed to the aqueous environment [19]. However, breaking lipid-protein and protein-protein interactions during the solubilization process can adversely affect the stability and function of membrane proteins. Here, we introduce some of the most recently developed detergents with enhanced features to improve the solubilization and stabilization process of membrane proteins. These advances in extraction and solubilizing methods have accelerated the production of eukaryotic membrane proteins for X-ray structural studies.

### 2.1.1 Maltose-neopentyl glycol (MNG) compounds

Maltose-neopentyl glycol (MNG) compounds were introduced in 2010 as a new class of amphiphiles suitable for direct extraction and solubilization of membrane proteins from lipid bilayers (Table 1) [24]. MNG has facilitated the purification and structure determination of several challenging membrane proteins, including GPCRs and ABC transporters (Table 2). The high efficacy of MNG in extracting and solubilizing membrane proteins is attributed to its ability to pack compactly during the micelle formation process

**Table 1** The most recently developed detergents for use in membrane protein crystallography.

| Detergent | Detergent type | Chain length | CMC (%, mM) | Reference |
|---|---|---|---|---|
| Decyl maltose neopentyl glycol (DMNG) | Nonionic | 10C | 0.036 mM 0.0034% | [24] |
| Undecyl maltose neopentyl glycol (UMNG) | Nonionic | 11C | – | Anatrace website |
| Lauryl maltose neopentyl glycol (LMNG) | Nonionic | 12C | 0.01 mM 0.001% | [24] |
| Octyl glucose neopentyl glycol (OGNG) | Nonionic | 8C | 1.02 mM 0.058% | [26] |
| NAPol | Nonionic | – | 2% | [27] |
| Calixarene | Anionic | 3C to 12C | 0.05 to 1.5 mM | [28] |
| Fluorinated octyl maltoside | Nonionic | 6F | 0.7 mM | [29] |

**Table 2** List of some membrane protein structures solved using novel detergents, see text.

| Detergent | Membrane protein | PDB code | Resolution (Å) | Reference |
|---|---|---|---|---|
| Maltose-neopentyl glycol (MNG) | $\beta_2$ adrenrgic receptor ($\beta_2$AR) | 3SN6 | 3.2 | [15] |
| LMNG | Agonist-$\beta_2$ adrenoceptor complex | 3PDS | 3.5 | [30] |
| LMNG | M3 muscarinic acetylcholine receptor | 4DAJ | 3.4 | [31] |
| LMNG | Neurotensin receptor NTS1 | 4GRV | 2.8 | [32] |
| LMNG | TatC | 4B4A | 3.5 | [33] |
| LMNG | TRPA1 ion channel | 3J9P | 4.24 | [34] |
| LMNG | OX2 orexin receptor bound to the insomnia drug suvorexant | 4S0V | 2.5 | [35] |
| LMNG | Rhodopsin-Arrestin complex | 4ZWJ | 3.3 | [36] |
| LMNG | ABC transporter PglK | 5C78 | 2.9 | [37] |
| LMNG | MFS transporter Ferroportin | 5AYN | 2.2 | [38] |
| LMNG | TRPA1 ion channel | 3J9P | 4.24 | [34] |
| LMNG | Influenza hemagglutinin | 6HJN | 3.3 | [39] |
| LMNG | Rhodopsin-transducin | 6OYA | 3.3 | [40] |
| LMNG/CHS mixture (8:1) | DUOX1-DUOXA1 | 6WXR 6WXU | 3.2 | [41] |
| LMNG/CHS mixture (8:1) | KCNQ1 potassium channel | 6UZZ 6V00 | 3.1 | [42] |
| DMNG | TRPV2 ion channel | 5HI9 | 4.4 | [43] |
| DMNG | $H^+/Ca^{2+}$ exchanger | 4KPP | 2.3 | [44] |
| OGNG/CHS | $K^+$ channel TREK-2 | 4BW5 | 3.2 | [45] |
| NAPol | TOM core complex | 5O8O | 6.8 | [46] |
| NAPol | Photosystem I supercomplex | Cryo-EM | 6.9 | [47] |
| NAPol | Photosystem II supercomplex | Cryo-EM | 5.8 | [48] |
| Calixarene | Potassium chloride cotransporter KCC2 | Cryo-EM | 15 | [49] |

(Table 1). This results in an exceptionally low critical micelle concentration (CMC) (as low as 11 nM) [25], improvement of protein stability (indicated by an increase in thermal stability of MNG-solubilized proteins), and an increase in water solubility of MNG micelles. Fig. 1A illustrates the structure

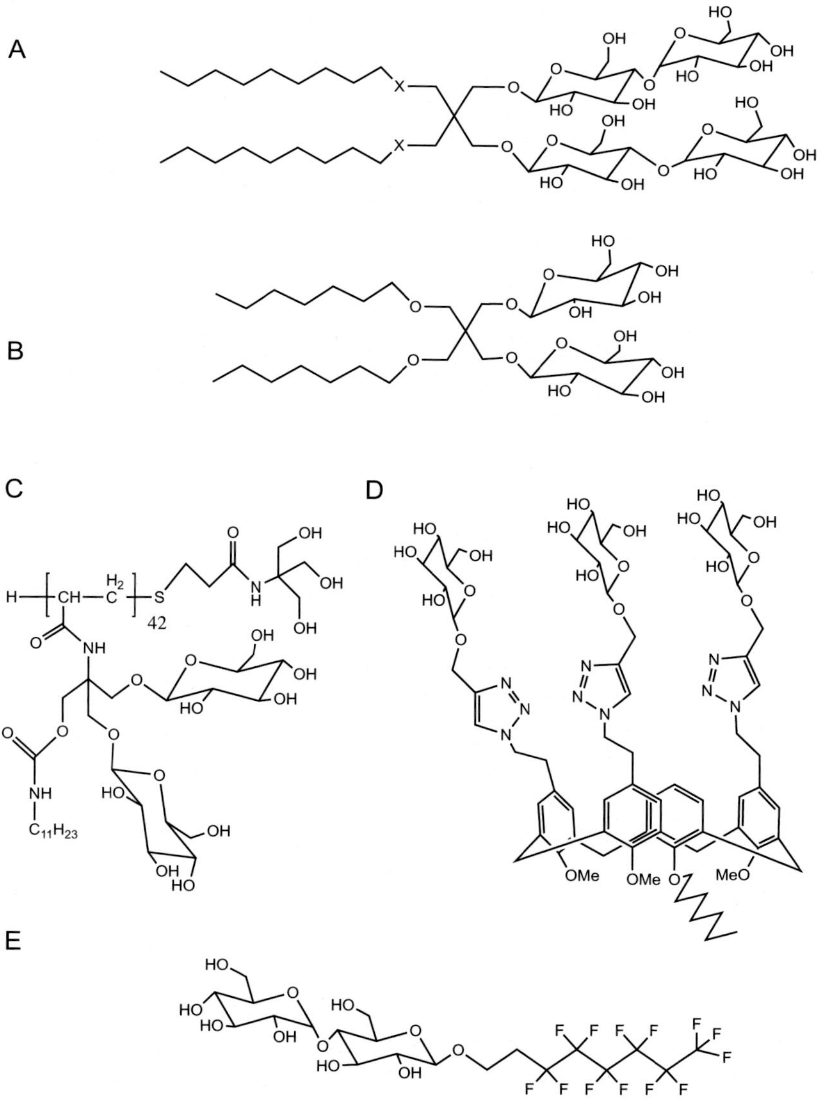


**Fig. 1** Chemical structures of some of the novel detergents developed for membrane proteins' purification and stabilization: (A) maltose neopentyl glycol (MNG), (B) glucose neopentyl glycol (GNG), (C) NAPol, (D) calixarene, and (E) fluorinated octyl maltoside.

of a representative MNG in which two hydrophilic heads, each made of a maltose unit, are connected to two *n*-decyl lipophilic chains via a quaternary carbon and variously an amide, ether, or aliphatic moiety [24]. In order to crystallize membrane proteins with different numbers of transmembranes (TMs) and intracellular and extracellular loop sizes, MNG amphiphiles with varying chain lengths have been generated. Lauryl maltose neopentyl glycol (LMNG), with 12 carbon chain lengths and more than 10 published structures, is the most successful member of the MNG class in determining the crystal structure of membrane proteins (Table 2). Decyl maltose neopentyl glycol (DMNG) (10 carbon chain length) occupies the second place in this class with two solved crystal structures (Table 2). Undecyl maltose neopentyl glycol (UMNG) is the most recent member of this class, which carries a chain composed of 11 carbons. UMNG was created recently to provide an intermediate chain length between LMNG and DMNG.

### 2.1.2 Glucose-neopentyl glycerol (GNG) compounds

The next class of detergents, glucose-neopentyl glycerols (GNGs), is developed based on MNG detergents, in which the maltose moiety is replaced with glucose (Fig. 1B) [26]. As a result, GNG amphiphiles form smaller micelles compared to their MNG counterparts and provide an extra hydrophilic surface area necessary to form crystal contacts (Table 1). However, lower efficacy in stabilizing membrane proteins, compared to the MNG class [26], is the main drawback of GNG and explains the limited number of solved protein structures in the presence of this detergent (Table 2). One possible strategy to exploit the benefits of both classes (MNG is more efficient in stabilizing membrane proteins, whereas GNG promotes crystallization by providing a larger surface area) is to extract and solubilize the target protein using the MNG class and transfer the MNG-solubilized protein into the GNG detergent immediately before setting up the crystal trays, for instance, during size exclusion chromatography (SEC).

### 2.1.3 Nonionic amphipols (NAPol)

It is crucial to retain membrane proteins extracted from lipid bilayers in a functional form throughout the purification process when the molecular mechanism of the target protein needs to be elucidated using X-ray crystallography. However, detergents can impact the activity of solubilized membrane proteins by forming protein-free micelles in the solution [50]. Free micelles can affect the stability of extracted proteins by disrupting protein-protein interactions and thus influencing the structural determination efforts by phase separation during crystallization. Nonionic amphipols (NAPol) are a new class of amphipols that stabilizes membrane proteins in the aqueous solution by forming small globular particles made of two glucose molecules and one undecyl chain with a diameter of about 6nm [27] (Fig. 1C). The amphipathic nature of amphipols enables them to stabilize membrane proteins while preserving the native state of membrane proteins, demonstrating a gentler alternative to detergents. Membrane proteins are usually transferred to NAPol after initial extraction and solubilization using conventional detergents. In a study elucidating the function of GPCRs, Granier's lab demonstrated that GPCRs reconstituted in NAPols are highly functional and show the same activity as in living cells [51]. Similarly, NAPol proved to be essential for determining the structure of the TOM complex, the central entry gate for precursor proteins into mitochondria [46]. NAPol has proved to be highly beneficial in stabilizing protein complexes and solving their structures using cryo-electron microscopy (cryo-EM) [47,48].

### 2.1.4 Calixarene

Calixarene is a new class of detergents based on the calix[4]arene scaffold, in which three negatively charged methylene-carboxylate groups are attached to one side of the scaffold, and the other side harbors a single hydrophobic chain, 1–12 carbon chain long (Fig. 1D) [28]. Depending on the length of the hydrophobic tail, the CMCs of these detergents vary from 1.5mM (3C) to 0.05mM (12C). Similar to other detergents, calixarene maintains the membrane proteins in solution by forming micellar structures; however, in some cases, it further improves the stability of purified membrane proteins by establishing a series of salt bridges with positively charged residues located on the intracellular loops of these proteins [28]. Potassium chloride cotransporter KCC2 [49] and influenza viral envelope matrix protein 2 (M2) [52] are a few examples that were successfully purified and solubilized in the functional form using this new class of detergents.

### 2.1.5 Fluorinated surfactants

Fluorinated surfactants are another class of surfactants specifically designed to improve the stability of extracted membrane proteins rather than de novo solubilizing membrane proteins from lipid bilayers. Fluorinated surfactants possess a similar structure to classical detergents, except that they carry a segment of fluorine in their hydrophobic tails (Fig. 1E) [29,53]. The fluorinated chain makes the hydrophobic tail lipophobic and therefore impermeable to the lipid bilayer and unable to solubilize membrane proteins [53]. As a result, fluorinated surfactants are not very common among membrane protein biochemists and are considered nonconventional surfactants. However, given that fluorinated surfactants are less efficient than detergents in breaking protein-protein and protein-lipid interactions, they are more capable of maintaining the essential lipids bound to the membrane proteins and, therefore, more likely to maintain the solubility and activity of purified membrane proteins, as has been shown for bacteriorhodopsin and cytochrome $b_6f$ complex [54] and ryanodine receptor [55].

### 2.1.6 Commercial detergent screen kits

Identifying the most stabilizing detergents for solubilization and crystallization purposes can be laborious, time-consuming, and costly. Therefore, to facilitate this process, commercial detergent screens have been developed. Hampton Research Detergent Screen is a Deep Well block composed of 96 mild detergent reagents ready to use in solubilization or crystallization efforts (https://hamptonresearch.com/product-Detergent-Screen-723.html). These detergents represent some of the most promising classes of detergents extracted from the literature and past membrane protein work. They can be supplemented into the protein solution to explore the impact of individual detergents on the solubility of protein of interest during the purification process or used as an additive just before setting up crystal trays. Using these commercially available detergent screens, we have identified some detergents vital for obtaining the first high-resolution crystal structures of the small multidrug resistance (SMR) transporters [56,57].

## 2.2 Membrane mimetics

As discussed earlier, detergents can have deleterious effects on the stability and activity of solubilized membrane proteins. Therefore, detergent alternatives have gained significant attention during the past 2 decades. Membrane mimicking systems, such as nanodiscs, styrene-maleic acid copolymer lipid particles (SMALPs), and the saposin-lipoprotein nanoparticle system (Salipro), have been developed to minimize or eliminate the need to use detergents for purification and solubilization purposes and, therefore, to better preserve the integrity and function of target membrane proteins. Numerous reports currently demonstrate the success of membrane mimicking systems in structural and functional studies of membrane proteins using nuclear magnetic resonance (NMR), cryo-EM, surface plasmon resonance (SPR), small-angle X-ray scattering (SAXS), and X-ray crystallography [58,59].

### 2.2.1 Nanodiscs

One of these artificial lipid environments that have proved to be an efficient tool in structural and functional studies of membrane proteins is the nanodisc. Nanodiscs are discoidal patches of lipid bilayers surrounded by two molecules of amphipathic (having both hydrophilic and hydrophobic regions) helical scaffold proteins, the so-called MSP [60]. Nanodiscs provide a more native-like environment, which can improve the stability and function of the reconstituted protein [61]. Nanodisc assembly occurs instinctively when detergent-solubilized membrane protein is mixed with lipid bilayers and MSP scaffold protein, and the solubilizing detergent is gradually removed during dialysis or using biobeads [62,63]. The ratio between lipids, MSP, and membrane protein is determined empirically and is highly important for monodisperse and functional nanodisc formation.

Nanodiscs have gained more momentum in the lipidic cubic phase (LCP) and cryo-EM than in X-ray crystallography. The first solved structure of a membrane protein embedded into a nanodisc was the TM protein bacteriorhodopsin [58]. Bacteriorhodopsin was first solubilized using *n*-dodecyl-β-D-maltoside (DDM) and then reconstituted into MSP1 and MSP1E3D1 (a variant of MSP1 that contains three additional helices and is suitable for generating larger nanodiscs) nanodiscs and employed for crystallization. Surprisingly, traditional crystallography did not generate any crystals from bacteriorhodopsin-embedded nanodiscs unless an LCP monoolein (1-oleoyl-rac-glycerol) environment was used; this maneuver yielded high-resolution (1.8–2.0 Å) diffracting crystals [58]. Another example is the cryo-EM structure of a multidrug-resistant efflux protein AdeB from *Acinetobacter baumannii* [64]. Similar to bacteriorhodopsin, AdeB was initially solubilized in DDM and subsequently reconstituted into MSP1E3D1 nanodiscs. The structure was solved to 2.98 Å resolution and unraveled the binding site for several drugs as substrates for this transporter [64].

### 2.2.2 Styrene maleic acid copolymer lipid particles (SMALPs)

We discussed the importance of nanodiscs in structural studies of membrane proteins; however, the main drawback of using nanodiscs is their inefficacy in extracting membrane proteins from lipid bilayers. Therefore, the need to use traditional methods using detergents for the initial stages of extraction and solubilization remains in place [59,65]. To overcome this issue, detergent-free approaches have been developed. SMALPs are synthetic polymers composed of styrene (hydrophobic) and maleic acid (hydrophilic). The amphipathic features of SMALPs enable them to solubilize membrane proteins by encapsulating them with a patch of the surrounding lipid bilayer. The use of this platform can significantly affect the stability and functionality of solubilized membrane proteins because some of the associated lipids essential for the activity of these proteins remain bound throughout the purification process [66,67]. One of the early works on structural studies of membrane proteins using SMALP was performed on *Escherichia coli* secondary transporter AcrB [65]. AcrB was extracted and solubilized using SMALP, negative-stained, and visualized under an electron microscope. Although a low-resolution (>15 Å) structure was obtained, the entire solubilization and data collection were carried out in less than 1 week, which was significantly faster than that of similar structures obtained using X-ray crystallography at that time. In a recent work, the structure of the same protein, AcrB, was determined at 3.2 Å resolution using cryo-EM [68]. This new structure revealed the presence of patches of the lipid bilayer within the AcrB TM domain, likely essential for the function of this multidrug transporter. The crystal structure of AcrB extracted in the presence of detergents was devoid of these lipid patches, demonstrating the importance of detergent-free systems in preserving the native structure of proteins [68].

### 2.2.3 Saposin-lipoprotein nanoparticle system (Salipro)

A more recently developed artificial membrane-like environment for structural and functional studies of membrane proteins is the saposin-lipoprotein nanoparticle system, Salipro [69]. Salipro is comparable to nanodiscs in several different aspects. Salipro, similar to nanodiscs, is composed of lipid patches surrounded by a scaffold protein, which in this case is saposin A instead of apolipoprotein A-1. Saposin A is a known lipid membrane modulator with lipid-binding characteristics [69]. The presence of six invariable cysteine residues that make disulfide bridges in the structure of saposin A makes it an unusually stable protein [70]. The self-assembly of Salipro discs occurs when the detergent-solubilized membrane protein is combined with lipids and saposin A in an empirically determined ratio, and the detergent is removed over a relatively short period of time. The number of saposin A molecules that encapsulate the lipid core can vary based on the size of the membrane protein. This enables saposin A to be highly flexible and form stable and homogeneous Salipro discs based on the size of the incorporated membrane protein [69]. This property of Salipro nanoparticles presents a significant advantage over nanodiscs, which have a fixed diameter based on the length of the scaffolding apolipoprotein belt. The membrane protein incorporated into Salipro disc nanoparticles maintains its solubility, monodispersity, and oligomeric state and remains stable over a wide range of temperatures (0–95°C) [69]. Although Salipro, similar to nanodiscs and SMALPs, is mainly designed for the structural determination of membrane proteins using Cryo-EM, it can also be used for the biochemical and biophysical characterization of these macromolecules an the aqueous solution, providing complementary tools for structural studies using X-ray crystallography.

# 3. Increasing the solubility and stability of proteins

## 3.1 Crystallization chaperones

Generating diffraction-quality crystals remains a major challenge in obtaining high-resolution structural information on target proteins. Incorporating fusion partners into the target protein has proved to be an efficient tool for improving the solubility and stability of proteins and therefore increasing the crystallization likelihood. One of the earliest fusion proteins employed for this purpose is the maltose-binding protein (MBP) [71]. The PDB database shows that MBP is the most successful fusion protein for crystallizing soluble proteins. MBP is the periplasmic portion of the ATP-binding cassette (ABC) maltose/maltodextrin transporter. The MBP tag is mostly cleaved through specific proteolytic sites engineered in the linker region between the tag and the protein of interest. However, during tag removal, several difficulties might originate, such as a decrease in the yield of the target protein, precipitation of the target protein, or production of an inactive protein. One possible solution to these problems is to retain the tag during the crystallization process. Nonetheless, multidomain proteins are typically less prone to form well-ordered, diffracting crystals, perhaps because of conformational heterogeneity originated by flexible linker regions [72]. Recently, it has been shown that generating a rigid construct with an MBP tag is a promising strategy for crystallizing alpha-helical proteins [73]. In this approach, MBP is added to the N-terminus of the target protein in such a way that the last helix of MBP and the first helix of the target protein form a continuous helix (Fig. 2A). This approach has been more successful by adding alanine linkers, which induce alpha-helical propensity between MBP and the target protein. MBP fusion protein structures deposited in PDB depict that although most of these structures comprise short interdomain linkers, some are crystallized using long linkers. In several other cases, a mutated version of MBP has been shown to facilitate crystal formation by lowering the surface entropy. Furthermore, overlapping the fused and unfused structures demonstrates that

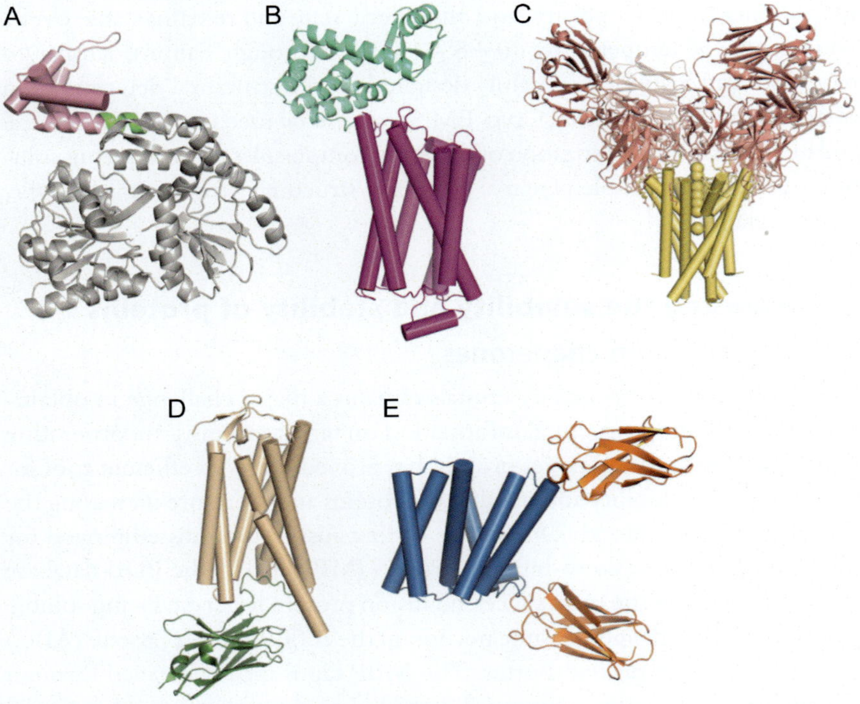


**Fig. 2** Crystallization chaperones developed for structural and functional studies of proteins: (A) crystal structure of hNLRP12-PYD (pink) bound to maltose-binding protein (gray) (PDB code 5H7N). The linker between the two fusion partners is shown in green. (B) Crystal structure of the human β2-adrenergic G protein-coupled receptor (purple) bound to T4 lysozyme (green cyan) (PDB code 2RH1). (C) Crystal structure of potassium channel KcsA (yellow) solved in the presence of Fab (pink) (PDB code 1K4C). (D) Crystal structure of angiotensin II type 1 receptor (wheat) stabilized using nanobody S1I8 (dark green) (PDB code 6DO1). (E) Crystal structure of the Gdx-Clo (dark blue) in the presence of the monobody (orange) (PDB code 6WK8). Protein macromolecules are shown in cylindrical shapes to make an easier distinction between the proteins and crystallization chaperones.

MBP does not contribute to the structural modifications of the target protein. Additionally, a mammalianized version of MBP (mMBP) was recently employed to produce highly posttranslationally modified eukaryotic proteins that are transiently expressed in mammalian cells. mMBP harbors mutations that improve both MBP solubility and affinity purification as well as enhance the crystallizability of MBP fusion proteins [74].

In the case of membrane proteins, crystallization chaperones have proved to be exceedingly efficient in i) improving the stability of membrane proteins and ii) expanding the crystallization surface area and subsequently enhancing the likelihood of crystallizing these proteins [75,76]. Membrane proteins are inherently dynamic, which enables them to fulfill their biological roles [77]. This flexibility, however, can increase conformational heterogeneity and prevent crystal contact formation. Even formed crystals might represent crystallography defects or diffract poorly. Crystallography chaperones can overcome this obstacle by locking the target protein into a fixed conformation, lowering conformational entropy and hence triggering uniform packing. In addition, crystallization chaperones further promote crystallization by masking highly unstable or flexible protein regions [75]. More importantly, chaperone fused to the membrane protein expands the hydrophilic surface area necessary for crystal contact formation. Membrane proteins are mainly composed of hydrophobic residues embedded in the lipid bilayer or detergent micelles upon solubilization, with a few hydrophilic loops present in the solvent, accessible for crystal contact formation. The small hydrophilic surface area available for crystal contact formation is one of the main reasons for the low crystallization rate of membrane proteins.

One of the successful examples of crystallography chaperones is T4 lysozyme. The application of T4 lysozyme as a crystallography chaperone is based on the idea that fusing an easy crystallizing partner to the target protein can facilitate the crystallization of the target protein. T4 lysozyme has been successfully used to solve the crystal structure of several GPCRs, including $\beta_2$-adrenergic receptor ($\beta_2$AR) [78], CXCR4 chemokine receptor [79], histamine H1 receptor [80], and dopamine D3 receptor [81]. In all these cases, T4 lysozyme was covalently linked to the protein by replacing an entire cytoplasmic loop (Fig. 2B). Although this strategy was crucial for the structure determination of these membrane proteins, in some cases, it impeded the functionality of the protein [82], raising the concern that introducing these changes might negatively impact the protein structure. In addition, the process of constructing T4 lysozyme fusion can be costly and laborious because the placement of T4 lysozyme on the target protein is important for a protein's solubility and functionality. This can only be determined empirically by generating numerous constructs and investigating the overexpression and thermal stability of each recombinant protein separately. Thus, the popularity of T4 lysozyme as a crystallization chaperone has drastically declined over the past decade.

Monoclonal antibodies specific to the target membrane protein have also been used as crystallization chaperones. Compared to the traditional antibody generation process in which antibodies are directly collected from plasma upon immunization, in the case of monoclonal antibodies, cells that produce antibodies and lymphocytes are harvested from the spleen or lymph nodes. They have an extended life expectancy through fusion with cancerous B-cells or myelomas. In this way, the resulting hybridoma cells are immortal and able to produce large amounts of monoclonal antibodies for many generations. Purified monoclonal antibodies are subjected to proteolysis to remove the fragment crystallizable region (Fc) and retain the antigen-binding fragment (Fab). Monoclonal antibodies can recognize a specific epitope on the target protein with extremely high affinity in the range of 20–200 pM (with an average of 66 pM) [83]. Using this rigorous process, monoclonal antibodies were efficient for stabilizing and solving the crystal structures of numerous proteins, including cytochrome *c* oxidase from *Paracoccus denitrificans* [84], $K^+$ channel KcsA [85], the ClC chloride channel [86], nitric oxide reductase [87], the SecYE protein-conducting channel [88], and so on (Fig. 2C). However, raising monoclonal antibodies specific to the understudy membrane protein can be time-consuming and expensive and may not be successful for every membrane protein.

A nanobody is an alternative crystallization chaperone to antibodies, which has proved to be beneficial in crystallizing membrane proteins, particularly GPCRs [89]. Nanobodies are single variable domains of camelid antibodies, which form the entire antigen-binding surface (Fig. 2D). Compared to conventional antibodies that are composed of two heavy chains and two light chains with both heavy and light chains contributing to form the antigen-binding surface, camelid antibodies are composed only of heavy chains forming a unique variable fragment, termed VHH or Nb. Nanobodies are composed of nine antiparallel β-strands connected by short loops (Fig. 2D). The antigen-binding interface of the nanobody consists of three loops designated as complementarity determining regions (CDRs). Out of these three loops, CDR3, which is the longest loop and plays the most critical role in antigen recognition, has been the subject of most randomization [90]. Randomization of CDRs enables nanobodies to conform to the epitope site on the surface of the target protein. Several different nanobody libraries have been developed to facilitate the screening and increase the number of binders. While previous generations of nanobodies were raised by immunization of camelids [91], new platforms are based on a fully synthetic library displayed on the yeast surface [90,92] or ribosome display [93]. The use of a fully synthetic library not only lowers the cost of nanobody generation and saves time substantially but also, importantly, addresses the immunological tolerance to self-antigens in the host [90]. Many of the medically important targets represent a high sequence similarity with their camel homologs; therefore, because of immunity to self-antigens, generating efficient antibodies is very challenging and, in many cases, impossible [94].

Enriching the target protein against the naïve library is performed during two rounds of magnetically activated cell sorting (MACS). MACS is a low-cost, highly efficient cell separation technique based on cell surface antigens. During each round of enrichment, the target protein is labeled with a distinct fluorophore, exposed to the nanobody expressing yeast library, followed by removing the excess of the target protein. High-affinity binders are isolated using antifluorophore magnetic microbeads. Before each selection round, it is crucial to ensure depleting the library from nonspecific nanobodies, which might bind to reagents, such as magnetic beads or fluorescent tags. The two commonly used fluorescent tags are Alexa Fluor 647 and fluorescein isothiocyanate (FITC). Using a distinct tag during each round of selection prevents enriching the fluorescent tags [90]. In the end, active clones are separated by fluorescence-activated cell sorting (FACS), which involves expressing the enriched library, staining with fluorescent-tagged protein, and sorting the cells based on the fluorescence of the tag attached to the antigen. The nanobody platform has aided the structure determination of several membrane proteins, including angiotensin II (AngII) type 1 receptor (AT1R) [94], KDEL receptor from Golgi [95], ABC exporter TM287/288 [96], mycobacterial ABC exporter IrtAB [97], and SARS-CoV-2 receptor-binding domain [98].

Monobodies present another distinct class of crystallization chaperones. They are synthetic binding proteins based on the human fibronectin type III domain (FN3), composed of a scaffold of seven antiparallel β-strands connected by three loops on each side of the protein (Fig. 2E) [99]. The general architecture of this scaffold is conserved among monobodies; however, the loops [99,100] or a combination of loops and β-sheets [101] are diversified to create new binders. Monobodies resemble antibodies by creating a binding surface for other proteins; however, they offer unique advantages over antibodies. Monobodies are small, monomeric proteins with about 100 residues and ~10 kDa, whereas antibodies are much larger, usually ~50 kDa. The small size of monobodies makes them ideal crystallization chaperones for small membrane proteins and structural studies of membrane proteins by NMR. An excellent example of small membrane proteins is the SMR family that formed high-resolution diffracting crystals only in the presence of a monobody (Fig. 2E) [56,102]. Furthermore, compared to antibodies, which require an oxidizing environment to form disulfide bonds and retain activity, monobodies do not form disulfide bonds and therefore can remain active under reducing conditions. This makes overexpression of monobodies in *E. coli* a feasible task [99]. Monobodies have played a significant role in expanding our understanding of the molecular mechanism of membrane proteins, particularly the mechanisms that regulate molecular recognition [103]. Monobodies tend to mimic the natural ligands and bind to the functional site on the target protein. The flexibility of the loops on the monobody scaffold enables them to enter and occupy the binding site, thereby providing a platform to compare the properties of the functional site in the presence of the natural ligand versus monobody [103]. In this regard, the crystal structure of two bacterial homologs of a dual topology fluoride ion channel (Fluc) was solved in the presence of three different monobodies [104]. The presence of monobodies was crucial for obtaining high-resolution diffracting crystals to unveil the unique two-pore architecture of these proteins while providing clear evidence of the dual-topology conformation within members of the Fluc family [104].

The selection process for high-affinity binders involves several rounds of sorting in which the biotinylated target protein is exposed to the phage display library carrying $\sim 2 \times 10^9$ clones. The target protein bound to its high-affinity binder is captured using streptavidin-coated magnetic beads, and nonspecific clones are washed away. The target protein bound to the monobody is released from the beads by cleaving the biotinylation linker using dithiothreitol or other reducing agents [105]. The recovered phage particles are amplified by infecting *E. coli* cells and using them during the next round of sorting. The desired phage clones with the highest affinities are sequenced and cloned in expression vectors for overexpression and purification.

In all these platforms, fusing a second soluble protein, which specifically binds the target membrane protein, crystallization chaperone, is the main key for expanding the surface area required to form crystal contacts, masking the disordered regions and subsequently increasing the likelihood of crystallizing the TM region. Similarly, one can use other intrinsic domains surrounding the TM region, mainly soluble domains such as cytoplasmic and extracellular domains, as pseudocrystallization chaperones to facilitate the crystallization of this region of the membrane protein of interest. Solving the crystal structure of the membrane-embedded domain of histidine kinase NarQ provides an interesting example of the efficacy of a pseudocrystallization chaperone. NarQ is a nitrate/nitrite sensor kinase composed of seven domains: the periplasmic sensor domain, TM domain, HAMP domain, signaling helix, GAF-like domain, DHp domain, and kinase CA domain [106]. In the absence of sensor and HAMP domains, the TM domain crystallized after 3 months and diffracted to 2.3 Å resolution [107]. It was observed from the structure that residues adjacent to the truncation site are largely disordered. Strikingly, the full-length protein composed of sensor-TM-HAMP crystallized faster, under 2 weeks, and diffracted to a higher resolution of 1.9 Å with well-ordered TM helices and loops [106]. Although the binding site in both proteins was fully ordered, the full-length structure was more beneficial in understanding the signaling process.

## 3.2 Thermostabilizing mutations

An alternative strategy to crystallization chaperones is the systematic mutagenesis of proteins to make them thermostable and, therefore, more amenable to crystallization [108]. There are numerous examples of stabilizing and crystallizing soluble proteins using this technique [109–113]. In this technique, known as surface entropy reduction (SER), multiple residues on the protein's surface are engineered to lower the conformational entropy and increase its propensity to crystallization [114]. The SERp server (http://services.mbi.ucla.edu/SER/) has been designed to identify such residues [115]. Compared to soluble proteins, membrane proteins are inherently more flexible and unstable, and generating diffracting crystals of these proteins poses a greater challenge. Therefore, generating the thermostable versions of these proteins appears to be essential for efficient crystallization.

Introducing a single-point mutation or a combination of single-point mutations in the structure of membrane proteins can cause an increase in the membrane protein rigidity, a decrease in conformational exchange, a reduction in site-specific tensions, and an increase in the number of ordered water molecules and therefore can result in an improvement in the thermostability of membrane proteins and subsequently an increase in the crystallization probability of target protein [116]. One of the main advantages of this strategy over the crystallization chaperones is the tendency of thermostabilized proteins to crystallize with different ligands. For example, the cocrystal structure of thermostabilized b1-adrenergic receptor (b1AR) with 12 different ligands has been solved, facilitating the detailed understanding of ligand binding and receptor activation [117–120]. On the other hand, the laborious and high cost of generating thermostabilized proteins makes this strategy less appealing to membrane protein crystallographers. Typically, each residue in the protein structure is mutated to alanine, whereas natural alanine is mutated into leucine, and most stabilizing mutants are combined together to identify a particular conformation, which increases the thermostability of the target protein and favors its crystallization [121].

The crystal structures of adenosine $A_{2A}$ receptor ($A_{2A}R$) [122], neurotensin receptor (NTSR1) [123], chemokine receptor (CCR5) [124], free fatty acid receptor (FFA1R) [125], corticotrophin release factor receptor (CRF1R) [126], and metabotropic glutamate receptor (mGlu5) [127] represent only a few examples of GPCRs, whose structures have been determined using this technique. For a complete list of thermostabilized GPCRs, please refer to [128] (or https://gpcrdb.org/construct/mutations). Although GPCRs are the main class of membrane proteins, which have been the subject of thermostabilization by mutagenesis, this technique is equally applicable to other classes of membrane proteins. This methodology has been essential for obtaining the high-resolution crystal structures of dopamine transporter [129], AMPA receptor ion channel [130], and multidrug transporter LmrP [131].

## 4. Assessing the homogeneity and purity of protein samples

The success of protein crystallization efforts is highly dependent on the level of protein purity and homogeneity. There are various methods to assess the quality of protein samples that are extensively used in protein production and crystallization pipelines.

### 4.1 UV-vis and fluorescence spectroscopy

This technique is not only used for the quantitative measurement of protein concentration but also used to detect nonprotein contaminates and protein aggregates. Amino acids such as tryptophan and tyrosine have an absorption maximum at 280 nm because of their aromatic nature, which can be exploited for measuring protein concentration using their extinction coefficients. In contrast, nucleic acids have absorption signatures at around 260 nm. A high 260/280 nm absorbance ($>0.57$) indicates the DNA/RNA contamination of protein samples. Large aggregates and particles in protein samples can also be detected through UV-vis spectroscopy (if their hydrodynamic radius is $>200$ nm). The samples without aggregates or large particles do not have any absorption beyond 320 nm, whereas the presence of the aggregates leads to scattering of light. Besides UV spectroscopy, fluorescence spectroscopy is also used to evaluate the degree of aggregation in a protein sample. Through excitation at 280 nm, the emission signal can be monitored at 280 and 340 nm, correlated with light scattering and intrinsic protein fluorescence, respectively. The ratio of the intensities at 280 and 340 nm ($I_{280}/I_{340}$) corresponds to the degree of aggregation of the sample. The $I_{280}/I_{340}$ is close to zero in the aggregation-free sample [132,133].

## 4.2 Size exclusion chromatography

SEC is another technique commonly used to assess the quality and homogeneity of protein samples. It is usually the last step of the protein purification process, which separates molecules according to their hydrodynamic size, often defined by their hydrodynamic radius. An SEC column is composed of a porous matrix of spherical particles (beads) that do not possess reactivity and adsorptive properties. In this technique, molecules larger than the pore size do not diffuse into the beads, so they elute early, and molecules that can penetrate the pores experience various migration times based on their size and thereby elute at different retention times. SEC is also useful for separating different oligomeric forms of a protein as well as large aggregates, which might interfere with the crystal formation. These species can be detected easily using traditional UV detectors [132]. It is crucial to remember that SEC can dilute a protein sample approximately tenfold and therefore alter the equilibration between various oligomerization states of a protein sample. In addition, although SEC possesses a porous matrix with no reactivity and absorptivity, in some cases, the protein might interact with the column matrix. Therefore, the size calculated based on running standard samples might not reflect the true size of the protein [134].

## 4.3 Dynamic light scattering

Dynamic light scattering (DLS) is a rapid and convenient technique to investigate the monodispersity of a protein sample and evaluate the presence of higher-order oligomers and soluble aggregates. DLS determines the size distribution of the particles, including proteins, polymers, micelles, vesicles, and so on, by measuring their Brownian motion. This technique measures the macromolecule's fluctuations in scattered light intensity as a result of diffusing particles. DLS shows the particle population at different diameters. If the system is monodispersed, only one population exists, whereas a polydispersed system would demonstrate multiple particle populations. The advantage of DLS is that it can be performed at various temperatures and times to assess protein stability and can be performed using various buffers during the optimization steps and sample preparation. The method requires a very small amount of protein sample (0.5–2 mL, 0.3–50 mg $mL^{-1}$) [19]. DLS can also be coupled with SEC to understand the size distribution of protein samples [132,135,136].

### 4.4 Size exclusion chromatography with multiangle light scattering (SEC-MALS)

Determining the oligomeric state of proteins is essential for structural studies. In some cases, the goal is to produce the monomeric form of a protein, and in some others, the oligomeric state is desirable because certain proteins are typically active in their oligomeric forms. On the other hand, nonnative oligomers can be detrimental to structural determination by X-ray crystallography. In addition, nonnative oligomers can lead to inaccuracies in binding measurements using functional assays such as isothermal titration calorimetry (ITC) and SPR. In these cases, SEC with multiangle light scattering (SEC-MALS) can be a useful technique to determine the exact molecular mass of proteins and separate various oligomeric states. This provides a significant advantage over SEC. The size estimation by SEC can be inaccurate because the retention time in this technique mainly relies on the hydrodynamic radius of macromolecules instead of their molecular mass. Also, as mentioned earlier, proteins and other macromolecules might interact with the matrix of the SEC column, or they can adopt various conformations making the measurement of their absolute size and molar mass by running standard proteins arduous. An alternative technique is to combine SEC with MALS and differential refractive index (dRI) detectors. In SEC-MALS, the SEC column is employed only to resolve different species in solution as they enter the MALS. The dRI detector calculates the concentration using changes in the refractive index because of the analyte. At the same time, the MALS detector can measure the proportion of light scattered by an analyte at multiple angles relative to the incident laser beam. This feature enables the measurement of molecular mass independent of elution time [132,137].

## 5. New crystallization methods

Protein crystallization is a multiparameter process. It is divided into three steps that begin with nucleation, then the crystal growth stage, and last, cessation of growth. These steps determine the number, size, and quality of obtained crystals. For a protein to precipitate from the solution, the system should be driven into a supersaturated state. The attainment of the supersaturation level and the rate of supersaturation can be explored on the basis of various crystallization techniques. For soluble proteins, there are four major approaches for obtaining supersaturation: vapor diffusion, batch, dialysis, and free interface diffusion (FID) [138]. In the case of membrane proteins, crystallization can be performed using LCP and bilayered bicelle crystallization methods. Each technique influences protein crystallization differently, even though the final chemical composition of the crystal system might be the same. The lack of generalized methods for high-quality crystal production is still a major bottleneck. The development of methods to control protein nucleation and crystal growth might significantly help to reduce this hurdle in the crystallization process. Now, we will discuss the most recently developed strategies to improve the crystallization process.

## 5.1 Automation of crystallization

Crystallization screening is the most time- and sample-consuming process in macromolecular crystallography. Because of numerous parameters to be tested, from precipitant, temperature, and pH to protein concentration, the crystallization trial is unlimited. Moreover, in some cases, crystallization is incorrectly considered an art rather than a science because of lower reproducibility and difficulty in obtaining well-diffracting crystals. Over the past decade, several automated platforms for high-throughput crystallization screening were developed to reduce manual work and increase reproducibility with a lower failure rate [139–141]. These developments include the steps from crystallization setup to drop observation. One such fully automated system is the protein crystallization and monitoring system (PXS), which has made an outstanding contribution to achieving high-throughput protein crystallization screening [139]. So far, the most commonly used crystallization method is vapor diffusion owing to its small sample volume requirement to optimize crystallization screening and easy harvesting of crystals. The drop size can vary from 50 to 1000 nL with varying parameters to set up the plates [142]. There are two types of automation robots used for setting up vapor diffusion experiments. Both robots can set up drops by mixing protein and precipitant volumes, but one can also transfer mother liquor from deep-well blocks to the plate. This avoids the extra step of setting prefilled plates by hand or another robot [143]. Similar automated systems are available for the batch crystallization process, where the protein and precipitant are dispensed simultaneously, either with or without oil [144]. Unlike vapor diffusion, this method provides slow equilibration of the drop. However, there is no endpoint in this experiment, and the drop continues to evaporate until it dries out. This method helps to understand the phase diagram of a protein in depth. Because of their unique setup, dialysis and the interface diffusion method cannot be automated.

LCP and bicelle crystallization are preferred methods for crystallizing membrane proteins, which otherwise cannot crystallize in an aqueous phase [145]. In the standard LCP technique, membrane protein is transferred into the LCP by mixing the protein solution with lipids such as monoacylglycerols (MAGs) in an optimum ratio that might vary with the protein sample. Because of their lower chemical stability, MAGs affect crystal growth and nucleation. Recently, various research groups have been designing and synthesizing new lipids to support crystallization by forming LCP [146–149]. Because LCP is a semisolid material with high viscosity, a specialized small-diameter syringe system is used to dispense the sample in a crystallization plate [19]. Unlike soluble proteins, there are specialized automatic dispensers such as Gryphon LCP (Art Robbins), NTX (Formulatix), and Mosquito LCP (TTP LabTech) that rapidly and accurately dispense LCP to a 96-well plate. Once LCP is dispensed, the aqueous solution is added over eight wells at a time, leading to approximately 10 s exposure of these drops before being covered by the aqueous phase. To improve this, Oryx LCP (Douglas Instruments) can be used, which dispenses one LCP bolus and then the aqueous solution over it at a time within 1 s. This system can also be used for the bicelle crystallization method setup. There are a few other improvised systems, such as NT8 and ProCrys Meso Plus (Zinner Analytic), with built-in humidifiers to avoid evaporation.

With improvements in data collection and processing at beamlines, there has been a huge demand to develop a fully automated crystallization setup for the structure determination process. Recently, a new database system, PXS2, was developed to integrate the crystallization setup database with a database for the diffraction data collection on the photon factory (PF) synchrotron beamlines [150]. This system will also enable in situ data collection along with a minimized sample volume up to 0.1 mL and improved resolution of captured images. This upgraded system will significantly improve the MX efficiency in structural biology research.

## 5.2 On-chip crystal growth

Microfluidic devices, because of low sample consumption, finely control mass transport properties, with a large surface-area-to-volume ratio, and have emerged as a viable technology for protein crystallization and in situ X-ray diffraction experiments [151]. Miniaturization of experiments at a small scale also facilitates good control over crystallization parameters such as temperature, concentration gradients, and convection [152,153]. These devices are also suitable for studying protein phase diagrams and understanding the crystallization process in depth by using protein samples of only a few nanoliters to microliters [153,154].

Various innovative approaches are reported that enable these devices to investigate soluble [155,156] and membrane protein crystallization [157,158]. In general, there are three types of microfluidic approaches employed to identify optimum conditions for protein crystallization: (i) valve-based systems, (ii) droplet-based systems, and (iii) well-based systems. Primarily, vapor diffusion, microbatch, and free interface methods have been used for crystallization setup on microfluidic devices. A review explained the implementation of traditional crystallization methods (microbatch, vapor diffusion, and FID) in microfluidic devices by using the three approaches mentioned earlier [159]. To perform vapor diffusion, both valve- and droplet-based approaches can be used. The first approach relies on a formulator module to create a mixture of protein and precipitant and on a two-phase injector to generate small droplets encapsulated in an immiscible carrier fluid [160]. The second droplet-based approach generates alternating protein trial droplets and salt droplets. A fluorinated carrier fluid transports and separates the droplets. Batch crystallization is performed similar to the droplet-based approach, as explained above, using a fluorinated carrier fluid. All components pass through different aqueous channels to meet at a junction and form a nanoliter volume droplet [161]. Because the carrier fluid is separated for all the aqueous solutions, no evaporation or loss of chemicals occurs, unlike the conventional crystallization method. In the case of FID, a valve-based formulator was developed to carefully manipulate the diffusion of fluids in nanoliter volumes [155]. Diffusion times between the sample and precipitant can be varied by changing the connecting channel length, which enables rational crystallization screening using on-chip FID [162].

Another crystallization method that has been explored is dialysis, which enables precise control over crystallization conditions by using a semipermeable membrane with a molecular weight cutoff (MWCO) smaller than the size of the protein of interest. There are different approaches to combining membranes and microfluidics. The most commonly used method is the direct integration of membranes into microfluidic devices by gluing or clamping. Microfluidic dialysis setups have been used for SAXS and fluorescence recovery after photobleaching (FRAP) [163,164]. Structural changes in protein dynamics can be studied by combining microfluidic dialysis with a SAXS/UV exposure cells, and all this is possible by using a small amount of protein sample [163]. Recently, reusable microfluidic dialysis setups for in situ serial X-ray crystallography experiments have been developed. These microfluidic chips were also used for screening crystallization conditions to investigate temperature-precipitant concentration phase diagrams [165]. Such experiments open the possibilities of on-chip serial crystallography experiments under dynamically controllable sample conditions.

So far, the well-based microfluidic approach is one of the most adapted platforms for on-chip crystallization. The most recent development is the use of a triple-gradient generator device to screen the crystallization conditions for three different proteins [166]. There are other microfluidic approaches, such as a capillary-based microfluidic device that not only enables soaking of crystals at room temperature and in situ data collection but also allows stable sample shipping to synchrotrons [167]. This work paves the way for room-temperature microfluidics-based sample delivery methods to facilitate the automated workflow of high-throughput protein-crystallography-based screening of compounds for drug discovery. Various groups have attempted room-temperature serial crystallography and in situ data collection using a microfluidic device [168–172]. Lower mosaicity and good isomorphism in crystallographic data have also been observed by using microfluidic devices. Moreover, triggering events (such as light, temperature, etc.) can be easily enabled by fluidic control to investigate protein dynamics that would otherwise be inaccessible or difficult to study [170]. To mitigate radiation damage, complete crystallographic data using a microfluidic device have been collected under cryogenic temperature using lysozyme protein crystals [173], thus providing a solution for data collection of complex proteins under cryogenic conditions instead of room temperature.

For membrane protein crystallization, there are additional factors such as the presence of detergents or viscous LCP that can be challenging to form on-chip in a small volume. Because of this, the microfluidic system has been limited to detergent-based membrane protein crystallization [157,174]. A hybrid droplet-based approach that uses nanoliter plugs to minimize sample consumption has been utilized in crystallization trials of a porin from *Rhodobacter capsulatus*. In situ data collection was performed using the same approach giving high-quality crystallographic data [157]. A microfluidic device to investigate the phase behavior of sarco(endo)plasmic reticulum

$Ca^{2+}$ adenosine triphosphatase (SERCA) has also been attempted previously, which showed the feasibility of membrane protein crystallization using microfluidic technology [174]. Droplet-based microfluidic systems have been suitable for handling viscous solutions such as LCP. This system was used to dispense nanoliter volume LCP droplets and later merged them with aqueous droplets to perform crystallization trials on bacteriorhodopsin (*Halobacterium salinarum*) [175]. They also developed a cyclodextrin-based host-guest chemistry approach in a microfluidic device. This simplified the process of protein concentration by removing free detergent micelles and thus affecting the packing of protein-detergent complexes. A time-controlled removal of loosely bound detergent molecules could also be enabled by this approach. Another microfluidic system with pneumatic valves was also used to form LCP on-chip with a droplet volume below 20 nL [176]. Implementation of current microfluidic devices is still limited to soluble proteins and should be further focused on studying membrane proteins. An improved imaging system, optimized modifications in terms of detergents, and a reliable database system can increase the number of solved membrane protein structures [159].

Microfluidic devices have emerged as a powerful tool in protein crystallization. The use of microchips has facilitated low sample and reagent consumption with a high density of experiments. This technology has revolutionized membrane protein crystallization by miniaturizing screening trials, a high operational speed, and good reproducibility. A step further was given by generating microchips with new materials such as a copolymer of cyclic olefin for in situ data collection from crystals obtained by counterdiffusion [177]. Furthermore, other advances in microfluidics are expected to mitigate radiation damage of protein crystals under the beam from a deeper understanding of the chemistry inside irradiated crystals [178].

## 6. New crystallization additives

Nucleation is the first step of the crystallization process that can be homogeneous or heterogeneous. A homogeneous process is a random process where multiple nuclei are formed with a simultaneous assembly of several protein molecules at a high supersaturation level. The energy barrier is low in this state, facilitating critical nuclei formation. If there is excessive supersaturation, it leads to unfavorable structural defects and excessive nucleation. To avoid this, heterogeneous nucleation is promoted to perform crystallization in a controlled manner.

The major challenge of protein crystallization is often the inability to obtain crystals. Even if crystals are obtained, other obstacles such as poor diffraction and reproducibility issues might occur. Various methods have been proposed to overcome this problem by controlling protein crystallization parameters (such as temperature, pH, concentration gradient, etc.), utilizing microgravity or electric field environments, or even developing advanced microfluidics platforms, as discussed previously [179]. Even though these methods have been an enormous success, they have some limitations. Thus, various nucleants came into existence, which induces heterogeneous nucleation and improves crystal diffraction quality in a very controlled manner [180]. Since then, there has been a massive debate on multiple substances treated as "universal" nucleants for protein nucleation. Some of these agents include minerals, microcrystals for seeding, natural nucleants (e.g., horsehair, human hair, cellulose), porous substances (e.g., silicon), and even charged surfaces (e.g., mica) [179].

## 6.1 Porous nucleants

Various porous nucleants have been developed, such as mesoporous bioactive gel glass, carbon-nanotube-based materials, and nanoporous gold nucleants. The most successful heterogeneous porous nucleant has been bioglass, also known as "Naomi's Nucleant" sold by Molecular Dimensions. Bioglass has a disordered pore distribution (2–10 nm) that promotes the nucleation of various difficult-to-crystallize proteins [181]. In principle, protein molecules would be trapped in pores, thereby encouraging them to form aggregates in crystalline order. Bioglass has been effective over a range of physical parameters such as different pH values, temperature, and varying isoelectric points of protein. Because of its flexibility, numerous target proteins have been crystallized, including a membrane protein [182]. Even though it is more convenient to use bioglass than silicon, its surface chemistry is not easy to control. Thus, carbon-nanotube-based nucleants with controlled pore distribution were developed [183]. Low-density or nonporous substances such as polystyrene divinylbenzene microspheres (SDB) are also effective nucleants [184]. They employ adsorption and desorption theory, where protein is first adsorbed at a high concentration and then desorbed at a low concentration. A higher-concentration region is favorable to nucleation, and a lower-concentration region allows crystal nuclei stability leading to the growth of high-quality protein crystals. The discovery of such effective porous materials has led to structural studies of various proteins by inducing nucleation. This has also set a trend for developing new nucleants for protein crystallization.

## 6.2 Molecularly imprinted polymers (MIPs)

Molecularly imprinted polymers (MIPs) are another such nucleants that produce molecularly selective sites via the polymerization of a functional monomer and a template biomolecule. There is an interaction between the functional monomer and template molecule via hydrogen bonding and weak Van der Waals forces. Once polymerized, the template molecule is trapped inside the polymer. The highly selective cavities remain after the removal of the template molecule. These cavities can remember the cognate template molecule and rebind to a noncognate molecule of a similar shape and size [185,186]. MIPs have been successfully used as a nonprotein nucleant in protein crystallization. They are very effective in increasing crystal hits and improving crystal quality. The first water-based MIP (HydroMIP), also known as 'smart material', was developed by Naomi Chayen in 2011 [187]. The most commonly used MIPs are acrylamide (AA), *N*-hydroxymethyl acrylamide (NHMA), and N-isopropyl acrylamide (NiPAm). To prepare MIPs, a functional monomer (i.e., AA) and a crosslinker (i.e., N,N0-methylenebisacrylamide) are dissolved in deionized water to form a pre-MIP solution. The solution is then polymerized by adding ammonium persulphate (APS) and *N,N,N, N*-tetramethylethyldiamine (TEMED) at room temperature. Last, formed gels are crushed, and the template protein molecule is removed, forming cavities. These cavities form a protein-rich phase when protein molecules migrate toward their surface. This overcomes the energy barrier for the first crystal nuclei formation step. The cavity surface is used as a support for protein crystal growth.

Various proteins such as lysozyme, catalase, hemoglobin, trypsin, alpha crustacyanin, and human macrophage migration inhibitory factor (MIF) have been tested for nucleation using MIPs [187,188]. Nucleation is induced by cognate MIP and also via other similar noncognate MIPs under metastable conditions. By using MIPs, the diffraction quality of the crystals can be improved with low protein consumption. MIPs can also be used in high-throughput automated crystallization trials and give high reproducibility [189]. To further improve the crystal quality, zwitterionic additives have been immobilized in MIPs that gave higher hits and well-diffracting single crystals of concanavalin A protein in a short time [190].

## 6.3 Crystallophore (Tb-Xo4)

Apart from getting high-quality protein crystals, solving the phase problem is another major issue in crystallography. For more than a decade, lanthanide complexes have been widely used to resolve the phase problem because of their large anomalous contribution of lanthanide ions [191]. These complexes can be inserted into protein crystals by (i) substituting $Ca^{2+}$ in calcium-binding proteins [192], (ii) by covalent grafting of a lanthanide binding tag [193], and (iii) cocrystallization of lanthanide complexes with the protein [194]. Previously, several such complexes such as macrocyclic (DOTA, DO3A, and HPDO3A), polydentate (DTPA-BMA), or tris-dipicolinate lanthanide complexes have been used for the structural determination of new proteins [195,196].

Recently, a new terbium (III) complex (cationic) has been developed that has all the properties such as nucleating, phasing, and luminescence to overcome major issues of macromolecular crystallography. Engilberge et al. named this type of complex as crystallophore (Xo4), which has been used for the structural determination of various proteins with known and unknown structures (such as lysozyme, thaumatin, malate dehydrogenase, and pb9 from the T5 phage tail) [197]. It has been proposed that nucleating properties of Tb-Xo4 could be because of SER or by a slight modification of the surface topology favoring a contact between protein molecules. This complex is highly stable under crystallization conditions and acts as a good phasing agent enabling the protein structure determination using single-wavelength anomalous diffraction (SAD) or multiwavelength anomalous diffraction (MAD) methods. The addition of this crystallophore also improved the diffraction quality and gave greater data completeness, as observed previously for an adhesion protein PitA [198]. The use of heavy atoms to solve phasing problems has recently become popular, and a new Gd(III) clathrochelate, a metal cage complex, has been designed by Castañeda et al. [199]. Tb-Xo4 has huge potential in facilitating the structural elucidation of new proteins with its capabilities to overcome issues such as phasing and crystal nucleation. In the future, it will be interesting to see this crystallophore's compatibility in phasing tools with serial crystallography.

## 6.4 Other nucleants

Natural nucleants such as hair are preferred because of their biocompatibility and easy availability. Mineral substances have also shown superior results. Previously, nucleation of four different proteins and crystal growth were observed using different mineral samples as nucleants [200]. However, they altered the crystal morphology and unit cell symmetry [201]. Porous substances such as silicon have also demonstrated crystal formation in the metastable zone for various proteins. The silicon pore size is similar to protein's molecular size in the crystalline form. Thus, these pores might trap protein molecules to induce crystal nucleation [202].

# 7. Advances in instrument and data-processing software

## 7.1 Automations in screening crystallization conditions

Following advances in X-ray radiation sources, technologies related to mounting and storing crystals, postdiffraction data processing programs, and obtaining phases are booming. One of these advances is automating the crystallization process from preparing conditions to sample dispensing. The focus of automation technologies for protein crystallization majored in two independent approaches: the robotic systems equipped with microdispensing heads and microfluidic chips with parallel valves [203]. Phoenix (Art Robbins Instruments, Inc.), Mosquito (TTP Labtech, Ltd.), Oryx Nano (Douglas Instruments, Ltd.), and NT8 (Formulatrix, Inc.) are some examples of the automated robots that can dispense and mix protein and crystallization solutions to the smallest volumes (nanoliter scales). High reliability, high throughput, low sample usage, and complete automation are prominent advantages of using these robots [204]. Microfluidic devices have also been developed in various designs to handle a small amount of liquids for different crystallization setups. In one of the designs, groups of microvalves were incorporated on a chip to combine the mixtures of proteins and crystallization solutions rapidly in a formulation chip or generate volume-defined microchambers for mixing nanoliter scales of proteins through an array of pneumatic valves [205].

## 7.2 Detecting protein crystals using an automated plate imager

The new instrumental advancement allows crystallization plates to be stored in a RockImager (Formulatrix, Bedford, MA, USA) for automated storage and imaging. In this approach, one can monitor the crystal growth while maintaining the plates at a constant temperature according to a schedule determined by the user. Periodic imaging of individual drops on each plate is performed to monitor the crystallization process. The RockImager system can store 1000 plates at once and image them according to the schedule. The images of drops, conditions for each drop, and the user experimental notes are all available on a server database pertinent to the imaging system. Using the RockMaker software, the users can monitor and score the drops from 0 to 9, starting from the clear drop and dust to crystals and marking the ones containing crystals as interesting drops. RockImager is also capable of distinguishing between protein and salt crystals. Tryptophan shows a high quantum yield when it gets excited at around 290 nm, which makes it useful as a fluorescent probe. On average, 1.09% of the residues in proteins constitute tryptophan, which provides a sufficient fluorescence signal for UV imaging. A protein crystal can be easily distinguished in UV images even when the crystals are buried under heavy precipitation. It must be noted that this technique has several limitations that have been reviewed thoroughly elsewhere [206,207].

## 7.3 Advances in synchrotron radiation instrumentation

Most X-ray crystal structures deposited in the PDB have been collected using synchrotrons because of the transition to high-brightness, third-generation synchrotron radiation sources. The European Synchrotron Radiation Facility (ESRF) was one of the earliest generations of these synchrotrons. Shortly afterward, Advanced Photon Source (APS) in USA and Spring 8 in Japan (Super Photon Ring 8GeV) launched their activities. The highly intense X-ray beams in ESRF, APS, and Spring 8 are achieved using X-ray undulators. Another essential feature of these undulators is a tunable X-ray wavelength that alters the magnet gap. New generations of synchrotrons offer much smaller, focused beams that allow the measurement of X-ray diffraction data for microcrystal samples [208]. Additionally, other experiments, such as optimized anomalous dispersion measurement or phasing, data collection for large unit cells, and high-resolution measurements, are now amenable through a much smaller number of crystals and much shorter recording times. Furthermore, robotics has become quite a standard method for handling samples, and in most synchrotron locations, one only needs to ship the crystals using a dry-shipper to the site, and all the handling can be performed remotely.

The detectors for synchrotron X-ray diffraction have also been advanced from photographic film to TV systems, imaging plates, charge-coupled devices (CCDs), and pixel detectors. These advancements have led to faster data collection at synchrotrons than home source X-ray generators [209–211].

## 7.4 In situ X-ray screening and data collection

As mentioned earlier, one of the methods to distinguish protein versus salt crystals is to use UV fluorescence and second-order harmonic generation techniques [206]. However, X-ray screening not only identifies protein versus salt crystals but also provides data collection-related information, including space group, unit cell, and diffraction quality. Therefore, in situ screening speeds up the crystallization screening pipeline to determine the diffraction-based optimal conditions and ligand binding state. This can be highly beneficial for drug discovery applications [212,213]. Also, in experiments such as serial crystallography, where many crystals are required, harvesting that many crystals become almost impossible, and an in situ experiment is a major tool [214]. In situ data collection is executed mostly at room temperature. The room-temperature data collection allows time-resolved experiments for probing chemical reactions in the crystals. The disadvantage of this method is that the data collection at room temperature increases the mosaicity of crystals during the data collection. New advances in in situ sample preparation, such as utilizing thin-film samples, enable the flash-cooling of in situ samples and data collection under cryogenic conditions [215].

## 7.5 In situ data collection using X-ray free-electron laser (XFEL)

The cryocooling of samples decreases the rate of secondary damage stemming from the ionizing radiation. However, it has been shown that X-ray damage can also be omitted by employing very short X-ray pulses comprising fewer than 100-fs intervals. The "diffraction-before-destruction" principle enables the collection of diffraction data from very small protein crystals before they deteriorate through extremely bright and short X-ray pulses [216]. Several X-ray free-electron laser (XFEL) sources have been established around the world, such as the linac coherent light source (LCLS) in Menlo Park, USA, Spring-8 Angstrom Coherent Laser (SACLA) in Harima, Japan, PAL-XFEL in Pohang, South Korea, the European XFEL in Hamburg, Germany, and SwissFEL in Villigen, Switzerland. These sources can generate coherent X-rays with energies up to ~13keV (25keV for some sources) and a peak brilliance of about 9–10 orders of magnitude stronger than a third-generation synchrotron [217].

XFEL provides an excellent platform for time-resolved experiments [218] to elucidate the mechanism of proteins and enzymes in subpicosecond time scales. Recently, Nogly et al. visualized the early step of photoisomerization of retinal in bacteriorhodopsin using a pump-probe technique. Their study initiated the excitation of BR microcrystals through optical laser pulses (pump), and the microcrystals were injected across the femtosecond X-ray pulses (probe). They elucidated the microenvironment of proteins that drives the photoisomerization of retinal during and after the excitation of the chromophore [219].

## 7.6 The computational tools available in protein crystallography

Since the early days of protein crystallography, the instruments for X-ray data collection and the software employed in data processing have undergone tremendous improvements. The data collection process has been programmed into an automatic process, whereas the accuracy of these processes depends upon the quality of the integration and scaling software. Three software packages are mainly employed for processing the diffraction data: iMosflm (part of the CCP4 suite), HKL2000, and HKL3000 (comprising the integration program Denzo as well as merging and scaling the program Scalepack) and the XDS suite (containing programs for both scaling and integration) [220].

DIALS is another software for data processing for synchrotrons and XFELs; however, it does not perform scaling or merging tasks. The software's automated data collection pipelines enable the user to acquire an optimal data collection strategy, collect high-quality data, and complete the dataset for each crystal. To process the more complex datasets, graphical user interfaces (GUIs) allow manual processing. All recent interfaces provide a general workflow of reading diffraction images, finding spots for indexing, indexing the spots, refining the crystal and detector parameters, integrating the diffraction maxima, and scaling, merging, or exporting the reflection files [221]. Two major protein crystallography software for data quality assessment, model and ligand building, phasing, refinements, and so on are Collaborative Computational Project Number 4 (CCP4) and Python-based Hierarchical Environment for Integrated Crystallography (Phenix) [222], which are extensively used in solving X-ray and cryo-EM protein structures. In addition, various model viewers such as Coot [219], PyMol, and UCSF Chimera [223] have been developed for visualizing protein structures and generating high-quality graphics.

# 8. Conclusions and future perspectives

The emergence of new technologies such as XFEL, neutron diffraction, and Micro-ED is expanding the toolkit necessary for obtaining the high-resolution structure of target proteins. Although XFEL requires nano-size protein crystals to be continuously delivered at room temperature, neutron diffraction uses perdeuterated millimeter-sized crystals, and Micro-ED requires micrometer-sized crystals under cryogenic conditions. This diversity in techniques encourages the demand for a better understanding and reproducibility of the biomolecular crystallization process. Meanwhile, new technological advances are introduced to meet the requirement of this diversity of crystal sizes for various diffraction techniques. However, with the advent of pulsed neutron sources such as ESS, the need for large crystals might reduce as the labeling methods for sample preparation are improving [224]. Devices capable of producing short pulses of X-rays are developed, which are far brighter than the radiation pulses generated by current synchrotrons. This will further revolutionize the field of structural biology, where one could get away with very small-size crystals. Other technological advances include cryo-electron tomography (cryo-ET), which enables the investigation of the molecular architecture of protein complexes within the native cell. These assist scientists in understanding fundamental key cell processes and building the image of inner cell structures. During the past decade, microfluidic devices (microchips) have also emerged as a powerful tool in protein crystallization. They enable the miniaturization of crystallization experiments using sample volumes much smaller than the sample volume necessary for current robotic systems. It is anticipated that microfluidic devices will play a central role in protein crystallization studies ranging from screening and crystal improvement to in situ data collection in the near future.

## References


[1] Kendrew JC, Bodo G, Dintzis HM, Parrish RG, Wyckoff H, Phillips DC. A three-dimensional model of the myoglobin molecule obtained by X-ray analysis. Nature 1958;181:662–6. https://doi.org/10.1038/181662a0.

[2] Perutz MF, Rossmann MG, Cullis AF, Muirhead H, Will G, North AC. Structure of haemoglobin: a three-dimensional Fourier synthesis at 5.5-A resolution, obtained by X-ray analysis. Nature 1960;185:416–22. https://doi.org/10.1038/185416a0.

[3] Crowther RA, Klug A. Structural analysis of macromolecular assemblies by image reconstruction from electron micrographs. Annu Rev Biochem 1975;44:161–82. https://doi.org/10.1146/annurev.bi.44.070175.001113.

[4] Deisenhofer J, Epp O, Miki K, Huber R, Michel H. X-ray structure analysis of a membrane protein complex. Electron density map at 3 A resolution and a model of the chromophores of the photosynthetic reaction center from *Rhodopseudomonas viridis*. J Mol Biol 1984;180:385–98. https://doi.org/10.1016/s0022-2836(84)80011-x.

[5] Abrahams JP, Leslie AG, Lutter R, Walker JE. Structure at 2.8 A resolution of F1-ATPase from bovine heart mitochondria. Nature 1994;370:621–8. https://doi.org/10.1038/370621a0.

[6] Doyle DA, Morais Cabral J, Pfuetzner RA, Kuo A, Gulbis JM, Cohen SL, Chait BT, MacKinnon R. The structure of the potassium channel: molecular basis of K + conduction and selectivity. Science 1998;280:69–77. https://doi.org/10.1126/science.280.5360.69.

[7] Cramer P, Bushnell DA, Kornberg RD. Structural basis of transcription: RNA polymerase II at 2.8 angstrom resolution. Science 2001;292:1863–76. https://doi.org/10.1126/science.1059493.

[8] Rasmussen SG, Choi HJ, Rosenbaum DM, Kobilka TS, Thian FS, Edwards PC, Burghammer M, Ratnala VR, Sanishvili R, Fischetti RF, Schertler GF, Weis WI, Kobilka BK. Crystal structure of the human beta2 adrenergic G-protein-coupled receptor. Nature 2007;450:383–7. https://doi.org/10.1038/nature06325.

[9] Krogh A, Larsson B, von Heijne G, Sonnhammer ELL. Predicting transmembrane protein topology with a hidden markov model: application to complete genomes 11 edited by F. Cohen. J Mol Biol 2001;305:567–80. https://doi.org/10.1006/jmbi.2000.4315.

[10] Rask-Andersen M, Almen MS, Schioth HB. Trends in the exploitation of novel drug targets. Nat Rev Drug Discov 2011;10:579–90. https://doi.org/10.1038/nrd3478.

[11] Santos R, Ursu O, Gaulton A, Bento AP, Donadi RS, Bologa CG, Karlsson A, Al-Lazikani B, Hersey A, Oprea TI, Overington JP. A comprehensive map of molecular drug targets. Nat Rev Drug Discov 2017;16:19–34. https://doi.org/10.1038/nrd.2016.230.

[12] Palczewski K, Kumasaka T, Hori T, Behnke CA, Motoshima H, Fox BA, Trong IL, Teller DC, Okada T, Stenkamp RE, Yamamoto M, Miyano M. Crystal structure of rhodopsin: a G protein-coupled receptor. Science 2000;289:739. https://doi.org/10.1126/science.289.5480.739.

[13] Cherezov V, Rosenbaum DM, Hanson MA, Rasmussen SGF, Thian FS, Kobilka TS, Choi H-J, Kuhn P, Weis WI, Kobilka BK, Stevens RC. High-resolution crystal structure of an engineered human beta2-adrenergic G protein-coupled receptor. Science (New York, NY) 2007;318:1258–65. https://doi.org/10.1126/science.1150577.

[14] Rasmussen SGF, Choi H-J, Rosenbaum DM, Kobilka TS, Thian FS, Edwards PC, Burghammer M, Ratnala VRP, Sanishvili R, Fischetti RF, Schertler GFX, Weis WI, Kobilka BK. Crystal structure of the human β2 adrenergic G-protein-coupled receptor. Nature 2007;450:383–7. https://doi.org/10.1038/nature06325.

[15] Rasmussen SG, DeVree BT, Zou Y, Kruse AC, Chung KY, Kobilka TS, Thian FS, Chae PS, Pardon E, Calinski D, Mathiesen JM, Shah ST, Lyons JA, Caffrey M, Gellman SH, Steyaert J, Skiniotis G, Weis WI, Sunahara RK, Kobilka BK. Crystal structure of the beta2 adrenergic receptor-Gs protein complex. Nature 2011;477:549–55. https://doi.org/10.1038/nature10361.

[16] Doyle DA, Cabral JM, Pfuetzner RA, Kuo A, Gulbis JM, Cohen SL, Chait BT, MacKinnon R. The structure of the potassium channel: molecular basis of K + conduction and selectivity. Science 1998;280:69. https://doi.org/10.1126/science.280.5360.69.

[17] Jiang Y, Lee A, Chen J, Cadene M, Chait BT, MacKinnon R. Crystal structure and mechanism of a calcium-gated potassium channel. Nature 2002;417:515–22. https://doi.org/10.1038/417515a.

[18] Congreve M, Dias JM, Marshall FH. Structure-based drug design for G protein-coupled receptors. In: Lawton G, Witty DR, editors. Progress in medicinal chemistry. Elsevier; 2014. p. 1–63 [Chapter 1].

[19] Kermani AA. A guide to membrane protein X-ray crystallography. FEBS J 2021;288:5788–804. https://doi.org/10.1111/febs.15676.

[20] Singh DB, Tripathi T. Frontiers in protein structure, function, and dynamics. 1st. Singapore: Springer Nature; 2020. p. 1–458.

[21] Tripathi T, Dubey VK. Advances in protein molecular and structural biology methods. 1st. Cambridge, MA, USA: Academic Press; 2022. p. 1–714.

[22] Katrusiak A. High-pressure crystallography. Acta Crystallogr Sect A: Found Crystallogr 2008;64:135–48. https://doi.org/10.1107/S0108767307061181.

[23] McIlwain BC, Kermani AA. Membrane protein production in *Escherichia coli*. In: Perez C, Maier T, editors. Expression, purification, and structural biology of membrane proteins. New York, NY: Springer US; 2020. p. 13–27.

[24] Chae PS, Rasmussen SGF, Rana RR, Gotfryd K, Chandra R, Goren MA, Kruse AC, Nurva S, Loland CJ, Pierre Y, Drew D, Popot J-L, Picot D, Fox BG, Guan L, Gether U, Byrne B, Kobilka B, Gellman SH. Maltose-neopentyl glycol (MNG) amphiphiles for solubilization, stabilization and crystallization of membrane proteins. Nat Methods 2010;7:1003–8. https://doi.org/10.1038/nmeth.1526.

[25] Chung KY, Kim TH, Manglik A, Alvares R, Kobilka BK, Prosser RS. Role of detergents in conformational exchange of a G protein-coupled receptor. J Biol Chem 2012;287:36305–11. https://doi.org/10.1074/jbc.M112.406371.

[26] Chae PS, Rana RR, Gotfryd K, Rasmussen SGF, Kruse AC, Cho KH, Capaldi S, Carlsson E, Kobilka B, Loland CJ, Gether U, Banerjee S, Byrne B, Lee JK, Gellman SH. Glucose-neopentyl glycol (GNG) amphiphiles for membrane protein study. Chem Commun (Camb) 2013;49:2287–9. https://doi.org/10.1039/c2cc36844g.

[27] Sharma KS, Durand G, Gabel F, Bazzacco P, Le Bon C, Billon-Denis E, Catoire LJ, Popot J-L, Ebel C, Pucci B. Non-ionic amphiphilic homopolymers: synthesis, solution properties, and biochemical validation. Langmuir 2012;28:4625–39. https://doi.org/10.1021/la205026r.

[28] Matar-Merheb R, Rhimi M, Leydier A, Huché F, Galián C, Desuzinges-Mandon E, Ficheux D, Flot D, Aghajari N, Kahn R, Di Pietro A, Jault J-M, Coleman AW, Falson P. Structuring detergents for extracting and stabilizing functional membrane proteins. PLoS One 2011;6, e18036. https://doi.org/10.1371/journal.pone.0018036.

[29] Frotscher E, Danielczak B, Vargas C, Meister A, Durand G, Keller S. A fluorinated detergent for membrane-protein applications. Angew Chem Int Ed Eng 2015;54:5069–73. https://doi.org/10.1002/anie.201412359.

[30] Rosenbaum DM, Zhang C, Lyons JA, Holl R, Aragao D, Arlow DH, Rasmussen SGF, Choi H-J, DeVree BT, Sunahara RK, Chae PS, Gellman SH, Dror RO, Shaw DE, Weis WI, Caffrey M, Gmeiner P, Kobilka BK. Structure and function of an irreversible agonist-β2 adrenoceptor complex. Nature 2011;469:236–40. https://doi.org/10.1038/nature09665.

[31] Kruse AC, Hu J, Pan AC, Arlow DH, Rosenbaum DM, Rosemond E, Green HF, Liu T, Chae PS, Dror RO, Shaw DE, Weis WI, Wess J, Kobilka BK. Structure and dynamics of the M3 muscarinic acetylcholine receptor. Nature 2012;482:552–6. https://doi.org/10.1038/nature10867.

[32] White JF, Noinaj N, Shibata Y, Love J, Kloss B, Xu F, Gvozdenovic-Jeremic J, Shah P, Shiloach J, Tate CG, Grisshammer R. Structure of the agonist-bound neurotensin receptor. Nature 2012;490:508–13. https://doi.org/10.1038/nature11558.

[33] Rollauer SE, Tarry MJ, Graham JE, Jääskeläinen M, Jäger F, Johnson S, Krehenbrink M, Liu S-M, Lukey MJ, Marcoux J, McDowell MA, Rodriguez F, Roversi P, Stansfeld PJ, Robinson CV, Sansom MSP, Palmer T, Högbom M, Berks BC, Lea SM. Structure of the TatC core of the twin-arginine protein transport system. Nature 2012;492:210–4. https://doi.org/10.1038/nature11683.

[34] Paulsen CE, Armache J-P, Gao Y, Cheng Y, Julius D. Structure of the TRPA1 ion channel suggests regulatory mechanisms. Nature 2015;520:511–7. https://doi.org/10.1038/nature14367.

[35] Yin J, Mobarec JC, Kolb P, Rosenbaum DM. Crystal structure of the human OX2 orexin receptor bound to the insomnia drug suvorexant. Nature 2015;519:247–50. https://doi.org/10.1038/nature14035.

[36] Kang Y, Zhou XE, Gao X, He Y, Liu W, Ishchenko A, Barty A, White TA, Yefanov O, Han GW, Xu Q, de Waal PW, Ke J, Tan MHE, Zhang C, Moeller A, West GM,

Pascal BD, Van Eps N, Caro LN, Vishnivetskiy SA, Lee RJ, Suino-Powell KM, Gu X, Pal K, Ma J, Zhi X, Boutet S, Williams GJ, Messerschmidt M, Gati C, Zatsepin NA, Wang D, James D, Basu S, Roy-Chowdhury S, Conrad CE, Coe J, Liu H, Lisova S, Kupitz C, Grotjohann I, Fromme R, Jiang Y, Tan M, Yang H, Li J, Wang M, Zheng Z, Li D, Howe N, Zhao Y, Standfuss J, Diederichs K, Dong Y, Potter CS, Carragher B, Caffrey M, Jiang H, Chapman HN, Spence JCH, Fromme P, Weierstall U, Ernst OP, Katritch V, Gurevich VV, Griffin PR, Hubbell WL, Stevens RC, Cherezov V, Melcher K, Xu HE. Crystal structure of rhodopsin bound to arrestin by femtosecond X-ray laser. Nature 2015;523:561–7. https://doi.org/10.1038/nature14656.

[37] Perez C, Gerber S, Boilevin J, Bucher M, Darbre T, Aebi M, Reymond J-L, Locher KP. Structure and mechanism of an active lipid-linked oligosaccharide flippase. Nature 2015;524:433–8. https://doi.org/10.1038/nature14953.

[38] Taniguchi R, Kato HE, Font J, Deshpande CN, Wada M, Ito K, Ishitani R, Jormakka M, Nureki O. Outward- and inward-facing structures of a putative bacterial transition-metal transporter with homology to ferroportin. Nat Commun 2015;6:8545. https://doi.org/10.1038/ncomms9545.

[39] Benton DJ, Nans A, Calder LJ, Turner J, Neu U, Lin YP, Ketelaars E, Kallewaard NL, Corti D, Lanzavecchia A, Gamblin SJ, Rosenthal PB, Skehel JJ. Influenza hemagglutinin membrane anchor. Proc Natl Acad Sci 2018;115:10112. https://doi.org/10.1073/pnas.1810927115.

[40] Gao Y, Hu H, Ramachandran S, Erickson JW, Cerione RA, Skiniotis G. Structures of the rhodopsin-transducin complex: insights into G-protein activation. Mol Cell 2019;75, 781-790.e783. https://doi.org/10.1016/j.molcel.2019.06.007.

[41] Sun J. Structures of mouse DUOX1-DUOXA1 provide mechanistic insights into enzyme activation and regulation. Nat Struct Mol Biol 2020;27:1086–93. https://doi.org/10.1038/s41594-020-0501-x.

[42] Sun J, MacKinnon R. Structural basis of human KCNQ1 modulation and gating. Cell 2020;180, 340-347.e349. https://doi.org/10.1016/j.cell.2019.12.003.

[43] Huynh KW, Cohen MR, Jiang J, Samanta A, Lodowski DT, Zhou ZH, Moiseenkova-Bell VY. Structure of the full-length TRPV2 channel by cryo-EM. Nat Commun 2016;7:11130. https://doi.org/10.1038/ncomms11130.

[44] Nishizawa T, Kita S, Maturana AD, Furuya N, Hirata K, Kasuya G, Ogasawara S, Dohmae N, Iwamoto T, Ishitani R, Nureki O. Structural basis for the counter-transport mechanism of a $H^{+}/Ca^{2+}$ exchanger. Science 2013;341:168–72. https://doi.org/10.1126/science.1239002.

[45] Dong YY, Pike AC, Mackenzie A, McClenaghan C, Aryal P, Dong L, Quigley A, Grieben M, Goubin S, Mukhopadhyay S, Ruda GF, Clausen MV, Cao L, Brennan PE, Burgess-Brown NA, Sansom MS, Tucker SJ, Carpenter EP. K2P channel gating mechanisms revealed by structures of TREK-2 and a complex with Prozac. Science 2015;347:1256–9. https://doi.org/10.1126/science.1261512.

[46] Bausewein T, Mills DJ, Langer JD, Nitschke B, Nussberger S, Kühlbrandt W. Cryo-EM structure of the TOM core complex from Neurospora crassa. Cell 2017;170. https://doi.org/10.1016/j.cell.2017.07.012. 693–700.e697.

[47] Kubota-Kawai H, Burton-Smith RN, Tokutsu R, Song C, Akimoto S, Yokono M, Ueno Y, Kim E, Watanabe A, Murata K, Minagawa J. Ten antenna proteins are associated with the core in the supramolecular organization of the photosystem I supercomplex in *Chlamydomonas reinhardtii*. J Biol Chem 2019;294:4304–14. https://doi.org/10.1074/jbc.RA118.006536.

[48] Burton-Smith RN, Watanabe A, Tokutsu R, Song C, Murata K, Minagawa J. Structural determination of the large photosystem II–light-harvesting complex II supercomplex of *Chlamydomonas reinhardtii* using nonionic amphipol. J Biol Chem 2019;294:15003–13. https://doi.org/10.1074/jbc.RA119.009341.

[49] Agez M, Schultz P, Medina I, Baker DJ, Burnham MP, Cardarelli RA, Conway LC, Garnier K, Geschwindner S, Gunnarsson A, McCall EJ, Frechard A, Audebert S, Deeb TZ, Moss SJ, Brandon NJ, Wang Q, Dekker N, Jawhari A. Molecular architecture of potassium chloride co-transporter KCC2. Sci Rep 2017;7:16452. https://doi.org/10.1038/s41598-017-15739-1.

[50] Tribet C, Audebert R, Popot JL. Amphipols: polymers that keep membrane proteins soluble in aqueous solutions. Proc Natl Acad Sci U S A 1996;93:15047–50. https://doi.org/10.1073/pnas.93.26.15047.

[51] Rahmeh R, Damian M, Cottet M, Orcel H, Mendre C, Durroux T, Sharma KS, Durand G, Pucci B, Trinquet E, Zwier JM, Deupi X, Bron P, Banères J-L, Mouillac B, Granier S. Structural insights into biased G protein-coupled receptor signaling revealed by fluorescence spectroscopy. Proc Natl Acad Sci 2012;109:6733–8. https://doi.org/10.1073/pnas.1201093109.

[52] Desuzinges Mandon E, Traversier A, Champagne A, Benier L, Audebert S, Balme S, Dejean E, Rosa Calatrava M, Jawhari A. Expression and purification of native and functional influenza A virus matrix 2 proton selective ion channel. Protein Expr Purif 2017;131:42–50. https://doi.org/10.1016/j.pep.2016.11.001.

[53] Park KH, Berrier C, Lebaupain F, Pucci B, Popot JL, Ghazi A, Zito F. Fluorinated and hemifluorinated surfactants as alternatives to detergents for membrane protein cell-free synthesis. Biochem J 2007;403:183–7. https://doi.org/10.1042/bj20061473.

[54] Breyton C, Chabaud E, Chaudier Y, Pucci B, Popot JL. Hemifluorinated surfactants: a non-dissociating environment for handling membrane proteins in aqueous solutions? FEBS Lett 2004;564:312–8. https://doi.org/10.1016/s0014-5793(04)00227-3.

[55] Willegems K, Efremov RG. Influence of lipid mimetics on gating of ryanodine receptor. Structure 2018;26:1303–1313.e1304. https://doi.org/10.1016/j.str.2018.06.010.

[56] Kermani AA, Macdonald CB, Burata OE, Ben Koff B, Koide A, Denbaum E, Koide S, Stockbridge RB. The structural basis of promiscuity in small multidrug resistance transporters. Nat Commun 2020;11:6064. https://doi.org/10.1038/s41467-020-19820-8.

[57] Kermani AA, Macdonald CB, Gundepudi R, Stockbridge RB. Guanidinium export is the primal function of SMR family transporters. Proc Natl Acad Sci U S A 2018;115:3060–5. https://doi.org/10.1073/pnas.1719187115.

[58] Nikolaev M, Round E, Gushchin I, Polovinkin V, Balandin T, Kuzmichev P, Shevchenko V, Borshchevskiy V, Kuklin A, Round A, Bernhard F, Willbold D, Büldt G, Gordeliy V. Integral membrane proteins can be crystallized directly from nanodiscs. Cryst Growth Des 2017;17:945–8. https://doi.org/10.1021/acs.cgd.6b01631.

[59] Broecker J, Eger BT, Ernst OP. Crystallogenesis of membrane proteins mediated by polymer-bounded lipid nanodiscs. Structure 2017;25:384–92. https://doi.org/10.1016/j.str.2016.12.004.

[60] Denisov IG, Grinkova YV, Lazarides AA, Sligar SG. Directed self-assembly of monodisperse phospholipid bilayer nanodiscs with controlled size. J Am Chem Soc 2004;126:3477–87. https://doi.org/10.1021/ja0393574.

[61] Denisov IG, Sligar SG. Nanodiscs for structural and functional studies of membrane proteins. Nat Struct Mol Biol 2016;23:481–6. https://doi.org/10.1038/nsmb.3195.

[62] Bayburt TH, Sligar SG. Membrane protein assembly into nanodiscs. FEBS Lett 2010;584:1721–7. https://doi.org/10.1016/j.febslet.2009.10.024.

[63] Goddard AD, Dijkman PM, Adamson RJ, dos Reis RI, Watts A. Reconstitution of membrane proteins: a GPCR as an example. Methods Enzymol 2015;556:405–24. https://doi.org/10.1016/bs.mie.2015.01.004.

[64] Su C-C, Morgan CE, Kambakam S, Rajavel M, Scott H, Huang W, Emerson CC, Taylor DJ, Stewart PL, Bonomo RA, Yu EW. Cryo-electron microscopy structure of an *Acinetobacter baumannii* multidrug efflux pump. mBio 2019;10, e01295-01219.-https://doi.org/10.1128/mBio.01295-19.

[65] Postis V, Rawson S, Mitchell JK, Lee SC, Parslow RA, Dafforn TR, Baldwin SA, Muench SP. The use of SMALPs as a novel membrane protein scaffold for structure study by negative stain electron microscopy. Biochim Biophys Acta Biomembr 2015;1848:496–501. https://doi.org/10.1016/j.bbamem.2014.10.018.
[66] Phillips R, Ursell T, Wiggins P, Sens P. Emerging roles for lipids in shaping membrane-protein function. Nature 2009;459:379–85. https://doi.org/10.1038/nature08147.
[67] Guo Y. Be cautious with crystal structures of membrane proteins or complexes prepared in detergents. Crystals 2020;10. https://doi.org/10.3390/cryst10020086.
[68] Qiu W, Fu Z, Xu GG, Grassucci RA, Zhang Y, Frank J, Hendrickson WA, Guo Y. Structure and activity of lipid bilayer within a membrane-protein transporter. Proc Natl Acad Sci 2018;115:12985. https://doi.org/10.1073/pnas.1812526115.
[69] Frauenfeld J, Löving R, Armache J-P, Sonnen AFP, Guettou F, Moberg P, Zhu L, Jegerschöld C, Flayhan A, Briggs JAG, Garoff H, Löw C, Cheng Y, Nordlund P. A saposin-lipoprotein nanoparticle system for membrane proteins. Nat Methods 2016;13:345–51. https://doi.org/10.1038/nmeth.3801.
[70] Bruhn H. A short guided tour through functional and structural features of saposin-like proteins. Biochem J 2005;389:249–57. https://doi.org/10.1042/BJ20050051.
[71] Kapust RB, Waugh DS. *Escherichia coli* maltose-binding protein is uncommonly effective at promoting the solubility of polypeptides to which it is fused. Protein Sci 1999;8:1668–74. https://doi.org/10.1110/ps.8.8.1668.
[72] Smyth DR, Mrozkiewicz MK, McGrath WJ, Listwan P, Kobe B. Crystal structures of fusion proteins with large-affinity tags. Protein Sci 2003;12:1313–22. https://doi.org/10.1110/ps.0243403.
[73] Jin T, Chuenchor W, Jiang J, Cheng J, Li Y, Fang K, Huang M, Smith P, Xiao TS. Design of an expression system to enhance MBP-mediated crystallization. Sci Rep 2017;7:40991. https://doi.org/10.1038/srep40991.
[74] Bokhove M, Al Hosseini HS, Saito T, Dioguardi E, Gegenschatz-Schmid K, Nishimura K, Raj I, de Sanctis D, Han L, Jovine L. Easy mammalian expression and crystallography of maltose-binding protein-fused human proteins. J Struct Biol 2016;194:1–7. https://doi.org/10.1016/j.jsb.2016.01.016.
[75] Lieberman RL, Culver JA, Entzminger KC, Pai JC, Maynard JA. Crystallization chaperone strategies for membrane proteins. Methods 2011;55:293–302. https://doi.org/10.1016/j.ymeth.2011.08.004.
[76] Koide S. Engineering of recombinant crystallization chaperones. Curr Opin Struct Biol 2009;19:449–57. https://doi.org/10.1016/j.sbi.2009.04.008.
[77] Zhou M, Robinson CV. Flexible membrane proteins: functional dynamics captured by mass spectrometry. Curr Opin Struct Biol 2014;28:122–30. https://doi.org/10.1016/j.sbi.2014.08.005.
[78] Rosenbaum DM, Cherezov V, Hanson MA, Rasmussen SG, Thian FS, Kobilka TS, Choi HJ, Yao XJ, Weis WI, Stevens RC, Kobilka BK. GPCR engineering yields high-resolution structural insights into beta2-adrenergic receptor function. Science 2007;318:1266–73. https://doi.org/10.1126/science.1150609.
[79] Wu B, Chien EYT, Mol CD, Fenalti G, Liu W, Katritch V, Abagyan R, Brooun A, Wells P, Bi FC, Hamel DJ, Kuhn P, Handel TM, Cherezov V, Stevens RC. Structures of the CXCR4 chemokine GPCR with small-molecule and cyclic peptide antagonists. Science (New York, NY) 2010;330:1066–71. https://doi.org/10.1126/science.1194396.
[80] Shimamura T, Shiroishi M, Weyand S, Tsujimoto H, Winter G, Katritch V, Abagyan R, Cherezov V, Liu W, Han GW, Kobayashi T, Stevens RC, Iwata S. Structure of the human histamine H1 receptor complex with doxepin. Nature 2011;475:65–70. https://doi.org/10.1038/nature10236.

[81] Chien EYT, Liu W, Zhao Q, Katritch V, Won Han G, Hanson MA, Shi L, Newman AH, Javitch JA, Cherezov V, Stevens RC. Structure of the human dopamine D3 receptor in complex with a D2/D3 selective antagonist. Science 2010;330:1091–5. https://doi.org/10.1126/science.1197410.
[82] Rosenbaum DM, Rasmussen SGF, Kobilka BK. The structure and function of G-protein-coupled receptors. Nature 2009;459:356–63. https://doi.org/10.1038/nature08144.
[83] Landry JP, Ke Y, Yu G-L, Zhu XD. Measuring affinity constants of 1450 monoclonal antibodies to peptide targets with a microarray-based label-free assay platform. J Immunol Methods 2015;417:86–96. https://doi.org/10.1016/j.jim.2014.12.011.
[84] Iwata S, Ostermeier C, Ludwig B, Michel H. Structure at 2.8 A resolution of cytochrome c oxidase from *Paracoccus denitrificans*. Nature 1995;376:660–9. https://doi.org/10.1038/376660a0.
[85] Zhou Y, Morais-Cabral JH, Kaufman A, MacKinnon R. Chemistry of ion coordination and hydration revealed by a K+ channel-Fab complex at 2.0 Å resolution. Nature 2001;414:43–8. https://doi.org/10.1038/35102009.
[86] Dutzler R, Campbell EB, Cadene M, Chait BT, MacKinnon R. X-ray structure of a ClC chloride channel at 3.0 Å reveals the molecular basis of anion selectivity. Nature 2002;415:287–94. https://doi.org/10.1038/415287a.
[87] Hino T, Matsumoto Y, Nagano S, Sugimoto H, Fukumori Y, Murata T, Iwata S, Shiro Y. Structural basis of biological $N_2O$ generation by bacterial nitric oxide reductase. Science 2010;330:1666–70. https://doi.org/10.1126/science.1195591.
[88] Tsukazaki T, Mori H, Fukai S, Ishitani R, Mori T, Dohmae N, Perederina A, Sugita Y, Vassylyev DG, Ito K, Nureki O. Conformational transition of Sec machinery inferred from bacterial SecYE structures. Nature 2008;455:988–91. https://doi.org/10.1038/nature07421.
[89] Ghosh E, Kumari P, Jaiman D, Shukla AK. Methodological advances: the unsung heroes of the GPCR structural revolution. Nat Rev Mol Cell Biol 2015;16:69–81. https://doi.org/10.1038/nrm3933.
[90] McMahon C, Baier AS, Pascolutti R, Wegrecki M, Zheng S, Ong JX, Erlandson SC, Hilger D, Rasmussen SGF, Ring AM, Manglik A, Kruse AC. Yeast surface display platform for rapid discovery of conformationally selective nanobodies. Nat Struct Mol Biol 2018;25:289–96. https://doi.org/10.1038/s41594-018-0028-6.
[91] Pardon E, Laeremans T, Triest S, Rasmussen SGF, Wohlkönig A, Ruf A, Muyldermans S, Hol WGJ, Kobilka BK, Steyaert J. A general protocol for the generation of nanobodies for structural biology. Nat Protoc 2014;9:674–93. https://doi.org/10.1038/nprot.2014.039.
[92] Uchański T, Zögg T, Yin J, Yuan D, Wohlkönig A, Fischer B, Rosenbaum DM, Kobilka BK, Pardon E, Steyaert J. An improved yeast surface display platform for the screening of nanobody immune libraries. Sci Rep 2019;9:382. https://doi.org/10.1038/s41598-018-37212-3.
[93] Zimmermann I, Egloff P, Hutter CAJ, Arnold FM, Stohler P, Bocquet N, Hug MN, Huber S, Siegrist M, Hetemann L, Gera J, Gmür S, Spies P, Gygax D, Geertsma ER, Dawson RJP, Seeger MA. Synthetic single domain antibodies for the conformational trapping of membrane proteins. Elife 2018;7, e34317. https://doi.org/10.7554/eLife.34317.
[94] Wingler LM, McMahon C, Staus DP, Lefkowitz RJ, Kruse AC. Distinctive activation mechanism for angiotensin receptor revealed by a synthetic nanobody. Cell 2019;176. https://doi.org/10.1016/j.cell.2018.12.006. 479-490.e412.
[95] Bräuer P, Parker JL, Gerondopoulos A, Zimmermann I, Seeger MA, Barr FA, Newstead S. Structural basis for pH-dependent retrieval of ER proteins from the Golgi by the KDEL receptor. Science 2019;363:1103. https://doi.org/10.1126/science.aaw2859.

[96] Hutter CAJ, Timachi MH, Hürlimann LM, Zimmermann I, Egloff P, Göddeke H, Kucher S, Štefanić S, Karttunen M, Schäfer LV, Bordignon E, Seeger MA. The extracellular gate shapes the energy profile of an ABC exporter. Nat Commun 2019;10:2260. https://doi.org/10.1038/s41467-019-09892-6.
[97] Arnold FM, Weber MS, Gonda I, Gallenito MJ, Adenau S, Egloff P, Zimmermann I, Hutter CAJ, Hürlimann LM, Peters EE, Piel J, Meloni G, Medalia O, Seeger MA. The ABC exporter IrtAB imports and reduces mycobacterial siderophores. Nature 2020;580:413–7. https://doi.org/10.1038/s41586-020-2136-9.
[98] Walter JD, Hutter CAJ, Zimmermann I, Earp J, Egloff P, Sorgenfrei M, Hürlimann LM, Gonda I, Meier G, Remm S, Thavarasah S, Plattet P, Seeger MA. Synthetic nanobodies targeting the SARS-CoV-2 receptor-binding domain. bioRxiv 2020. https://doi.org/10.1101/2020.04.16.045419.
[99] Koide A, Bailey CW, Huang X, Koide S. The fibronectin type III domain as a scaffold for novel binding proteins 11 edited by J. Wells. J Mol Biol 1998;284:1141–51. https://doi.org/10.1006/jmbi.1998.2238.
[100] Karatan E, Merguerian M, Han Z, Scholle MD, Koide S, Kay BK. Molecular recognition properties of FN3 monobodies that bind the Src SH3 domain. Chem Biol 2004;11:835–44. https://doi.org/10.1016/j.chembiol.2004.04.009.
[101] Koide A, Wojcik J, Gilbreth RN, Hoey RJ, Koide S. Teaching an old scaffold new tricks: monobodies constructed using alternative surfaces of the FN3 scaffold. J Mol Biol 2012;415:393–405. https://doi.org/10.1016/j.jmb.2011.12.019.
[102] Kermani AA, Burata OE, Koff BB, Koide A, Koide S, Stockbridge RB. Crystal structures of bacterial small multidrug resistance transporter EmrE in complex with structurally diverse substrates. eLife 2022;11:e76766. https://doi.org/10.7554/eLife.76766.
[103] Sha F, Salzman G, Gupta A, Koide S. Monobodies and other synthetic binding proteins for expanding protein science. Protein Sci 2017;26:910–24. https://doi.org/10.1002/pro.3148.
[104] Stockbridge RB, Kolmakova-Partensky L, Shane T, Koide A, Koide S, Miller C, Newstead S. Crystal structures of a double-barrelled fluoride ion channel. Nature 2015;525:548–51. https://doi.org/10.1038/nature14981.
[105] Koide A, Gilbreth RN, Esaki K, Tereshko V, Koide S. High-affinity single-domain binding proteins with a binary-code interface. Proc Natl Acad Sci U S A 2007;104:6632–7. https://doi.org/10.1073/pnas.0700149104.
[106] Gushchin I, Melnikov I, Polovinkin V, Ishchenko A, Yuzhakova A, Buslaev P, Bourenkov G, Grudinin S, Round E, Balandin T, Borshchevskiy V, Willbold D, Leonard G, Buldt G, Popov A, Gordeliy V. Mechanism of transmembrane signaling by sensor histidine kinases. Science 2017;356. https://doi.org/10.1126/science.aah6345.
[107] Gushchin I, Melnikov I, Polovinkin V, Ishchenko A, Gordeliy V. Crystal structure of a proteolytic fragment of the sensor histidine kinase NarQ. Crystals 2020;10. https://doi.org/10.3390/cryst10030149.
[108] Deller MC, Kong L, Rupp B. Protein stability: a crystallographer's perspective. Acta Crystallogr F Struct Biol Commun 2016;72:72–95. https://doi.org/10.1107/S2053230X15024619.
[109] Chung IYW, Li L, Tyurin O, Gagarinova A, Wibawa R, Li P, Hartland EL, Cygler M. Structural and functional study of Legionella pneumophila effector RavA. Protein Sci 2021;30:940–55. https://doi.org/10.1002/pro.4057.
[110] Zhang Y, Prach LM, O'Brien TE, DiMaio F, Prigozhin DM, Corn JE, Alber T, Siegel JB, Tantillo DJ. Crystal structure and mechanistic molecular modeling studies of *Mycobacterium tuberculosis* diterpene cyclase Rv3377c. Biochemistry 2020;59:4507–15. https://doi.org/10.1021/acs.biochem.0c00762.

[111] Andaleeb H, Ullah N, Falke S, Perbandt M, Brognaro H, Betzel C. High-resolution crystal structure and biochemical characterization of a GH11 endoxylanase from *Nectria haematococca*. Sci Rep 2020;10:15658. https://doi.org/10.1038/s41598-020-72644-w.
[112] Sato Y, Tsuchiya H, Yamagata A, Okatsu K, Tanaka K, Saeki Y, Fukai S. Structural insights into ubiquitin recognition and Ufd1 interaction of Npl4. Nat Commun 2019;10:5708. https://doi.org/10.1038/s41467-019-13697-y.
[113] Kung WW, Ramachandran S, Makukhin N, Bruno E, Ciulli A. Structural insights into substrate recognition by the SOCS2 E3 ubiquitin ligase. Nat Commun 2019;10:2534. https://doi.org/10.1038/s41467-019-10190-4.
[114] Cooper DR, Boczek T, Grelewska K, Pinkowska M, Sikorska M, Zawadzki M, Derewenda Z. Protein crystallization by surface entropy reduction: optimization of the SER strategy. Acta Crystallogr D Biol Crystallogr 2007;63:636–45. https://doi.org/10.1107/s0907444907010931.
[115] Goldschmidt L, Cooper DR, Derewenda ZS, Eisenberg D. Toward rational protein crystallization: a web server for the design of crystallizable protein variants. Protein Sci 2007;16:1569–76. https://doi.org/10.1110/ps.072914007.
[116] Vaidehi N, Grisshammer R, Tate CG. How can mutations thermostabilize G-protein-coupled receptors? Trends Pharmacol Sci 2016;37:37–46. https://doi.org/10.1016/j.tips.2015.09.005.
[117] Christopher JA, Brown J, Doré AS, Errey JC, Koglin M, Marshall FH, Myszka DG, Rich RL, Tate CG, Tehan B, Warne T, Congreve M. Biophysical fragment screening of the β1-adrenergic receptor: identification of high affinity arylpiperazine leads using structure-based drug design. J Med Chem 2013;56:3446–55. https://doi.org/10.1021/jm400140q.
[118] Warne T, Edwards PC, Leslie AG, Tate CG. Crystal structures of a stabilized β1-adrenoceptor bound to the biased agonists bucindolol and carvedilol. Structure 2012;20:841–9. https://doi.org/10.1016/j.str.2012.03.014.
[119] Warne T, Moukhametzianov R, Baker JG, Nehmé R, Edwards PC, Leslie AGW, Schertler GFX, Tate CG. The structural basis for agonist and partial agonist action on a β(1)-adrenergic receptor. Nature 2011;469:241–4. https://doi.org/10.1038/nature09746.
[120] Warne T, Serrano-Vega MJ, Baker JG, Moukhametzianov R, Edwards PC, Henderson R, Leslie AG, Tate CG, Schertler GF. Structure of a beta1-adrenergic G-protein-coupled receptor. Nature 2008;454:486–91. https://doi.org/10.1038/nature07101.
[121] Robertson N, Jazayeri A, Errey J, Baig A, Hurrell E, Zhukov A, Langmead CJ, Weir M, Marshall FH. The properties of thermostabilised G protein-coupled receptors (StaRs) and their use in drug discovery. Neuropharmacology 2011;60:36–44. https://doi.org/10.1016/j.neuropharm.2010.07.001.
[122] Lebon G, Warne T, Edwards PC, Bennett K, Langmead CJ, Leslie AG, Tate CG. Agonist-bound adenosine A2A receptor structures reveal common features of GPCR activation. Nature 2011;474:521–5. https://doi.org/10.1038/nature10136.
[123] Krumm BE, White JF, Shah P, Grisshammer R. Structural prerequisites for G-protein activation by the neurotensin receptor. Nat Commun 2015;6:7895. https://doi.org/10.1038/ncomms8895.
[124] Tan Q, Zhu Y, Li J, Chen Z, Han GW, Kufareva I, Li T, Ma L, Fenalti G, Li J, Zhang W, Xie X, Yang H, Jiang H, Cherezov V, Liu H, Stevens RC, Zhao Q, Wu B. Structure of the CCR5 chemokine receptor-HIV entry inhibitor maraviroc complex. Science (New York, NY) 2013;341:1387–90. https://doi.org/10.1126/science.1241475.
[125] Srivastava A, Yano J, Hirozane Y, Kefala G, Gruswitz F, Snell G, Lane W, Ivetac A, Aertgeerts K, Nguyen J, Jennings A, Okada K. High-resolution structure of the

human GPR40 receptor bound to allosteric agonist TAK-875. Nature 2014; 513:124–7. https://doi.org/10.1038/nature13494.
[126] Hollenstein K, Kean J, Bortolato A, Cheng RK, Doré AS, Jazayeri A, Cooke RM, Weir M, Marshall FH. Structure of class B GPCR corticotropin-releasing factor receptor 1. Nature 2013;499:438–43. https://doi.org/10.1038/nature12357.
[127] Doré AS, Okrasa K, Patel JC, Serrano-Vega M, Bennett K, Cooke RM, Errey JC, Jazayeri A, Khan S, Tehan B, Weir M, Wiggin GR, Marshall FH. Structure of class C GPCR metabotropic glutamate receptor 5 transmembrane domain. Nature 2014;511:557–62. https://doi.org/10.1038/nature13396.
[128] Munk C, Mutt E, Isberg V, Nikolajsen LF, Bibbe JM, Flock T, Hanson MA, Stevens RC, Deupi X, Gloriam DE. An online resource for GPCR structure determination and analysis. Nat Methods 2019;16:151–62. https://doi.org/10.1038/s41592-018-0302-x.
[129] Penmatsa A, Wang KH, Gouaux E. X-ray structure of dopamine transporter elucidates antidepressant mechanism. Nature 2013;503:85–90. https://doi.org/10.1038/nature12533.
[130] Chen L, Dürr KL, Gouaux E. X-ray structures of AMPA receptor-cone snail toxin complexes illuminate activation mechanism. Science 2014;345:1021–6. https://doi.org/10.1126/science.1258409.
[131] Debruycker V, Hutchin A, Masureel M, Ficici E, Martens C, Legrand P, Stein RA, McHaourab HS, Faraldo-Gómez JD, Remaut H, Govaerts C. An embedded lipid in the multidrug transporter LmrP suggests a mechanism for polyspecificity. Nat Struct Mol Biol 2020;27:829–35. https://doi.org/10.1038/s41594-020-0464-y.
[132] Raynal B, Lenormand P, Baron B, Hoos S, England P. Quality assessment and optimization of purified protein samples: why and how? Microb Cell Factories 2014;13:180. https://doi.org/10.1186/s12934-014-0180-6.
[133] Pignataro MF, Herrera MG, Dodero VI. Evaluation of peptide/protein self-assembly and aggregation by spectroscopic methods. Molecules 2020;25:4854. https://doi.org/10.3390/molecules25204854.
[134] McMullan D, Canaves JM, Quijano K, Abdubek P, Nigoghossian E, Haugen J, Klock HE, Vincent J, Hale J, Paulsen J, Lesley SA. High-throughput protein production for X-ray crystallography and use of size exclusion chromatography to validate or refute computational biological unit predictions. J Struct Funct Genom 2005;6:135–41. https://doi.org/10.1007/s10969-005-2898-1.
[135] Wilson WW. Monitoring crystallization experiments using dynamic light scattering: assaying and monitoring protein crystallization in solution. Methods 1990;1:110–7. https://doi.org/10.1016/S1046-2023(05)80154-9.
[136] Al-Ghobashy MA, Mostafa MM, Abed HS, Fathalla FA, Salem MY. Correlation between dynamic light scattering and size exclusion high performance liquid chromatography for monitoring the effect of pH on stability of biopharmaceuticals. J Chromatogr B Anal Technol Biomed Life Sci 2017;1060:1–9. https://doi.org/10.1016/j.jchromb.2017.05.029.
[137] Some D, Amartely H, Tsadok A, Lebendiker M. Characterization of proteins by size-exclusion chromatography coupled to multi-angle light scattering (SEC-MALS). J Vis Exp 2019;, e59615. https://doi.org/10.3791/59615.
[138] Russo Krauss I, Merlino A, Vergara A, Sica F. An overview of biological macromolecule crystallization. Int J Mol Sci 2013;14:11643–91. https://doi.org/10.3390/ijms140611643.
[139] Hiraki M, Kato R, Nagai M, Satoh T, Hirano S, Ihara K, Kudo N, Nagae M, Kobayashi M, Inoue M. Development of an automated large-scale protein-crystallization and monitoring system for high-throughput protein-structure analyses. Acta Crystallogr D Biol Crystallogr 2006;62:1058–65.

[140] Sugahara M, Asada Y, Shimizu K, Yamamoto H, Lokanath NK, Mizutani H, Bagautdinov B, Matsuura Y, Taketa M, Kageyama Y. High-throughput crystallization-to-structure pipeline at RIKEN SPring-8 Center. J Struct Funct Genom 2008;9:21–8.
[141] Gorrec F, Löwe J. Automated protocols for macromolecular crystallization at the MRC laboratory of molecular biology. J Vis Exp 2018.
[142] Shaw Stewart P, Mueller-Dieckmann J. Automation in biological crystallization. Acta Crystallogr F: Struct Biol Commun 2014;70:686–96.
[143] Bard J, Ercolani K, Svenson K, Olland A, Somers W. Automated systems for protein crystallization. Methods 2004;34:329–47.
[144] Shah AK, Liu Z-J, Stewart PD, Schubot FD, Rose JP, Newton MG, Wang B-C. On increasing protein-crystallization throughput for X-ray diffraction studies. Acta Crystallogr D Biol Crystallogr 2005;61:123–9.
[145] Landau EM, Rosenbusch JP. Lipidic cubic phases: a novel concept for the crystallization of membrane proteins. Proc Natl Acad Sci 1996;93:14532–5.
[146] Salvati Manni L, Zabara A, Osornio YM, Schöppe J, Batyuk A, Plückthun A, Siegel JS, Mezzenga R, Landau EM. Phase behavior of a designed cyclopropyl analogue of monoolein: implications for low-temperature membrane protein crystallization. Angew Chem Int Ed 2015;54:1027–31.
[147] Borshchevskiy V, Moiseeva E, Kuklin A, Büldt G, Hato M, Gordeliy V. Isoprenoid-chained lipid β-XylOC16 + 4—a novel molecule for in meso membrane protein crystallization. J Cryst Growth 2010;312:3326–30.
[148] Misquitta LV, Misquitta Y, Cherezov V, Slattery O, Mohan JM, Hart D, Zhalnina M, Cramer WA, Caffrey M. Membrane protein crystallization in lipidic mesophases with tailored bilayers. Structure 2004;12:2113–24.
[149] Ishchenko A, Peng L, Zinovev E, Vlasov A, Lee SC, Kuklin A, Mishin A, Borshchevskiy V, Zhang Q, Cherezov V. Chemically stable lipids for membrane protein crystallization. Cryst Growth Des 2017;17:3502–11.
[150] Kato R, Hiraki M, Yamada Y, Tanabe M, Senda T. A fully automated crystallization apparatus for small protein quantities. Acta Crystallogr F: Struct Biol Commun 2021;77.
[151] Perry SL, Guha S, Pawate AS, Bhaskarla A, Agarwal V, Nair SK, Kenis PJ. A microfluidic approach for protein structure determination at room temperature via on-chip anomalous diffraction. Lab Chip 2013;13:3183–7.
[152] Junius N, Oksanen E, Terrien M, Berzin C, Ferrer J-L, Budayova-Spano M. A crystallization apparatus for temperature-controlled flow-cell dialysis with real-time visualization. J Appl Crystallogr 2016;49:806–13.
[153] Laval P, Giroux C, Leng J, Salmon J-B. Microfluidic screening of potassium nitrate polymorphism. J Cryst Growth 2008;310:3121–4.
[154] Dhouib K, Malek CK, Pfleging W, Gauthier-Manuel B, Duffait R, Thuillier G, Ferrigno R, Jacquamet L, Ohana J, Ferrer J-L. Microfluidic chips for the crystallization of biomacromolecules by counter-diffusion and on-chip crystal X-ray analysis. Lab Chip 2009;9:1412–21.
[155] Hansen CL, Skordalakes E, Berger JM, Quake SR. A robust and scalable microfluidic metering method that allows protein crystal growth by free interface diffusion. Proc Natl Acad Sci 2002;99:16531–6.
[156] Talreja S, Kim DY, Mirarefi AY, Zukoski CF, Kenis PJ. Screening and optimization of protein crystallization conditions through gradual evaporation using a novel crystallization platform. J Appl Crystallogr 2005;38:988–95.
[157] Li L, Mustafi D, Fu Q, Tereshko V, Chen DL, Tice JD, Ismagilov RF. Nanoliter microfluidic hybrid method for simultaneous screening and optimization validated with crystallization of membrane proteins. Proc Natl Acad Sci 2006;103:19243–8.
[158] Du W, Li L, Nichols KP, Ismagilov RF. SlipChip. Lab Chip 2009;9:2286–92.

[159] Li L, Ismagilov RF. Protein crystallization using microfluidic technologies based on valves, droplets, and SlipChip. Annu Rev Biophys 2010;39:139–58.

[160] Lau BT, Baitz CA, Dong XP, Hansen CL. A complete microfluidic screening platform for rational protein crystallization. J Am Chem Soc 2007;129:454–5.

[161] Zheng B, Gerdts CJ, Ismagilov RF. Using nanoliter plugs in microfluidics to facilitate and understand protein crystallization. Curr Opin Struct Biol 2005;15:548–55.

[162] Hansen CL, Sommer MO, Quake SR. Systematic investigation of protein phase behavior with a microfluidic formulator. Proc Natl Acad Sci 2004;101:14431–6.

[163] Skou M, Skou S, Jensen TG, Vestergaard B, Gillilan RE. In situ microfluidic dialysis for biological small-angle X-ray scattering. J Appl Crystallogr 2014;47:1355–66.

[164] Scrimgeour J, Cho JK, Breedveld V, Curtis J. Microfluidic dialysis cell for characterization of macromolecule interactions. Soft Matter 2011;7:4762–7.

[165] Junius N, Jaho S, Sallaz-Damaz Y, Borel F, Salmon J-B, Budayova-Spano M. A microfluidic device for both on-chip dialysis protein crystallization and in situ X-ray diffraction. Lab Chip 2020;20:296–310.

[166] Li Y, Xuan J, Hu R, Zhang P, Lou X, Yang Y. Microfluidic triple-gradient generator for efficient screening of chemical space. Talanta 2019;204:569–75.

[167] Sui S, Mulichak A, Kulathila R, Mcgee J, Filiatreault D, Saha S, Cohen A, Song J, Hung H, Selway J, Kirby C, Shrestha OK, Weihofen W, Fodor M, Xu M, Chopra R, Perry SL. A capillary-based microfluidic device enables primary high-throughput room-temperature crystallographic screening. J Appl Crystallogr 2021;54:1034–46. https://doi.org/10.1107/s1600576721004155.

[168] Heymann M, Opthalage A, Wierman JL, Akella S, Szebenyi DM, Gruner SM, Fraden S. Room-temperature serial crystallography using a kinetically optimized microfluidic device for protein crystallization and on-chip X-ray diffraction. IUCrJ 2014;1:349–60.

[169] Gicquel Y, Schubert R, Kapis S, Bourenkov G, Schneider T, Perbandt M, Betzel C, Chapman HN, Heymann M. Microfluidic chips for in situ crystal X-ray diffraction and in situ dynamic light scattering for serial crystallography. J Vis Exp 2018.

[170] Perry SL, Guha S, Pawate AS, Henning R, Kosheleva I, Srajer V, Kenis PJ, Ren Z. In situ serial Laue diffraction on a microfluidic crystallization device. J Appl Crystallogr 2014;47:1975–82.

[171] Pinker F, Brun M, Morin P, Deman A-L, Chateaux J-FO, Oliéric V, Stirnimann C, Lorber B, Terrier N, Ferrigno R. ChipX: a novel microfluidic chip for counter-diffusion crystallization of biomolecules and in situ crystal analysis at room temperature. Cryst Growth Des 2013;13:3333–40.

[172] Sui S, Wang Y, Dimitrakopoulos C, Perry SL. A graphene-based microfluidic platform for electrocrystallization and in situ X-ray diffraction. Crystals 2018;8:76.

[173] Maeki M, Pawate AS, Yamashita K, Kawamoto M, Tokeshi M, Kenis PJ, Miyazaki M. A method of cryoprotection for protein crystallography by using a microfluidic chip and its application for in situ X-ray diffraction measurements. Anal Chem 2015;87:4194–200.

[174] Sommer MO, Larsen S. Crystallizing proteins on the basis of their precipitation diagram determined using a microfluidic formulator. J Synchrotron Radiat 2005;12:779–85.

[175] Li L, Fu Q, Kors CA, Stewart L, Nollert P, Laible PD, Ismagilov RF. A plug-based microfluidic system for dispensing lipidic cubic phase (LCP) material validated by crystallizing membrane proteins in lipidic mesophases. Microfluid Nanofluid 2010;8:789–98.

[176] Perry SL, Roberts GW, Tice JD, Gennis RB, Kenis PJ. Microfluidic generation of lipidic mesophases for membrane protein crystallization. Cryst Growth Des 2009;9:2566–9.

[177] Pinker F, Brun M, Morin P, Deman A-L, Chateaux J-F, Oliéric V, Stirnimann C, Lorber B, Terrier N, Ferrigno R, Sauter C. ChipX: a novel microfluidic chip for counter-diffusion crystallization of biomolecules and in situ crystal analysis at room temperature. Cryst Growth Des 2013;13:3333–40. https://doi.org/10.1021/cg301757g.

[178] Garman EF. Radiation damage in macromolecular crystallography: what is it and why should we care? Acta Crystallogr D Biol Crystallogr 2010;66:339–51.
[179] Zhou R-B, Cao H-L, Zhang C-Y, Yin D-C. A review on recent advances for nucleants and nucleation in protein crystallization. CrystEngComm 2017;19:1143–55. https://doi.org/10.1039/c6ce02562e.
[180] Saridakis E, Chayen NE. Towards a 'universal' nucleant for protein crystallization. Trends Biotechnol 2009;27:99–106.
[181] Nanev CN, Saridakis E, Chayen NE. Protein crystal nucleation in pores. Sci Rep 2017;7:35821. https://doi.org/10.1038/srep35821.
[182] Khurshid S, Saridakis E, Govada L, Chayen NE. Porous nucleating agents for protein crystallization. Nat Protoc 2014;9:1621–33.
[183] Shah UV, Williams DR, Heng JY. Selective crystallization of proteins using engineered nanonucleants. Cryst Growth Des 2012;12:1362–9.
[184] Guo Y-Z, Sun L-H, Oberthuer D, Zhang C-Y, Shi J-Y, Di J-L, Zhang B-L, Cao H-L, Liu Y-M, Li J. Utilisation of adsorption and desorption for simultaneously improving protein crystallisation success rate and crystal quality. Sci Rep 2014;4:1–8.
[185] Zhou T, Zhang K, Kamra T, Bülow L, Ye L. Preparation of protein imprinted polymer beads by pickering emulsion polymerization. J Mater Chem B 2015;3:1254–60.
[186] Vasapollo G, Sole RD, Mergola L, Lazzoi MR, Scardino A, Scorrano S, Mele G. Molecularly imprinted polymers: present and future prospective. Int J Mol Sci 2011;12:5908–45.
[187] Saridakis E, Khurshid S, Govada L, Phan Q, Hawkins D, Crichlow GV, Lolis E, Reddy SM, Chayen NE. Protein crystallization facilitated by molecularly imprinted polymers. Proc Natl Acad Sci 2011;108:11081–6.
[188] Saridakis E, Chayen NE. Imprinted polymers assisting protein crystallization. Trends Biotechnol 2013;31:515–20.
[189] Khurshid S, Govada L, El-Sharif HF, Reddy SM, Chayen NE. Automating the application of smart materials for protein crystallization. Acta Crystallogr D Biol Crystallogr 2015;71:534–40.
[190] Xing Y, Hu Y, Jiang L, Gao Z, Chen Z, Chen Z, Ren X. Zwitterion-immobilized imprinted polymers for promoting the crystallization of proteins. Cryst Growth Des 2015;15:4932–7.
[191] Clares MP, Serena C, Blasco S, Nebot A, del Castillo L, Soriano C, Domènech A, Sánchez-Sánchez AV, Soler-Calero L, Mullor JL. Mn (II) complexes of scorpiand-like ligands. A model for the MnSOD active centre with high in vitro and in vivo activity. J Inorg Biochem 2015;143:1–8.
[192] Weis WI, Kahn R, Fourme R, Drickamer K, Hendrickson WA. Structure of the calcium-dependent lectin domain from a rat mannose-binding protein determined by MAD phasing. Science 1991;254:1608–15.
[193] Silvaggi NR, Martin LJ, Schwalbe H, Imperiali B, Allen KN. Double-lanthanide-binding tags for macromolecular crystallographic structure determination. J Am Chem Soc 2007;129:7114–20.
[194] Girard É, Stelter M, Vicat J, Kahn R. A new class of lanthanide complexes to obtain high-phasing-power heavy-atom derivatives for macromolecular crystallography. Acta Crystallogr D Biol Crystallogr 2003;59:1914–22.
[195] Pompidor G, d'Aleo A, Vicat J, Toupet L, Giraud N, Kahn R, Maury O. Protein crystallography through supramolecular interactions between a lanthanide complex and arginine. Angew Chem 2008;120:3436–9.
[196] Stelter M, Molina R, Jeudy S, Kahn R, Abergel C, Hermoso JA. A complement to the modern crystallographer's toolbox: caged gadolinium complexes with versatile binding modes. Acta Crystallogr D Biol Crystallogr 2014;70:1506–16.
[197] Engilberge S, Riobé F, Di Pietro S, Lassalle L, Coquelle N, Arnaud C-A, Pitrat D, Mulatier J-C, Madern D, Breyton C. Crystallophore: a versatile lanthanide complex

for protein crystallography combining nucleating effects, phasing properties, and luminescence. Chem Sci 2017;8:5909–17.
[198] Kumari Yadav R, Krishnan V. The adhesive PitA pilus protein from the early dental plaque colonizer *Streptococcus oralis*: expression, purification, crystallization and X-ray diffraction analysis. Acta Crystallogr F: Struct Biol Commun 2020;76:8–13.
[199] Prieto-Castañeda A, Martínez-Caballero S, Agarrabeitia AR, García-Moreno I, Moya SDL, Ortiz MJ, Hermoso JA. First lanthanide complex for de novo phasing in native protein crystallography at 1 Å radiation. ACS Appl Bio Mater 2021;4:4575–81. https://doi.org/10.1021/acsabm.1c00305.
[200] McPherson A, Shlichta P. Heterogeneous and epitaxial nucleation of protein crystals on mineral surfaces. Science 1988;239:385–7. https://doi.org/10.1126/science.239.4838.385.
[201] Akella SV, Mowitz A, Heymann M, Fraden S. Emulsion-based technique to measure protein crystal nucleation rates of lysozyme. Cryst Growth Des 2014;14:4487–509. https://doi.org/10.1021/cg500562r.
[202] Chayen NE, Saridakis E, El Bahar R, Nemirovsky Y. Porous silicon: a nucleation-inducing material for protein crystallization. Acta Crystallogr Sect A: Found Crystallogr 2002;58:c224. https://doi.org/10.1107/s0108767302093972.
[203] Stewart PS, Mueller-Dieckmann J. Automation in biological crystallization. Acta Crystallogr F Struct Biol Commun 2014;70:686–96. https://doi.org/10.1107/s2053230x14011601.
[204] Zhu Y, Zhu L-N, Guo R, Cui H-J, Ye S, Fang Q. Nanoliter-scale protein crystallization and screening with a microfluidic droplet robot. Sci Rep 2014;4:5046. https://doi.org/10.1038/srep05046.
[205] Dauter Z, Wlodawer A. Progress in protein crystallography. Protein Pept Lett 2016;23:201–10. https://doi.org/10.2174/0929866523666160106153524.
[206] Desbois S, Seabrook SA, Newman J. Some practical guidelines for UV imaging in the protein crystallization laboratory. Acta Crystallogr Sect F: Struct Biol Cryst Commun 2013;69:201–8. https://doi.org/10.1107/S1744309112048634.
[207] Weber P, Pissis C, Navaza R, Mechaly AE, Saul F, Alzari PM, Haouz A. High-throughput crystallization pipeline at the crystallography core facility of the Institut Pasteur. Molecules 2019;24:4451.
[208] Owen RL, Juanhuix J, Fuchs M. Current advances in synchrotron radiation instrumentation for MX experiments. Arch Biochem Biophys 2016;602:21–31. https://doi.org/10.1016/j.abb.2016.03.021.
[209] Yamamoto M, Hirata K, Yamashita K, Hasegawa K, Ueno G, Ago H, Kumasaka T. Protein microcrystallography using synchrotron radiation. IUCrJ 2017;4:529–39. https://doi.org/10.1107/s2052252517008193.
[210] Duke EMH, Johnson LN. Macromolecular crystallography at synchrotron radiation sources: current status and future developments. Proc R Soc A: Math Phys Eng Sci 2010;466:3421–52. https://doi.org/10.1098/rspa.2010.0448.
[211] Cohen AE. A new era of synchrotron-enabled macromolecular crystallography. Nat Methods 2021;18:433–4. https://doi.org/10.1038/s41592-021-01146-y.
[212] Martiel I, Olieric V, Caffrey M, Wang M. Chapter 1: Practical approaches for in situ x-ray crystallography: from high-throughput screening to serial data collection. United Kingdom: Royal Society of Chemistry; 2018.
[213] Maveyraud L, Mourey L. Protein X-ray crystallography and drug discovery. Molecules 2020;25. https://doi.org/10.3390/molecules25051030.
[214] Broecker J, Morizumi T, Ou W-L, Klingel V, Kuo A, Kissick DJ, Ishchenko A, Lee M-Y, Xu S, Makarov O, Cherezov V, Ogata CM, Ernst OP. High-throughput in situ X-ray screening of and data collection from protein crystals at room temperature and under cryogenic conditions. Nat Protoc 2018;13:260–92. https://doi.org/10.1038/nprot.2017.135.

[215] Cipriani F, Röwer M, Landret C, Zander U, Felisaz F, Márquez JA. CrystalDirect: a new method for automated crystal harvesting based on laser-induced photoablation of thin films. Acta Crystallogr D Biol Crystallogr 2012;68:1393–9. https://doi.org/10.1107/s0907444912031459.
[216] Johansson LC, Stauch B, Ishchenko A, Cherezov V. A bright future for serial femtosecond crystallography with XFELs. Trends Biochem Sci 2017;42:749–62. https://doi.org/10.1016/j.tibs.2017.06.007.
[217] Liu H, Lee W. The XFEL protein crystallography: developments and perspectives. Int J Mol Sci 2019;20. https://doi.org/10.3390/ijms20143421.
[218] Martin-Garcia JM. Protein dynamics and time resolved protein crystallography at synchrotron radiation sources: past, present and future. Crystals 2021;11:521.
[219] Nogly P, Weinert T, James D, Carbajo S, Ozerov D, Furrer A, Gashi D, Borin V, Skopintsev P, Jaeger K, Nass K, Båth P, Bosman R, Koglin J, Seaberg M, Lane T, Kekilli D, Brünle S, Tanaka T, Wu W, Milne C, White T, Barty A, Weierstall U, Panneels V, Nango E, Iwata S, Hunter M, Schapiro I, Schertler G, Neutze R, Standfuss J. Retinal isomerization in bacteriorhodopsin captured by a femtosecond x-ray laser. Science 2018;361, eaat0094. https://doi.org/10.1126/science.aat0094.
[220] Jain D, Lamour V. Computational tools in protein crystallography. Methods Mol Biol 2010;673:129–56. https://doi.org/10.1007/978-1-60761-842-3_8.
[221] Powell HR. X-ray data processing. Biosci Rep 2017;37. https://doi.org/10.1042/bsr20170227.
[222] Adams PD, Afonine PV, Bunkóczi G, Chen VB, Davis IW, Echols N, Headd JJ, Hung LW, Kapral GJ, Grosse-Kunstleve RW, McCoy AJ, Moriarty NW, Oeffner R, Read RJ, Richardson DC, Richardson JS, Terwilliger TC, Zwart PH. PHENIX: a comprehensive Python-based system for macromolecular structure solution. Acta Crystallogr D Biol Crystallogr 2010;66:213–21. https://doi.org/10.1107/s0907444909052925.
[223] Pettersen EF, Goddard TD, Huang CC, Couch GS, Greenblatt DM, Meng EC, Ferrin TE. UCSF Chimera—a visualization system for exploratory research and analysis. J Comput Chem 2004;25:1605–12. https://doi.org/10.1002/jcc.20084.
[224] Aggarwal S, von Wachenfeldt C, Fisher SZ, Oksanen E. A protocol for production of perdeuterated OmpF porin for neutron crystallography. Protein Expr Purif 2021;188, 105954.

# CHAPTER 12

# Spectroscopic methods to study protein folding kinetics: Methodology, data analysis, and interpretation of the results

**Akram Shirdel**[a] **and Khosrow Khalifeh**[b,c]
[a]Department of Biochemistry, Faculty of Biological Sciences, Tarbiat Modares University, Tehran, Iran
[b]Department of Biology, Faculty of Sciences, University of Zanjan, Zanjan, Iran
[c]Department of Biotechnology, Research Institute of Modern Biological Techniques, University of Zanjan, Zanjan, Iran

## 1. Introduction

The enzymatic activity of a protein was first observed with the catalytic ability of starch degradation in 1833 [1]; however, its true nature as a protein was reported in 1926 [2]. In 1951, Pauling [3] discovered two different ordered arrangements of amino acids known as the α-helix and the β-sheet. In 1955, the sequence of insulin was determined by Sanger [4], which showed that proteins are made up of a linear string of covalent bonds (peptide bonds) and that the amino acid sequence of peptide chains is unique for individual proteins. In 1958, the first crystallographic three-dimensional (3D) structure of protein myoglobin was resolved by Kendrew et al. [5], showing that its structure was not entirely regular; instead, it consists of helices, sheets, and coils collectively arranged into a well-defined compact tertiary structure. In 1961, Anfinsen et al. [6] observed that ribonuclease A (RNase A) spontaneously folds back into its correct shape upon changing the environmental conditions from unfolded to the folded state. In 1968, Ramachandran and Sasisekharan [7] had intensively discussed the structural aspects relevant to the conformation in the immediate neighborhood of residues in a protein chain. They considered a protein chain as some peptide units linked to one another at α-carbon atoms. They assumed that a variety of allowed conformations could exist for a pair of linked peptide units for the angles of rotation about the two bonds, including $\varphi$ and $\psi$ dihedral angles. After the description of geometrical aspects of the polypeptide chain and using the concept of dihedral angles, Cyrus Levinthal [8] noted that proteins could explore

*Advanced Spectroscopic Methods to Study Biomolecular Structure and Dynamics*
https://doi.org/10.1016/B978-0-323-99127-8.00008-8

through an astronomical number of possible conformational states while folding randomly. Hence, it should take a long time for a protein to fold up into its native structure; however, proteins fold very quickly in a time scale of seconds or less. Levinthal concluded his paradox by suggesting that the random sampling of all possible conformations is not an effective method for a protein to achieve the correct folded state. Accordingly, there should be mechanisms or laws governing protein folding and stability. At the same time, between 1968 and 1970, Tanford [9–11] posed the principles of thermodynamics and kinetics of protein folding. Five years after the publication of the Levinthal paradox, Anfinsen [12] stated that the amino acid sequence of a protein contains all required information to direct an unfolded state of a protein toward its native tertiary structure and that the primary structure of a protein reflects the state of the lowest conformational energy under native conditions. The proposition of the protein folding problem by Levinthal, followed by Tanford's data and Anfinsen's theory, led researchers to a multitude of investigations, including theoretical, computational, and experimental studies to unravel the mechanism of protein folding.

Generally, the theoretical study refers to the statistical analysis of the sequence and structural data deposited into bioinformatics databases. These studies aim to establish empirical rules between the amino acid sequences and the 3D structure of proteins [13–18]. For example, as a preliminary work and 2 years after the publication of Anfinsen's theory, Chou and Fasman statistically analyzed known protein structures that were solved with X-ray crystallography and determined the relative frequencies of all natural amino acids in the context of secondary structural elements. Using these data, they prepared a set of probability parameters for the occurrence of individual amino acids in each secondary structural element. The Chou-Fasman parameter for each amino acid in a given secondary structure represents the probability of finding these residues in that secondary structural element. The Chou-Fasman model was then used in secondary structure prediction algorithms to predict the probability of forming a helix, a strand, or a turn by a given sequence of a protein [13,15]. However, to increase the prediction accuracy, this method was updated to the Garnier-Osguthorpe-Robson (GOR) method [19,20]. Similar to Chou-Fasman's analysis, the frequency distribution of all 400 dipeptides was determined in the main structural classes of proteins and different secondary structural elements [16,17]. It was observed that the majority of dipeptides were randomly distributed in different secondary structures of proteins, whereas some residues in representative dipeptides prefer to select or reject their neighbors according to the conformation of the first and second positions of the dipeptide [17].

To use the physicochemical properties of a given sequence to predict the respective structure, several attempts have been made to determine the propensities of amino acid side chains for the hydrophobic/hydrophilic environment. Therefore, the suitability of amino acids for an aqueous or hydrophobic environment can be used for a quantitative description of the degree to which fragments of a given protein tend to be exposed to a polar or hydrophobic medium [21–29]. The resulting models were used in other predictive algorithms, including the identification of surface-exposed as well as transmembrane regions in the protein structure [21–23], identification of potentially antigenic regions in proteins [24,28], prediction of α-helices in proteins [25], determination of the surface accessibility of amino acids in globular proteins [26,27], and determination of the chain flexibility [29] in proteins.

Since the 1970s, the thermodynamic stability and the kinetics of protein folding have been intensively studied using a variety of experimental methods [30–38]. These studies led to the presentation of different models for protein folding, including nucleation-growth [36,39–43], framework model [36,44–47], diffusion-collision model [48], hydrophobic collapse model [49], and the possibility of sequential or parallel folding pathways [50,51]. Reviews containing a detailed description of protein folding models are available [52–54]. In the early 1990s, the association of rapid mixing techniques and site-directed mutagenesis directed protein engineering studies toward a detailed description of protein folding and stability using the concept of $\varphi$-value analysis. This strategy made it possible to compare the kinetics of refolding and thermodynamic stability of the wild type (WT) and mutants of a given protein to determine the effect of a specific interaction on the protein folding pathway and the stability of the folded state. In a useful series of articles by Fersht et al. [55–59], the assumptions behind this strategy and interpretation of data were provided. Chevron plot analysis was also used to characterize the structural elements in the folding pathway of proteins [60–63]. Different aspects of the protein folding pathway have been reviewed in classic articles [64,65]. In this chapter, we provide an overview of the principle and methodological aspects of protein folding kinetics. This article complements our recent publication on the thermodynamics of protein folding and stability [66].

## 2. Kinetics of protein folding

### 2.1 Principle

The kinetics of a reaction represents the changes in the concentration of the reactants and products over time. In a simple two-state unfolding reaction

$N \rightarrow D$, where a native protein (N) unfolds into the ensemble of denatured states (D), a kinetic study involves the determination of the rate constant of the reaction that reflects the changes of the population of N or D over time according to the following equations:

$$k = \frac{\partial D}{\partial t} \uparrow \quad (1)$$

$$k = \frac{\partial N}{\partial t} \downarrow \quad (2)$$

where $k$ is the unfolding rate constant obtained from the kinetics studies.

## 2.2 Methodology

As noted previously, in thermodynamic studies, we have sufficient time to achieve an equilibrium condition where the structural changes (changes in the population of the unfolded and folded states) can be monitored with conventional spectroscopic measurements [66]. However, to monitor the structural changes of proteins, immediately after initiation of folding/unfolding reaction, measuring the rate of the reactions by simple mixing of reagents in the observation cell of a spectrophotometer would not provide reliable results for certain reasons. First, no signal is produced in fast reactions; second, for reactions whose kinetics takes a time scale of minutes or more, the speed of mixing reagents and the speed of signal observation do not occur in a reasonable time scale. However, thanks to the capabilities of spectroscopic methods attached to the stopped-flow apparatus, it is now possible to monitor the early events of folding reactions by rapid filling of the observation cell with reagents [67–69]. A schematic representation of a simple stopped-flow apparatus is shown in Fig. 1. Stopped flow contains two or more (up to four) syringes that are filled with reagents and can inject defined volumes of reagents into the observation cell under the control of microprocessors. After hitting the syringe's plunger to a mechanical stop, the flow of reagents is stopped, and the reaction is initiated by mixing the reagents in the observation cell, whereas simultaneous monitoring of the changes in the signal is performed by a detector of the spectrophotometer. It is noticeable that the temperature jump experiments are also possible with the stopped-flow instrument to monitor the unfolding of the protein upon a sudden temperature jump from a starting temperature to a final one using an appropriate accessory for the stopped-flow apparatus. However, in the current chapter, we focus our attention on chemical unfolding experiments.

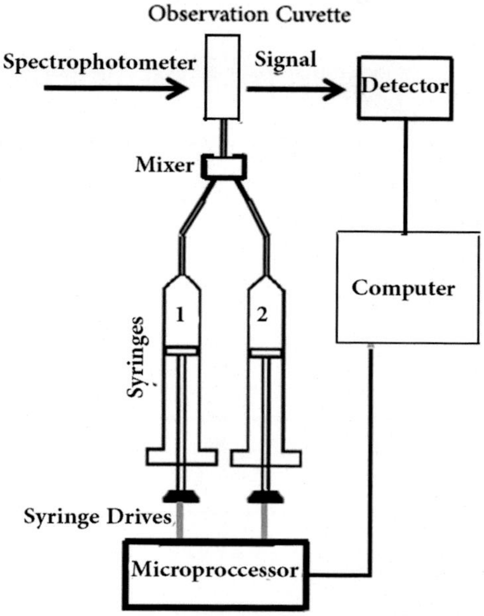


**Fig. 1** A schematic representation of a stopped-flow apparatus. Two Hamilton syringes containing different solutions are pushed by driving systems based on stepping motors. Using preprogrammed sequence driving, the reagents are directed toward the observation cell through the narrow tubes. The mixing of reagents in observation cells occurs under highly turbulent flow conditions, particularly in conditions where solutions have different viscosities. Hence, a mixer is positioned before the observation cell to inject the fully homogeneous mixture into the observation cell.

## 2.3 Data analysis and interpretation

### *2.3.1 Obtaining the experimental rate constant*

Fig. 2 shows the kinetics of the unfolding and refolding of lipase in the absence (A) and presence (B) of proline measured by a BioLogic μ-SFM-20 fluorescence-detected stopped-flow instrument, in which refolding and unfolding processes were monitored by recording the fluorescence signal at 340 nm upon excitation of fluorophores at 290 nm [70].

To gain quantitative insights into the refolding and unfolding curves, data from Fig. 2 should initially be fitted into a sum of exponentials [71] according to the following equation:

$$A(t) = \sum_{i=1}^{n} A_i \exp(\pm k_i t) + at + c \quad (3)$$

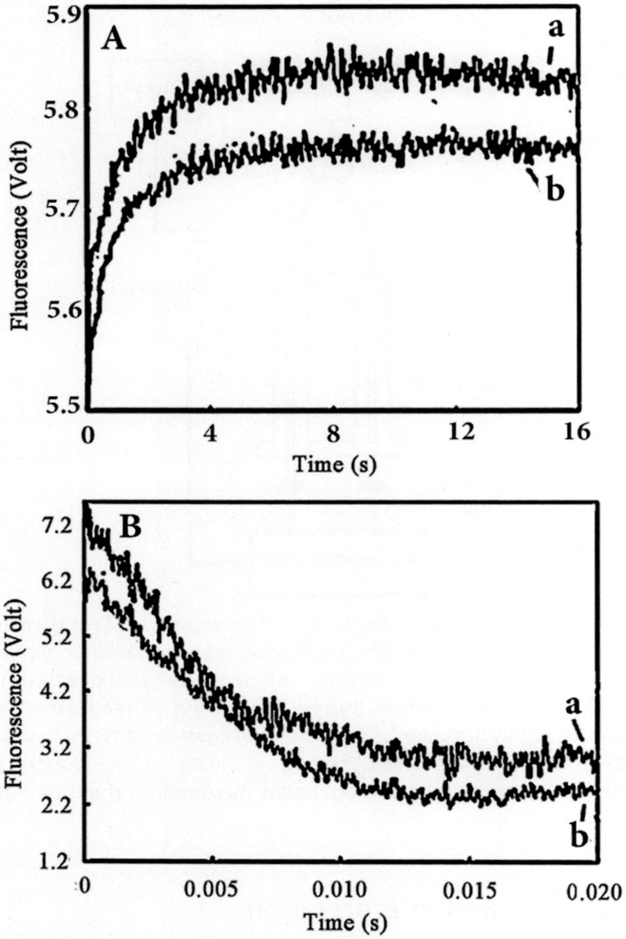


**Fig. 2** Stopped-flow mixing experiment. Typical kinetic traces of the (A) unfolding and (B) refolding reactions of *Pseudomonas fluorescence* lipase in the absence (a) and presence (b) of proline. The unfolding reaction was triggered by mixing 1 volume of protein in 100 mM Tris-HCl buffer, pH 7.5, with 6 volumes of the unfolding buffer (7 M concentration of urea) containing 0 and 4.2 M proline. Therefore, unfolding was initiated at 0 and 0.6 M concentrations of proline. The refolding was performed by a 7-fold dilution of the unfolded protein in 6 M urea into refolding buffers at two concentrations of proline. The reaction was monitored by measuring the fluorescence signal. *(Figure reproduced with permission from Hakiminia F, Ranjbar B, Khalifeh K, Khajeh K. Kinetic and thermodynamic properties of pseudomonas fluorescence lipase upon addition of proline. Int J Biol Macromol 2013;55:123–6.)*

where A(t) is the value of the spectroscopic signal at a given time t, a is the slope, and c is the offset of the kinetic trace corresponding to the baseline, $k_i$ is the rate constant of the *i*th kinetic phase, and $A_i$ is the amplitude of the *i*th phase.

This analysis provides numerical values for the observed rate constants and their corresponding amplitudes. If fitting a one-exponential model to the data does not provide a reliable rate constant, then the multiexponential model should be tested for obtaining more than one rate constant. Theoretically, each kinetic curve can be fitted into Eq. (3) by setting $i=6$, which results in obtaining six rate constants and the respective amplitudes; however, more than three kinetic phases are biologically meaningless and are just mathematical manipulation. As an example, unfolding kinetic traces of Fig. 2 were fitted into a double-exponential function providing slow and fast rate constants, demonstrating that the unfolded proteins are partitioned into two sets of the population which refold from two different pathways, whereas the refolding kinetic curves were fitted into a single-exponential function.

### 2.3.2 Transition state

The fitting of kinetics data into single-exponential or multiexponential functions is the starting point for mechanistic investigation of protein folding pathways. Similar to thermodynamic studies [66], the kinetic experiments of the refolding and unfolding should be performed at a range of denaturant concentrations, and the experimental evidence should be extrapolated to zero denaturant concentration to predict the properties of the kinetics elements under conditions where no denaturant is present.

To examine the kinetics of the refolding and unfolding reactions, we first consider a simple reversible two-state process in which the rate of the reaction is described in terms of a single rate constant for the unfolding ($k_D$) and folding ($k_N$) reactions (Fig. 3). From a thermodynamic point of view, only the population of the denatured and native states determines the equilibrium constant and the Gibbs free energy of the reaction under equilibrium conditions. In contrast, from a kinetic point of view, the rate of interconversion of N and D is determined by the characteristics of the high-energy activated structure, namely, the transition state, in addition to the properties of the native and denatured states [11].

It is worth mentioning that the transition states for protein molecules differ from those in small organic molecules, as in simple chemical reactions,

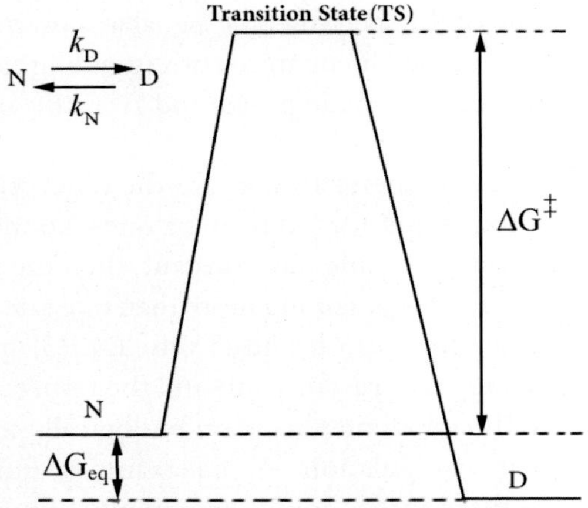


**Fig. 3** Schematic representation of the energy profile of a simple two-state unfolding reaction of a given protein. The energy barrier of the reaction ($\Delta G^{\ddagger}$) is calculated from the observed rate constant ($k_{obs}$) using Eq. (4). The energy difference between the native and denatured states is calculated by the observed equilibrium constant (**K**) using the thermodynamic equation $\Delta G = -RT \ln \mathbf{K}$.

the transition state involves covalent bond alterations that lead to large energy changes. However, in reactions involving protein molecules, low-energy noncovalent bonds are made and broken so that the energy levels are small compared with $k_B T$ [57]. Therefore, in protein folding studies, the transition state determines the pathway of the reaction, where the type and pattern of specific interactions in the structure of the transition state may be locally changed upon changing the experimental conditions.

The Eyring equation can then be used for the conversion of the rate constant into the energy of the transition state [72]:

$$k = \frac{k_B T}{h} \exp\left(\frac{-\Delta G^{\ddagger}}{RT}\right) \tag{4}$$

### 2.3.3 Modeling of the folding/unfolding pathway

In analogy with the treatment of equilibrium constants [66], the logarithm of the rate constant for the refolding and unfolding of small proteins in urea ($k$) is proportional to the concentration of urea [11,73,74]:

$$\log k_{obs} = \log k^{H_2O} + m[\text{urea}] \tag{5}$$

where $k_{obs}$ is the experimental rate constant at a given urea concentration, obtained by fitting the experimental refolding/unfolding curve into Eq. (3).

$k^{H_2O}$ is the theoretical rate constant at zero concentration of the denaturant, obtained by extrapolating the linear line to zero concentration of the denaturant.

This situation is observed under conditions where the kinetics obeys the two-state folding, and no detectable intermediate states are present during protein unfolding or refolding. The linear dependence of the rate constants of refolding and unfolding reactions to the denaturant concentration based on Eq. (5) is shown in Fig. 4A and B, respectively. As shown in the figure, both refolding and unfolding kinetics data can be combined into a single V-shape curve known as a chevron plot (Fig. 4C). The midpoint of the chevron plot, where the rate of refolding and unfolding reaction is equal, reflects the equilibrium constant of the reaction. The entire process of

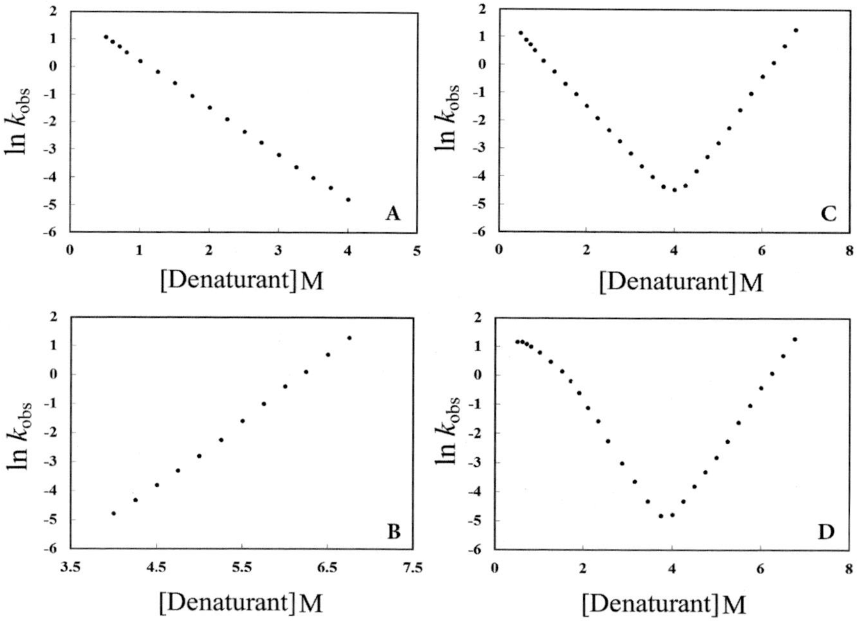


**Fig. 4** Linear dependence of the logarithm of the rate constants of the (A) folding and (B) unfolding to denaturant concentration. (C and D) Chevron plots as the logarithm of the rate constants of the folding and unfolding reactions. Both refolding and unfolding arms of the chevron plot in (C) have a linear dependence on the denaturant concentration, whereas in the refolding arm of the chevron plot in (D), there is an intermediate state in the low concentration of the denaturant that is less sensitive to urea. Each data point is derived from fitting a separate refolding/unfolding reaction performed by the stopped-flow apparatus.

refolding/unfolding, as shown in the chevron plot (Fig. 4C), can be analyzed with a nonlinear least-square fitting procedure. In the case of a two-state reversible mechanism, the chevron plot can be fitted into Eq. (6) [75]:

$$\log k_{obs} = \log\left\{\left(k_N^{H_2O}\exp\left(\frac{m_N[\text{urea}]}{RT}\right) + k_D^{H_2O}\exp\left(\frac{m_D[\text{urea}]}{RT}\right)\right\}\right. \tag{6}$$

where $k_N^{H_2O}$ and $k_D^{H_2O}$ are the rate constants for refolding and unfolding reactions at 0 M denaturant concentration, respectively. The dependence of $\log(k_N^{H_2O})$ and $\log(k_D^{H_2O})$ on [denaturant] is given by $m_N$ and $m_D$, respectively. The relative values of $m_N$ and $m_D$ provide an estimate of how much surface area is buried in the transition state structure during the folding reaction [11]. The ratio $\frac{m_N}{m_N + m_D}$, which is referred to $\theta_m$, indicates the surface area buried in the structure of the transition state upon folding [75].

It is important to note that some representative small proteins fold by a two-state mechanism, whereas there are other mechanisms involving the presence of intermediates in the folding pathway of proteins [43,76–80]. Generally, intermediate states are trapped in local energy minima; however, they are less stable than folded and unfolded structures in the protein folding process. On the other hand, transition states are located at energy maxima and have a very short half-life. According to the degree of stability, thermodynamic intermediates are those structures that are trapped under particular conditions and can be detected directly by conventional spectroscopic techniques. However, kinetic intermediates are short-lived, unstable structures that appear during the protein folding process and can be predicted by modeling of the experimental evidence. Kinetic intermediates can also be classified as on-pathway (Reaction i) and off-pathway (Reaction ii) intermediates [65,81].

$$N \underset{K_{IN}}{\overset{k_{NI}}{\rightleftarrows}} I \underset{k_{DI}}{\overset{k_{ID}}{\rightleftarrows}} D \tag{i}$$

$$N \underset{K_{DN}}{\overset{k_{ND}}{\rightleftarrows}} D \underset{k_{ID}}{\overset{k_{DI}}{\rightleftarrows}} I \tag{ii}$$

In complicated reactions, more than one kinetic intermediate is observed. Here, we focus on the presence of one on-pathway intermediate, which can be predicted by modeling of experimental evidence as discussed below.

As mentioned above, the presence of rollover in the refolding/unfolding arm of the chevron plot demonstrates that an intermediate is formed in the protein folding process. For instance, as shown in Fig. 4D, an apparent deviation from the linear dependence to denaturant concentration is observed at the low concentration of the denaturant, indicating the presence of a molecular species in the refolding pathway that is less sensitive to the denaturant when compared with the native state. In other words, according to Reaction (i) and Fig. 5, it can be assumed that the formation of the intermediate state (I) is essential for the creation of the native state, and the rate of formation of the I from D ($k_{DI}$) occurs much more quickly than the formation of N from I ($K_{IN}$). This leads to the accumulation of the intermediate state because of a rapid equilibrium between the D and I, so that the equilibrium constant can be defined as $\mathbf{k}_{DI} = \frac{[I]}{[D]}$ for this steady-state condition. Considering these assumptions, the chevron plot of Fig. 4D can be fitted to Eq. (7) [60,61,79]:

$$\log k_{obs} = \log\left(\left(\frac{\mathbf{k}_{DI}^{H_2O}\exp\left(-\frac{m_{DI}[urea]}{RT}\right)}{\left(1+\mathbf{k}_{DI}^{H_2O}\exp\left(-\frac{m_{DI}[urea]}{RT}\right)\right)}\right)\right) \times k_{IN}^{H_2O}$$

$$\exp\left(-\frac{m_{IN}^{\ddagger}[urea]}{RT}\right) + \left(k_{NI}^{H_2O}\exp\left(\frac{m_{NI}^{\ddagger}[urea]}{RT}\right)\right)\Bigg)\Bigg) \quad (7)$$

where $\mathbf{k}_{DI}^{H_2O}$ is the equilibrium constant between the unfolded and intermediate states at zero denaturant concentration. $k_{ij}$ is the rate constant of conversion of $i$ to $j$ extrapolated to the absence of the denaturant, and $m_{ij}$ is the difference in exposed surface area between states $i$ and $j$.

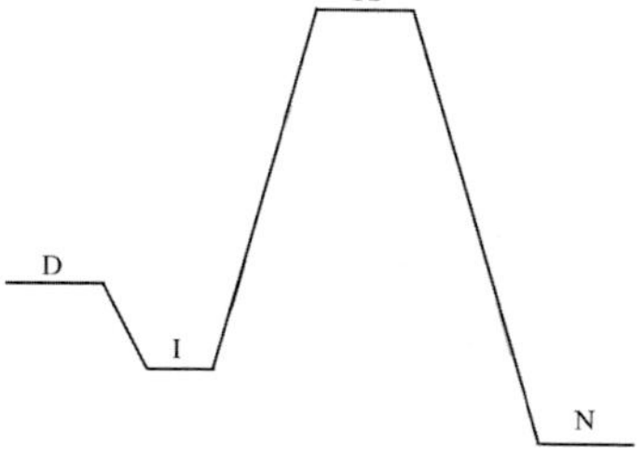


**Fig. 5** Free-energy profile for a protein folding based on Reaction (i) via a major intermediate (*I*) and a transition state (TS). The energy level of N (the right-hand side of the figure) is calculated from equilibrium data, and the energy levels of I and TS are calculated from kinetic studies.

The free energy of the transition state between I and N ($\Delta G^{\ddagger-I}$) can be calculated by the Eyring equation (Eq. 4) using the numerical value of $K_{IN}$.

$$k_{IN} = \frac{k_B T}{h} \exp\left(\frac{-\Delta G^{\ddagger-I}}{RT}\right) \tag{8}$$

Regarding the effect of all rate constants on the accumulation of the I state at the steady-state condition and considering the conversion of I to N as the rate-limiting step, the free energy of the intermediate state as a reflection of its population is determined by the following familiar thermodynamic equation:

$$\Delta G_{DI} = -RT \ln \mathbf{K}_{DI} \tag{9}$$

Finally, calculated parameters should be normalized by considering the unfolded state as the reference (Fig. 5) [74].

## 3. Applications in protein engineering

Protein engineering strategies make it possible to rationally design and construct a variety of mutants, each chosen to manipulate a specific interaction of a particular residue [59,82,83]. Protein folding studies on WT protein and its mutants can provide useful information to investigate the role of any particular residue and its corresponding interactions on the structure of transition and intermediate states, as well as the stability of the folded state. Using thermodynamics and kinetics data of WT and its mutant, as depicted in Fig. 5, the difference energy diagram between the WT and mutants can be plotted. Fig. 6 shows the free energy diagram for the structural elements on the folding pathway of different mutants of barnase [84]. Comparing the free-energy profile of a given protein and its mutants makes it possible to graphically represent the role of a specific residue on a particular position of a protein on the structure of the transition state as well as the stability of the intermediate and folded states of the protein. In another strategy known as $\varphi$-value analysis, protein engineering is used for rational site-directed mutagenesis, followed by kinetics and thermodynamic studies on the folding/unfolding reactions of WT and mutants of a given protein [55,84–86].

Based on the data in Fig. 7, the parameter $\varphi$ for the unfolding reaction is defined as the following relationship [57]:

$$\varphi = \frac{\Delta G^{\ddagger} - \Delta G_N}{\Delta G_D - \Delta G_N} \tag{10}$$

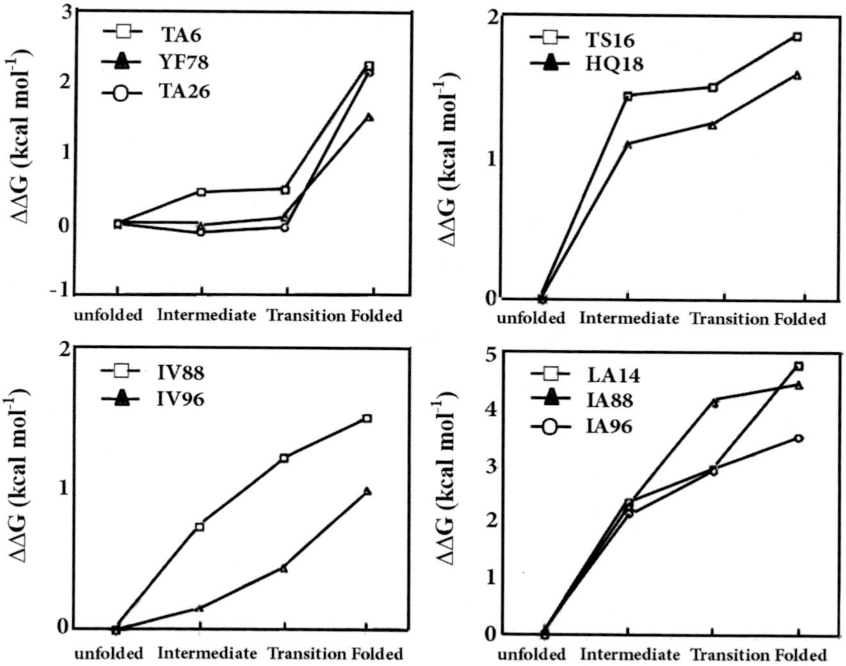


**Fig. 6** Difference energy diagram. The energy of structural elements in the folding pathway of barnase and some representative mutants was measured, and differences in energies between individual mutants and the WT protein were calculated according to Fig. 5. *(Figure reproduced with permission from Matouschek A, Kellis JT, Serrano L, Bycroft M, Fersht AR. Transient folding intermediates characterized by protein engineering. Nature 1990;346:440–5.)*

where $\Delta G^{\ddagger}$, $\Delta G_N$, and $\Delta G_D$ are the differences in free energy of the transitions and folded and unfolded states, respectively, between the WT and mutant protein. It should be noted that the values of free energy in zero denaturant concentration as standard states are included in Eq. (10).

Similar values for $\Delta G^{\ddagger}$ and $\Delta G_D$ ($\varphi = 1$) demonstrate that the site of mutation is exposed to the solvent in the transition state to the same extent as in the unfolded state for both WT and the mutant [85]. In other words, in this condition, the region of the protein containing the mutation is as unfolded in the transition state structure as in the unfolded state for both WT and the mutant [55]. On the other hand, in situations where the values of $\Delta G^{\ddagger}$ and $\Delta G_N$ are the same ($\varphi = 0$), it can be said that the position of protein containing the mutation is as completely folded in the structure of the transition state as it is in the folded state for both WT and the mutant,

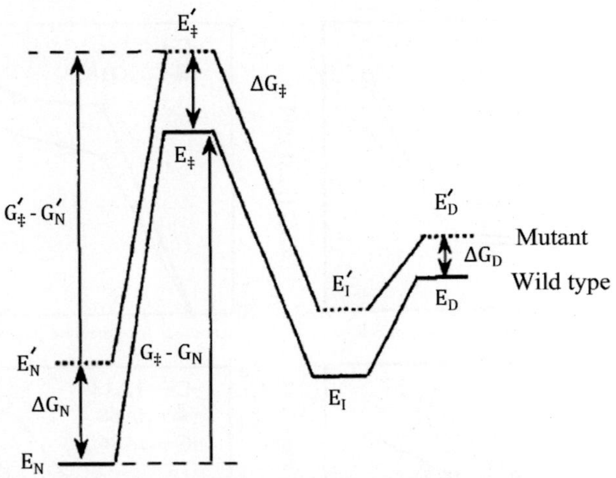


**Fig. 7** $\varphi$-value analysis. Free-energy profiles for the folding of WT and the mutant of a given protein according to Reaction (i), and the assumption of the presence of one on-pathway intermediate is used for $\varphi$-value analysis. Comparing the values of $\Delta G^{\ddagger}$, $\Delta G_N$, and $\Delta G_D$ according to Eq. (8) allows determining whether interactions of a given residue in the structure of the transition state are the same as those in folded or unfolded states during protein unfolding. *(Figure reproduced with permission from Fersht AR, Matouschek A, Serrano L. The folding of an enzyme. I. Theory of protein engineering analysis of stability and pathway of protein folding. J Mol Biol 1991;224:771–82.)*

whereas other regions are unaffected by the mutation. There could be an alternative condition when the residue targeted for mutation makes a new interaction in the transition state containing the same energy as in the folded state [55].

## 4. Conclusions

Despite passing more than a half-century after the proposition of the protein folding problem by Levinthal and the availability of a plethora of research articles, this field is still a challenging area that has attracted the attention of researchers from various disciplines [87,88]. This review as a comprehensive chapter yields several key methods accompanied by manipulating experimental shreds of evidence in protein folding studies and will be especially useful for researchers in different backgrounds who are interested in works with proteins. Two main research trends represent massive opportunities and are needed to be considered in this field. The first one includes updating the current in vitro models on protein folding. The second one is extending the models into physiological conditions addressing how to promote the native folding and prevent the misfolding and aggregation processes.

## References

[1] Payen FW, Persoz JF. Memoir on diastase, the principal products of its reactions, and their applications to the industrial arts. Ann Chim Phys 1833;53:73–92.
[2] Sumner JB. Enzyme urease. J Biol Chem 1926;69:435–41.
[3] Pauling L, Corey R. Atomic coordinates and structure factors for two helical configurations of polypeptide chains. Proc Natl Acad Sci U S A 1951;37:235–40.
[4] Sanger F, Thompson EO, Kitai R. The amide groups of insulin. Biochem J 1955;59:509–18.
[5] Kendrew JC, Bodo G, Dintzis HM, Parrish RG, Wyckoff H, Phillips DC. A three-dimensional model of the myoglobin molecule obtained by X-ray analysis. Nature 1958;181:662–6.
[6] Anfinsen CB, Haber E, Sela M, White FH. The kinetics of formation of native ribonuclease during oxidation of the reduced polypeptide chain. Proc Natl Acad Sci U S A 1961;47:1309–14.
[7] Ramachandran GN, Sasisekharan V. Conformation of polypeptides and proteins. Adv Protein Chem 1968;23:283–437.
[8] Levinthal C. Are there pathways for protein folding? J Chim Phys 1968;65:44–5.
[9] Tanford C. Protein denaturation. Part A. Characterization of the denatured state. Adv Protein Chem 1968;23:121–217.
[10] Tanford C. Protein denaturation. Part B. The transition from native to denatured state. Adv Protein Chem 1968;23:218–75.
[11] Tanford C. Protein denaturation. Part C. Theoretical models for the mechanism of denaturation. Adv Protein Chem 1970;24:1–95.
[12] Anfinsen CB. Principles that govern the folding of protein chains. Science 1973;181:223–30.
[13] Chou PY, Fasman GD. Prediction of protein conformation. Biochemistry 1974;13:222–45.
[14] Go N. Theoretical studies of protein folding. Ann Rev Biophys Bioeng 1983;12:183–210.
[15] Chou PY, Fasman GD. Corrections: conformational parameters for amino acids in helical, β-sheet, and random coil regions calculated from proteins (Biochemistry (1974) 13(2) (211–222) (10.1021/bi00699a001)). Biochemistry 1975;14:196.
[16] Ghadimi M, Heshmati E, Khalifeh K. Distribution of dipeptides in different protein structural classes: an effort to find new similarities. Eur Biophys J 2018;47:31–8.
[17] Ghadimi M, Khalifeh K, Heshmati E. Neighbor effect and local conformation in protein structures. Amino Acids 2017;49:1641–6.
[18] Mirny L, Shakhnovich E. Evolutionary conservation of the folding nucleus. J Mol Biol 2001;308:123–9.
[19] Robson B, Pain RH. Analysis of the code relating sequence to conformation in globular proteins: the distribution of residue pairs in turns and kinks in the backbone chain. Biochem J 1974;141:899–904.
[20] Garnier J, Osguthorpe DJ, Robson B. Analysis of the accuracy and implications of simple methods for predicting the secondary structure of globular proteins. J Mol Biol 1978;120:97–120.

[21] Kyte J, Doolittle RF. A simple method for displaying the hydropathic character of a protein. J Mol Biol 1982;157:105–32.
[22] Engelman D. Identifying nonpolar transbilayer helices in amino acid sequences of membrane proteins. Annu Rev Biophys Biomol Struct 1986;15:321–53.
[23] Eisenberg D, Schwarz E, Komaromy M, Wall R. Analysis of membrane and surface protein sequences with the hydrophobic moment plot. J Mol Biol 1984;179:125–42.
[24] Hopp TP, Woods KR. Prediction of protein antigenic determinants from amino acid sequences. Immunology 1981;78:3824–8.
[25] Cornette JL, Cease KB, Margalit H, Spouge JL, Berzofsky JA, DeLisi C. Hydrophobicity scales and computational techniques for detecting amphipathic structures in proteins. J Mol Biol 1987;195:659–85.
[26] Rose GD, Geselowitz AR, Lesser GJ, Lee RH, Zehfus H. Hydrophobicity of amino acid residues in globular proteins. Science 2016;229:834–8.
[27] Janin J. Surface and inside volumes in globular proteins. Nature 1979;277:491–2.
[28] Kolaskar AS, Tongaonkar PC. A semi-empirical method for prediction of antigenic determinants on protein antigens. FEBS Lett 1990;276:172–4.
[29] Karplus PA, Schulz GE. Prediction of chain flexibility in proteins. Naturwissenschaften 1985;72:212–3.
[30] Pace CN, Grimsley GR, Thomson JA, Barnett BJ. Conformational stability and activity of ribonuclease T1 with zero, one, and two intact disulfide bonds. J Biol Chem 1988;263:11820–5.
[31] Pace CN, Laurents DV, Thomson JA. pH dependence of the urea and guanidine hydrochloride denaturation of ribonuclease A and ribonuclease Tl. Biochemistry 1990;29:2564–72.
[32] Privalov PL, Gill SJ. Stability of protein structure and hydrophobic interaction. Adv Protein Chem 1988;39:191–234.
[33] Santoro MM, Bolen DW. Unfolding free-energy changes determined by the linear extrapolation method. 1. Unfolding of phenylmethanesylfonyl alpha-chymotrypsin using different denaturants. Biochemistry 1988;27:8063–8.
[34] Ptitsyn OB. Protein folding: hypotheses and experiments. J Protein Chem 1987;6:273–93.
[35] Pace CN, Hebert EJ, Shaw KL, Schell D, Both V, Krajcikova D, Sevcik J, Wilson KS, Dauter Z, Hartley RW, Grimsley GR. Conformational stability and thermodynamics of folding of ribonucleases Sa, Sa2 and Sa3. J Mol Biol 1998;279:271–86.
[36] Ptitsyn OB. Structures of folding intermediates. Curr Opin Struct Biol 1995;5:74–8.
[37] Fersht AR, Bycroft M, Horovitz A, Kellis Jr JT, Matouschek A, Serrano L. Physical-organic molecular biology: pathway and stability of protein folding. Philos Trans R Soc L B Biol Sci 1991;332:171–6.
[38] Veitshans T, Klimov DK, Thirumalai D. Protein folding kinetics: time scales, pathways, and energy landscapes in terms of sequence dependent properties. Fold Des 1996;2:1–22.
[39] Wetlaufer DB. Nucleation, rapid folding, and globular intrachain regions in proteins. Proc Natl Acad Sci U S A 1973;70:697–701.
[40] Kim P, Baldwin R. Specific intermediates in the folding reactions of small proteins and the mechanism of protein folding pathways. Annu Rev Biochem 1982;51:459–89.
[41] Vallée-Bélisle A, Michnick SW. Multiple tryptophan probes reveal that ubiquitin folds via a late misfolded intermediate. J Mol Biol 2007;374:791–805.
[42] Kuwajima K. Protein folding: in vitro. Curr Opin Biotechnol 1992;3:462–7.
[43] Matagne A, Chung EW, Ball LJ, Radford SE, Robinson CV, Dobson CM. The origin of the α-domain intermediate in the folding of hen lysozyme. J Mol Biol 1998;277:997–1005.

[44] Udgaonkar JB, Baldwin RL. NMR evidence for an early framework intermediate on the folding pathway of ribonuclease A. Nature 1988;335:700–4.
[45] Ptitsyn OB, Rashin AA. A model of myoglobin self-organization. Biophys Chem 1975;3:1–20.
[46] Kim S P, Baldwin L R. Intermediates in the folding reactions of small proteins. Annu Rev Biochem 1990;59:631–60.
[47] Brown JE, Klee WA. Helix-coil transition of the isolated amino terminus of ribonuclease. Biochemistry 1971;10:470–6.
[48] Karplus M, Weaver DL. Protein folding dynamics: the diffusion-collision model and experimental data. Protein Sci 1994;3:650–68.
[49] Baldwin RL. How does protein folding get started? Trends Biochem Sci 1989;14:291–4.
[50] Wallace LA, Matthews CR. Sequential vs. parallel protein-folding mechanisms: experimental tests for complex folding reactions. Biophys Chem 2002;101–102:113–31.
[51] Aksel T, Barrick D. Direct observation of parallel folding pathways revealed using a symmetric repeat protein system. Biophys J 2014;107:220–32.
[52] Daggett V, Fersht AR. Is there a unifying mechanism for protein folding? Trends Biochem Sci 2003;28:18–25.
[53] Mirny L, Shakhnovich E. Protein folding theory: from lattice to all-atom models. Annu Rev Biophys Biomol Struct 2001;30:361–96.
[54] Jackson SE. How do small single-domain proteins fold? Fold Des 1998;3:81–91.
[55] Fersht AR, Matouschek A, Serrano L. The folding of an enzyme. I. Theory of protein engineering analysis of stability and pathway of protein folding. J Mol Biol 1991;224:771–82.
[56] Serrano L, Kellis JT, Cann P, Matouschek A, Fersht AR. The folding of an enzyme. II. Substructure of barnase and the contribution of different interactions to protein stability. J Mol Biol 1992;224:783–804.
[57] Serrano L, Matouschek A, Fersht AR. The folding of an enzyme. III. Structure of the transition state for unfolding of barnase analysed by a protein engineering procedure. J Mol Biol 1992;224:805–18.
[58] Fersht AR. Protein folding and stability: the pathway of folding of barnase. FEBS Lett 1993;325:5–16.
[59] Fersht AR, Serrano L. Principles of protein stability derived from protein engineering experiments. Curr Opin Struct Biol 1993;3:75–83.
[60] Zarrine-Afsar A, Davidson AR. The analysis of protein folding kinetic data produced in protein engineering experiments. Methods 2004;34:41–50.
[61] Hakiminia F, Khalifeh K, Sajedi RH, Ranjbar B. Determination of structural elements on the folding reaction of mnemiopsin by spectroscopic techniques. Spectrochim Acta A Mol Biomol Spectrosc 2016;158:49–55.
[62] Inaba K, Kobayashi N, Fersht AR. Conversion of two-state to multi-state folding kinetics on fusion of two protein foldons. J Mol Biol 2000;302:219–33.
[63] Batey S, Scott KA, Clarke J. Complex folding kinetics of a multidomain protein. Biophys J 2006;90:2120–30.
[64] Englander SW, Mayne L. The nature of protein folding pathways. Proc Natl Acad Sci U S A 2014;111:15873–80.
[65] Baldwin RL. On-pathway versus off-pathway folding intermediates. Fold Des 1996;1:1–8.
[66] Shirdel SA, Khalifeh K. Thermodynamics of protein folding: methodology, data analysis and interpretation of data. Eur Biophys J 2019;48:305–16.
[67] Gomez-Hens A, Perez-Bendito D. The stopped-flow technique in analytical chemistry. Anal Chim Acta 1991;242:147–77.

[68] Biro FN, Zhai J, Doucette CW, Hingorani MM. Application of stopped-flow kinetics methods to investigate the mechanism of action of a DNA repair protein. J Vis Exp 2010;2–8.
[69] Hilczer M, Barzykin AV, Tachiya M. Theory of the stopped-flow method for studying micelle exchange kinetics. Langmuir 2001;17:4196–201.
[70] Hakiminia F, Ranjbar B, Khalifeh K, Khajeh K. Kinetic and thermodynamic properties of pseudomonas fluorescence lipase upon addition of proline. Int J Biol Macromol 2013;55:123–6.
[71] Takano K, Higashi R, Okada J, Mukaiyama A, Tadokoro T, Koga Y, Kanaya S. Proline effect on the thermostability and slow unfolding of a hyperthermophilic protein. J Biochem 2009;145:79–85.
[72] Eyring H. The activated complex and the absolute rate of chemical reactions. Chem Rev 1935;17:65–77.
[73] Matouschek A, Kellis JT, Serrano L, Fersht AR, Kellis Jr JT, Serrano L, Fersht AR. Mapping the transition state and pathway of protein folding by protein engineering. Nature 1989;340:122–6.
[74] Fersht AR, Matouschek A, Bycroft M, Kellis Jr JT, Serrano L. Physical-organic molecular biology: pathway and stability of protein folding. Philos Trans R Soc L B Biol Sci 1991;63:187–94.
[75] Kuhlman B, Luisi DL, Evans PA, Raleigh DP. Global analysis of the effects of temperature and denaturant on the folding and unfolding kinetics of the N-terminal domain of the protein L9. J Mol Biol 1998;284:1661–70.
[76] Dobson CM, Evans PA, Radford SE. Understanding how proteins fold: the lysozyme story so far. Trends Biochem Sci 1994;19:31–7.
[77] Baldwin RL. Pulsed H/D-exchange studies of folding intermediates. Curr Opin Struct Biol 1993;3:84–91.
[78] Kiefhaber T, Labhardt AM, Baldwin RL. Direct NMR evidence for an intermediate preceding the rate-limiting step in the unfolding of ribonuclease A. Nature 1995;375:513–5.
[79] Khorasanizadeh S, Peters LD, Roder H. Evidence for a three-state model of protein folding from kinetic analysis of ubiquitin variants with altered core residues. Nat Struct Biol 1996;3:193–205.
[80] Roder H, Colón W. Kinetic role of early intermediates in protein folding. Curr Opin Struct Biol 1997;7:15–28.
[81] Matthews CR. Pathways of protein folding. Annu Rev Biochem 1993;62:653–83.
[82] Porebski BT, Buckle AM. Consensus protein design. Protein Eng Des Sel 2016;29:245–51.
[83] Coombs GS, Corey DR. In: Angeletti RH, editor. Proteins analysis and design. London: Academic Press; 1998.
[84] Matouschek A, Kellis JT, Serrano L, Bycroft M, Fersht AR. Transient folding intermediates characterized by protein engineering. Nature 1990;346:440–5.
[85] Matouschek A, Kellis JT, Serrano L, Fersht AR. Mapping the transition state and pathway of protein folding by protein engineering. Nature 1989;340:122–6.
[86] Fersht AR, Sato S. phi-Value analysis and the nature of protein-folding transition states. Proc Natl Acad Sci U S A 2004;101:7976–81.
[87] Singh D, Tripathi T, editors. Frontiers in protein structure, function, and dynamics. 1st. Singapore: Springer; 2020. p. 1–458.
[88] Tripathi T, Dubey V, editors. Advances in protein molecular and structural biology methods. USA: Academic Press; 2022. p. 1–714.

CHAPTER 13

# Spectroscopic methods to study the thermodynamics of biomolecular interactions

**Bharti and Maya S. Nair**
Department of Biosciences and Bioengineering, Indian Institute of Technology Roorkee, Roorkee, Uttarakhand, India

## 1. Introduction

Biomolecules are the elementary components of biological systems involved in fundamental processes such as cell division, growth, morphogenesis, and development. They are categorized into macromolecules such as proteins, carbohydrates, lipids, nucleic acids, small molecules such as primary and secondary metabolites, and plant-derived natural products [1]. The interaction among biomolecules is essential for all the physical and metabolic processes involved in the living system. Understanding the biomolecular interactions and their complexities is the main focus of current research to gain more insight into cell growth and development, disease and its progression, drug targeting, and delivery systems. Binding reactions involve complex energetics involving different energy profiles from a free state to a complex form. A broad analysis of the equilibrium binding of a macromolecule to another molecule can give the following information: (i) model-dependent intrinsic binding constants and model-independent macroscopic binding constants, (ii) the equilibrium binding parameters dependent on the solution variables such as pH, concentration, buffer, or temperature, (iii) the stoichiometry of the interacting macromolecules, and (iv) the presence and extent of cooperativity [2]. A detailed insight into the biomolecular interactions provides structural, molecular, thermodynamic, and kinetic properties. Several structural and molecular techniques are used to probe the conformational dynamics of biomolecules and their interactions. These techniques cover various biophysical and biochemical studies, X-ray crystallography, and nuclear magnetic resonance (NMR) studies assisted with bioinformatics tools (Fig. 1). Thermodynamic studies play a pivotal role in elucidating the molecular forces driving the interactions and binding phenomenon, with

*Advanced Spectroscopic Methods to Study Biomolecular Structure and Dynamics*
https://doi.org/10.1016/B978-0-323-99127-8.00001-5

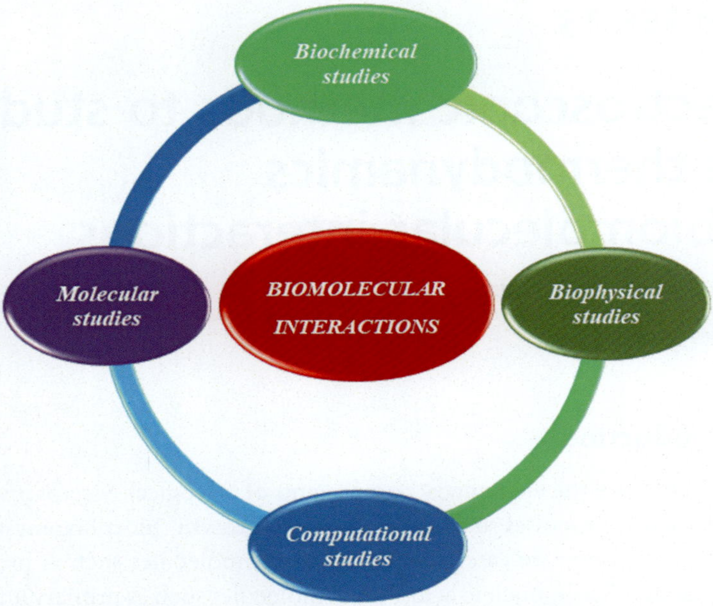


**Fig. 1** Schematic representation of different approaches in the studies of biomolecular interactions.

structure-activity relationship. Calorimetric techniques such as differential scanning calorimetry (DSC) and isothermal calorimetry (ITC) are universal techniques and are widely used to study the thermodynamics of biomolecular interactions. Different spectroscopic techniques are vastly utilized in thermodynamic studies. Their ease of use, rapidness, and low sample requirements make them suitable for studying many molecular systems. Spectroscopic methods provide the advantage of labeling and tracking specific binding molecules in a crowded mixture. At the same time, in DSC/ITC, it is not possible to distinguish specific binding molecules within the crowded environment [3].

This chapter focuses on (i) the forces involved in biomolecular interactions, (ii) how characterizing thermodynamic parameters involved in interactions provides information about the nature of interactions, and (iii) methods that experimentally determine the thermodynamics of interactions. We present various spectroscopic techniques that provide thermodynamic information involved in the interaction and examples of their applications.

## 2. Overview of biomolecular forces

Biological processes are associated with a series of biomolecular interactions, generally followed by conformational changes in the molecules. These forces are measured by considering the macroscopic parameters such as viscosity, pressure, volume, and temperature data. A summary of these interactions is described below.

### 2.1 Hydrogen bonding

Hydrogen bonding is a weak and noncovalent interaction. It is a type of dipole–dipole bond that describes the attraction between a hydrogen atom and an electronegative element such as nitrogen, oxygen, or fluorine [4,5]. The donor molecule is the one that provides hydrogen, whereas the acceptor molecule is the one that has a lone pair that participates in hydrogen bonding. Hydrogen bonding between complementary base pairing in the DNA double helix provides excellent stability to the structure, whereas in protein, the interaction between groups such as —NH, —C=O, and —OH is important in predicting regular helical or sheet conformation. Hydrogen bonds are relatively weak in biological systems, with strengths ranging from 5 to 30 kJ/mol [6]. The energy required to break a hydrogen bond between two groups in vacuum is in the range of 25 kJ/mol for a peptide hydrogen bond [5,7]. However, in water, such exposed groups are more prone to create new hydrogen bonds with nearby water molecules, canceling the impact. Thus, the de facto hydrogen bond strength between groups in an aqueous environment may be closer to zero, and the entire interaction will be entropy-driven [8]. As hydrogen bonds are weak, they form or break readily during binding events, protein folding/unfolding, or structural changes. Hydrogen bonding can thus be turned on and off in biological systems with energies within the range of thermal variations and is one of the essential variables in macromolecular interaction and biological activity. Thus, hydrogen bonding is concrete because of its orientations, lengths, and angular preferences. Because of these characteristics, hydrogen bonds play a critical role in controlling specific interactions in biological recognition processes.

### 2.2 Hydrophobic interactions

Hydrophobic interactions are relatively more robust than the other interactions. These interactions are manifested as repulsive forces between nonpolar groups and water as these groups are poorly soluble in water.

Hydrophobic groups in water generally clump together because of their collective repulsion from water rather than any particular direct affinity for one another [9,10]. When nonpolar molecules are present in an aqueous solution, hydrogen bonds between water molecules are broken. Thus, heat is introduced into the system when bonds are broken; this is an endothermic reaction. Water molecules deformed by the presence of the nonpolar molecules will establish additional hydrogen bonds and form an ice-like cage structure surrounding the nonpolar molecules termed a clathrate cage [11]. This orientation increases the ordered structure of the system while decreasing the total entropy ($\Delta S < 0$) of the system [12]. Because the new hydrogen bonds can partially or wholly compensate for the hydrogen bonds broken by the presence of the nonpolar molecule, the change in enthalpy ($\Delta H$) of the system can be negative, zero, or positive. Thus, the overall change in Gibb's energy, $\Delta G$, turns negative (around $-20$ to $-40$ kJ/mol). Such interactions are essential in protein folding and keeping protein stable and folded [13]. In the case of binding events, the binding free energy is heavily influenced by interactions between ligands and proteins' hydrophobic side chains. In protein, residues such as Tyr, Phe, and Trp have aromatic rings that undergo stacking interactions with aromatic ligands. Water and other polar groups also resist the hydrophobic residues, giving rise to a resultant attraction of the ligand's nonpolar groups. Several studies showed that the hydrophobic interactions are measured by the sum of the hydrophobic surface buried upon ligand binding, which is the structural parameter that relates well with the binding free energy and is well known in various protein-protein and protein-ligand reactions [14,15].

## 2.3 Van der Waals interactions

It is a distance-dependent interaction that includes attractive and repulsive forces between atoms or molecules [16]. Slight changes in electron density distribution in one group of molecules fluctuate the surrounding electrostatic field, influencing nearby groups. This fluctuation induces an equivalent and opposite dipole in a nearby group where the transient dipoles attract each other. The strength of interaction depends on the distance and frequency of polarizability of molecules, which decreases with the reciprocal of the sixth power of the separation distance. Van der Waals interactions are exceedingly weak as opposed to covalent bonds or electrostatic contacts (0.1–4 kJ/mol). However, the enormous number of contacts that develop during biomolecule recognition events contribute significantly to total free energy. They are thus thought to render the enthalpy of interactions without any considerable change in the entropy component [17].

## 2.4 Electrostatic interactions

These forces are also known as the charge-charge or ionic interactions. The electrostatic attraction exists when the two strong opposite charges come closer. These interactions are specific and involved in biochemical reactions and interactions such as DNA stability, protein folding, and its function and flexibility. The force between two charged particles is

$$V(r) = \frac{1}{D}\frac{Q_1 Q_2}{r} \tag{1}$$

where $V$ = force between two charged molecules, $D$ = dielectric constant of the solvent; $Q_1$ and $Q_2$ are the magnitudes of the charged molecule, and $r$ = distance between the charged molecules.

The strength of interaction depends on the dielectric constant. Water has a high dielectric constant $D$ (~80), reducing the intensity of interaction, whereas the hydrophobic interior of the proteins and membranes has a low dielectric constant (~3–4), strengthening the intensity of these interactions. These interactions contribute to the binding energies and enthalpy of the biomolecular interactions. However, the entropic contribution is also seen when there is a displacement of counterions during complex formation. For proteins, interactions between ionized amino acid side chains with positive and negative groups (such as Arg, His, Lys, and Glu side chain) and the dipole of the ligand moiety or water molecule contribute significantly to the enthalpy change associated with a binding event [18].

## 2.5 Configurational entropy

It is the entropy associated with the different configurations or states of the biomolecule and their interactions. Generally, the free-energy change in noncovalent interactions between biomolecules is dominated by configurational entropy change [19]. It destabilizes the complex formation and is quantified by the Boltzmann's equation:

$$S = R \ln W \tag{2}$$

where $W$ = number of conformations and $R$ = gas constant.

To determine the conformational entropy, the molecule's potential conformations are first divided into a finite number of states, each characterized by a particular structural feature associated with the energy. In the case of proteins, backbone dihedral angles and side-chain rotamers are utilized as parameters, whereas the base-pairing pattern can be used in the case of nucleic acids [20]. Generally, during a binding event, there is a high unfavorable increase in entropy coming from the many conformational states available for the unbound systems [21].

### 2.6 Bonded interactions

These forces are involved in forming covalent bonds between adjacent atoms of biomolecules. Bending, rotating, and stretching the bonds require work and alter the system's total energy, thus changing binding partners' energies [17]. Stretching or bending a covalent bond is back-breaking labor requiring energy that is typically outside the scope for thermal motions. It is considered that covalent bonds in proteins adopt low energy conformation with the lowest stress.

## 3. Thermodynamics overview

Thermodynamics is a science that deals with energy. It is the quantitative study of energy conversions in living systems and associated biological reactions from one form to another. The principles covered under thermodynamics are its laws, Gibb's free energy, and chemical kinetics. The first law of thermodynamics states that energy cannot be created or destroyed. It may be changed or transformed into another form according to the second law of thermodynamics, which happens all the time in a living system. All chemical reactions such as folding, unfolding, and stability of macromolecules, macromolecule interaction, and the intricate properties of macromolecular assemblies and biological cells are governed by the principal laws of thermodynamics. Therefore, it becomes necessary to evaluate the free energy in the bulk and on the surface.

The part of the energy in a system that is not used for the entropic configurations or not used in any thermal agitations represents the free energy, generally termed Gibb's free energy $G$. It is a measure of the usable energy in the system and is defined as.

$$G = H - TS \tag{3}$$

where $H =$ enthalpy (joule), $T =$ temperature (kelvin), and $S =$ entropy (joule per kelvin).

Enthalpy of a thermodynamic system is the sum of the system's internal energy ($U$) and the product of its pressure ($P$) and volume ($V$). Entropy S is a physical quality usually linked with randomness or uncertainty in the system. Therefore, Gibb's free energy $G$ is

$$G = U - TS + PV \tag{4}$$

where $U$ = internal energy, $T$ = temperature in kelvin, $S$ = entropy, $P$ = pressure, and $V$ = volume.

Gibb's free energy change ($\Delta G = G_{\text{final}} - G_{\text{initial}}$) provides information about the spontaneity of a reaction at constant temperature and pressure. If G is negative, the reaction can happen on its own without added energy. The reaction is nonspontaneous if G is positive [22]. $\Delta G$ is the difference of free energies left to do work. Using $\Delta G$ as a parameter, biochemical reactions and processes can be grouped into three: (1) exergonic ($\Delta G < 0$), (2) endergonic ($\Delta G > 0$), and (3) processes (reactions) in equilibrium ($\Delta G = 0$). Because the energy of the products is lower than the energy of the reactants, exergonic reactions are favorable and spontaneous [23]. Pressure, temperature, pH, denaturants, ligand concentration, and other solvent changes may cause a biomolecule to switch between conformational states, and these transitions are associated with thermodynamic parameters. The enthalpy-driven forces are bonded interaction, electrostatic interaction, van der Waals interaction, and hydrogen bonding, whereas the entropy-driven forces are the hydrophobic forces and solvent effect.

The crucial thermal parameter is the heat capacity difference ($\Delta C_p$). It is the amount of heat given to a system to bring about a unit change in its temperature [24]. With the help of $\Delta C_p$, other thermal variables such as enthalpy and entropy can be determined. The enthalpy is the total heat absorbed or emitted by a system at constant pressure, and the enthalpy change ($\Delta H$) represents the internal energy changes because of intermolecular bonding. The entropy change $\Delta S$ hints at the number of degrees of freedom of the components involved in the binding reaction. These parameters can be determined using the following equation:

$$\Delta H = \int_0^T \Delta Cp \cdot dT + \Delta H(0) \tag{5}$$

$$\Delta S = \int_0^T \left(\frac{\Delta Cp}{T}\right) dT \tag{6}$$

where $\Delta H$ = enthalpy change and $\Delta H(0)$ = enthalpy change for the process at 0 K.

Eq. (5) portrays how enthalpy change accounts for the differences in heat energy when a specific molecular state is created. Eq. (6) demonstrates how the entropy change is a measure to disperse that energy among the different molecular energy levels [17]. The Gibbs free energy is the most critical parameter determining the biomolecular equilibration state. Thus, it indicates the direction in which the reaction progresses and the amount of work required to make them proceed.

$$\Delta G = \Delta H - T\Delta S \tag{7}$$

Eq. (7) shows that the free-energy change depends on the change in enthalpy associated with bond formation (negative $\Delta H$), bond breakage or disruption (positive $\Delta S$), and entropy change associated with the number of configurations available to the system. When different energy states are available to a system, they will be populated to different extents as a function of temperature. The probability of each state being occupied at a particular temperature is given by Boltzmann statistics.

Under standard conditions (1 M concentration, 1 atm partial pressure, 25°C, pH 7), the standard free-energy change $\Delta G^{\circ}$ is

$$\Delta G^{\circ} = -RT \ln K_{eq} \tag{8}$$

where $R$ = gas constant and $K_{eq}$ = equilibrium constant. The free-energy change for a reaction starting at any value of $K$ is

$$\Delta G = \Delta G^{\circ} - (-RT \ln K) = \Delta G^{\circ} + RT \ln K$$

$$\Delta G = -RT \ln \left( \frac{K_{eq}}{K} \right) \tag{9}$$

If $K_{eq} = 1$, then $\Delta G = 0$, that is, when the concentration of reactants is equal to the concentration of products. If $K > 1$, $\Delta G$ is negative and the reaction occurs spontaneously, and if $K < 1$, $\Delta G$ is positive and the reaction requires input energy to proceed under standard conditions [25].

## 3.1 Thermodynamics of protein folding, binding reactions, and interactions

A protein's native structure is governed by its amino acid sequence, and its function depends on its ability to fold into its native, globular structure [26]. The folding process of protein has a concise time scale of folding from milliseconds to seconds and results in a single minimum free-energy state, which is thermodynamically stable [27]. Protein folding is an entropically driven process and is contributed by hydrophobic forces. Apart from this, other interactions such as hydrogen bonding, van der Waals forces, and electrostatic interactions; conformational entropy; and the physical environment (pH, buffer, ionic strength) all have a role in the native protein's folding and stability [28]. During the folding process, protein adopts many intermediate states before being folded into the lowest energy conformation. The protein always exists in folded (native) and unfolded (denatured) conformations during equilibrium. The stability of the folded conformation depends on Gibb's free energy compared to the unfolded state [28,29]. Therefore, for the equilibrium condition [30],

$$K = \frac{D}{N} \tag{10}$$

where $K$ = equilibrium constant, $D$ = concentration of the denatured state, and $N$ = concentration of the native state.

The energetics of the unfolding reaction can be determined using the equilibrium constant:

$$\Delta G_{eq} = -RT \ln K = -RT \ln\left(\frac{D}{N}\right) \tag{11}$$

where $R$ = universal gas constant, $T$ = temperature in kelvin, and $\Delta G_{eq}$ = Gibbs free energy of the unfolding reaction and is the sum of enthalpy and entropy change.

For a given protein's unfolding reaction at equilibrium, the value of $\Delta G_{eq}$ is positive when the population of the native state is larger than that of the denatured state and thus represents the stability of the protein in the equilibrium state. However, in the refolding process, the negative value of $\Delta G_{eq}$ means the stability of protein as the native conformation is the final state of the process [30].

For a reversible reaction between a macromolecule (A) and ligand (B), the reaction can be represented as

$$\mathrm{A} + \mathrm{B} \rightleftarrows \mathrm{AB}$$

$$K = \frac{[AB]}{[A][B]} \tag{12}$$

In terms of energetics,

$$\Delta G = -RT \ln\left(\frac{[AB]}{[A][B]}\right)$$

$$\Delta G = -RT \ln K \tag{13}$$

or

$$\Delta G = -RT \ln K_a = RT \ln K_d \tag{14}$$

where $K = K_a$ is the equilibrium affinity constant and $K_d$ is the dissociation constant. Further, the binding parameter, $\vartheta$, is the ratio of bound ligand $[B_b]$ to total macromolecule $[A_t]$ concentration.

$$\vartheta = \frac{[B_b]}{[A_t]} = \frac{[AB]}{[A_t]} = \frac{[AB]}{[AB] + [A]} = \frac{K_a[B]}{1 + K_a[B]} = \frac{[B]}{K_d + [B]} \tag{15}$$

The $K_a$ or $K_d$ can be determined experimentally using UV, fluorescence, circular dichroism (CD), and NMR spectroscopic techniques [29]. For the thermodynamic characterization, enthalpy change associated with the change in heat capacity can be obtained using Eq. (7).

The sign and magnitude of the enthalpic and entropic contributions result from the interactions such as ionic, hydrogen, hydrophobic, or van der Waals force, which fret about the strength and specificity of the interactions between the partners. Enthalpically, the most significant contribution comes from hydrogen bonding. Therefore, the sign reflects whether the hydrogen bond network between the reacting species, including the solvent, has been redistributed in a net favorable (negative) or unfavorable (positive) manner. The relative degree of the disorder can be attributed to the entropy change. This change in entropy gives an account of the degree of freedom of the components involved in the binding interaction. The release of water molecules into the bulk solvent is a source of favorable entropy. As a result, hydrophobic interactions have a low enthalpy (positive or negative) and high entropy.

## 3.2 Enthalpy contributions

A change in enthalpy describes the energy changes linked with specific, non-covalent interactions. During the binding of two molecules, many bonds form and break, which results in enthalpy changes. It gives an account of the loss of hydrogen bonds between protein and the solvent, the ligand and solvent, van der Waals interactions, salt bridge formation, protein-ligand bonds, the intramolecular solvent, the hydrogen bonding network, conformational changes at the protein surface, and reorganization near the protein surface [7,29]. The alterations may be favorable or unfavorable, depending on the system.

## 3.3 Cooperativity

The contributions of enthalpy and entropy are interlinked. An increase in enthalpy because of tighter binding may directly affect entropy by limiting the mobility of the interacting molecules [31]. Cooperativity is generally linked to noticeable structural changes in the macromolecule, which can be positive or negative. If the macromolecule has more than one site of binding of the ligand, different conditions may apply. When the occupation of one site favors the binding to another site, it is called cooperative binding. Conversely, the occupation of one site causes a decrease in the binding affinity of a second one; it is called anticooperative. However, in the case of noncooperative interactions, the binding affinity of the remaining ligand is not affected, and binding pockets remain independent. The binding energy associated with protein-ligand interactions can be cooperative, implying that it differs from the sum of the individual contributions of binding free energies of protein and the ligand [18].

Positive cooperativity leads to tightening of the target molecule and shortening of interatomic distances, making interactions enthalpically favorable. An example of such an interaction is the binding of biotin-streptavidin [32]. Streptavidin is a homotetramer with a high affinity toward biotin and has one of the strongest noncovalent bonds in nature. Each monomer unit of streptavidin has a β-barrel structure which then associates to form its tetrameric quaternary structure. At the end of each β-barrel, there is a binding site of biotin. Biotin binding is confined to the active site of streptavidin. It forms eight hydrogen bonds to the conserved residues of β-barrel and has van der Waals interactions with nonpolar groups. Many X-ray crystallography and other biophysical studies showed shape complementarity and extensive hydrogen bond networking between the biotin and binding sites of streptavidin, which is the reason for strong cooperativity [18,33]. Many studies suggested that protein dynamics is interrelated to the binding events. In some cases, the dynamics of proteins increased far away from the ligand-binding site, compensating for the unfavorable entropic contribution to ligand binding [34,35].

## 4. Methods for binding constant and thermodynamics study

### 4.1 Differential scanning calorimetry and isothermal titration calorimetry

Conventionally, methods such as DSC and ITC are directly employed to study the energetics of DNA and proteins. Both techniques offer a broad application base for biomolecule binding, stability, and melting studies. ITC measures the heat change by titrating one binding partner with another in a solution at a constant temperature [36]. The resultant isotherm produced gives information about the type of reaction, stoichiometry ($n$), heat capacity change ($\Delta C_p$), association constant ($K_a$), entropy ($\Delta S$), enthalpy ($\Delta H$), and free-energy change ($\Delta G$) of the reaction [37]. DSC is another extensively used method to study the energetics of the conformational transition as a function of temperature. This method measures the heat capacity difference between a macromolecule solution and the same volume of the solvent alone. Other thermodynamics parameters that govern the conformational equilibrium between two states of a macromolecule and their interactions can be derived from the heat capacity profile [38,39]. ITC and DSC alone and together can provide much valuable information about macromolecule interaction, majorly in a moderate binding regime. Additional biophysical, biochemical, and structural analysis techniques are required to understand their energetics better.

This chapter focuses on the energetic study employing various spectroscopic techniques. The following sections discuss the details of different spectroscopic methods to characterize biomolecular interactions in terms of the thermodynamic parameters.

## 4.2 Spectroscopic techniques for thermodynamics studies

This section discusses the principle and use of different spectroscopic techniques to study the binding phenomenon and thermodynamics analysis of biomolecular interactions, as shown in Fig. 2. The characteristic signal produced from the macromolecule or the ligand (absorbance, fluorescence, CD signals, NMR linewidth, chemical shift, relaxation, etc.) varying at every titration point as complex formation takes place is monitored. Generally, either a fixed concentration of macromolecule titrated with the increasing ligand concentration, referred to as 'normal' titration, or a fixed concentration of ligand with the increasing concentration of macromolecule, referred to as

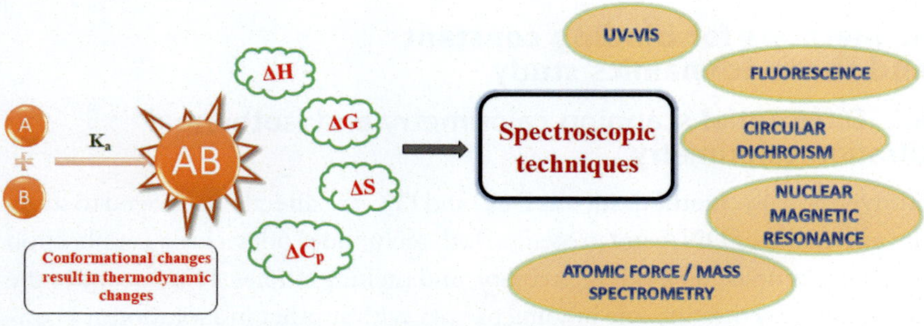


**Fig. 2** Diagrammatic illustration of energetics and spectroscopic techniques for portraying biomolecular interactions.

'reverse' titration, is performed. This protocol depends on whether the signal monitoring is from the macromolecule or the ligand.

### 4.2.1 Ultraviolet-visible (UV-vis) spectroscopy

Ultraviolet-visible (UV-vis) spectroscopy is a technique for determining how molecules interact with electromagnetic radiation. The molecule absorbs the energy of the radiation to jump from the lower energy level (ground state) to a higher energy level (excited state). The resultant spectrum generated shows the absorbance of molecules as a function of wavelength. In aromatic systems, molecules containing electrons absorb light in the near-UV (150–400 nm) or visible (400–800 nm) regions. This technique works on the principle of Lambert-Beer's law. It states that the absorbance (A) is directly proportional to the solute concentration and is related to the intensity of light before and after passing through the sample [40].

$$A = \varepsilon\, cl \tag{16}$$

where $\varepsilon$ = molar absorption coefficient (L/mol/cm), $c$ = concentration of solute, and $l$ = path length in centimeters.

Proteins show absorption maxima at 280 nm because of aromatic amino acids Trp and Tyr, which are sensitive to their microenvironment. Variations in the environment bring about absorbance intensity or shift changes. Nucleic acids show absorbance around 260 nm because of the $\pi$-$\pi^*$ transitions of the pyrimidine and purine rings of the bases. The biomolecular reaction follows the host-guest complexation process of supramolecular chemistry [40,41]. For the quantitative analysis of binding, knowledge about the binding constant is essential. The main forces involved in the binding can be assessed by the signs and magnitudes of the thermodynamic parameters such as enthalpy change ($\Delta H$) and entropy change ($\Delta S$) with the help of the van't Hoff equation. According to van't Hoff, under standard enthalpy change ($\Delta H$), change in the equilibrium constant $K$ is related to the change in the temperature $T$ for a given biochemical reaction as

$$\ln K = -\frac{\Delta H}{RT} + \frac{\Delta S}{R} \tag{17}$$

A simple equilibrium model has been used for measuring the quantitative parameters with high reliability using the following equation [42]:

$$a \cdot H + b \cdot G \leftrightarrow C \tag{18}$$

where $H$ = host, $G$ = guest, $C$ = complex, and $a$ and $b$ = stoichiometry of the host and guest, respectively.

$$K = \frac{[C]}{[H]^a \cdot [G]^b} \tag{19}$$

$$[H]_t = [H] + a \cdot [C]; [G]_t = [G] + b \cdot [C]$$

where $[H]_t$ = total concentration of the host molecule; $[G]_t$ = total concentration of the guest molecule; and $[H]$, $[G]$, and $[C]$ are the concentrations of the host, guest, and complex, respectively, at equilibrium.

$K$ can be determined using the equation

$$K = \frac{[C]}{\left([H]_t - a \cdot [C]^a\right) \cdot \left([G]_t - b \cdot [C]^b\right)} \tag{20}$$

There are various ways of calculating stoichiometry, such as the continuous variation method [43], slope ratio method [44], and mole ratio method [45]. The most widely used method is the continuous variation method. Many studies on DNA-drug /Protein-drug/DNA-protein systems utilize UV-vis spectroscopy to study the binding mode and thermodynamic parameters.

One example of such interaction is between human serum albumin (HSA) and beta-blocker drugs, Atenolol and Metoprolol. HSA is a single-chain protein involved in controlling [46] the osmotic pressure and pH of the blood. In this study, Arastou et al. [47] showed the binding of HSA with these drugs [Atenolol (Atn) and Metoprolol (Met)] to understand the behavior and mechanism of drugs in the bloodstream in the body. Using the van't Hoff relation (Eq. 17), they derived the thermodynamic parameters at 298 K and gave an insight into the interactions and energies involved, as summarized in Table 1. The values of $-\Delta H$ and $-\Delta S$ showed that the van der Waals force and/or hydrogen bonds play a vital role in the interaction and make the complex stable. Data reported at different temperatures showed higher $\Delta H$ values ($|\Delta H|$) than $|T\Delta S|$, arguing that the process is enthalpically governed.

**Table 1** HSA-Atn and HSA-Met thermodynamic parameters [47].

| System | Binding constant $K_b$ ($M^{-1}$) | $-\Delta H$ (kJ/mol) | $-\Delta G$ (kJ/mol) | $-\Delta S$ (J/K/mol) |
|---|---|---|---|---|
| HSA-Atn | $3.1 \times 10^3$ | 72.74 | 64.05 | 29.18 |
| HSA-Met | $2.4 \times 10^3$ | 112.23 | 99.75 | 42.31 |

In another study, Afrin et al. [48] examined the interaction of ticlopidine, an antiplatelet drug, with calf thymus DNA (CT DNA) using different spectroscopic techniques. UV-vis absorption data espied a binding constant of $11 \times 10^3$ $M^{-1}$, and the free energy $\Delta G$ was obtained as $-5.5$ kcal/mol. They investigated the thermal melting profiles of CT DNA in the absence and presence of the drug, and no significant changes in the melting temperatures $T_m$ were observed. Using different spectroscopic and calorimetric results, the authors identified a minor groove binding mode. The reaction was mainly entropically driven (from ITC data). The entropic contributions were primarily from the desolvation energy and hydrophobic interactions.

### 4.2.2 Fluorescence spectroscopy

From basic research to clinical applications, fluorescence spectroscopy has become one of the most potent and commonly used tools in the biological sciences. It can detect extracellular and intracellular molecular interactions and enzymatic processes in real time, a significant advantage over other techniques for studying biomolecular interactions. An intrinsic or extrinsic fluorophore emits light at a longer wavelength after absorbing light at a particular wavelength. This emitted light is called fluorescence. Many biomolecular interactions are of great importance. Generally, the binding of one molecule to another may lead to a decrease in its fluorescence intensity, and this phenomenon is known as quenching. Many processes such as excited-state reactions, molecular rearrangements, energy transfer, ground-state complex creation, and collisional quenching can cause quenching [49]. Quenching can be collisional or static.

The equilibrium constant at various temperatures can be determined from the van't Hoff equation (Eq. 17). These data can further be plotted on a graph between $1/T$ on the x-axis and the log of $K_{eq}$ ($\ln K_{eq}$) on the $y$-axis, as shown in Fig. 3. A linear fit of the data to the van't Hoff equation estimates the change in enthalpy and entropy of a biochemical reaction [50]. The intercept and slope of the plot are used to get the change in enthalpy ($\Delta H$) and entropy ($\Delta S$), respectively. If there is no significant change in enthalpy, then free-energy change $\Delta G$ can be obtained from Eq. (13).

Interaction of the drug tetrandrine (TETD) to human and bovine serum albumins (HSA and BSA) showed that TETD showed a high binding affinity to HSA with binding constants ranging from $10^3$ to $10^5$ $M^{-1}$ [51]. The $\Delta G$ value observed for the TETD-HSA complex was $-33.90$ kJ/mol and that for TETD-BSA was $-39.30$ kJ/mol, indicating spontaneous binding

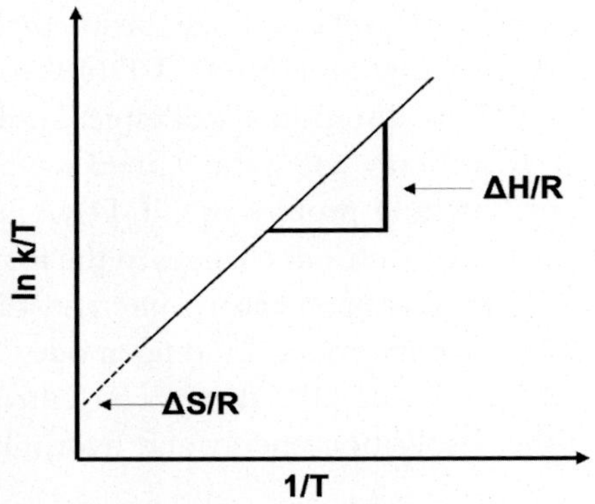


**Fig. 3** A typical van't Hoff plot showing the correlation between thermodynamic parameters.

processes. However, the positive enthalpy, $\Delta H = 59.15$ kJ/mol, and the positive entropy, $\Delta S = 317.58$ J/mol/K, values for the interaction of TETD with HSA were linked to hydrophobic interactions. TETD-BSA interaction resulted in $\Delta H = -27.61$ kJ/mol and $\Delta S = 22.77.58$ J/mol/K, indicating an exothermic interaction and pointing to the role of an increased entropy in the interaction.

Another study conducted to study the energetics of the DNA and drug system is the interaction of β-resorcylic acid (BR) with the CT DNA. BR is a secondary metabolite and possesses antibacterial properties. Hussain et al. showed that BR is a groove binder with the binding constant in the order of $10^3$ $M^{-1}$. The free-energy *(ΔG)* value obtained, −5.30 kcal/mol, indicated an exothermic reaction, and the negative enthalpy (Δ*H*), −20.37 kcal/mol, and the negative entropy (Δ*S*), −50.41 cal/K/mol, support the presence of hydrogen bonding and van der Waals interactions in binding of BR to CT DNA [52].

Recently, the association of lead sulfide quantum dots (PbS QDs) with HSA was described by Wang et al. using the fluorescence quenching method. Using the van't Hoff equation, they calculated the thermodynamic parameters and analyzed the driving forces involved. From the negative enthalpy and entropy values obtained, it was shown that the interaction between the PbS QD and HSA was enthalpically driven with significant contributions from a covalent bond, hydrogen bond, and van der Waal interactions. They validated the measured values with 25 other reported values in the literature and determined the enthalpy-entropy compensation. Their results are similar to the values obtained from protein-protein interaction, hinting that nanoparticle–protein interactions resemble that of protein-protein interactions [53].

In addition to the different fluorescence methods, fluorescence correlation spectroscopy (FCS) can also be used to evaluate membrane proteins' thermodynamic details. This method will monitor the fluorescence intensity fluctuations of a few fluorescent molecules diffusing through a small focal volume. Posokhov et al. [54] demonstrated FCS as a successful tool in characterizing the energetics of pH-triggered transmembrane insertion of annexin $B_{12}$. They obtained comparable free-energy values for final transmembrane insertion and interfacial insertion, ~10–12 kcal/mol. They concluded that electrostatic and hydrophobic interactions are involved in the insertion of annexin into the membrane. Similar studies using FCS are reported in the literature [55,56].

### 4.2.3 Circular dichroism (CD) spectroscopy

CD spectroscopy is the differential absorption spectroscopy used to study the conformation of biomolecules in the solution. It measures the difference in absorption by a chiral molecule when right and left circularly polarized light passes through it. It is an excellent technique for secondary structure determination of biomolecules.

$$\Delta A = A_L - A_R \tag{21}$$

where $\Delta A$ = difference in absorbance, $A_L$ = absorbance because of left circularized polarized light, and $A_R$ = absorbance because of right circularized polarized light [57].

The main applications of CD spectroscopy are (i) to determine the secondary structures of protein and nucleic acids; (ii) to study the conformational changes in protein or DNA [58,59] under the effect of pH, temperature, salt, ions, or any other environmental factor; (iii) stability studies between native and mutant forms; (iv) monitoring protein-protein, protein-DNA, DNA-drug, or protein-drug interactions; and (v) monitoring the kinetics of macromolecular reactions. Chromophores in an asymmetrical environment contribute to CD signals. In protein, the n-$\pi^*$ and $\pi$-$\pi^*$ transitions (around 220 and 190 nm, respectively) of the amide backbones are the chromophores giving dominant CD peaks. The CD signals from these chromophores correspond to the various secondary structures adopted by the protein. The aromatic amino acids (Trp, Tyr, Phe) have the characteristic absorption band in the near UV range of 250–300 nm [60]. For nucleic acids, the CD signals from the electronic transitions of the bases are observed. CD signals are sensitive to nucleic acid conformation, particularly within the 180–320 nm wavelength range [58].

CD is mainly used to study protein/DNA conformation as unfolding or a change leads to significant differences in the near-UV and far-UV CD spectra. Many tiny globular proteins have a two-state unfolding mechanism, with only the native/folded (N) and denatured/unfolded (D) states considerably populated at equilibrium. The thermodynamics of unfolding, namely, the van't Hoff enthalpy ($\Delta H$) and entropy ($\Delta S$), the midpoint of the unfolding transition ($T_m$), and the free energy ($\Delta G$), can be determined using differences in CD as a function of temperature at particular wavelengths. The heat capacity change of unfolding ($\Delta C_p$) can be determined from CD data if the apparent $T_m$ of unfolding changes as a function of protein concentration or is affected by salt or pH changes [61].

The conformational stability of a protein can be defined in terms of the free energy of unfolding using Eq. (13). The unfolding thermodynamics of a molecule with temperature can be determined by measuring the unfolding constant. The fraction of the folded state at any temperature $\alpha$ is

$$\alpha = [N]/([N] + [D]) \tag{22}$$

$$\alpha = \frac{K}{1 + K} \text{ or } \alpha = (\theta_t - \theta_D)/(\theta_N - \theta_D) \tag{23}$$

$$[\theta_t] = \alpha(\lfloor\theta_N\rfloor - [\theta_D]) + [\theta_D] \tag{24}$$

where $\theta_t$ = observed ellipticity at any temperature and $\theta_D$ and $\theta_N$ = ellipticity observed of unfolded and fully folded forms, respectively.

The variations in CD at a single wavelength as a function of temperature $T$ are used to follow the folding process [61,62]. The unfolding of a monomer with temperature can be tracked using the Gibbs-Helmholtz equation below [63,64].

$$\Delta G = \Delta H \left(\frac{1 - T}{T_m}\right) - \Delta C_p(T_m - T) + T \ln \left(\frac{T}{T_m}\right) \tag{25}$$

$\Delta C_p$ is the heat capacity change when moving from the folded to the unfolded state, and $T_m$ is the temperature at which $\alpha = 0.5$. The initial parameters such as $\Delta H$, $T_m$, $[\theta_N]$, and $[\theta_D]$ are measured and used in fitting the curves. This calculation can also be used for other hetero-oligomers [65].

When one macromolecule interacts with another molecule (small/macromolecule), changes are usually observed in the CD signal of either one or both the components. The binding constant can be determined by the direct or indirect method. The primary criterion for a direct approach is that the sum of spectra of the components (*A* and *B*) should differ from that of the complex (*AB*) at a specific wavelength [57]. If this condition is satisfied, the equilibrium dissociation constant, $K_d$, can be estimated from the titration of *A* with *B*. For any mixture of *A* and *B* at a 1:1 ratio, the observed CD signal is given by

$$Signal = \Delta\varepsilon_A A_T + \Delta\varepsilon_B B_T + (\Delta\varepsilon_{AB} - \Delta\varepsilon_A - \Delta\varepsilon_B)$$
$$= \left(\frac{(A_T + B_T + K_d) - (A_T + B_T + K_d)^2) - 4\,A_T B_T)^{0.5}}{2}\right) \quad (26)$$

where $A_T$ and $B_T$ = total concentrations of *A* and *B*, $K_d$ = dissociation constant, and $\Delta\varepsilon_{AB}$ = measured ellipticity difference of both components.

The other method to determine the binding constant is the indirect mode. This method is relevant when there is no optimum signal to quantify binding interaction. In many cases, when a ligand binds to the native state of the protein, then it becomes more stable. This is because in equilibrium, the concentrations of native and denatured forms of the protein are equal. When complex formation occurs, some of the native form of protein gets engaged in binding, whereas some in the denatured state refold again to maintain the equilibrium. Thus, the equilibrium dissociation constant can be calculated using the increase in the stability in the presence of the ligand. Therefore, for this thermal unfolding, experiments can be used to study the effect of ligand binding on the midpoint for unfolding using the equation [66]

$$\Delta T_m = \frac{T_m T_m^* R}{\Delta H_{Tm^*}} \ln\left(1 + \frac{[L]}{Kd}\right) \quad (27)$$

where $\Delta T_m$ = difference in the midpoint of the unfolding curve, $T_m$ and $T_m^*$ = midpoint values in the absence and presence of the ligand, $\Delta H_{Tm^*}$ = enthalpy of unfolding in the presence of the ligand, and $[L]$ = free ligand concentration [67].

Free energy can be calculated as given below:

$$\Delta G_L - \Delta G = RT \ln\left(1 + \frac{[L]}{Kd}\right) \quad (28)$$

where $\Delta G_L$ and $\Delta G$ = free energies of unfolding in the presence and absence of the ligand, respectively.

CD spectroscopy is an excellent approach to monitor the conformational changes of the molecule upon a change in environment or using thermal or chemical denaturation experiments and further gain insight into energetics. Changes in CD signals were monitored when lysozyme was subjected to thermal and guanidine hydrochloride (GdnHCl) denaturation in the presence of tri-*N*-acetyl glucosamine ($(GlcNAc)_3$). $(GlcNAc)_3$ is the trisaccharide carbohydrate that binds to lysozyme. They evaluated the free-energy change during denaturation for lysozyme under different concentrations of $(GlcNAc)_3$. They found that in the presence of $3 \times 10^{-4}$ M of $(GlcNAc)_3$, the stability of lysozyme increased by 495 kcal/mol under GdnHCl denaturation. The findings of this work indicate that binding small molecules/ligands to specific sites in the native conformation of protein enhances the thermal stability of protein under different denaturing conditions [68].

Another example is the binding of calmodulin (CaM) to its binding targets (CaMBTs). CaMBTs are specific sequences present in the proteins that have an affinity to CaM. CaM is a calcium-binding protein that helps in many biological processes. Dunlap et al. studied and determined the $\Delta G$ value of CaM with different CaMBT sequences using the CD technique. Their finding suggested that the affinity of all peptide sequences is almost similar as their binding energy lies in the range of 35–49 kJ/mol. They also demonstrated that the binding of different sequences to CaM is entropically driven [69].

Gondeau et al. studied the formation and stability of a DNA triple helix in the absence and presence of salt. They found that in the presence of $MgCl_2$, an oligonucleotide sequence 5′-d($G_4A_4G_4[T_4]C_4T_4C_4[T_4]G_4T_4G_4$) adopts an intramolecular parallel triplex structure in contrast to a duplex structure it adopts in the absence of $MgCl_2$. The CD spectrum of this oligonucleotide was found to be different from that of the antiparallel triplex-forming oligonucleotide sequence 5′-d($G_4T_4G_4[T_4]G_4A_4G_4[T_4]C_4T_4C_4$). Melting studies on the parallel triplex-forming sequence showed a biphasic transition, triplex to duplex and duplex to coil. The stability of this sequence was found to be lower than that of the antiparallel triplex forming sequence. They also obtained the $\Delta H$ and $\Delta S$ values for the triplex to duplex transition as 37 kcal/mol and 124 eu, respectively. These values are much lower (for the Hoogsteen transition) than those for a Watson-Crick duplex to coil transition. The authors accounted for these lower values to the higher phosphate-phosphate electrostatic repulsion in the Hoogsteen part of the triplex. The lower value of the entropy was attributed to the solvent effect. The positive value is associated with the removal of more water molecules during triplex formation than in duplex formation [70].

### 4.2.4 Nuclear magnetic resonance (NMR) spectroscopy

NMR spectroscopy is widely used to study the structure and properties of different molecules. The finite nuclear magnetic moments of elements, such as $^{1}H$, $^{13}C$, $^{15}N$, and $^{19}F$, are the basis of NMR spectroscopy. When placed in a magnetic field and applying a radiofrequency (RF) field, these NMR-active nuclei will absorb the energy corresponding to the particular nuclei's precession frequency. These elements' distinct nuclear spin states become quantized in an external magnetic field ($B_o$). The energies of these states are proportionate to their projections onto $B_o$. It is an excellent method for studying three-dimensional structural and dynamical features of systems and their interactions because the differences in energies are proportional to the strength of the magnetic field and depend on the chemical environment of the interacting element [71]. These methods consist of one-, two-, and three-dimensional NMR approaches. One-dimensional NMR helps in identifying the spin systems present in the sample. Another important NMR parameter arises from the indirect coupling between the nuclear spins through bonds, leading to the splitting of spectral lines. During the nonradiative relaxation of nuclei, many interactions come into play. The systematic analysis of these events helps measure a molecular system's kinetic and thermodynamic processes. NMR spectra provide real-time analysis of the reaction process and information about the molecular structure. In the NMR spectrum, the areas of the spectral lines are directly proportional to the number of molecular species present. Researchers are generally using NMR relaxation studies to gain insight into the thermodynamics of DNA-ligand, protein-ligand, or DNA-protein binding.

Advancements in NMR techniques made it possible to analyze ligand-induced conformational changes, investigate bound water molecules' locations and dynamic behavior, and calculate conformational entropy. Nuclear Overhauser effects (NOEs) are beneficial for studying the structure of three-dimensional macromolecules in solution [71]. NOE effectively elaborates noncovalent binding interactions between macromolecules as it does not occur through chemical bonds. Chemical shift mapping is another helpful NMR approach for understanding the binding thermodynamics [72]. Chemical shift changes show how the electronic environment affects nuclear magnetic energy levels in a molecule. Several factors such as electron density, electronegativity, and aromaticity can influence these shifts. As a result, chemical shift changes occur in the binding or interaction interface zone when two molecules come in contact. Thus, chemical shifts help locate the crucial binding sites and modes of binding.

The rate of a chemical reaction usually depends on several factors such as the concentration of different species, pressure, and temperature. The equilibrium process between the product and the reactant is referred to as the chemical exchange in NMR. For a reaction

$$A \underset{K_{-1}}{\Leftarrow \overset{K_1}{\Rightarrow}} B$$

the population difference of any molecular species from equilibrium at any given time can be written as [73]

$$\Delta[A](t) = \left([A] - [A]_{eq}\right) \tag{29}$$

The time dependence can be written as

$$\Delta[A](t) = \Delta[A](0)e^{-k_x t} \tag{30}$$

where $k_x = k_1 + k_{-1}$.

At equilibrium, the rate of the forward reaction is equal to the rate of backward reaction and

$$k_{eq} = \frac{k_1}{k_{-1}} = \frac{[B]_{eq}}{[A]_{eq}} \tag{31}$$

Using the Arrhenius equation [74], the temperature dependence of the reaction rate can be written as

$$k_{ex} = A \ \exp^{\left(-\frac{\Delta E}{RT}\right)} \tag{32}$$

where $A$ = preexponential factor, $R$ is the gas constant (8.314J/Kmol), $T$ = sample's temperature in kelvin, and $\Delta E$ = activation energy to the transition state.

The modified form of the Arrhenius equation by Eyring allows to define $\Delta E$ in terms of Gibbs energy ($\Delta G$), enthalpy ($\Delta H$), and entropic ($\Delta S$) contributions of the transition state as follows:

$$k_{ex} = \frac{k_B T}{h} \exp^{\frac{\Delta G^{\circ}}{RT}} = \frac{k_B T}{h \exp^{\left(-\frac{\Delta H}{RT} + \frac{\Delta S}{R}\right)}} \tag{33}$$

where $k_B$ = Boltzmann constant ($1.3807 \times 10^{-23}$J/K), $h$ = Planck's constant ($6.62697 \times 10^{-34}$ Js), and $\Delta G^{\circ}$ = Gibbs energy at standard conditions. The temperature dependence of the difference in the Gibbs energy between the reactant and product is given by modifying Eq. (9), which becomes

$$K_{eq} = \exp^{\left(-\frac{\Delta G}{RT}\right)} \tag{34}$$

Experimental data at different temperatures help to calculate the thermodynamic parameters. The plot where inverse temperature ($1/T$) is plotted against the natural logarithm of ($k_{eq}/T$), known as the Eyring plot (Fig. 4), is used to extract the parameters. The intercept and the slope signify the value of $\Delta S$ and $\Delta H$ [75,76].

Choosing the right type of experiment for different exchange processes involved (slow, fast, intermolecular, and intramolecular exchanges) and measuring over a wide range of temperatures can provide the reaction rates, kinetic energy barrier, entropy, and enthalpy information [73]. The relaxation rates of the backbone and side chains of proteins can be determined using isotopically labeled proteins ($^{15}N$, $^{13}C$, and $^{2}H$).

NMR has a time scale range of approximately $10^{-12}$ to $10^{5}$ s, allowing it to detect most of the protein dynamics. Conformational changes linked with ligand binding happen on slow time scales (microseconds to milliseconds), leading to slower motions than protein backbone and side-chain motions (picoseconds to nanoseconds). Studies on different ligand-enzyme systems showed that many binding events that lower the rapid motions can enhance, decrease, or have no effect on slow motions [71,77,78]. Multiple time scale dynamics of ligand-protein complexes have been studied using NMR relaxation techniques [18]. Their findings reveal that despite substantial conformational changes taking place on a slow time scale, 'rapid' protein mobility (picoseconds to nanoseconds) is critical in all elements of the binding event. The longitudinal relaxation rate (R1), the transverse relaxation rate (R2), and the NOE are the three relaxation rates that are routinely measured.

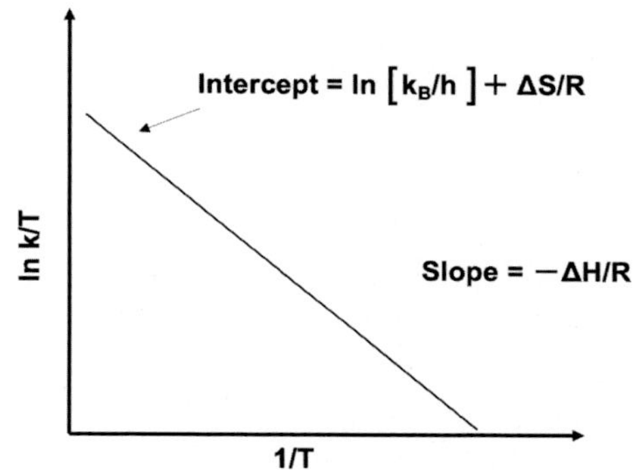


**Fig. 4** A typical eyring plot for the determination of thermodynamic parameters.

The spectral density function, $J(\omega)$, is directly connected to these relaxation rates. The amplitude of the fluctuating magnetic field at the frequency $\omega$ is proportional to this function. Molecular mobility in an external magnetic field, strongly related to nuclear spin relaxation, causes such fluctuating magnetic fields. Many studies showed that biomolecular association is mainly contributed by the perturbations in configurational entropy. The NMR relaxation studies often measure these changes by estimating order parameter $S^2$. The order parameter $S^2$ measures the internal motion or measures the spatial fluctuation of a bond vector and hence can be correlated with the configurational entropy [79,80].

The study conducted by Bracken and his team showed the dynamics of the N-terminal leucine zipper domain of yeast transcription factor GCN4 to the DNA [81]. In the absence of DNA, the basic region adopts different temporary structures. When DNA is bound, the N-terminal region adopts a stable helical structure. As a result, an unfavorable input from the change in conformational entropy of the protein backbone was observed. The overall rotational correlation time, $\tau_M$, was calculated using spectral density function ($J(\omega)$), which was further used to calculate the order parameter $S^2$ using the equation

$$S^2 = 5(J(0) - J(\omega_N)(1 - \omega_{N^2}\tau_{N^2})2\omega_{N^2}\tau_{M^3} \tag{35}$$

The average value of $\tau_M$ estimated for the coiled-coil domain was used to determine $S^2$ for residues in the basic region, as shown in Figs. 5 and 6. The observed entropy $\Delta S_b$ is in the order of ~0.6 kJ/mo/K, which is close to the already reported calorimetric observation. The observed free energy of binding $\Delta G$ calculated was 150–180 kJ/mol [81].

Tzeng et al. used the NMR technique and conventional thermodynamic data to analyze how the conformation and internal dynamics are influenced when DNA binds to many variants of the catabolite activator protein (CAP) DNA binding domain. They showed that the binding of DNA to CAP variants, CAP-S62F-cAMP$_2$ and CAP-T127L/S128I, resulted in configurational entropy changes of −15 and −9 kcal/mol, respectively. These entropy changes are correlated to the higher degree of freedom of the residues upon binding to DNA. However, the binding of DNA to wild-type WT-CAP-cAMP$_2$ resulted in an unfavorable configurational entropy (~15 kcal/mol) attributed to the increased rigidity of residues. Their result suggests that fast and slow internal dynamics can play a role in the DNA/ligand binding process [79,80].

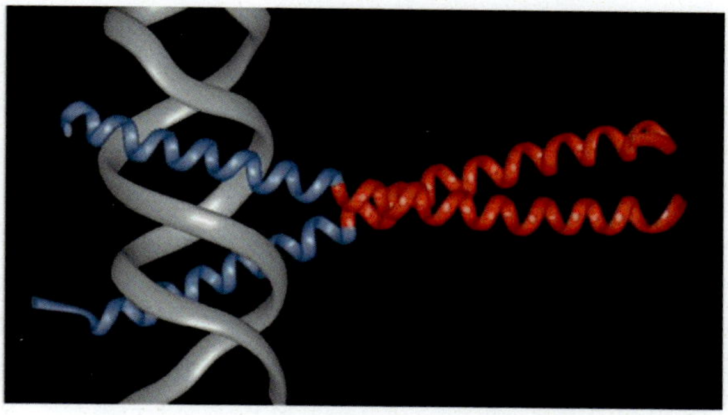

10 20
MKDPAALKRARNTEAARRSRARKLQRMKQ
30 40 50
LEDKVEELLSKNYHLENEVARLKKLVGER

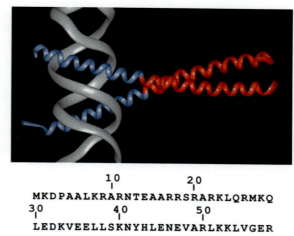

**Fig. 5** Structure and sequence of GCN4-58; the ribbon shows the DNA binding domain complexed with DNA. *(Figure adapted with permission from Bracken C, Carr PA, Cavanagh J, Palmer AG. Temperature dependence of intramolecular dynamics of the basic leucine zipper of GCN4: implications for the entropy of association with DNA. J Mol Biol 1999;285(5):2133–46, https://doi.org/10.1006/jmbi.1998.2429.)*

### *4.2.5 Atomic force spectroscopy*

Other spectroscopic techniques are also used for investigating biomolecules and their interactions. One of the techniques is atomic force spectroscopy (AFS). It acquires information at a single-molecule resolution on biorecognition

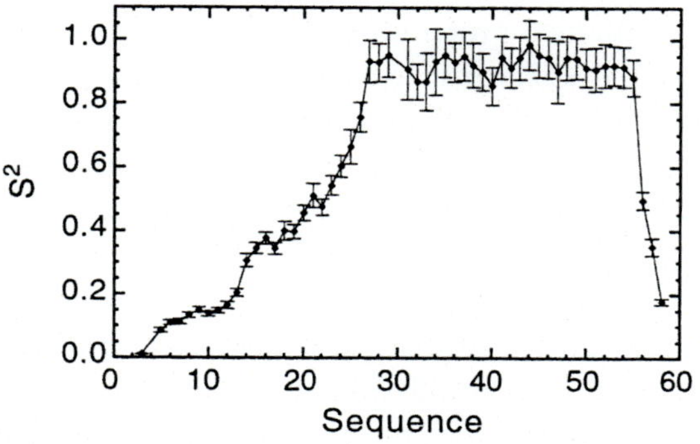


**Fig. 6** Amplitudes of intramolecular motions in GCN4-58. Using spectral density mapping, the order parameter $S^2$ was calculated at 310K. *(Figure adapted with permission from Bracken C, Carr PA, Cavanagh J, Palmer AG. Temperature dependence of intramolecular dynamics of the basic leucine zipper of GCN4: implications for the entropy of association with DNA. J Mol Biol 1999;285(5):2133–46, https://doi.org/10.1006/jmbi.1998.2429.)*

processes [82]. This technique does not need any labeling of molecules and measures the forces between the molecules under physiological conditions. Thus, it provides complete details of the kinetic and thermodynamic behavior of the interaction system. AFS has been used to study several biological structures and processes, including paired interactions between ligands and receptors or antibodies and antigens, protein unfolding, conformational changes, molecular stretching, membrane elasticity, cell deformation, and cell adhesion [83]. In AFS, gold, silicone, and glass-coated substrates are used with heterofunctional groups containing linker groups for substrate binding at one end and biomolecule binding sites at another. Biomolecules can be immobilized directly or indirectly to the substrate. A tip containing a ligand is paired with a substrate to measure the force between them. Some examples where AFS is used are the interactions between HSA and Anti-HSA, p53 and azurin, biotin and streptavidin, and so on [84].

### 4.2.6 Mass spectrometry

Mass spectrometry (MS) is a well-recognized analytical tool for probing a variety of classes of compounds, including proteins, protein-protein/ligand complexes, nucleic acids, protein/ligand-nucleic acid complexes, and so on. With the advent of technology, the recently developed 'native mass spectrometric technique' can determine the thermodynamic parameters $\Delta H$, $\Delta S$, and $\Delta G$ for protein-ligand interactions [85,86]. The technique is based on the assumption that when biomolecules are in their compact, folded native state, they house fewer charges, resulting in a higher mass to charge ratio ($m/z$) than unfolded proteins. Any change in the conformation state displays a change in the charge state distribution (CSD) in the resulting mass spectrum. This technique is suitable for investigating the protein-ligand interaction as the intermolecular interactions are retained in the gas phase and do not require immobilization [87,88]. The MS methods have the advantage of being fast and having small sample requirements over the other techniques described earlier.

Cong et al. [89] described the native mass spectrometric technique in determining the binding thermodynamics of a few membrane protein–lipid and protein-ligand systems. With the help of a constructed device to control and change the temperatures of the analyte and air surrounding the mass spectrometer, they analyzed the binding thermodynamics using van't Hoff analysis. They obtained the $\Delta H$, $\Delta S$, and $\Delta G$ values and compared the results with ITC, DSC, and SPR data, which were in good agreement. The concept of the MS-based analysis is shown in Fig. 7.

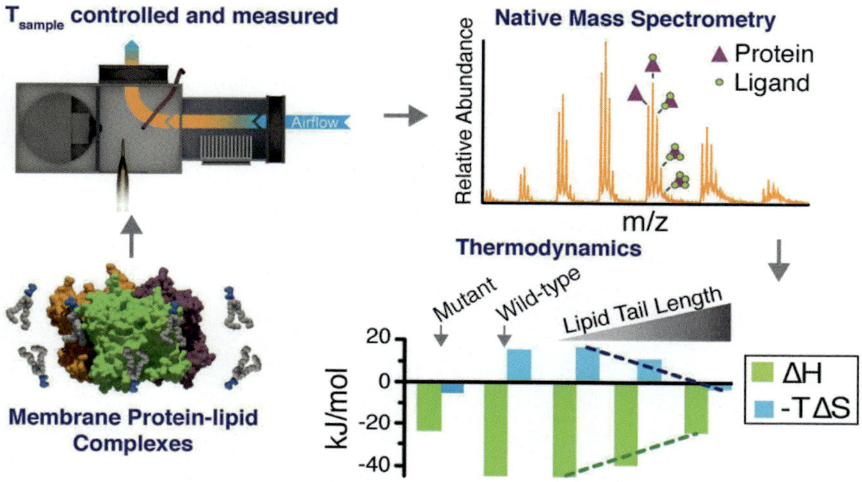


**Fig. 7** Schematic of the method to get the binding thermodynamics using native MS. Membrane protein-lipid complexes are introduced into the apparatus, which controls the sample temperature. Mass spectra of the protein-ligand complex will be obtained for different ligand concentrations or temperatures. Mass spectra were deconvoluted to get the mole fraction of free and bound states and globally fit to a binding model to obtain the equilibrium binding constant using which binding thermodynamics were determined through van't Hoff analysis. *(Figure adapted with permission from Cong X, Liu Y, Liu W, Liang X, Russell DH, Laganowsky A. Determining membrane protein-lipid binding thermodynamics using native mass spectrometry. J Am Chem Soc 2016;138(13):4346–49, https://doi.org/10.1021/jacs.6b01771.)*

Native MS was used to determine the energetics of lipid binding events to modulate the binding of the ammonia channel, AmtB, with the membrane regulatory protein, GlnK [90]. Cong et al. showed that the lipids binding to AmtB allosterically modulate the activity of membrane protein, GlnK. The observed value of $\Delta G$ through native MS was $-34.45\,\text{kJ/mol}$. They also determined the changes in enthalpy ($\Delta H$), heat capacity ($\Delta C_p$), and entropy ($T\Delta S$) using the nonlinear van't Hoff equation. The values observed were $\Delta H = -47.50\,\text{kJ/mol}$, $T\Delta S = 13.05\,\text{kJ/mol}$, and $\Delta C_p = -10.24$ kJ/mol/K. Their finding suggests conformational changes upon lipid binding events, which are regulated allosterically.

SPROX (stability of proteins from rates of oxidation) is another MS-based approach to measure the energetics, stability of proteins, and protein-ligand complex. In this technique, the extent of oxidation of proteins with added hydrogen peroxide by a denaturant is calculated as a function of its concentration using electrospray/matrix-assisted laser

**Table 2** Different spectroscopic techniques used to characterize the thermodynamic properties.

| Molecular system | Type of spectroscopic technique | Thermodynamic parameters | Reference |
|---|---|---|---|
| BSA and EAR8-II-THP (peptide-drug complex) | Fluorescence quenching studies | $K_{SV} = 1689\,M^{-1}$<br>$\Delta G = -18\,kJ/mol$<br>$\Delta H = -14.35\,kJ/mol$<br>$\Delta S = 12.71\,J/mol$ | [94] |
| SS-DNA and replication protein (RPA) | Fluorescence correlation | $K_d = 2.61 \pm 0.80 \times 10^{-10}\,M$<br>$\Delta G = -54.7 \pm 0.75\,kJ/mol$<br>$\Delta H = -66.5 \pm 8.9\,kJ/mol$<br>$T\Delta S = -11.8\,kJ/mol$ | [95] |
| L-Arabinose binding protein (ABP) and 1-deoxy-D-galactose | NMR | $K_d = 17.1 \pm 5\,mM$<br>$\Delta G = -15$ to $-24\,kJ/mol$ | [96] |
| DNA (chicken blood) and dexamethasone sodium phosphate (DPS) | UV-vis | $K_b = 8.43 \pm 0.29 \times 10^4\,M^{-1}$<br>$\Delta G = -28.10 \pm 1.15\,kJ/mol$<br>$\Delta H = -65.07 \pm 3.45\,kJ/mol$<br>$\Delta S = 0.31\,kJ/K\,mol$ | [97] |
| CT-DNA and ticlopidine | CD, fluorescence quenching studies | $K_b = 1.12 \times 10^3\,M^{-1}$<br>Number of binding sites, $n = 0.940$<br>$\Delta G = -5.0\,kcal/mol$ | [48] |
| CT DNA and antileukemic drug, Imanitib | CD, fluorescence | $K_b = 6.62 \pm 0.09 \times 10^3\,L/mol$<br>$\Delta G = -21.80 \pm 0.33\,kJ/mol$<br>$\Delta H = 95.29 \pm 1.43\,kJ/mol$<br>$\Delta S = 0.393 \pm 0.01\,kJ/K\,mol$ | [98] |

| | | | |
|---|---|---|---|
| Histamine and recombinant histamine binding protein | NMR, ITC | $K_d = 2.5 \pm 1.4$ nM<br>$\Delta G = -49.1 \pm 1.4$ kJ/mol<br>$\Delta H = -58.3 \pm 1.2$ kJ/mol<br>$T\Delta S = -9.3 \pm 2.5$ kJ/mol | [99] |
| Mouse major urinary protein (MUP) and pheromone, 2-s-butyl-4,5-dihydrothiazole | NMR relaxation studies | Favorable entropic contribution<br>$\Delta S = \sim 50$ kJ/mol | [100] |
| HSA and cleviprex | Fluorescence | $K_{SV} = 2.220 \times 10^4$ L/mol<br>$\Delta G = -26.66$ kJ/mol<br>$\Delta H = -18.30$ kJ/mol<br>$\Delta S = 28.94$ J/mol/K | [101] |
| Salomon testes DNA and Actinomycin D | UV-vis, fluorescence, CD | $\Delta G = -7.6$ kcal/mol<br>$\Delta H = 0.9$ kcal/mol<br>$\Delta S = 31.0$ cal/degmol | [102] |
| Poly d(AT) and Daunomycin | UV-vis, fluorescence, CD | $\Delta G = -9.3$ kcal/mol<br>$\Delta H = -8.9$ kcal/mol<br>$\Delta S = 1.3$ cal/degmol | [102] |
| Poly d(AT) and Netropsin | UV-vis, fluorescence, CD | $\Delta G = -12.3$ kcal/mol<br>$\Delta H = -9.2$ kcal/mol<br>$\Delta S = 10.3$ cal/degmol | [102] |
| CT DNA and Caffeic acid | UV, fluorescence | $\Delta G = -4.55$ kcal/mol<br>$\Delta H = -4.69$ kcal/mol<br>$\Delta S = -0.48$ cal/mol/K | [103] |
| CT DNA and Mesalamine (5-ASA) | UV, fluorescence | $\Delta G = -13.70$ kJ/mol<br>$\Delta H = -55.23$ kJ/mol<br>$\Delta S = -139.33$ J/mol/K | [104] |

*Continued*

**Table 2** Different spectroscopic techniques used to characterize the thermodynamic properties—cont'd

| Molecular system | Type of spectroscopic technique | Thermodynamic parameters | Reference |
|---|---|---|---|
| CT DNA and Citral | UV, fluorescence | $\Delta G = 11.64$ kJ/mol<br>$\Delta H = -8.06$ kJ/mol<br>$\Delta S = 3.58$ kJ/mol/K | [105] |
| CT DNA and Telmisartan | UV | $\Delta G = -19.59$ kJ/mol<br>$\Delta H = -183$ kJ/mol<br>$\Delta S = -550$ J/mol/K | [106] |
| Human hemoglobin protein (HMG) and metoprolol tartrate (MPT) | Fluorescence, CD | $\Delta G = -30.36$ kJ/mol<br>$\Delta H = -46.30$ kJ/mol<br>$\Delta S = -53.44$ J/mol/K | [107] |
| Human hemoglobin protein (HMG) and guaifenesin (GF) | Fluorescence, CD | $\Delta G = -30.29$ kJ/mol<br>$\Delta H = -38.44$ kJ/mol<br>$\Delta S = -27.34$ J/mol/K | [107] |
| Deoxytetranucleotide d(ApCpGpT) and ethidium bromide (EB) | NMR | $\Delta G = -6.6$ kcal/mol<br>$\Delta H = -22.7$ kcal/mol<br>$\Delta S = -54.0$ cal/mol/K | [108] |
| SS-DNA and Lamotrigine (LMT) | UV | $K = 1.19 \times 10^6$ $M^{-1}$<br>$\Delta G = -34.08$ kJ/mol | [109] |
| DS-DNA and Lamotrigine (LMT) | UV | $K = 6.93 \times 10^5$ $M^{-1}$<br>$\Delta G = -32.76$ kJ/mol | [109] |
| HSA and Morpholine (CMR) | Fluorescence | $\Delta G = -31.35$ kJ/mol<br>$\Delta H = 355.9$ kJ/mol<br>$\Delta S = 1.299$ kJ/mol/K | [110] |

| | | | |
|---|---|---|---|
| HSA and N-l-naphthyllaurohydroxamic acid (HA) | Fluorescence | $\Delta G = -5.34$ kJ/mol<br>$\Delta H = -12.28$ kJ/mol<br>$\Delta S = 17.87$ J/mol/K | [111] |
| HSA and Exenatide | Fluorescence, CD | $\Delta G = -25.33$ kJ/mol<br>$\Delta H = -45.30$ kJ/mol<br>$\Delta S = -66.67$ J/mol/K | [112] |
| Mannose binding protein (MBP) and Maltose | MS | $\Delta G = -34.9$ kJ/mol<br>$\Delta H = -14.1$ kJ/mol<br>$T\Delta S = -20.8$ kJ/mol/K | [89] |
| Lysozyme and NAG3 | MS | $\Delta G = -28.4$ kJ/mol<br>$\Delta H = -28.4$ kJ/mol<br>$T\Delta S = -0.01$ kJ/mol/K | [89] |
| Double-stranded oligonucleotide and Fluorophore-Texas Red (F) and Quencher-Iowa Black RQ (Q)<br>F-CGTACACATGC3′<br>5′-GCATGTGTACG-Q | Fluorescence | $\Delta G = -15.9$ kcal/mol<br>$\Delta H = -84.0$ kcal/mol<br>$\Delta S = -219.5$ kcal/mol/K | [113] |
| Double-stranded oligonucleotide and Fluorophore-Texas Red (F) and Quencher-Iowa Black RQ (Q)<br>F-CATACTACAAATA-3′<br>5′-TATTTGTAGTATG-Q | UV melting studies | $\Delta G = -14.8$ kcal/mol<br>$\Delta H = -90.1$ kcal/mol<br>$\Delta S = -242.8$ kcal/mol/K | [113] |

deposition/ionization mass spectrometry [91]. Oxidation rate was used to determine the folding free energy ($\Delta G_f$). The authors demonstrated the technique for ubiquitin and RNaseA and compared the values with values obtained from conventional methods.

## 5. Conclusions

Researchers need to understand the molecular recognition underlying many biological processes. Macromolecules, especially proteins, interact with a multitude of binding partners while they perform their functions. The association between these interacting partners drives the metabolic processes. Understanding the forces and energies involved in the biological world will help us better understand the processes involved. This knowledge can shed light on controlling these functions and lead us to design molecules for the betterment of life. Thermodynamic properties of the binding interactions can provide the minute details of the nature of interaction and the driving forces. The thermodynamic stability of a molecule/complex is an essential parameter that is to be determined during a chemical/biochemical reaction. The enthalpy and entropic contributions in the inculcation of this binding reaction will educate one about the forces involved. Based on the forces involved, such as hydrophobic, hydrophilic, covalent, noncovalent, or electrostatic interaction, one can get a better perception of the processes.

This chapter discussed different spectroscopic techniques utilized by the scientific community and a few applications where these techniques are successfully employed in deriving information. An extensive summary of different spectroscopic techniques used in understanding the thermodynamic properties of various systems is provided in Table 2. Spectroscopic techniques, such as UV-vis, fluorescence, CD, and NMR, utilize the interaction of electromagnetic radiation with the sample without causing much damage to the test sample [92,93]. These techniques can be used without any external agent (e.g., a fluorescence tag) in a majority of the cases. However, in the case of a crowded environment, tagging/labeling the components will help resolve the reacting partners. The more general calorimetric approaches using ITC and DSC methods lack these properties. The sample cannot be recovered after their use in these techniques. The spectroscopic process is fast, and the data are in good agreement with the corresponding calorimetric data and thus are reliable. With the advancement in technologies in the spectroscopic world, highly complex systems can be studied spectroscopically in the near future.

## Acknowledgments

Bharti acknowledges the Council of Scientific and Industrial Research (CSIR), India, for the Ph.D. assistantship (09/143(0909)/2018-EMR-I). Maya S Nair acknowledges the support of the Council of Scientific and Industrial Research (CSIR), India, for a grant 37(1732)/19/EMR-II.

## References

[1] Bowler TD. Treatise on basic philosophy, volume 4, ontology II: a world of systems. Int J Gen Syst 1980;6(1):51–3. https://doi.org/10.1080/03081078008934780.
[2] Lohman TM, Bujalowski W. Thermodynamic methods for model-independent determination of equilibrium binding isothermsa for protein-DNA interactions: spectroscopic approaches to monitor binding. Methods Enzymol 1991;208:258–90. https://doi.org/10.1016/0076-6879(91)08017-C.
[3] Puglisi R, Brylski O, Alfano C, Martin SR, Pastore A, Temussi PA. Quantifying the thermodynamics of protein unfolding using 2D NMR spectroscopy. Commun Chem 2020;3(1):100. https://doi.org/10.1038/s42004-020-00358-1.
[4] McNaught AD, Wilkinson A. Hydrogen bond. In: The IUPAC compendium of chemical terminology, vol. 1077. Research Triangle Park, NC: International Union of Pure and Applied Chemistry (IUPAC); 2014. p. 2899. https://doi.org/10.1351/goldbook.H02899.
[5] Rose GD, Wolfenden R. Hydrogen bonding, hydrophobicity, packing, and protein folding. Annu Rev Biophys Biomol Struct 1993;22(1):381–415. https://doi.org/10.1146/annurev.bb.22.060193.002121.
[6] Emsley J. Very strong hydrogen bonding. Chem Soc Rev 1980;9(1):91. https://doi.org/10.1039/cs9800900091.
[7] Lazaridis T, Archontis G, Karplus M. Enthalpic contribution to protein stability: insights from atom-based calculations and statistical mechanics. Adv Protein Chem 1995;47:231–306. Academic Press https://doi.org/10.1016/S0065-3233(08)60547-1.
[8] Kresheck GC, Klotz IM. Thermodynamics of transfer of amides from an apolar to an aqueous solution. Biochemistry 1969;8(1):8–12. https://doi.org/10.1021/bi00829a002.
[9] Kauzmann W. Some factors in the interpretation of protein denaturation. Adv Protein Chem 1959;1–63. https://doi.org/10.1016/S0065-3233(08)60608-7.
[10] Krimm S. The hydrophobic effect: formation of micelles and biological membranes. J Polym Sci, Polym Lett Ed 1980;18(10):687. https://doi.org/10.1002/pol.1980.130181008.
[11] Atkins P, De Paula J. Physical chemistry. 8th ed. New York: Oxford University Press; 2006.
[12] Zimmerman J. Physical chemistry for the biosciences: Chang, Raymond. Biochem Mol Biol Educ 2005;33(5):382. https://doi.org/10.1002/bmb.2005.49403305383.
[13] Garrett RH, Grisham CM. Biochemistry. 3rd ed. Belmont, CA: Thomson Brooks/Cole; 2005.
[14] Bissantz C, Kuhn B, Stahl M. A medicinal chemist's guide to molecular interactions. J Med Chem 2010;53(14):5061–84. https://doi.org/10.1021/jm100112j.
[15] Perozzo R, Folkers G, Scapozza L. Thermodynamics of protein-ligand interactions: history, presence, and future aspects. J Recept Signal Transduction 2004;24(1–2):1–52. https://doi.org/10.1081/RRS-120037896.

[16] Wang J, Wolf RM, Caldwell JW, Kollman PA, Case DA. Development and testing of a general amber force field. J Comput Chem Jul. 2004;25(9):1157–74. https://doi.org/10.1002/jcc.20035.
[17] Cooper A. Thermodynamics of protein folding and stability. In: Allen G, editor. Protein: a comprehensive treatise, vol. 2. Stamford, CT: JAI Press Inc.; 1999. p. 217–70.
[18] Bronowska AK. Thermodynamics of ligand-protein interactions: implications for molecular design. In: Thermodynamics—interaction studies—solids, liquids and gases. InTech; 2011. p. 1–49. https://doi.org/10.5772/19447.
[19] Fleck M, Zagrovic B. Configurational entropy components and their contribution to biomolecular complex formation. J Chem Theory Comput 2019;15(6):3844–53. https://doi.org/10.1021/acs.jctc.8b01254.
[20] Doig AJ, Sternberg MJE. Side-chain conformational entropy in protein folding. Protein Sci 1995;4(11):2247–51. https://doi.org/10.1002/pro.5560041101.
[21] Mills JL, Liu G, Skerra A, Szyperski T. NMR structure and dynamics of the engineered fluorescein-binding lipocalin FluA reveal rigidification of β-barrel and variable loops upon enthalpy-driven ligand binding. Biochemistry 2009;48(31):7411–9. https://doi.org/10.1021/bi900535j.
[22] Bergethon PR. Philosophy and practice of biophysical study. In: The physical basis of biochemistry. New York, NY: Springer New York; 2010. p. 5–22. https://doi.org/10.1007/978-1-4419-6324-6_2.
[23] Djordjevic IB. Fundamentals of biological thermodynamics, biomolecules, cellular genetics, and bioenergetics. In: Quantum biological information theory. Cham: Springer International Publishing; 2016. p. 75–142. https://doi.org/10.1007/978-3-319-22816-7_3.
[24] David H, Robert R, Jearl W. Fundamental of physics. 10th ed. John Wiley & Sons; 2013.
[25] Serdyuk IN, Zaccai NR, Zaccai J. Methods in molecular biophysics. Cambridge: Cambridge University Press; 2007. https://doi.org/10.1017/CBO9780511811166.
[26] Creighton TE. Protein folding. Biochem J 1990;270(1):1–16. https://doi.org/10.1042/bj2700001.
[27] Hao M, Scheraga HA. Statistical thermodynamics of protein folding: sequence dependence. J Phys Chem 1994;98(39):9882–93. https://doi.org/10.1021/j100090a024.
[28] Gonzalez-Diaz PF, Siguenza CL. Protein folding and cosmology. J Phys Condens Matter 1997;22(41):415106. https://doi.org/10.1088/0953-8984/22/41/415106.
[29] Garcia-Fuentes L, Ramiro T, Sanz IQ-S, Baro C. Thermodynamics of molecular recognition by calorimetry. In: Thermodynamics—physical chemistry of aqueous systems. InTech; 2011. https://doi.org/10.5772/21306.
[30] Akram S, Khosrow S. Thermodynamics of protein folding : methodology, data analysis and interpretation of data. Eur Biophys J 2019;48(4):305–16. https://doi.org/10.1007/s00249-019-01362-7.
[31] Dunitz JD. Win some, lose some: enthalpy-entropy compensation in weak intermolecular interactions. Chem Biol 1995;2(11):709–12. https://doi.org/10.1016/1074-5521(95)90097-7.
[32] Williams MA, Ladbury JE. Protein science encyclopedia. Weinheim, Germany: Wiley-VCH Verlag GmbH & Co. KGaA; 2008. https://doi.org/10.1002/9783527610754.
[33] Liu F, Zhang JZH, Mei Y. The origin of the cooperativity in the streptavidin-biotin system: a computational investigation through molecular dynamics simulations. Sci Rep 2016;6(1):27190. https://doi.org/10.1038/srep27190.
[34] Evans DA, Bronowska AK. Implications of fast-time scale dynamics of human DNA/RNA cytosine methyltransferases (DNMTs) for protein function. Theor Chem Accounts 2010;125(3–6):407–18. https://doi.org/10.1007/s00214-009-0681-2.

[35] MacRaild CA, Daranas AH, Bronowska A, Homans SW. Global changes in local protein dynamics reduce the entropic cost of carbohydrate binding in the arabinose-binding protein. J Mol Biol 2007;368(3):822–32. https://doi.org/10.1016/j.jmb.2007.02.055.
[36] Jelesarov I, Bosshard HR. Isothermal titration calorimetry and differential scanning calorimetry as complementary tools to investigate the energetics of biomolecular recognition. J Mol Recognit 1999;12(1):3–18. https://doi.org/10.1002/(SICI)1099-1352(199901/02)12:1<3::AID-JMR441>3.0.CO;2-6.
[37] Leavitt S, Freire E. Direct measurement of protein binding energetics by isothermal titration calorimetry. Curr Opin Struct Biol 2001;11(5):560–6. https://doi.org/10.1016/S0959-440X(00)00248-7.
[38] Gill P, Moghadam TT, Ranjbar B. Differential scanning calorimetry techniques: applications in biology and nanoscience. J Biomol Tech 2010;21(4):167–93.
[39] Prenner E, Chiu M. Differential scanning calorimetry: an invaluable tool for a detailed thermodynamic characterization of macromolecules and their interactions. J Pharm Bioallied Sci 2011;3(1):39. https://doi.org/10.4103/0975-7406.76463.
[40] Schmid F. Biological macromolecules: UV-visible spectrophotometry. In: eLS. Wiley; 2001. p. 1–4. https://doi.org/10.1038/npg.els.0003142.
[41] Li Y, He W, Liu J, Sheng F, Hu Z, Chen X. Binding of the bioactive component Jatrorrhizine to human serum albumin. Biochim Biophys Acta Gen Subj 2005;1722(1):15–21. https://doi.org/10.1016/j.bbagen.2004.11.006.
[42] Hirose K. A practical guide for the determination of binding constants. J Incl Phenom Mol Recognit Chem 2001;39(3–4):193–209.
[43] Tsuchida R. A spectrographic method for the study of unstable compounds in equilibrium. Bull Chem Soc Jpn 1935;10(1):27–39. https://doi.org/10.1246/bcsj.10.27.
[44] Harvey AE, Manning DL. Spectrophotometric studies of empirical formulas of complex ions 1. J Am Chem Soc 1952;74(19):4744–6. https://doi.org/10.1021/ja01139a005.
[45] Yoe JH, Jones AL. Colorimetric determination of iron with disodium-1,2-dihydroxybenzene-3,5-disulfonate. Ind Eng Chem Anal Ed 1944;16(2):111–5. https://doi.org/10.1021/i560126a015.
[46] Maciążek-Jurczyk M, Sułkowska A, Bojko B, Równicka J, Sułkowski WW. Fluorescence analysis of competition of phenylbutazone and methotrexate in binding to serum albumin in combination treatment in rheumatology. J Mol Struct 2009;924–926:378–84. https://doi.org/10.1016/j.molstruc.2008.12.023.
[47] Raoufi A, Ebrahimi M, Bozorgmehr MR. Determination of thermodynamics constant of interaction among of atenolol and metoprolol with human serum albumin: spectroscopic and molecular modeling approaches. Russ J Phys Chem A 2021;95(6):1269–76. https://doi.org/10.1134/S0036024421140181.
[48] Afrin S, et al. Molecular spectroscopic and thermodynamic studies on the interaction of antiplatelet drug ticlopidine with calf thymus DNA. Spectrochim Acta A Mol Biomol Spectrosc 2017;186:66–75. https://doi.org/10.1016/j.saa.2017.05.073.
[49] Lakowicz JR. Principles of fluorescence spectroscopy. Boston, MA: Springer US; 2006. https://doi.org/10.1007/978-0-387-46312-4.
[50] Hino S, Ichikawa T, Kojima Y. Thermodynamic properties of metal amides determined by ammonia pressure-composition isotherms. J Chem Thermodyn 2010;42(1):140–3. https://doi.org/10.1016/j.jct.2009.07.024.
[51] Cheng Z, Liu R, Jiang X. Spectroscopic studies on the interaction between tetrandrine and two serum albumins by chemometrics methods. Spectrochim Acta A Mol Biomol Spectrosc 2013;115:92–105. https://doi.org/10.1016/j.saa.2013.06.007.
[52] Hussain I, Fatima S, Siddiqui S, Ahmed S, Tabish M. Exploring the binding mechanism of β-resorcylic acid with calf thymus DNA: insights from multi-spectroscopic,

thermodynamic and bioinformatics approaches. Spectrochim Acta A Mol Biomol Spectrosc 2021;260:119952. https://doi.org/10.1016/j.saa.2021.119952.
[53] Wang Q, Chen W, Liu X-Y, Liu Y, Jiang F. Thermodynamic implications and time evolution of the interactions of near-infrared PbS quantum dots with human serum albumin. ACS Omega 2021;6(8):5569–81. https://doi.org/10.1021/acsomega.0c05974.
[54] Posokhov YO, Rodnin MV, Lu L, Ladokhin AS. Membrane insertion pathway of annexin B12: thermodynamic and kinetic characterization by fluorescence correlation spectroscopy and fluorescence quenching. Biochemistry 2008;47 (18):5078–87. https://doi.org/10.1021/bi702223c.
[55] Ladokhin AS. Fluorescence spectroscopy in thermodynamic and kinetic analysis of pH-dependent membrane protein insertion. Methods Enzymol 2009;19–42. https://doi.org/10.1016/S0076-6879(09)66002-X.
[56] Kyrychenko A, Posokhov YO, Vargas-uribe M, Ghatak C, Rodnin MV, Ladokhin AS. Reviews in Fluorescence 2016. Cham: Springer International Publishing; 2017. https://doi.org/10.1007/978-3-319-48260-6.
[57] Martin SR, Schilstra MJ. Circular dichroism and its application to the study of biomolecules. Methods Cell Biol 2008;84(07):263–93. https://doi.org/10.1016/S0091-679X(07)84010-6.
[58] Bishop GR, Chaires JB. Characterization of DNA structures by circular dichroism. In: Current protocols in nucleic acid chemistry. Hoboken, NJ, USA: John Wiley & Sons, Inc; 2002. p. 7.11.1–8. https://doi.org/10.1002/0471142700.nc0711s11.
[59] Masino L, Martin SR, Bayley PM. Ligand binding and thermodynamic stability of a multidomain protein, calmodulin. Protein Sci 2000;9(8):1519–29. https://doi.org/10.1110/ps.9.8.1519.
[60] Woody RW. Circular dichroism. Methods Enzymol 1995;246:34–71. https://doi.org/10.1016/0076-6879(95)46006-3.
[61] Greenfield NJ. Determination of the folding of proteins as a function of denaturants, osmolytes or ligands using circular dichroism. Nat Protoc 2006;1(6):2733–41. https://doi.org/10.1038/nprot.2006.229.
[62] Miles AJ, Wallace BA. Synchrotron radiation circular dichroism spectroscopy of proteins and applications in structural and functional genomics. Chem Soc Rev 2006;35 (1):39–51. https://doi.org/10.1039/B316168B.
[63] Greenfield NJ. Using circular dichroism spectra to estimate protein secondary structure. Nat Protoc 2006;1(6):2876–90. https://doi.org/10.1038/nprot.2006.202.
[64] Ohgushi M, Wada A. 'Molten-globule state': a compact form of globular proteins with mobile side-chains. FEBS Lett 1983;164(1):21–4. https://doi.org/10.1016/0014-5793(83)80010-6.
[65] Greenfield NJ. Circular dichroism analysis for protein-protein interactions. Methods Mol Biol 2004;261:55–78. https://doi.org/10.1385/1-59259-762-9. 05.
[66] Schellman JA. Macromolecular binding. Biopolymers 1975;14(5):999–1018. https://doi.org/10.1002/bip.1975.360140509.
[67] Mayhood TW, Windsor WT. Ligand binding affinity determined by temperature-dependent circular dichroism: cyclin-dependent kinase 2 inhibitors. Anal Biochem 2005;345(2):187–97. https://doi.org/10.1016/j.ab.2005.07.032.
[68] Pace CN, McGrath T. Substrate stabilization of lysozyme to thermal and guanidine hydrochloride denaturation. J Biol Chem 1980;255(9):3862–5. https://doi.org/10.1016/S0021-9258(19)85604-1.
[69] Dunlap TB, Kirk JM, Pena EA, Yoder MS, Creamer TP. Thermodynamics of binding by calmodulin correlates with target peptide α-helical propensity. Proteins: Struct Funct Bioinf 2013;81(4):607–12. https://doi.org/10.1002/prot.24215.

[70] Gondeau C. Circular dichroism and UV melting studies on formation of an intramolecular triplex containing parallel T*A:T and G*G:C triplets: netropsin complexation with the triplex. Nucleic Acids Res 1998;26(21):4996–5003. https://doi.org/10.1093/nar/26.21.4996.

[71] Boehr DD, Dyson HJ, Wright PE. An NMR perspective on enzyme dynamics. Chem Rev 2006;106(8):3055–79. https://doi.org/10.1021/cr050312q.

[72] Meyer B, Peters T. NMR spectroscopy techniques for screening and identifying ligand binding to protein receptors. Angew Chem Int Ed 2003;42(8):864–90. https://doi.org/10.1002/anie.200390233.

[73] Krishnan V. Molecular thermodynamics using nuclear magnetic resonance (NMR) spectroscopy. Inventions 2019;4(1):13. https://doi.org/10.3390/inventions4010013.

[74] Logan SR. The origin and status of the Arrhenius equation. J Chem Educ 1982;59(4):279. https://doi.org/10.1021/ed059p279.

[75] Bain AD. Chemical exchange in NMR. Prog Nucl Magn Reson Spectrosc 2003;43(3–4):63–103. https://doi.org/10.1016/j.pnmrs.2003.08.001.

[76] Eyring H. The activated complex in chemical reactions. J Chem Phys 1935;3(2):107–15. https://doi.org/10.1063/1.1749604.

[77] Kleckner IR, Foster MP. An introduction to NMR-based approaches for measuring protein dynamics. Biochim Biophys Acta Protein Proteomics Aug. 2011;1814(8):942–68. https://doi.org/10.1016/j.bbapap.2010.10.012.

[78] Grutsch S, Brüschweiler S, Tollinger M. NMR methods to study dynamic allostery. PLoS Comput Biol 2016;12(3). https://doi.org/10.1371/journal.pcbi.1004620, e1004620.

[79] Tzeng S, Kalodimos CG. Protein activity regulation by conformational entropy. Nature 2012;488(7410):236–40. https://doi.org/10.1038/nature11271.

[80] Sapienza PJ, Lee AL. Using NMR to study fast dynamics in proteins: methods and applications. Curr Opin Pharmacol 2010;10(6):723–30. https://doi.org/10.1016/j.coph.2010.09.006.

[81] Bracken C, Carr PA, Cavanagh J, Palmer AG. Temperature dependence of intramolecular dynamics of the basic leucine zipper of GCN4: implications for the entropy of association with DNA. J Mol Biol 1999;285(5):2133–46. https://doi.org/10.1006/jmbi.1998.2429.

[82] Butt HJ, Cappella B, Kappl M. Force measurements with the atomic force microscope: technique, interpretation and applications. Surf Sci Rep 2005;59(1–6):1–152. https://doi.org/10.1016/j.surfrep.2005.08.003.

[83] Oberhauser AF, Hansma PK, Carrion-Vazquez M, Fernandez JM. Stepwise unfolding of titin under force-clamp atomic force microscopy. Proc Natl Acad Sci 2001;98(2):468–72. https://doi.org/10.1073/pnas.98.2.468.

[84] Bizzarri AR, Cannistraro S. The application of atomic force spectroscopy to the study of biological complexes undergoing a biorecognition process. Chem Soc Rev 2010;39(2):734–49. https://doi.org/10.1039/B811426A.

[85] Loo JA. Studying noncovalent protein complexes by electrospray ionization mass spectrometry. Mass Spectrom Rev 1997;16(1):1–23. https://doi.org/10.1002/(SICI)1098-2787(1997)16:1<1::AID-MAS1>3.0.CO;2-L.

[86] Marcoux J, Robinson CV. Twenty years of gas phase structural biology. Structure 2013;21(9):1541–50. https://doi.org/10.1016/j.str.2013.08.002.

[87] Gülbakan B, Barylyuk K, Zenobi R. Determination of thermodynamic and kinetic properties of biomolecules by mass spectrometry. Curr Opin Biotechnol 2015;31:65–72. https://doi.org/10.1016/j.copbio.2014.08.003.

[88] Fenn JB, Mann M, Meng CK, Wong SF, Whitehouse CM. Electrospray ionization for mass spectrometry of large biomolecules. Science 1989;246(4926):64–71. https://doi.org/10.1126/science.2675315.

[89] Cong X, Liu Y, Liu W, Liang X, Russell DH, Laganowsky A. Determining membrane protein-lipid binding thermodynamics using native mass spectrometry. J Am Chem Soc 2016;138(13):4346–9. https://doi.org/10.1021/jacs.6b01771.
[90] Cong X, Liu Y, Liu W, Liang X, Laganowsky A. Allosteric modulation of protein-protein interactions by individual lipid binding events. Nat Commun 2017;8 (1):2203. https://doi.org/10.1038/s41467-017-02397-0.
[91] West GM, Tang L, Fitzgerald MC. Thermodynamic analysis of protein stability and ligand binding using a chemical modification- and mass spectrometry-based strategy. Anal Chem 2008;80(11):4175–85. https://doi.org/10.1021/ac702610a.
[92] Tripathi T, Dubey V, editors. Advances in protein molecular and structural biology methods. 1st. USA: Academic Press; 2022. p. 1–714.
[93] Singh D, Tripathi T, editors. Frontiers in protein structure, function, and dynamics. 1st. Singapore: Springer; 2022. p. 1–458.
[94] Sadatmousavi P, Kovalenko E, Chen P. Thermodynamic characterization of the interaction between a peptide-drug complex and serum proteins. Langmuir 2014;30 (37):11122–30. https://doi.org/10.1021/la502422u.
[95] Schubert F, Zettl H, Häfner W, Krauss G, Krausch G. Comparative thermodynamic analysis of DNA-protein interactions using surface plasmon resonance and fluorescence correlation spectroscopy. Biochemistry 2003;42(34):10288–94. https://doi.org/10.1021/bi034033d.
[96] MacRaild CA, Daranas AH, Bronowska A, Homans SW. Global changes in local protein dynamics reduce the entropic cost of carbohydrate binding in the arabinose-binding protein. J Mol Biol 2007;368(3):822–32. https://doi.org/10.1016/j.jmb.2007.02.055.
[97] Shah A, Khan AM, Usman M, Qureshi R, Siddiq M, Shah SS. Thermodynamic characterization of dexamethasone sodium phosphate and its complex with DNA as studied by conductometric and spectroscopic techniques. J Chil Chem Soc 2009;54 (2):134–7. https://doi.org/10.4067/S0717-97072009000200007.
[98] Hegde AH, Seetharamappa J. Fluorescence and circular dichroism studies on binding and conformational aspects of an anti-leukemic drug with DNA. Mol Biol Rep 2014;41(1):67–71. https://doi.org/10.1007/s11033-013-2838-2.
[99] Syme NR, Dennis C, Bronowska A, Paesen GC, Homans SW. Comparison of entropic contributions to binding in a 'Hydrophilic' versus 'Hydrophobic' ligand-protein interaction. J Am Chem Soc 2010;132(25):8682–9. https://doi.org/10.1021/ja101362u.
[100] Zídek L, Novotny MV, Stone MJ. Increased protein backbone conformational entropy upon hydrophobic ligand binding. Nat Struct Biol 1999;6 (12):1118–21. https://doi.org/10.1038/70057.
[101] Wang X, et al. Studies on the competitive binding of cleviprex and flavonoids to plasma protein by multi-spectroscopic methods: a prediction of food-drug interaction. J Photochem Photobiol B Biol 2017;175(July):192–9. https://doi.org/10.1016/j.jphotobiol.2017.08.037.
[102] Marky LA, Snyder JG, Remeta DP, Breslauer KJ. Thermodynamics of drug-DNA interactions. J Biomol Struct Dyn 1983;1(2):487–507. https://doi.org/10.1080/07391102.1983.10507457.
[103] Sarwar T, Ishqi HM, Rehman SU, Husain MA, Rahman Y, Tabish M. Caffeic acid binds to the minor groove of calf thymus DNA: a multi-spectroscopic, thermodynamics and molecular modelling study. Int J Biol Macromol 2017;98:319–28. https://doi.org/10.1016/j.ijbiomac.2017.02.014.
[104] Shahabadi N, Fili SM, Kheirdoosh F. Study on the interaction of the drug mesalamine with calf thymus DNA using molecular docking and spectroscopic techniques. J Photochem Photobiol B Biol 2013;128:20–6. https://doi.org/10.1016/j.jphotobiol.2013.08.005.

[105] Alam MF, Varshney S, Khan MA, Laskar AA, Younus H. In vitro DNA binding studies of therapeutic and prophylactic drug citral. Int J Biol Macromol 2018;113:300–8. https://doi.org/10.1016/j.ijbiomac.2018.02.098.

[106] Dong X, Lou Y, Zhou K, Shi J. Exploration of association of telmisartan with calf thymus DNA using a series of spectroscopic methodologies and theoretical calculation. J Mol Liq 2018;266:1–9. https://doi.org/10.1016/j.molliq.2018.06.057.

[107] Duman O, Tunç S, Kancı Bozoğlan B. Characterization of the binding of metoprolol tartrate and guaifenesin drugs to human serum albumin and human hemoglobin proteins by fluorescence and circular dichroism spectroscopy. J Fluoresc 2013;23 (4):659–69. https://doi.org/10.1007/s10895-013-1177-y.

[108] Davies DB, Djimant LN, Baranovsky SF, Veselkov AN. 1H-NMR determination of the thermodynamics of drug complexation with single-stranded and double-stranded oligonucleotides in solution: ethidium bromide complexation with the deoxytetranucleotides 5′-d(ApCpGpT), 5′-d(ApGpCpT), and 5′-d(TpGpCpA). Biopolymers 1997;42(3):285–95. https://doi.org/10.1002/(SICI)1097-0282(199709)42:3<285::AID-BIP2>3.0.CO;2-I.

[109] Morawska K, Popławski T, Ciesielski W, Smarzewska S. Interactions of lamotrigine with single- and double-stranded DNA under physiological conditions. Bioelectrochemistry 2020;136:107630. https://doi.org/10.1016/j.bioelechem.2020.107630.

[110] Fulya K, Taskın TT, Halil D, Zeynel S, Elmas G. Synthesis of new morpholine containing 3-amido-9-ethylcarbazole derivative and studies on its biophysical interactions with calf thymus DNA/HSA. J Biomol Struct Dyn 2021;39(5):1561–71. https://doi.org/10.1080/07391102.2020.1734093.

[111] Agrawal R, Thakur Y, Tripathi M, Khursheed M. Elucidating the binding propensity of naphthyl hydroxamic acid to human serum albumin (HSA): multi-spectroscopic and molecular modeling approach. J Mol Struct 2019;1184:1–11. https://doi.org/10.1016/j.molstruc.2019.01.067.

[112] Śliwińska-hill U, Wiglusz K. The interaction between human serum albumin and antidiabetic agent—exenatide: determination of the mechanism binding and effect on the protein conformation by fluorescence and circular dichroism techniques—Part I. J Biomol Struct Dyn 2020;38(8):2267–75. https://doi.org/10.1080/07391102.2019.1630007.

[113] You Y, Tataurov AV, Owczarzy R. Measuring thermodynamic details of DNA hybridization using fluorescence. Biopolymers 2011;95(7):472–86. https://doi.org/10.1002/bip.21615.

CHAPTER 14

# Spectroscopic methods to detect and analyze protein oligomerization, aggregation, and fibrillation

**Kummari Shivani*, Amrita Arpita Padhy*, Subhashree Sahoo, Varsha Kumari, and Parul Mishra**
Department of Animal Biology, School of Life Sciences, University of Hyderabad, Telangana, India

## 1. Introduction

The folding of proteins into discrete conformations provides a solid foundation for the diverse functions served by these macromolecules in a cell. The information to acquire the native conformation is deeply embedded in the amino acid sequence of a protein [1]. The three-dimensional native conformations include the tertiary and quaternary structures that result from the interactions between the side chains of amino acids of one or more polypeptide chains. The functional requirements of each protein dictate whether it exists as a monomer or an oligomer [2,3]. Although both forms are abundantly present in cells, oligomerization of protein helps regulate or maximize its functions [4,5]. In hyperthermophilic organisms, the oligomers of some proteins are thermodynamically more stable than their monomeric counterparts [6]. A regulatory role of oligomerization is well documented in hemoglobin (Hb), where the tetrameric form is preferred for positive cooperativity of its oxygen-binding function [7]. Further, oligomeric forms are also evolutionarily selected owing to their enhanced stability [8,9].

Monomeric proteins display a smooth folding pathway with their native conformation representing the lowest energy state acquired during protein folding. Contrary to this, the oligomers or large polypeptides have a rugged energy landscape indicating an ensemble competing for low-energy conformational states. The nonnative interactions in these states often stabilize the partially folded or unfolded proteins that exist as aggregates. Amyloid fibrils

* Equal contribution.

*Advanced Spectroscopic Methods to Study Biomolecular Structure and Dynamics*
https://doi.org/10.1016/B978-0-323-99127-8.00016-7

are such insoluble aggregate species of proteins that are highly polymorphic in their structures but represent the kinetically trapped metastable folding intermediates. Protein aggregation is often initiated by misfolding of the native protein, followed by fibrillization, a relatively slow process that takes hours or weeks to yield the fibrils. Understanding the kinetics of folding, misfolding, and aggregation is critical for the structural and functional analysis of proteins. Accumulation of misfolded and aggregated proteins is associated with many neurodegenerative diseases such as Huntington's, Alzheimer's, Parkinson's, and prion diseases. Interestingly, the partially folded intermediates are the primary pathogenic forms in these disease conditions and can either transform into mature fibrils or refold back into monomers. This dynamic nature and conformational heterogeneity impose a challenge for the structural understanding of intermediate species in the unfolding pathway of aggregation-prone proteins [10].

Although physiological conditions can be used to analyze the native structures of proteins, characterization of folding intermediates and aggregates is possible by stabilizing these structures using variable experimental conditions such as temperature, pH, or chemical denaturants. Fluorescence-based methods, electron microscopy, or atomic force microscopy are frequently used to analyze the monomeric, oligomeric, and aggregate conformations [11]. Based on the fluorescence intensity of the fluorophore, the monomeric and oligomeric conformations can be quantified and distinguished. Imaging techniques coupled with fluorophore tagging can also inform the spatial and temporal distribution of oligomers and monomers. Because transient changes in cellular conditions can trigger unfolding and aggregation, these techniques can demonstrate the effects of altering protein concentrations, pH, or temperatures on the stability of the conformers. Although it is possible to learn about the different sizes of oligomers or aggregates from single-molecule imaging techniques, low concentrations of proteins employed in these experiments limit the identification of oligomers that are stable only at higher protein concentrations. Furthermore, secondary structural elements or their transitions cannot be inferred from image analysis of oligomers or aggregates. Spectroscopic techniques can help overcome these limitations. They are highly reproducible and commonly used to evaluate the unfolding kinetics of soluble proteins, which is unachievable using high-resolution methods. Although many spectroscopic methods such as circular dichroism (CD) and nuclear magnetic resonance (NMR) study protein oligomerization or aggregation in vitro, infrared (IR) spectral analysis by Fourier transform infrared spectroscopy (FTIR) has demonstrated the kinetics of inclusion body formation within actively growing *Escherichia coli* cells [12,13]. Nonpathological protein aggregates have gained attention for their benefits in the food industry and as drug delivery nanoparticles. Collagen, gelatin, and many milk proteins can denature and self-assemble to form nanoparticles with therapeutic values [14]. Understanding the structural details of these soluble proteins and their aggregated forms is imperative for their utilization in the field of nanomedicine. Analysis of oligomers or aggregates by spectroscopic methods provides a better understanding of the kinetics of protein aggregation impacted by changes in the cellular environment or solution properties. In this chapter, we discuss different spectroscopic approaches that have been successfully applied to monitor various protein assemblies to translate their functional modalities in biomedical research, therapeutics, and fundamental protein science.

## 2. UV-visible spectroscopy

UV-visible (UV-Vis) spectroscopy is a common spectroscopic technique for the assessment of protein conformations of soluble or aggregating properties. Being a nondestructive, quick, and yet very sensitive method for spectral analysis, it offers a flexible platform for detecting variable protein aggregate species. With recent advancements, the shift from conventional detectors to diode-array detectors has been successfully combined with this technique to renew its utilization in the structural analysis of proteins.

### 2.1 Principle

It is based on the absorption of electromagnetic radiation by macromolecules such as proteins. The light energy of 100–400 kJ/mol in the near UV-Vis region is absorbed by a chromophore and causes a transition of electrons from their ground state to an excited state, resulting in an absorption spectrum at different wavelengths. It follows Beer-Lambert's law, which states that when monochromatic light passes through a protein sample, the intensity of transmitted light decreases exponentially as the thickness and concentration of absorbing species increase.

$$A = \log_{10}(I_0/I) = kl$$

where $A$ is the absorbance of molecules at a specific wavelength, $I_0$ is the intensity of incident light, $I$ is the intensity of light transmitted from a sample, $l$ is the optical pass of the sample, and $k$ is the chromophore concentration proportionality constant.

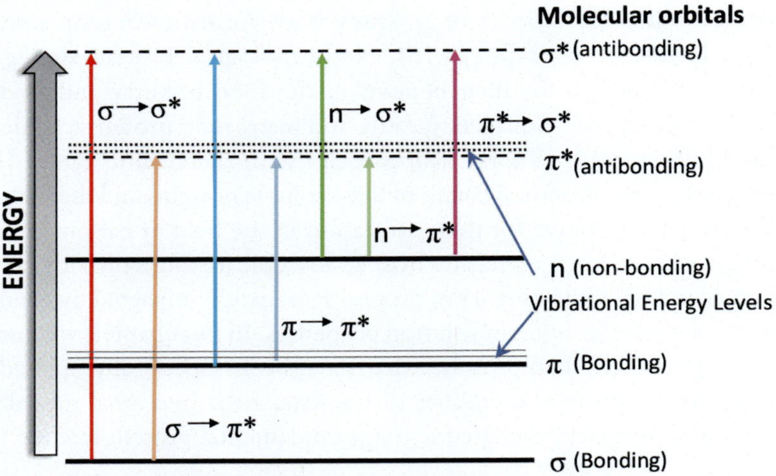


**Fig. 1** Electronic transitions observed in UV-Vis and CD spectroscopy within the molecular orbitals of different energy levels when the electromagnetic radiations interact with organic molecules. The solid lines represent the bonding and nonbonding orbitals, whereas the dashed lines represent the antibonding orbitals.

In the UV-Vis region, the absorption of electromagnetic radiation causes the molecules to jump from pi ($\pi$) bonding orbitals to pi antibonding ($\pi^*$) orbitals and from nonbonding ($n$) orbitals to pi antibonding ($\pi^*$) orbitals (Fig. 1). This characteristic feature is attributed by chromophores to quantify the proteins, which is an indirect method of quantifying the protein molecules. Monomeric or soluble proteins absorb in the range of 280–320 nm, whereas an absorption maximum beyond 320 nm is exhibited by aggregated particles. The shifts in wavelength are indicative of structural changes occurring in the protein. Hence, UV-Vis spectroscopy can be used to estimate protein concentrations. Proteins containing aromatic amino acids can be identified by spectral peaks at 280 nm, whereas nucleic acid can be detected at 260 nm. Other buffer components such as dithiothreitol (DTT) broaden the peak, showing an absorption maximum near 250 nm or below. Absorption values at 260/280 nm should be carefully recorded as having a ratio close to 0.57 indicates a noncontaminated protein sample. Samples dissolved in aqueous buffers are preferred, with sample buffers not absorbing light in the region of the spectrum of interest. Solute absorption is directly proportional to its concentration, and hence, this method is well suited for quantitative studies of protein solutions. Deuterium or tungsten lamps are generally used as sources of light. Although the conventional method implies the use of a monochromator through which light passes to the cuvette containing the protein sample, which is further directed to a photomultiplier tube, present-day spectrophotometers utilize polychromators and silicon photodiodes as a more reliable and cost-effective arrangement for versatile applications (Fig. 2).

This spectroscopic method includes two different approaches for analyzing protein oligomerization and aggregation, one being an indirect way to evaluate the protein absorption spectra through intrinsic and extrinsic chromophores and the other comprising direct detection of aggregating protein.

## 2.2 Indirect detection of protein conformation changes

Chromophores form the basis of indirect determination of structural changes in protein aggregates as they absorb light of different wavelengths based on protein composition. The protein amide bonds absorb light in the far ultraviolet (UV) range of 180–230 nm, with two significant peak transitions observed at 195 and 220 nm. The aromatic residues (Phe, Trp, Tyr) and cysteine absorb in the near UV range of 240–295 nm, with Trp and Tyr having an absorption maximum at 280 nm, whereas a weak absorption spectrum for Phe is observed at 260 nm. Although these aromatic residues do not absorb light above 310 nm, with oxidation or ionization reactions persisting in these aromatic residues, changes in the absorption spectrum are frequently observed with the formation of distinct bands at 320 nm. Indirect changes in protein conformation have been analyzed during the binding of lysozyme with a planar molecule, crystal violet, in the presence of azoTAB. At low concentrations (<2 mM) of azoTAB, crystal violet has an absorption maximum of 590 nm. Increasing the concentration of the azoTAB from 2 to 5 mM causes the unfolding of lysozyme and binding of the crystal violet to the hydrophobic interiors of lysozyme, which decreases the polarity of the microenvironment around crystal violet and changes its absorption maxima to 600 nm [15].

The second derivative of spectra (a strategy to separate overlapping peaks in a UV-Vis spectrum) is used as a powerful tool to monitor conformational alterations in protein structure with prominent transitions observed in peak positions along with its accurate intensity to quantify proteins. This strategy has been applied to study the self-assembly of proteins, one being a case of assessment of 33-mer gliadin peptide self-assembled oligomers at several concentrations. This study employed dynamic light scattering (DLS) and UV-Vis spectroscopy to detect 33-mer peptide self-assembly in aqueous

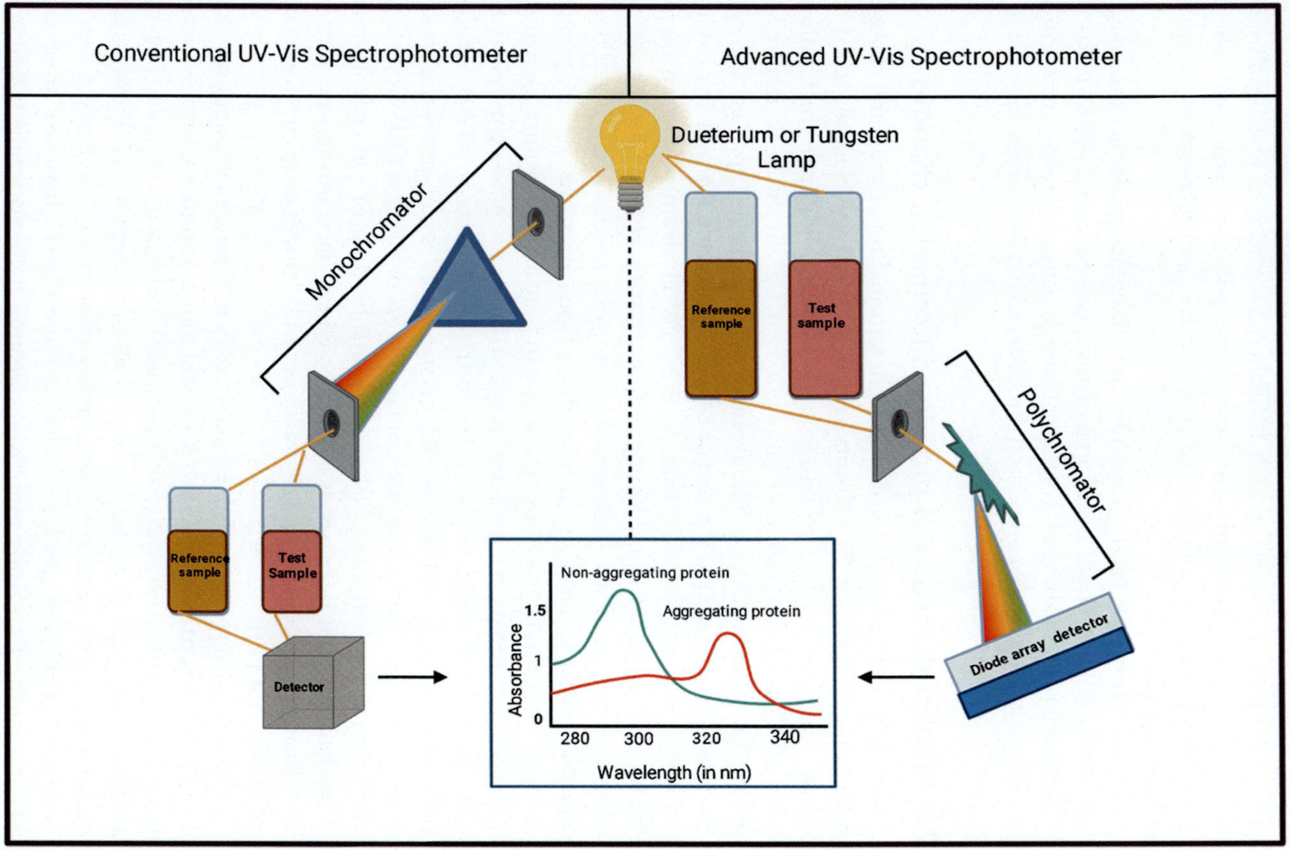


**Fig. 2** Schematic representation of the instrumentation design of conventional and advanced UV-Vis spectrophotometers.

solutions at pH 7 over a wide range of concentrations from <125 to 615 μM [16]. Recent studies analyzing second-derivative spectra of human serum albumin (HSA) observed a notable increase in the presence of disordered structures as seen by far UV circular dichroism (far-UV CD), indicating a structural characteristic of amyloid fibrils [17]. Another amyloid fibrillation study assessed the changes in the original microenvironment of the Ac-FFC-NH2 molecules (minimalist β-amyloids) because of aggregation with analysis of maximum absorbance of the second derivative of the UV-Vis spectrum in the region of 250–310 nm [18].

Besides the absorption properties of amide bonds and amino acids, folding properties of proteins can also be followed with the help of cofactors such as FADH, NADH, or heme groups, which act as extrinsic chromophores and absorb radiation in near UV and visible regions of the spectrum. The shift in the absorption maximum in the wavelength range specific to these cofactors can be attributed to the changes in the microenvironment from polar to hydrophobic upon interaction with their specific proteins. The aggregation propensity of Hb in the presence of glycans has been recently studied using UV-Vis spectroscopy [19]. A similar study was performed to analyze the conformational variations of bovine Hb at increasing concentrations of glyoxal to study its heterogeneous properties in high glycemic conditions [20]. At 70% glyoxal, a blue shift of 20 nm of the Soret band showcased alterations in the steric pattern of Hb with aggregation propensities. Extrinsic probes such as Congo red, Thioflavin-T (ThT), and so on show a shift in absorbance peak from 490 to 540 nm upon binding to amyloidogenic and prion proteins [21,22].

## 2.3 Direct detection of protein conformational changes

A direct assessment of aggregation can be obtained by evaluating the decrease in the intensity of incident light because of the scattering effect of protein particles that are formed in a pure protein sample. The turbidity of a solution can be measured to analyze the aggregation of proteins where turbidity ($\tau$) is the parameter for the total scattering of light in different directions and is defined as optical density (optical density = scattering + absorbance). It is measured at wavelengths longer than 320 nm as signals beyond 320 nm arise as a result of light scattering exclusively by large protein aggregates with hydrodynamic radii of >200 nm. The aggregation kinetics of the C-terminal domain of TDP-43 protein was monitored over time using turbidity as a characteristic to compare the same samples under different conditions. Another relevant example was seen in the case of Hb aggregates with an increase in turbidity at all hours as induced by bovine serum albumin (BSA) acting as a molecular crowder at different concentrations [23]. This approach was also applied to investigate the role of dietary monosodium glutamate (MSG) in protein aggregation. Kinetics of aggregation was studied with a constant BSA concentration of 5 μM, titrated with different concentrations of MSG. An increase in MSG concentration enhanced the BSA aggregation as denoted by an increase in absorbance at 600 nm as a function of time [24]. Another study monitored the aggregation behavior of BSA depending on BSA nanoparticle size using turbidity measurements at 450 nm at different pH and temperatures [25]. Several other studies include tubulin polymerization kinetics upon interaction with α-synuclein fibrils [26], plasminogen-induced transthyretin amyloidogenesis [27], β-amyloid aggregation in the presence of zinc [28], and detection of insoluble dihydrofolate reductase protein aggregates, which were based on UV-based turbidity assays at 330–350 nm for qualitative analysis of protein aggregates under the influence of varying factors.

The existence of oligomers in solution has been studied through a different approach using aggregation index (*AI*) as a parameter to check the aggregation state of any sample. Absorbance values at $A_{280}$, $A_{320}$, and $A_{340}$ are used to calculate the aggregation index of the different samples using the equation below:

$$AI_{340} = 100 \times [A_{340}/(A_{280} - A_{340})] \text{ OR}$$

$$AI_{320} = 100 \times [A_{320}/(A_{280} - A_{320})]$$

The range of aggregation index varies from less than 3 to more than 30, with values $<3$ being significant for a transparent sample solution, values between 3 and 30 indicating a moderate level of aggregates in solution, and values of absorbance index $>30$ indicating large amounts of aggregates.

Moreover, UV-Vis studies, along with two other widely used spectroscopic methods, CD and fluorescence spectroscopy, have been used to study the fibrillar network of HSA in the presence of metal ions, with copper (II) ions inducing a faster rate of fibrillation in protein solutions as evident from ThT-stained fibrils excited at a near-UV range [29]. Overall, UV-Vis spectroscopy is a simple tool to analyze insoluble protein species in solution, with details on protein-ligand binding reactions, enzyme catalysis, and structural transitions.

## 2.4 Estimating the concentration of proteins

Many colorimetric assays such as Bradford, Lowry, or Bicinchoninic acid have been developed to measure total protein concentrations, but they depend on the amino acid composition of the protein of interest. The errors associated with these assays can be avoided by using UV-Vis absorbance estimations of the protein of interest. This requires a pure protein whose extinction coefficient can be calculated based on its amino acid composition. It is best suited for proteins with a known number of Tyr and Trp, which absorb light at 280 nm. For other proteins, FTIR-based protein quantifications can be used to construct calibration curves of protein samples.

## 2.5 Advantages and limitations

Although it is a quick, versatile, and nondestructive technique suitable for quantitative and qualitative analysis of protein aggregates, it holds some limitations. Distinguishing transitions in the secondary structures, be it α-helix, β-sheet, or random coil, is limited through this approach as each structural conformation exhibits a different UV spectrum in the far-UV region. High-overlapping signals are also produced in this region because of oxygen or inorganic molecules, thereby affecting the accuracy of absorbance. As these limitations persist in the indirect approach, direct approaches are considered more reliable because of their easy application in aggregating systems in a productive and cost-effective manner [30]. This method also provides high time resolution over a vast range of time scales that can be efficiently enforced in detecting early intermediates in molecular dynamic studies, which is crucial for understanding the initiation stages of disease biology.

# 3. Circular dichroism spectroscopy

CD is a valuable tool for characterizing conformational changes in the optically active compounds having chiral centers such as proteins, nucleic acids, glycosides, and so on. It is widely used to study secondary and tertiary structures in native proteins and folding intermediates.

## 3.1 Principle

CD is based on the absorption of defined components of electromagnetic waves. Both electric and magnetic vectors in these waves oscillate in all directions perpendicular to the direction of propagation of light. The linearly polarized light comprises oscillating electric vectors confined to one plane, whereas the circularly polarized light forms when two vectors differ in phase by a quarter-wave. These vectors can rotate either clockwise (right hand) or counterclockwise (left hand) in a plane perpendicular to the direction of the wave. The optical element in the spectroscope converts the linearly polarized light to the right and left circularly polarized lights (CPLs), which are then differentially absorbed by the chiral molecules (such as proteins) to finally trace an elliptical path. The unequal absorption of left and right polarized light can be represented by the Beer-Lambert law as

$$\Delta A = A_l - A_r = \Delta \varepsilon Cl$$

where $\Delta A$ is the absorption signal, $A_l$ is the absorption of the left circularly polarized light, $A_r$ is the absorption of the left circularly polarized light, $\Delta\varepsilon$ is the difference between the molar extinction coefficient of left and right CPL, $C$ is the concentration (mg mL$^{-1}$), and $l$ is the pathlength (cm).

CD is represented as molar ellipticity ($\theta$) (deg $\times$ cm$^2$ dmol$^{-1}$) as a function of wavelength. The CD signal is positive when the left CPL is absorbed more than the right CPL and negative when the left CPL is absorbed to a lesser extent.

$$\theta = 3298.2\Delta\varepsilon$$

The CD spectrometer is similar to the classical UV-Vis spectrometer, with the only difference being the presence of a light polarizer and a photoelastic modulator (PEM) between the sample and the monochromator. The polarizer generates a linearly polarized light that is further converted into circularly polarized light by the PEM. The transmitted light from the sample then goes to a detector, which is connected to an amplifier and a modulator where the data are processed [30]. CD spectra in the visible and ultraviolet regions indicate different secondary structural elements in the proteins. Spectra can be obtained in the far-UV range (190–250 nm) or the near-UV range (250–320 nm). At 220 and 190 nm, electronic transitions occur from $n \rightarrow \pi^*$ and $\pi \rightarrow \pi^*$, respectively, because of absorption by the amide bonds in the protein backbone. The transition by the amide bond specifies β-sheet structures. The electronic transitions in the chiral centers of a protein have distinct fingerprints for different secondary structural elements. The α-helical structures show a positive ellipticity at 190 nm and negative ellipticity at 222 and 208 nm, whereas β-sheets show a negative ellipticity at 218 nm and a positive ellipticity at 195 nm. The random coil conformations show low values for negative ellipticity at ~200 nm (Fig. 3).

In the case of near-UV CD, aromatic residues and cysteines in the side chains act as chromophores to monitor the tertiary structure [31]. A positive ellipticity in the range of 270–290 nm indicates a folded conformation of the protein, and a change in the absorption maximum in this wavelength range for a protein with a mutation in the aromatic residues suggests the

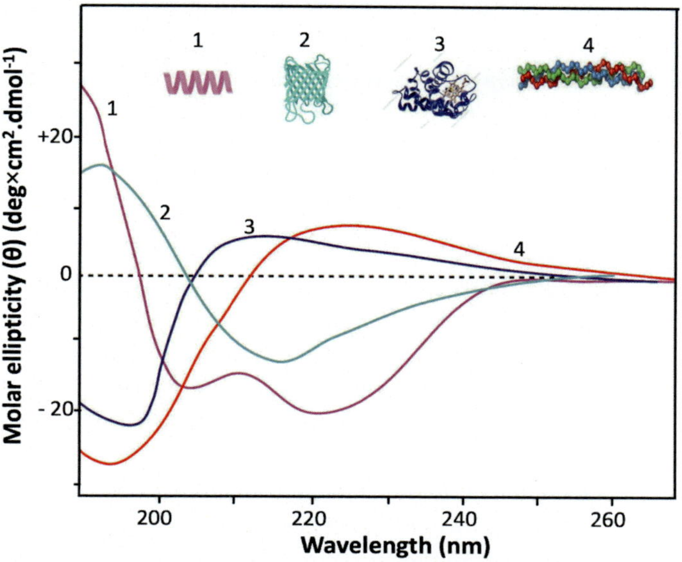


**Fig. 3** Representative characteristic spectral output for different secondary structural elements of proteins analyzed by far-UV CD spectroscopy. The α-helix (1, pink) shows negative ellipticity at 222 and 208 nm and a positive ellipticity at 192 nm. The β-sheet (2, green) shows a negative ellipticity at 192 nm and a positive ellipticity at 195 nm, whereas the random coil (3, blue) shows a negative ellipticity at 195 nm and a positive ellipticity at 212 nm. The collagen triple helix (4, red) shows a negative ellipticity at 198 nm and a positive ellipticity at 220 nm.

contribution of those residues to the tertiary conformation relative to the wild-type protein. Sample preparation for CD includes using aqueous buffers with or without cosolvents. As numerous chemical bonds have strong absorption in the far-UV range, choosing buffers with minimum absorption is crucial. Phosphate buffers are commonly used for CD spectral analysis. The signal-to-noise ratio is relatively lower at wavelengths below 210 nm because of the cumulative absorbance by the buffer and peptide chromophores as well as the composition material of the optical cell. Rectangular quartz cuvettes of a path length of 1–10 mm with all four sides made of the same material are preferred. In addition, light scattering effects from any undissolved protein or suspended particles in the solution, such as lipids present in membrane samples, may also contribute to this challenge. The reproducibility of spectral data depends on the accurate quantification and purity of protein samples. CD buffers should be filtered using a 0.2 μm filter to prevent light scattering by particle contamination in buffers. Before the experiments, the protein should be desalted and dialyzed into the CD buffer, and concentration should be estimated using the molar extinction coefficient of the protein. After the experiment, the spectrum of the buffer should be subtracted from the spectrum of protein for baseline corrections.

## 3.2 Far-UV CD

CD spectra obtained in the far-UV region provide in situ and real-time monitoring of changes in secondary structures of polypeptides during thermal unfolding, misfolding, or aggregation. Sometimes, the heat-induced unfolded proteins cannot refold because of their inherent propensity to aggregate and form fibrillar structures containing β-sheets [32]. Far-UV CD is useful for evaluating conformational changes of both amyloidogenic and nonamyloidogenic proteins.

Using time-dependent CD signal measurements, a study on α-synuclein, an intrinsically disordered protein associated with various synucleinopathies, demonstrated the gradual conversion of monomeric disordered α-synuclein into a right-handed twisted β-sheet structure, which has a higher aggregation propensity [33]. The shift from a disordered to a β-sheet structure is confirmed by the spectral minimum obtained at 218 nm [34]. Another time-dependent CD experiment revealed the gradual conversion of disordered regions to amyloid-β ordered oligomers. The amyloid-β peptide fragment containing the first 42 residues ($A\beta_{1-42}$) self-assembles to form small soluble oligomers that are toxic to cells. It also forms insoluble protofibrils and fibrils, which possess a high amount of β-sheet structures. The slow transition of the random coil to the β-sheet conformation of this peptide was monitored with the increase in minimum at 215 nm and maximum at 195 nm [35]. Transthyretin (TTR), which binds to and transports thyroxine and holo-retinol binding protein, is involved in amyloidosis. CD was used to characterize the structure and conformation of wild-type TTR tetramers, dimers, and oligomers. For the small oligomers, the signal was obtained between the dimer and native tetramer. A positive ellipticity at 197 nm and a negative ellipticity at 214 nm confirmed the presence of β-sheet conformation [36]. The CD signal for *E. coli* regulatory ATPase variant A revealed that at low pH and high temperature, the protein was unfolded, and more amounts of β-strands were observed that contribute to the formation of fibrillar aggregates [37]. For some other proteins such as insulin, glucagon, hen egg white lysozyme, and serum amyloid A, far-UV CD was used to study the formation of the amyloid fibrillar structure and conformational changes in their secondary structures at acidic, basic, or neutral pH [38–40].

CD not only is used to elucidate the conformational changes in amyloidogenic proteins but also provides characteristic features of several non-amyloidogenic proteins. The thermal unfolding and aggregation of the wild-type homotetrameric enzyme asparaginase 2 from *E. coli* and its mutants were characterized using far-UV CD by observing the change in turbidity or light scattering. Additionally, far-UV CD and light scattering were used to monitor the temperature-dependent irreversible denaturation and fusion of lipoproteins with an increase in the size of the particles [41]. CD has also been used to study the self-assembly property of surface protein A present in the alveolus. The aggregation process was indicated by the decrease in the negative ellipticity because of increased $Ca^{2+}$ and $Na^{+}$ concentrations. The presence of NaCl and low pH decreases the negative ellipticity and shifts it from 207 to 209 nm [42]. Far-UV CD was also used to detect the loss of secondary structure and aggregation of therapeutic monoclonal antibodies (IgG), whereas high-throughput thermal scanning CD analyzed their stability. CD also confirmed the aggregation mechanism in different buffer environments; the CD signal remained at 218 nm in the phosphate buffer, indicating the β-sheet structure, whereas the CD signal changed when the antibody was present in the citrate buffer. Additionally, far-UV CD revealed that the $Ca^{2+}$ ions promote the nonfibrillar aggregation of superoxide dismutase1 (SOD1), a protein involved in the pathology of amyotrophic lateral sclerosis (ALS), by increasing the β-sheet content in the protein [43]. Furthermore, CD was able to detect the self-aggregation properties of β-lactalbumin in a concentration-dependent manner. The continued heating of the protein at 80°C results in the formation of fibrils with the stabilization of α-helix and an increase in the amounts of β-sheet structures [44].

## 3.3 Near-UV CD

Near-UV CD has been used to detect the changes in the tertiary structure of proteins, considering the aromatic and cysteine side chains for evaluating it. To identify the conformational differences between the wild type and mutant α-synuclein, both near-UV CD and far-UV CD were performed. Measuring the UV signals at 280 nm revealed that the mutated forms of α-synuclein have a higher aggregation tendency than the wild type [45]. Frataxin is an essential protein whose loss of function causes Friedreich's Ataxia. The tertiary structures of Frataxin variants are characterized by CD. In the near-UV CD region, the lower magnitude signals for oligomers indicate the conformational heterogeneity among subunits with molten globule-like dynamics [46,47]. The variants of hypoxanthine phosphoribosyl transferase, an important enzyme of *Trypanosoma cruzi*, have also been studied by changes in CD spectra. The signal change at 220 nm with an increase in temperature was used to evaluate the aggregate formation. Minute changes in the absorption spectra of aromatic amino acids obtained from near-UV CD demonstrate the shift in the protonation state of the phenolic side chain [48]. CD was also applied to measure the photostability of water-soluble porcine lens protein. Exposure of this protein to UV-C light (254 nm) generates reactive oxygen species in solution, resulting in the conformational change in the protein. The signal obtained from near UV-CD at 295 nm disappeared, and a strong decrease in signal at 265 nm was observed [49].

### 3.4 Advantages and limitations

Although CD spectroscopy can provide limited structural information, it holds few advantages over high-resolution structural techniques such as crystallography, NMR, and electron microscopy. CD is a cost-effective, convenient, and rapid method with a low sample requirement (~20 μg). It allows the study of both secondary and tertiary structures of proteins in solutions similar to physiological conditions. Further, as CD is nondestructive, the sample can be recovered and used multiple times. Another advantage of CD is its high tolerance to a wide range of pH and temperatures in determining secondary structures. Synchrotron-based high-intensity light sources have been developed for faster data collection and more detailed information from low amounts of samples. However, because of the spectral diversity of the β-structure, this tool often fails to provide acceptable results on α/β-mixed or β-structure-rich proteins [50]. Further, in contrast to high-resolution structural methods that provide atomic-level information, CD spectroscopy provides the cumulative contributions of all amino acids that maintain the structural integrity of a protein.

## 4. Fluorescence spectroscopy

Fluorescence spectroscopy is a highly versatile and widely used technique for conformational analysis of natural and synthetic macromolecules [51]. This technique enables obtaining valuable insights into the structural and kinetic aspects of soluble and aggregate-forming proteins [52].

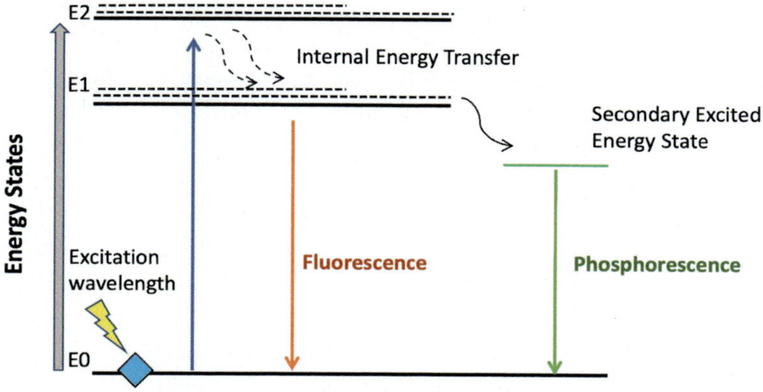


**Fig. 4** Schematic representation of photoluminescence phenomena illustrating absorption, fluorescence, and phosphorescence. E0 represents the lowest energy or ground state of the fluorophore shown as a blue diamond. E1 and E2 are the higher energy states for the fluorophore.

## 4.1 Principle

The phenomenon of fluorescence underlies the principle of this spectroscopic method. The method involves using a fluorophore that absorbs incident photons (quantum of electromagnetic radiation), and its electrons transit from a lower energy state to excited energy states (excited energy state 2), as shown in Fig. 4. Excited states are also associated with an increase in vibrational energy (energy associated with bond length and bond angle), which in turn can be lost because of molecular collisions and heat transfer. This relaxation in the vibrational energy results in the migration of electrons to a lower energy level (excited energy state 1). Because of the dissipation of some of the absorbed energy, the emitted photons have a lower energy and a longer wavelength than the absorbed photon, which is detected as fluorescence light. It is important to note that the delay time between excitation and emission of light is between $10^{-6}$ and $10^{-8}$ s. Sometimes, electrons in the lower excited energy state (state 1) undergo an internal conversion in the electronic spin (singlet to triplet state) to a secondary excited energy state of lower energy. A longer delay ($>10^{-6}$ s) in the return of these electrons to the lower energy state is observed as either a nonradiative process or phosphorescence light.

## 4.2 Intrinsic and extrinsic fluorophores

Fluorophores generate the fluorescence signals and hence are the central component of fluorescence spectroscopy. Fluorophores are classified as intrinsic if they self-fluoresce and extrinsic if they are made to fluoresce by associating with a fluorometric reagent. Naturally occurring fluorophores such as aromatic amino acids, NADH, flavins, and so on are intrinsic fluorophores. However, fluorescent molecules (covalently associated with the protein of interest) such as red fluorescent protein (RFP), green fluorescent protein (GFP), or yellow fluorescent protein (YFP) or dyes such as safranin, rhodamine 123, fluorescein, and DAPI (4′,6-diamidino-2-phenylindole) are extrinsic fluorophores. Studies have led to the identification of structure-specific intrinsic factors within the amyloid fibrils that are independent of aromatic residues in the polypeptide chain, which enable label-free fluorescence detection. Extrinsic probes include a wide array of chemicals and their derivatives that can be used for targeted studies of amyloids. ThT, Congo red, curcumin, dyes such as oxazine and thiophene, and their derivatives are commonly used to study insoluble amyloid-β aggregates. However, oligomeric amyloid-β aggregates are extensively studied via boron dipyrromethene (BODIPY) derivatives, 1-anilino-8-naphthalene-sulfonate (ANS), DCVJ, 9-(dicyanovinyl) julolidine, and so on [53].

## 4.3 Sample properties

Considering the sensitivity of fluorescence investigations, certain technical aspects should be carefully evaluated. First, samples with general absorption values of 0.05 or less should be taken for experimental measurements as turbid solutions lead to uneven absorption and generate distorted emission spectra. Another way to subdue this inner filter effect is to use a cuvette with a shorter path length. Second, to avoid instrument-dependent differences in intensities, signal calibration using Raman peaks should be performed [54]. All solvents scatter light observed as Raman peaks at wavelengths that differ from their excitation wavelength. For instance, at an excitation wavelength of 350 nm, water shows a Raman peak at 398 nm. The shift in energy between the excitation wavelength and the Raman peak is solvent-dependent, making the Raman peak a useful calibration tool for instrumentation error. The choice of the fluorophore, sample size, and experimental conditions such as temperature, viscosity, solvent polarity, solvent pH, and inner filter effects should be considered to achieve an optimum fluorescence signal. Autofluorescence by solvents should also be avoided by altering the solvent conditions.

## 4.4 Steady-state fluorescence

The detection and quantification of amyloids, particularly Aβ peptides, are most commonly achieved by steady-state fluorescence, wherein the binding of extrinsic fluorophores to the Aβ peptide fibrils increases the fluorescence emission. In contrast, the oligomeric state (number of subunits present in a complex) of proteins can be deciphered using steady-state anisotropy.

ThT is routinely used as an amyloid probe as it can accommodate itself in the cross-β structure of most fibrils. Although the interaction of ThT with the amyloid core is not fully understood, it serves as a prominent probe to study variants of amyloids, including Aβ and Alzheimer's-related peptides [55]. Acting as a molecular rotor, ThT dissipates its excitation energy via the relative torsional motion of its benzothiazole and aminobenzene rings, resulting in a short fluorescence lifetime and low quantum yields. The binding of ThT to amyloids results in fluorescence enhancement because of its restricted rotation [56]. Distinct surface structural features of amyloid fibrils can also be detected and differentiated based on the binding of ThT to the exposed β-sheet regions. The tyrosine derivative (tyrosinate) has also been used to study aggregates of human islet amyloid polypeptide (hIAPP) and Aβ by employing steady-state fluorescence [57]. A typical steady-state fluorescence curve shows a characteristic red shift (the excitation wavelength is 412 nm, and the emission wavelength is between 440 and 450 nm) and a lower fluorescence lifetime for ThT bound to amyloid fibrils (Fig. 5).

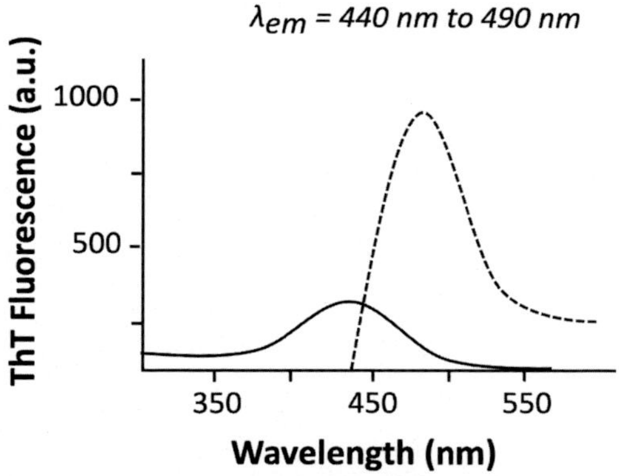


**Fig. 5** Graph representing shifts in ThT emission intensities from the unbound/free form in aqueous solution (solid line) and the bound state in amyloid-containing solution (dashed line). The solid line represents the fluorescence emission $\lambda_{em} = 440$ nm, whereas the dashed line represents the emission at $\lambda_{em} = 490$ nm.

The extent of the red shift might differ among the variants of amyloid fibrils. It is essential to estimate the soluble monomeric Aβ concentrations before and after aggregation (by separating the fibrillar material from the monomeric solution) to eliminate differences that might arise from the amount of fibrils present in the samples to be evaluated. The amount of fibrillar structures in proteins can be correlated to their ThT emission intensities [55,58].

## 4.5 Fluorescence anisotropy

Fluorescence anisotropy or polarization provides information on the molecular rotation, size, shape, and flexibility of a fluorophore. The anisotropy value enhances with the increase in protein self-assembly (Fig. 6A) and is measured by using the following equation for fluorescence anisotropy values:

$$r = \left(I_{\parallel} - I_{\perp}\right) / \left(I_{\parallel} + 2I_{\perp}\right)$$

Here, anisotropy ($r$) is measured as the ratio of fluorescence intensity difference between the emission polarizer oriented parallel ($I_{\parallel}$) and perpendicular ($I_{\perp}$), respectively, to total intensity of polarized light emitted by the sample. It can measure either the intrinsic fluorescence of a protein or an extrinsic molecule whose fluorescence properties change on binding to a protein. Fluorescence anisotropy is dimensionless and independent of the fluorophore concentration.

Briefly, plane polarized light is bombarded onto an isotonic solution, which results in excitation of the fluorophore with transition movements parallel to the direction of polarization. Changes in the polarity of emitted light provide information about molecular size as the rotational diffusion of the fluorophore brings about the depolarization of the emitted light. Thus, fluorescence anisotropy is sensitive to the rotational diffusion of a fluorophore or factors affecting it. Accurate measurement of anisotropic differences between the free protein and protein-associated complex requires careful consideration of the fluorescence lifetime of the fluorophore and the rotational correlation time (time taken by a molecule to rotate by one radian angle) of the target molecule to be similar. The fluorescence anisotropy approach employs a fluorescent entity being bombarded by a plane polarized light, and the emission signal is analyzed to determine its degree of polarization in both planes. It also facilitates the detection of molecular orientation and mobility. Being an intensive property, anisotropy measurements are independent of the amount of fluorophore and thus offer an

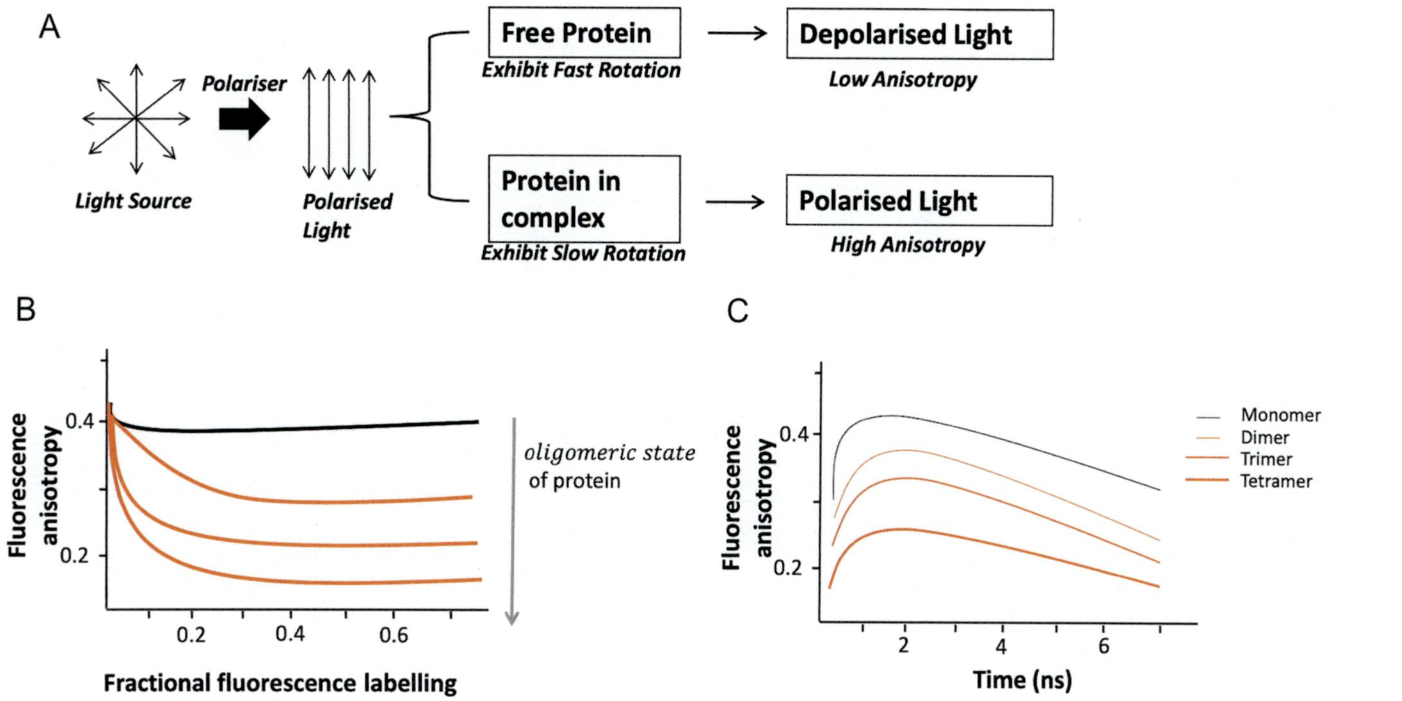


**Fig. 6** Fluorescence anisotropy: (A) Changes in fluorescence polarization or anisotropy of a fluorophore upon excitation by a polarized light source. (B) Decreasing fluorescence anisotropy indicates the transition of protein conformation from a monomer to an oligomer state. (C) Decay in fluorescence anisotropy followed over a short period of time for different oligomeric states.

advantage as fluorescence lifetime is also a critical parameter [54,59–61]. Fig. 6B shows that the anisotropic values decrease with enhanced stoichiometry when measured for a constant concentration of a protein.

The discrimination of the Aβ oligomer from the monomeric and fibrillar forms has been demonstrated in a rapid and straightforward fluorescence anisotropy-based assay using a pyrene fluorophore [60]. Recent studies have also performed a stoichiometric evaluation of oligomeric states such as dimeric leucine-zipper GCN4 transcription factor and its trimeric variant, GCN4-pII, using steady-state fluorescence anisotropy, with the use of minimal datasets from time-resolved measurements of reference proteins [59].

## 4.6 Time-resolved fluorescence spectroscopy

When a fluorophore is irradiated by UV, visible, or near-IR light, the change in fluorescence over a short period of time (picoseconds) is measured as time-resolved fluorescence spectroscopy (TRFS). The internal dynamics in protein structure can be observed by TRFS combined with steady-state fluorescence, that is, TRFS. The extremely rapid movements of tryptophan residues in a protein upon changes in its microenvironment can provide information about the folding or binding kinetics of a protein. Morphological distinctions in amyloid fibrils such as $A\beta_{1-40}$ and $A\beta_{1-42[48]}$ and inhibitory effects of molecular compounds on human lysozyme (HL) amyloid fibrillation [62], β-casein [63], and TTR [64] are well-studied examples through both steady-state and TRFS methods, which measure the difference in fluorescence quantum yields (brightness of a fluorophore). Although the ThT fluorophore can help determine the overall transition from monomers to fibrils, bis(triphenylphosphonium) tetraphenylethene (TPE-TPP) can differentiate between the numerous aggregates, including the toxic Aβ aggregates [65]. Specific fibril association in amyloid aggregates, such as differential strand organization in β-sheet aggregates, has been demonstrated by the time-resolved absorption technique owing to their considerable sensitivity toward conformational changes in protein and peptides [66]. Critical conformational transitions depicting early oligomerization in synuclein protein using TRFS have been reviewed elsewhere [67]. A recent study also characterized various states of protein fibrillation using ThT and real-time fluorescence measurements [68]. This technique has been widely used to study protein conformations and binding kinetics. A characteristic graph representing the decay of fluorescence as a function of time is shown in Fig. 6C.

## 4.7 Fluorescence correlation spectroscopy

Fluorescence correlation spectroscopy (FCS) is a powerful technique employed to quantitatively detect the molecular dynamics (aggregate size and shape in general) by measuring the fluctuations in fluorophore intensity in a confined detection volume, typically using a confocal laser microscope [69]. Along with molecular diffusion, it provides information about the chemical reaction kinetics when fluctuations from a limited number of molecules are studied in a confined space rather than the whole sample. Diffusion or chemical reaction rates and the intensity of fluorophore in the system are the two aspects of molecular behavior facilitated by FCS studies. Its ability to measure dynamic parameters without perturbing the system from the steady-state condition, sensitivity, chemical selectivity to lower fluorophore concentrations, the requirement of low sample volumes [70], and wide time scale measurements (subnanoseconds to nanoseconds) are some of the advantages offered by FCS [71]. Numerous complementary methods based on FCS include fluorescence cross-correlation spectroscopy (FCCS) [72], single-color fluorescence lifetime cross-correlation spectroscopy (sc-FLCCS) [73], scanning FCS [74], spot variation fluorescence correlation spectroscopy (sv-FCS) [75], fluorescence resonance energy transfer-fluorescence correlation spectroscopy (FRET-FCS) [76], and single-molecule fluorescence microscopy. The values can be plotted in terms of autocorrelation function because of the rapid increase or decrease in fluorescence intensity of molecules. The following parameters obtained from the graph are then used to quantify the autocorrelation function value [77]. The autocorrelation function describes the correlation function, $G(\tau)$, which measures the time scale of signal fluctuations and is calculated by the equation below, where $\delta F(t)$ represents the deviation from the mean signal and $F(t)$ corresponds to the average signal over time ($t$) with a delay time ($\tau$).

$$G(\tau) = \frac{<\delta F(t) \times \delta F(t+\tau)>}{<F(t)>^2}$$

Parameters of an ideal fluorophore for FCS studies include high intrinsic brightness, that is, a high extinction coefficient, and a good quantum yield. The fluorophore should be able to label all the subunits uniformly. Further, the fluorophore dye should not be altered by protein assemblies during incubation in the sample [78,79]. Care should be taken to avoid nonspecific attachment of the protein solution to the coverslip to facilitate accurate particle count. As FCS requires only a few fluorescent particles for detection, expression levels of protein under investigation should be optimized. A sample concentration below the nanomolar concentration is sufficient to facilitate the acquisition of autocorrelation functions. Autofluorescence of entities including phenol red, flavins, flavoproteins, dead and/or stressed cells should be optimized in the initial stages of the FCS experiment [80].

## 4.8 Advantages and limitations

Fluorescence spectroscopy facilitates highly specific analysis of proteins of interest through efficient tagging of proteins with fluorescent tags. A low concentration of samples (nanomolar range) can be used for rapid data acquisition with high sensitivity. Structural and functional characteristics of fluorescent proteins can be obtained either under physiological conditions or altered cellular environments such as altered temperature, pH, and osmolarity. As opposed to UV-Vis spectroscopy, which depends on specific requirements of specialized quartz cuvettes, fluorescence spectroscopy acquired high-quality data with inexpensive and simpler apparatuses such as sample holders with round-bottom cuvettes.

Although an increasing number of extrinsic fluorophores have enhanced the applications of this spectroscopic technique, the short lifespan of a fluorophore and its susceptibility toward autofluorescence continue to impose a limitation on its unbiased utility in studying protein structures.

# 5. Infrared spectroscopy

## 5.1 Principle

The elucidation of structural and mechanistic details of intermediate and toxic forms of protein aggregates is crucial for understanding the perplexing etiology underlying several neurodegenerative and nonneurodegenerative diseases. IR spectroscopy is highly sensitive to the secondary structure of proteins and has been applied to study protein dynamics, unfolding, misfolding, and aggregation. IR spectroscopy involves the absorption of the electromagnetic radiation in the IR region, along with vibrations and oscillations generated by different functional groups in the studied analyte that are selective to changes in the dipole moment. Most protein molecules, except for homonuclear diatomic molecules, exhibit IR absorption. The IR spectrum consists of three characteristic bands, providing information about the protein's secondary structures; these are amide I (1700–1600 $cm^{-1}$), amide II (~1550 $cm^{-1}$), and amide III (1400–1200 $cm^{-1}$) bands (Fig. 7).

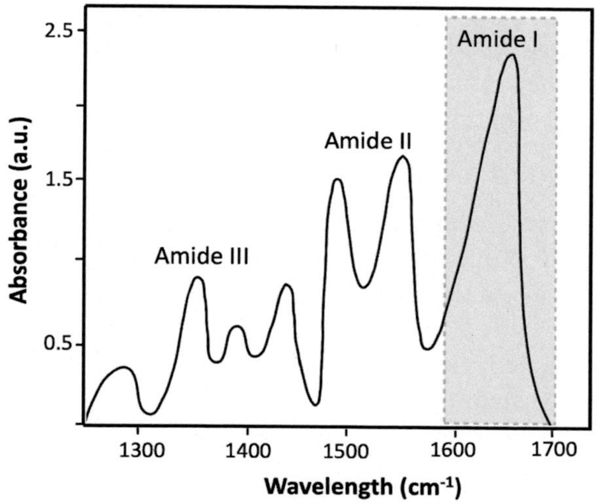


**Fig. 7** Spectrum obtained by IR spectroscopy showing wavelengths at which characteristic amide bands are observed for specific secondary structures.

Out of these, amide I possess the highest sensitivity toward the protein backbone. It arises from the C=O stretching vibration, and hence, it is used widely to analyze protein α-helix, β-sheet, turn, and disordered conformations.

FTIR is a widely used form of IR spectroscopy. In contrast to other spectroscopic techniques that require the excitation of protein samples with monochromatic light, FTIR involves a beam of light consisting of many wavelengths of different frequencies. The beam splits into discrete wavelengths when it falls on mirrors configured at different angles and is constantly moved by a motor. This shift of mirrors causes a single wavelength of light to fall on the sample, and its absorption pattern is then recorded, which is followed by another wavelength of light exciting the sample by wave interference. These different absorptions are converted into spectral patterns by an algorithm called Fourier Transform. This technique is frequently used to study biological molecules because of its high signal-to-noise ratio, quick data accomplishment, and favorable digital subtraction attributes. Sample preparation includes the use of demountable cells made with a variety of insoluble materials such as ZnSe, CdTe, $CaF_2$, and $BaF_2$ with spacers without wrinkles or pits to minimize leakage. The instrument stability is often inferred by averaging numerous scans (over 1000 scans). Following the removal of instrumental noise of the amide I band curve, the relationship between the peak position and the type of secondary structures is considered to estimate the secondary structure content. The quantitation of the secondary structures is subjected to a fraction of each secondary structure proportional to band intensities.

$$\textit{IR absorption intensity} \propto (\delta\mu/\delta qi)^2$$

Here, $\mu$ is the dipole moment of a molecule and qi is the *i*th normal coordinate. The oscillation of charged groups leads to shifts of charges that, in turn, result in an increased value of $(\delta\mu/\delta qi)^2$, thus exhibiting larger absorption.

FTIR is implied extensively in the analysis of different kinds of β-structures and therefore is widely used to characterize amyloidogenic proteins. Table 1 shows the strong signals corresponding to variable protein structures in the amide I region [81,82].

## 5.2 FTIR spectroscopy to analyze amyloid aggregates

Several studies have used FTIR for the quantitative and qualitative spectral analysis of variable species of protein aggregates. A comparative study between FTIR combined with attenuated total reflectance (ATR) and standard aggregation analysis methods including AFM, ThT fluorescence, and ANS fluorescence highlighted the tremendous potential of FTIR in detecting oligomeric intermediates during the amyloid fibrillation from hen egg white lysozyme solution. The qualitative analysis was prominent from changes in amide I and III regions, shifting toward β-sheet bands. The deconvoluted amide I region of FTIR spectra showed structural proximity between oligomeric intermediates and mature fibrils compared to monomers [83]. A novel synergic, in silico, in vitro, and in vivo validation study

**Table 1** IR marker bands for different protein conformations.

| Protein conformation | Average range of IR marker band intensity in the amide I region (in $cm^{-1}$) |
|---|---|
| α-Helix | 1659–1647 |
| Antiparallel β-sheet | 1695–1612 |
| Parallel β-sheet | 1633–1618 |
| β-Turn | 1678–1664 |
| Random coil | 1660–1639 |
| Unstructured/disordered/loops | 1649–1641 |

on the structural dynamics of A$\beta_{1-42}$ oligomers found in Alzheimer's disease (AD) involving FTIR indicated the spontaneous formation of nanosized barrels during early aggregation of A$\beta_{1-42}$, exhibiting its role as a toxic intermediate in the pathology of AD [84]. Time-resolved IR spectra of A$\beta_{1-42}$ showed a major band shift with an increase in the size of A$\beta_{1-42}$ oligomers [85]. Contradictory studies surround the IR spectral studies of α-synuclein, with one displaying β-sheets typical of fibrils with maximum absorption at 1628 cm$^{-1}$, whereas the other deconvolution analysis of FTIR spectra of α-synuclein with its three variants indicated the existence of antiparallel β-sheets in fibrillar structures [86,87]. Human Islet amyloid peptide (IAPP) aggregation inhibition using small molecules such as rhodamine derivatives and resveratrol was shown with the change in amide I band in ATR-FTIR spectra [88,89].

## 5.3 FTIR spectroscopy to analyze nonamyloid aggregates

FTIR has also been utilized in structural studies of nonamyloid aggregates. It helped in understanding the effects of pressure and heat on the unfolding of myoglobin. The temperature-induced unfolding of myoglobin resulted in IR bands specific to intermolecular antiparallel β-sheet aggregates. In contrast, pressure and low temperatures yielded a partially unfolded state with stable core helices, suggesting a difference in folding pathways under varying conditions [90]. A study was conducted on papain kept in cold storage, which identified structural rearrangements with an increase in 40% of intermolecular β-sheets, detected at a frequency of 1621 cm$^{-1}$ by FTIR [91]. Another study analyzed the secondary structure of inclusion bodies formed because of the expression of recombinant human growth hormone (h-GH) and human interferon-α-2b (IFNα-2b). Depending on protein expression levels, residual native-like conformations existed within these insoluble protein aggregates to varying extents, as determined by protein secondary structures using FTIR-microspectroscopy [92]. Several other studies have focused on residual structures embedded in protein aggregates to control the conformation quality of inclusion bodies rather than just targeting aggregate clearance (reviewed in Ref. [93]).

## 5.4 Combinatorial studies with FTIR

Recently, a more sensitive 2D-IR spectroscopy has been used to distinguish oligomeric intermediates from stable amyloid fibrils. A unique 1610 cm$^{-1}$ transition in 2D-IR spectra was seen for mature A$\beta_{1-42}$ fibrils, whereas no such transition was marked in other aggregating species [94]. A combinatorial study using 2D-IR, AFM, and FTIR demonstrated the effect of buffer conditions on synuclein fibrillar structures of α-synuclein protein. Although low-salt buffers contained loosely packed β-sheet structures, high-salt buffers contained tightly packed antiparallel β-sheet fibrillar conformations [95]. Recent research is directed toward developing an immune IR sensor to quantify the Aβ peptide secondary structure in the cerebrospinal fluid (CSF) and blood plasma of AD patients. With discrimination based on amide I band frequency downshift to the β-sheet structure in AD patients compared to disease controls, it holds a novel and promising strategy for label-free diagnosis of AD with a 90% accuracy for CSF samples [96].

## 5.5 Advantages and limitations

It is characterized as one of the most simple, reliable, fast, and cheap spectral analysis methods to eliminate complex spectral processing to establish undeviating amyloid assessment. Low sample requirements (10–100 μg) with a short measuring time make it a convenient technique for protein structural analysis. However, the strong water bending absorption band overlaps with the protein I amide band and thus masks the protein signal at $1630\,cm^{-1}$. Hence, isotopic labeling strategies are carried out in buffers, converting $H_2O$ to $D_2O$ (deuterated water) to shift the water bending peak from 1630 to $1290\,cm^{-1}$, thus facilitating appropriate amide I region measurements.

Compared to FTIR, attenuated total reflectance (ATR-FTIR) is an advantageous spectroscopic method with simpler sample preparations with minimal absorption by the solvents. Attaching an ATR accessory in FTIR with a sample placed over an ATR crystal with a high refractive index enhances the analysis of protein thin films. Repeated reflections on the crystal-sample interface in ATR-FTIR allow the sample to absorb more incident IR light, resulting in increased sensitivity and greater signal amplification [97,98]. Nevertheless, because of diffraction limits in conventional IR spectroscopic methods, low-defined resolution restricts the track change in single molecules. To overcome this challenge at a single molecule level, novel techniques such as nano-FTIR and AFM-IR have been established for studies with abnormal proteins. The nano-FTIR technique allows studying the IR characteristics of a sample at nanoscale resolution. Scattering-type scanning near-field optical microscopy (s-SNOM), a unique combination of

AFM with optical imaging and IR spectroscopy, offers a combination of nanometric spatial resolution of AFM with high chemical sensitivity of IR. This technique was first utilized to explore the structure of single insulin fibrils [99]. IR has been primarily implemented either individually or in combination with other biophysical and biochemical methods as a convenient and versatile approach for monitoring the structural and topological dynamics of aggregating proteins to unravel the unclear mechanism behind complicated cytotoxic and neurotoxic diseases.

# 6. Dynamic light scattering spectroscopy

## 6.1 Principle

DLS is a noninvasive method that measures the size distribution profiles of particles in solution. DLS is based on the Brownian motion of macromolecules in a solution. On collision with the solvent molecules, the proteins acquire movement relative to their size, where smaller molecules move faster than larger molecules. The Stokes-Einstein equation (below) describes the relationship between particle speed and particle size:

$$D = k_B T / 6\pi\eta R$$

Here, $D$ is the translational diffusion coefficient (speed of the particles), $k_B$ is the Boltzmann constant, $T$ is the temperature, $\eta$ is the solvent viscosity, and $R$ is the hydrodynamic radius. This equation is used considering that the particle is globular like spherical proteins. For a smaller sample size, the Brownian motion of molecules appears as fluctuations in the intensity of the scattered light, which further depends on the size and shape of the molecule. The rate of Brownian motion of molecules and solvent viscosity also depends on the temperature, so the temperature equilibration during the experiment is of paramount importance. The sample is exposed to monochromatic light, which scatters upon interacting with the sample molecules. The detector then captures the scattered light and transforms it into a particle size distribution (PSD) curve. Because the scattering of light is being analyzed, the size distribution given by DLS is an intensity-weighted distribution [100]. The spectrometer used to measure DLS consists of three major components: laser, sample holder, and light detector.

DLS samples should be dispersed in a liquid phase (solvent). When choosing a solvent, there are two key considerations to keep in mind. First, the solvent must effectively disperse the particles, and second, the solvent should not change the properties of the samples, such as dissolving the particles. The commonly used solvents are MilliQ water, toluene, methanol, ethanol, and glycerol. All the monodisperse particles are identical in size, shape, and mass, which results in one narrow peak in the PSD curve. On the other hand, the polydisperse particles are not uniform in these parameters. As the procedures for calculating the hydrodynamic radius distribution vary depending on whether the samples are monodispersed or polydispersed, it is critical to understand the polydispersity of the samples. The cumulant method (which determines the particle size and its width) is used to measure the monodisperse samples, whereas the CONTIN algorithm is used for the polydisperse and heterogeneous samples [101].

To track the movement of the particles, the scattered light is measured over time. The intensity of the scattered light does not remain constant throughout the time as it fluctuates. Smaller particles move faster, resulting in rapid fluctuations compared to the larger particles; as a result, the larger particles display higher amplitudes between the maximum and minimum scattering intensities. This information generates a correlation function. Later, an exponential decay of the correlation function is seen because the particle starts moving. The correlation function, in general, describes how long a particle stays in the same place in the sample. The correlation function is linear and nearly constant at the start, indicating that the particle is still in the same position, but it decreases later, indicating that the particle is moving. The size-dependent movement information is included in the decay of the correlation function. The decay is an indirect measure of the time that the particles take to alter their relative positions. As small particles move quickly, their decay occurs quickly; larger particles move slowly, so their decay takes a much longer time.

DLS is a quantitative method that can be used to study the protein oligomers, aggregates, and fibril formation across diverse systems. The studies related to aggregation of polyglutamine, Aβ, and lysozyme aggregation [102] are a few examples.

## 6.2 Protein aggregation analysis by DLS

The oligomeric status of the HIV-1 Nef protein was estimated by DLS. The hydrodynamic radius of full-length and truncated protein species was measured in pure protein samples under varying concentrations. It was observed that at low protein concentrations, Nef protein versions were monomeric, whereas they have different oligomeric states at higher protein concentrations [101]. The rate of aggregation of tau protein was estimated with automated static and DLS [103]. Initially, pure solutions of tau proteins were formulated with and without seeds. The purpose of the addition of seeds was to cause instantaneous aggregation. Early on, both solutions predominantly contained monomeric proteins with approximately 10 nm radius, stable at 37°C for more than 14 h. After 3 h, no noteworthy changes were exhibited by the solution containing pure tau monomers, whereas the solution formulated with seeds reported a decrease in 10 nm monomers and an increase in the aggregate population of approximately 200 nm. The different-sized aggregates of insulin analogs such as lispro, aspart, and glulisine were studied via a combinatorial approach using DLS and Raman spectroscopy [104]. The three insulin analogs were incubated at 37°C for 30 days without any preservatives. At several intervals throughout the incubation, samples were taken out and analyzed using DLS to determine the aggregate sizes. Both the intensity percent versus size and the volume percent versus size were investigated in the characterization. The intensity distribution detected smaller quantities of macroscopic aggregates, and the volume distribution detected the relative amounts of different-sized aggregates. The findings showed that the levels of aggregates formed during incubation for aspart and glulisine were low, even though the aggregate sizes were relatively large [105].

## 6.3 Advantages and limitations

DLS can accurately measure particle size, and these measurements are carried out with utmost precision in a short time. The measure of particle concentration is free from calibration, and by using the viscosity information of the liquid, the mean size of the particle can be calculated. DLS uses as little as 3 μL of samples for analysis and can determine the molecular weight and hydrodynamic radius of proteins. The molecules in the solution can be measured without using any external probe. This technique can be used to carry out experiments with different sample buffers and under variable temperatures. However, alterations in temperature and solvent viscosity can bring about changes in the values obtained by DLS; hence, both the temperature and solvent viscosity must remain constant throughout the process of measurement. Being a low-resolution method, it fails to differentiate between monomers and dimers. The large aggregates present in small amounts also affect the final measured values. Before taking the measurements, the concentration range has to be optimized to obtain authentic results as the DLS signals alter with changes in the concentration and size of macromolecules.

Overall, DLS is a quick method for measuring changes in the size of single molecules compared to complex mixtures.

# 7. Raman spectroscopy

## 7.1 Principle

Raman spectroscopy is also based on the Raman effect to analyze the structural features of a sample. Proteins in a sample scatter some amount of monochromatic light falling on them, but most of the light gets absorbed. When the frequency of the scattered light is different from the frequency of the incident monochromatic light, then the scattering is called Raman scattering or Raman effect, whereas a similar frequency of incident and scattered light is observed as Rayleigh scattering. Because frequency is directly related to the energy of light, it is noteworthy to understand frequency differences in the incident and scattered light. The energy of the incident photons is used to excite the electrons of interacting molecules to higher vibrational states. While returning to the ground level, if the electrons release the same amount of energy absorbed, then the frequency of the photon released is the same as the frequency of the photon absorbed (Rayleigh scattering). However, if the electrons return to a different vibrational level than their original levels, then the energy and frequency of the incident photon are different from those of the emitted photons giving rise to Raman scattering. The inelastic collision between the monochromatic light and the sample molecules gives rise to Raman spectra. If the frequency of the incident light is higher than that of the scattered light, the Raman spectrum displays Stokes lines. However, if the frequency of the incident light is lower than that of the scattered light, the Raman spectrum displays anti-Stokes lines. The scattered light is measured at a right angle to the incident light. The Stokes bands are measured using conventional Raman spectroscopy, whereas anti-Stokes bands are measured using fluorescence samples. The Raman scattering is affected by the wavelength of incident radiation. A Raman spectrum is a shift in intensity versus wavelength. A nondispersive Raman spectrometer uses an interferometer in contrast to the dispersive Raman spectrometer, which uses a grating or prism. Changes in the intensity of scattered light and shifts in peaks in amide band I (Fig. 7) are measured to study the secondary structural elements and the changes during oligomerization.

The ability to generate label-free and nondestructive spectral information with minimal sample preparation is one of the main advantages of Raman microspectroscopy in biological studies. Water is the best solvent to dissolve the sample. The formalin-fixed, paraffin-embedded (FFPE) tissues are extensively studied using Raman microspectroscopy. Formalin crosslinks proteins, which affect the spectral peaks associated with proteins between 1500 and 1700 $cm^{-1}$. Moreover, paraffin exhibits prominent signals in the fingerprint and higher-wavenumber spectrum regions. Dewaxing the sample or digital dewaxing can remove the strong paraffin peaks from the Raman spectrum. Snap-frozen or fresh tissue analysis can overcome these limitations. Raman microspectroscopy can easily evaluate liquid samples. The spectral artifacts caused by bubbles and surface tension can be eliminated using an immersion objective or a microfluidic device. A combined drop coating deposition Raman spectroscopy (DCDRS), also known as drop deposition Raman spectroscopy, is another method for biofluid analysis in which small amounts of biofluid are placed on a flat surface and allowed to dry [106]. Substrates that produce minimal or no background interference are chosen. Because calcium fluoride, quartz, and fused silica show minimal background interference, they are frequently used as Raman substrates.

A Raman spectrum is a unique fingerprint of a sample. The following inferences can be drawn from the Raman spectrum: (i) The peak intensity indicates the amount of a particular molecule, (ii) the peak shift identifies stress and strain states, (iii) the peak width determines the degree of crystallinity, and (iv) the polarization state provides the information about the crystal symmetry and orientation.

## 7.2 Protein analysis by deep UV Raman spectroscopy (DUVRR)

The resonant effect of deep UV Raman spectroscopy (DUVRR) allows it to selectively visualize nucleotide bases and aromatic amino acids in cells. When a molecule's electronic transition energy corresponds to the photon energy of Raman excitation light in the resonant Raman condition, the molecule's Raman scattering intensity is increased by up to $10^6$ compared to the nonresonant Raman scattering. Raman scattering from nucleotide bases and aromatic amino acids is preferentially amplified in DUVRR of cells because other biological substances in cells, such as lipids and carbohydrates, are not in the resonant state at DUV.

The application of two-dimensional correlation spectroscopy (2DCoS) paired with DUVRR spectroscopy was pioneered by the Lednev group. Shashilov et al. investigated hen egg white lysozyme aggregation using 2DCoS and DUVRR. At different time points of protein aggregation, they observed a link between the Cα-H band and the amide I band in the

DUVRR spectra. The emergence of a disordered protein secondary structure was attributed to the melting of the α-helix. As a result, the newly generated β-sheet emerges from the chaotic lysozyme. It was found that centrifugation could not remove β-sheet-rich species that emerged early in the lysozyme fibrillation process. They also discovered that supernatants from lysozyme samples incubated for 48 hours could seed protein fibrillation, thus removing the lag phase. It was concluded that the generated β-sheet-rich species found in the supernatants of the incubated samples could be nuclei [107].

## 7.3 Protein analysis by surface-enhanced Raman spectroscopy (SERS)

Surface-enhanced Raman spectroscopy (SERS) is a surface-sensitive technique for enhancing Raman scattering of adsorbate molecules on rough metal surfaces. The enhancement factor ranges from $10^{10}$ to $10^{11}$, implying that the method can detect single molecules [108].

SERS was used to identify insulin oligomers [109]. To accomplish this, sample aliquots were taken at various stages of insulin aggregation and centrifuged at pH 1.6 (at 65°C) to remove filaments and short fragments of fibrils. The sample supernatants were then mixed with 90 nm Au nanoparticles and analyzed with SERS. It was presumed that insulin dimers and higher-molecular-weight oligomers adsorb to the gold surface in nearly equal proportions. As prefibrillar oligomers were β-sheet-rich species, each adsorbed oligomer had a distinct "SERS fingerprint," specifically the amide I band in the 1665–1680 $cm^{-1}$ range. The number of SERS spectra in this spectral area that contained the amide I band for all sample aliquots taken at different states of insulin aggregation were counted. The amount of insulin oligomers increased more than twice after 1 h of incubation and then gradually decreased with time. This decrease in oligomer concentration is kinetically coupled to their conversion into fibrils. Hence, SERS can be used to detect rare protein species such as prefibrillar oligomers.

## 7.4 Advantages and limitations

Raman spectroscopy is a noninvasive technique that can be used for both solid and liquid samples. The sample preparation is very minimal, and there is minimal interference from the solvent water. Raman spectra are obtained quickly, within seconds. However, the Raman effect is very weak, so the detection requires highly sensitive and optimized instrumentation. The fluorescence of impurities or the sample can hide the Raman spectrum.

# 8. NMR spectroscopy

## 8.1 Principle

NMR spectroscopy is a promising technique for detecting the type and content of specific structural elements involved in protein folding pathways. Atomic nuclei of some isotopes ($^{1}$H, $^{13}$C, $^{15}$N, $^{19}$F, and $^{31}$P) possess angular spin along with electrical charge and behave like magnets when placed in a powerful and uniform external magnetic field. In the presence of an external magnetic field, they align themselves in either the opposite (spin states of higher energy levels) or the same direction (spin states of lower energy levels) to the external field, thereby utilizing the energy of interaction between the spinning nuclei and external field. Furthermore, the orientations are in accordance with the applied radiofrequency irradiation to the sample. At a wavelength resonating with radiofrequency waves, the energy is absorbed or emitted to transfer the nuclei from lower to higher energy levels or vice versa, respectively. The return of energy levels back to their original state is referred to as spin relaxation. The absorption of energies during these transitions is measured in the NMR spectroscopy method. These energies are also affected by the chemical environment of each atom as the electrons surrounding the nuclei create local magnetic fields and shield the protons from the external magnetic field, resulting in NMR signals at higher magnetic field strengths. Even the same kind of nuclei like $^{1}$H or $^{13}$C can have different NMR frequencies owing to differences in their electronic distributions, which is referred to as "chemical shift". Chemical shift refers to the difference in parts per million (ppm) between the resonance frequency of the observed proton and trimethylsilane (TMS) hydrogens used as a reference. This criterion can denote specific conformations of native and nonnative states of proteins. Hence, the implementation of different radiofrequency pulse sequences (a few microseconds) helps obtain information about variable protein structure dynamics, folding, kinetics, and conformation changes within a protein during its self-assembly. The resulting NMR spectra contain peaks (resonance signals) with different chemical shifts, linewidth, *J-Coupling*, cross-peaks, and the nuclear Overhauser effect to analyze the nature of atoms of the molecule of interest.

Based on differences in sample preparation, instrumentation, and structural results, NMR spectroscopy can be divided into solution-state NMR and solid-state NMR. Both these methods provide information about secondary, tertiary, and quaternary protein structures and have been successfully enforced in amyloid and protein aggregation studies.

## 8.2 Solution-state NMR

It is a high-resolution technique that analyzes three-dimensional structures of soluble protein samples with several distinct parameters, including chemical shifts and relaxation times. $^{1}H$, $^{13}C$, and $^{15}N$ are the commonly used isotopes in protein structure for solution-state NMR studies as interactions between these nuclei result in high-resolution structures of small proteins (less than 60 kDa). NMR relaxation time constants ($T_1$ and $T_2$) and rates ($R_1$ and $R_2$) are of great interest in this approach as they convey detailed information on the dynamic and aggregation events of proteins. The time taken by the excited nuclei (with different radiofrequency pulses) to relax and return to equilibrium is closely monitored with consideration of other factors such as the mobility of protein, internuclear distances, and so on to extract details on protein structure dynamics. Previous studies exploited the heteronuclear spin relaxation rates (correlation between different nuclei), chemical shifts, and translation diffusion coefficients of solution-state NMR combined with hydro-NMR to determine the weak association constants involved in the transient formation of oligomers of tyrosine phosphatase at different protein concentrations. These results revealed the formation of tetramers by the interaction of two dimers in solution for the first time [110]. Another study also used NMR relaxation measurements to study the kinetic and structural characterization of oligomeric states of the Aβ protein. Quantitative analysis of exchange between monomeric and oligomeric states at multiple magnetic fields and different peptide concentrations was performed to establish that a 3% fraction of oligomeric peptides were involved in free exchange with monomers [111]. NMR data in integration with the Rosetta symmetric docking method was used to determine the solution structures of nonintertwined and intertwined symmetric oligomers of several bacterial and viral proteins [104]. Dark-state exchange saturation transfer (DEST) NMR, a novel variant of solution-state NMR, revealed that exchange species between monomer and protofibrillar species of amyloid-β consist of low-populated species that are in direct contact with the surface and hence may contain the critical nucleus for fibril formation. Furthermore, the higher $R_2$ values of $^{15}N$ for Aβ42 than its close variant Aβ40 indicated the greater propensity of Aβ42 in aggregation and fibrillation as compared to Aβ40 [112]. Water-proton NMR, another new method for protein samples in solution, utilizes T2 of water protons to quantify protein assemblies. This rapid and noninvasive approach has been exploited in aggregation studies of several proteins (BSA, insulin, and γ-globulin, etc.) by monitoring the T2 of water as it linearly increased with protein aggregation in low and high magnetic fields [113–115].

However, the heterogeneous nature of amyloidogenic oligomers makes it difficult to study such oligomers by solution-state NMR as it requires higher protein concentrations in the sample. Earlier studies on aggregating proteins such as α-synuclein and amyloid-β used smaller fragments rather than full protein at nonphysiological pH in the presence of detergents or fluorinated alcohols to disintegrate peptides. As these conditions were not relevant to the physiological environment, recent studies have begun using full-length proteins diluted in aqueous solutions at neutral pH.

## 8.3 Solid-state NMR

This approach is primarily used for studying insoluble and noncrystalline protein aggregates that cannot be evaluated by X-ray crystallography. Comparatively, it is a better technique than solution-state NMR to analyze higher-order fibrillar aggregates at a higher resolution. Furthermore, solid-state NMR is less susceptible to dynamic structural changes than solution-state NMR and provides good quality data for the fibrillar species. Cross-polarization (CP), high-power proton decoupling, and magic angle spinning (MAS) are some of the parameters evaluated for obtaining high-resolution solid-state NMR spectra. CP-MAS experiments provide isotropic chemical shift values that can be utilized to assess site-specific secondary structures.

Extensive research using solid-state NMR has been implemented in the structural characterization of amyloid-β protein aggregates. Another comparative study utilized chemical shift and other parameters of solid-state NMR to characterize the difference in fibrillar structures between purely synthetic β-amyloid fibrils and Alzheimer's patient-derived brain fibrillar samples [116]. Qualitative differences were examined between variants of amyloid-β peptide fibrils to understand the correlation between the variation of AD phenotypes [108]. Recent studies used sensitivity-enhanced solid-state NMR for structural characterization of pico- to nanomolar amounts of brain-derived and synthetic Aβ42 fibrils that exhibited a novel polymorph structure [117]. Solid-state NMR was also used to probe the core molecular structure of amyloid fibrils formed by residues 106–126 of the human prion protein [118]. A recent three-dimensional study of tau-fibrils using solid-state NMR suggests that the P2 region of tau contributes to its amyloid formation as it loses its flexibility upon the formation of amyloids [119]. Interestingly, solid-state NMR combined with cryo-electron microscopy (cryo-EM) was used to evaluate the amyloid fibril structure of the RIPK3 RHIM-containing C-terminal domain (CTD involved in necroptosis [120]). $^{19}F$ NMR can be utilized to directly probe oligomer/fibril assembly because of its high sensitivity and extended chemical shift range. The position and intensity of $^{19}F$ for fluorinated proteins can provide real-time information about oligomer formation during fibril building. The $^{19}F$ signal of human islet associated polypeptide (IAPP) aggregation revealed that fibril assembly is a two-state process (monomer to fibrils) with no identifiable intermediate species [121]. Aβ40, on the other hand, has at least six oligomeric species, each with its own NMR fingerprint [122], and the octamer generation has been linked to increased prion protein amyloidogenicity [123].

### 8.4 Advantages and disadvantages

The most challenging problem associated with NMR includes the interpretation of structures of high-molecular-weight proteins. High-molecular-weight species show extensive crowding of peaks having a similar spectral width but an increased line width, resulting in low signal intensity and low resolution in the spectra. One way to lower resonance overlap could be the expansion of NMR spectra from one to multiple dimensions or retrieving data at higher magnetic fields for better resolution [124]. Solution-state NMR holds a disadvantage of requiring low protein concentrations of 0.1–1 mM, and hence, isotope labeling of proteins becomes necessary. This restricts the determination of higher-order oligomers and fibrils with higher protein concentrations. Moreover, time relaxation analysis and resonance assessment are very time-consuming for high-molecular proteins in solution-state NMR. As these issues do not persist in solid-state NMR, along with no isotropic tumbling in solid-state samples, it has emerged as a widely used technique in the structural determination of highly toxic species of protein aggregates [125].

## 9. Conclusion

This chapter highlights the applicability of spectroscopic methods to analyze the structural properties of oligomeric, aggregate, or fibrillar forms of proteins.

The feasibility to not only elucidate the structural transitions in protein structures under varying conditions but also quantify the native forms has conferred distinct advantages to spectroscopic techniques in scientific research. Although high-resolution methods such as cryo-EM or other microscopic techniques fetch detailed structural information about a protein molecule, they lack the ability to explore the effects of changes in microenvironments on the folding, self-assembly, or aggregation of proteins. Spectroscopic methods have been applied to study the folding or unfolding pathways for many proteins under complex experimental conditions. Such information has expanded our understanding of these processes and has immense applications in the pharmaceutical industry.

## Acknowledgments

This work is supported by SERB (ECR/2017/003431), MoE-STARS/STARS-1/PID (STARS1/634), and HGK-IYBA (BT/11/IYBA/2018/08) grants to PM. PM duly acknowledges the support of the Ramalingaswami Fellowship, Department of Biotechnology, GOI.

## References

[1] Boucher JI, Cote P, Flynn J, Jiang L, Laban A, Mishra P, et al. Viewing protein fitness landscapes through a next-gen lens. Genetics 2014;198(2):461–71. https://doi.org/10.1534/genetics.114.168351.
[2] Mishra P, Bhakuni V. Self-assembly of bacteriophage-associated hyaluronate lyase (HYLP2) into an enzymatically active fibrillar film. J Biol Chem 2009;284 (8):5240–9. https://doi.org/10.1074/jbc.M806730200.
[3] Flynn JM, Mishra P, Bolon DNA. Mechanistic asymmetry in Hsp90 dimers. J Mol Biol 2015;427(18):2904–11. https://doi.org/10.1016/j.jmb.2015.03.017.
[4] Mishra P, Akhtar MS, Bhakuni V. Unusual structural features of the bacteriophage-associated hyaluronate lyase (hylp2). J Biol Chem 2006;281(11):7143–50. https://doi.org/10.1074/jbc.M510991200.
[5] Mishra P, Bolon DNA. Designed Hsp90 heterodimers reveal an asymmetric ATPase-driven mechanism in vivo. Mol Cell 2014;53(2):344–50. https://doi.org/10.1016/j.molcel.2013.12.024.
[6] Thoma R, Hennig M, Sterner R, Kirschner K. Structure and function of mutationally generated monomers of dimeric phosphoribosylanthranilate isomerase from *Thermotoga maritima*. Structure 2000;8:265–76.
[7] Ahmed MH, Ghatge MS, Safo MK. Hemoglobin: structure, function and allostery. Subcell Biochem 2020;94:345–82.
[8] Goodsell DS, Olson AJ. Structural symmetry and protein function. Annu Rev Biophys Biomol Struct 2000;29:105–53.
[9] Starr TN, Flynn JM, Mishra P, Bolon DNA, Thornton JW. Pervasive contingency and entrenchment in a billion years of Hsp90 evolution. Proc Natl Acad Sci U S A 2018;115 (17):4453–8. https://doi.org/10.1073/pnas.1718133115.
[10] Singh DB, Tripathi T. Frontiers in protein structure, function, and dynamics. 1st. Singapore: Springer; 2020. p. 1–458. Place Published.

[11] Tripathi T, Dubey VK. Advances in protein molecular and structural biology methods. 1st. USA: Academic Press; 2022. p. 1–716. Place Published.
[12] Ami D, Natalello A, Gatti-Lafranconi P, Lotti M, Doglia SM. Kinetics of inclusion body formation studied in intact cells by FT-IR spectroscopy. FEBS Lett 2005; 579:3433–6.
[13] Curtis-Fisk J, Spencer RM, Weliky DP. Native conformation at specific residues in recombinant inclusion body protein in whole cells determined with solid-state NMR spectroscopy. J Am Chem Soc 2008;130:12568–9.
[14] Kianfar E. Protein nanoparticles in drug delivery: animal protein, plant proteins and protein cages, albumin nanoparticles. J Nanobiotechnol 2021;19:159.
[15] Hamill AC, Wang SC, Lee Jr CT. Probing lysozyme conformation with light reveals a new folding intermediate. Biochemistry 2005;44:15139–49.
[16] Herrera MG, Benedini LA, Lonez C, Schilardi PL, Hellweg T, Ruysschaert JM, Dodero VI. Self-assembly of 33-mer gliadin peptide oligomers. Soft Matter 2015; 11:8648–60.
[17] Maciazek-Jurczyk M, Janas K, Pozycka J, Szkudlarek A, Rogoz W, Owczarzy A, Kulig K. Human serum albumin aggregation/fibrillation and its abilities to drugs binding. Molecules 2020;25.
[18] Sequeira MA, Herrera MG, Dodero VI. Modulating amyloid fibrillation in a minimalist model peptide by intermolecular disulfide chemical reduction. Phys Chem Chem Phys 2019;21:11916–23.
[19] Hasan S, Naeem A. Consequence of macromolecular crowding on aggregation propensity and structural stability of haemoglobin under glycating conditions. Int J Biol Macromol 2020;162:1044–53.
[20] Iram A, Alam T, Khan JM, Khan TA, Khan RH, Naeem A. Molten globule of hemoglobin proceeds into aggregates and advanced glycated end products. PLoS One 2013;8, e72075.
[21] Inouye H, Kirschner DA. Alzheimer's beta-amyloid: insights into fibril formation and structure from Congo red binding. Subcell Biochem 2005;38:203–24.
[22] Hudson SA, Ecroyd H, Kee TW, Carver JA. The thioflavin T fluorescence assay for amyloid fibril detection can be biased by the presence of exogenous compounds. FEBS J 2009;276:5960–72.
[23] Siddiqui GA, Naeem A. Aggregation of globular protein as a consequences of macromolecular crowding: a time and concentration dependent study. Int J Biol Macromol 2018;108:360–6.
[24] Ahanger IA, Bashir S, Parray ZA, Alajmi MF, Hussain A, Ahmad F, Hassan MI, Islam A, Sharma A. Rationalizing the role of monosodium glutamate in the protein aggregation through biophysical approaches: potential impact on neurodegeneration. Front Neurosci 2021;15, 636454.
[25] Ghadami SA, Ahmadi Z, Moosavi-Nejad Z. The albumin-based nanoparticle formation in relation to protein aggregation. Spectrochim Acta A Mol Biomol Spectrosc 2021;252, 119489.
[26] Oikawa T, Nonaka T, Terada M, Tamaoka A, Hisanaga S, Hasegawa M. Alpha-synuclein fibrils exhibit gain of toxic function, promoting tau aggregation and inhibiting microtubule assembly. J Biol Chem 2016;291:15046–56.
[27] Mangione PP, Verona G, Corazza A, Marcoux J, Canetti D, Giorgetti S, Raimondi S, Stoppini M, Esposito M, Relini A, Canale C, Valli M, Marchese L, Faravelli G, Obici L, Hawkins PN, Taylor GW, Gillmore JD, Pepys MB, Bellotti V. Plasminogen activation triggers transthyretin amyloidogenesis in vitro. J Biol Chem 2018;293:14192–9.
[28] Barykin EP, Petrushanko IY, Kozin SA, Telegin GB, Chernov AS, Lopina OD, Radko SP, Mitkevich VA, Makarov AA. Phosphorylation of the amyloid-Beta peptide inhibits zinc-dependent aggregation, prevents Na,K-ATPase inhibition, and reduces cerebral plaque deposition. Front Mol Neurosci 2018;11:302.

[29] Pandey NK, Ghosh S, Dasgupta S. Fibrillation in human serum albumin is enhanced in the presence of copper(II). J Phys Chem B 2010;114:10228–33.
[30] Pignataro MF, Herrera MG, Dodero VI. Evaluation of peptide/protein self-assembly and aggregation by spectroscopic methods. Molecules 2020;25.
[31] Ranjbar B, Gill P. Circular dichroism techniques: biomolecular and nanostructural analyses—a review. Chem Biol Drug Des 2009;74:101–20.
[32] Zhang H, Zheng X, Kwok RTK, Wang J, Leung NLC, Shi L, Sun JZ, Tang Z, Lam JWY, Qin A, Tang BZ. In situ monitoring of molecular aggregation using circular dichroism. Nat Commun 2018;9:4961.
[33] Sahoo S, Padhy AA, Kumari V, Mishra P. Role of ubiquitin-proteasome and autophagy-lysosome pathways in α-synuclein aggregate clearance. Mol Neurobiol 2022. https://doi.org/10.1007/s12035-022-02897-1.
[34] Skamris T, Marasini C, Madsen KL, Fodera V, Vestergaard B. Early stage alpha-synuclein amyloid fibrils are reservoirs of membrane-binding species. Sci Rep 2019;9:1733.
[35] Tonali N, Dodero VI, Kaffy J, Hericks L, Ongeri S, Sewald N. Real-time BODIPY-binding assay to screen inhibitors of the early oligomerization process of Abeta1–42 peptide. Chembiochem 2020;21:1129–35.
[36] Dasari AKR, Hughes RM, Wi S, Hung I, Gan Z, Kelly JW, Lim KH. Transthyretin aggregation pathway toward the formation of distinct cytotoxic oligomers. Sci Rep 2019;9:33.
[37] Chan SW, Yau J, Ing C, Liu K, Farber P, Won A, Bhandari V, Kara-Yacoubian N, Seraphim TV, Chakrabarti N, Kay LE, Yip CM, Pomes R, Sharpe S, Houry WA. Mechanism of amyloidogenesis of a bacterial AAA+ chaperone. Structure 2016;24:1095–109.
[38] Mawhinney MT, Williams TL, Hart JL, Taheri ML, Urbanc B. Elucidation of insulin assembly at acidic and neutral pH: characterization of low molecular weight oligomers. Proteins 2017;85:2096–110.
[39] Gelenter MD, Smith KJ, Liao SY, Mandala VS, Dregni AJ, Lamm MS, Tian Y, Xu W, Pochan DJ, Tucker TJ, Su Y, Hong M. The peptide hormone glucagon forms amyloid fibrils with two coexisting beta-strand conformations. Nat Struct Mol Biol 2019;26:592–8.
[40] Brudar S, Hribar-Lee B. The role of buffers in wild-type HEWL amyloid fibril formation mechanism. Biomolecules 2019;9.
[41] Benjwal S, Verma S, Rohm KH, Gursky O. Monitoring protein aggregation during thermal unfolding in circular dichroism experiments. Protein Sci 2006;15:635–9.
[42] Ruano ML, Garcia-Verdugo I, Miguel E, Perez-Gil J, Casals C. Self-aggregation of surfactant protein A. Biochemistry 2000;39:6529–37.
[43] Leal SS, Cardoso I, Valentine JS, Gomes CM. Calcium ions promote superoxide dismutase 1 (SOD1) aggregation into non-fibrillar amyloid: a link to toxic effects of calcium overload in amyotrophic lateral sclerosis (ALS)? J Biol Chem 2013;288:25219–28.
[44] Hu Y, He C, Woo MW, Xiong H, Hu J, Zhao Q. Formation of fibrils derived from whey protein isolate: structural characteristics and protease resistance. Food Funct 2019;10:8106–15.
[45] Narhi L, Wood SJ, Steavenson S, Jiang Y, Wu GM, Anafi D, Kaufman SA, Martin F, Sitney K, Denis P, Louis JC, Wypych J, Biere AL, Citron M. Both familial Parkinson's disease mutations accelerate alpha-synuclein aggregation. J Biol Chem 1999;274:9843–6.
[46] Faraj SE, Venturutti L, Roman EA, Marino-Buslje CB, Mignone A, Tosatto SC, Delfino JM, Santos J. The role of the N-terminal tail for the oligomerization, folding and stability of human frataxin. FEBS Open Bio 2013;3:310–20.

[47] Castro IH, Bringas M, Doni D, Noguera ME, Capece L, Aran M, Blaustein M, Costantini P, Santos J. Relationship between activity and stability: design and characterization of stable variants of human frataxin. Arch Biochem Biophys 2020;691, 108491.
[48] Valsecchi WM, Cousido-Siah A, Defelipe LA, Mitschler A, Podjarny A, Santos J, Delfino JM. The role of the C-terminal region on the oligomeric state and enzymatic activity of *Trypanosoma cruzi* hypoxanthine phosphoribosyl transferase. Biochim Biophys Acta 1864;2016:655–66.
[49] Honisch C, Donadello V, Hussain R, Peterle D, De Filippis V, Arrigoni G, Gatto C, Giurgola L, Siligardi G, Ruzza P. Application of circular dichroism and fluorescence spectroscopies to assess photostability of water-soluble porcine lens proteins. ACS Omega 2020;5:4293–301.
[50] Micsonai A, Wien F, Kernya L, Lee YH, Goto Y, Refregiers M, Kardos J. Accurate secondary structure prediction and fold recognition for circular dichroism spectroscopy. Proc Natl Acad Sci U S A 2015;112:E3095–103.
[51] Frieden C. Protein aggregation processes: in search of the mechanism. Protein Sci 2007;16:2334–44.
[52] Munishkina LA, Fink AL. Fluorescence as a method to reveal structures and membrane-interactions of amyloidogenic proteins. Biochim Biophys Acta 2007;1768:1862–85.
[53] Lee D, Kim SM, Kim HY, Kim Y. Fluorescence chemicals to detect insoluble and soluble amyloid-beta aggregates. ACS Chem Nerosci 2019;10:2647–57.
[54] James NG, Jameson DM. Steady-state fluorescence polarization/anisotropy for the study of protein interactions. Methods Mol Biol 2014;1076:29–42.
[55] Lindberg DJ, Wranne MS, Gilbert Gatty M, Westerlund F, Esbjorner EK. Steady-state and time-resolved Thioflavin-T fluorescence can report on morphological differences in amyloid fibrils formed by Abeta(1–40) and Abeta(1–42). Biochem Biophys Res Commun 2015;458:418–23.
[56] Stsiapura VI, Maskevich AA, Kuzmitsky VA, Uversky VN, Kuznetsova IM, Turoverov KK. Thioflavin T as a molecular rotor: fluorescent properties of thioflavin T in solvents with different viscosity. J Phys Chem B 2008;112:15893–902.
[57] Singh A, Khatun S, Nath Gupta A. Simultaneous detection of tyrosine and structure-specific intrinsic fluorescence in the fibrillation of Alzheimer's associated peptides. Chemphyschem 2020;21:2585–98.
[58] Kuznetsova IM, Sulatskaya AI, Maskevich AA, Uversky VN, Turoverov KK. High fluorescence anisotropy of thioflavin T in aqueous solution resulting from its molecular rotor nature. Anal Chem 2016;88:718–24.
[59] Heckmeier PJ, Agam G, Teese MG, Hoyer M, Stehle R, Lamb DC, Langosch D. Determining the stoichiometry of small protein oligomers using steady-state fluorescence anisotropy. Biophys J 2020;119:99–114.
[60] Matveeva EG, Rudolph A, Moll JR, Thompson RB. Structure-selective anisotropy assay for amyloid Beta oligomers. ACS Chem Nerosci 2012;3:982–7.
[61] Ojha N, Rainey KH, Patterson GH. Imaging of fluorescence anisotropy during photoswitching provides a simple readout for protein self-association. Nat Commun 2020;11:21.
[62] Jin L, Gao W, Liu C, Zhang N, Mukherjee S, Zhang R, Dong H, Bhunia A, Bednarikova Z, Gazova Z, Liu M, Han J, Siebert HC. Investigating the inhibitory effects of entacapone on amyloid fibril formation of human lysozyme. Int J Biol Macromol 2020;161:1393–404.
[63] Wang J, Zhu H, Gan H, Meng Q, Du G, An Y, Liu J. The effect of heparan sulfate on promoting amyloid fibril formation by beta-casein and their binding research with multi-spectroscopic approaches. J Photochem Photobiol B 2020;202, 111671.
[64] Xu J, Wei Y, Yang W, Yang L, Yi Z. The mechanism and conformational changes of polybrominated diphenyl ethers to TTR by fluorescence spectroscopy, molecular simulation, and quantum chemistry. Analyst 2018;143:4662–73.

[65] Das A, Gupta A, Hong Y, Carver JA, Maiti S. A spectroscopic marker for structural transitions associated with amyloid-beta aggregation. Biochemistry 2020;59:1813–22.
[66] Hanczyc P, Mikhailovsky A, Boyer DR, Sawaya MR, Heeger A, Eisenberg D. Ultrafast time-resolved studies on fluorescein for recognition strands architecture in amyloid fibrils. J Phys Chem B 2018;122:8–18.
[67] Sahay S, Krishnamoorthy G, Maji SK. Site-specific structural dynamics of alpha-synuclein revealed by time-resolved fluorescence spectroscopy: a review. Methods Appl Fluoresc 2016;4, 042002.
[68] Rovnyagina NR, Budylin GS, Vainer YG, Tikhonova TN, Vasin SL, Yakovlev AA, Kompanets VO, Chekalin SV, Priezzhev AV, Shirshin EA. Fluorescence lifetime and intensity of thioflavin T as reporters of different fibrillation stages: insights obtained from fluorescence up-conversion and particle size distribution measurements. Int J Mol Sci 2020;21.
[69] Elson EL. Fluorescence correlation spectroscopy: past, present, future. Biophys J 2011;101:2855–70.
[70] Slaughter BD, Schwartz JW, Li R. Mapping dynamic protein interactions in MAP kinase signaling using live-cell fluorescence fluctuation spectroscopy and imaging. Proc Natl Acad Sci U S A 2007;104:20320–5.
[71] Elson EL. Introduction to fluorescence correlation spectroscopy-brief and simple. Methods 2018;140-141:3–9.
[72] Schwille P, Meyer-Almes FJ, Rigler R. Dual-color fluorescence cross-correlation spectroscopy for multicomponent diffusional analysis in solution. Biophys J 1997;72:1878–86.
[73] Stefl M, Herbst K, Rubsam M, Benda A, Knop M. Single-color fluorescence lifetime cross-correlation spectroscopy in vivo. Biophys J 2020;119:1359–70.
[74] Petrasek Z, Schwille P. Precise measurement of diffusion coefficients using scanning fluorescence correlation spectroscopy. Biophys J 2008;94:1437–48.
[75] Ruprecht V, Wieser S, Marguet D, Schutz GJ. Spot variation fluorescence correlation spectroscopy allows for superresolution chronoscopy of confinement times in membranes. Biophys J 2011;100:2839–45.
[76] Torres T, Levitus M. Measuring conformational dynamics: a new FCS-FRET approach. J Phys Chem B 2007;111:7392–400.
[77] Sahoo B, Drombosky KW, Wetzel R. Fluorescence correlation spectroscopy: a tool to study protein oligomerization and aggregation in vitro and in vivo. Methods Mol Biol 2016;1345:67–87.
[78] Sanchez SA, Chen Y, Muller JD, Gratton E, Hazlett TL. Solution and interface aggregation states of Crotalus atrox venom phospholipase A2 by two-photon excitation fluorescence correlation spectroscopy. Biochemistry 2001;40:6903–11.
[79] Sanchez SA, Brunet JE, Jameson DM, Lagos R, Monasterio O. Tubulin equilibrium unfolding followed by time-resolved fluorescence and fluorescence correlation spectroscopy. Protein Science 2004;13:81–8.
[80] Gunther G, Jameson DM, Aguilar J, Sanchez SA. Scanning fluorescence correlation spectroscopy comes full circle. Methods 2018;140-141:52–61.
[81] Wilkosz N, Czaja M, Seweryn S, Skirlinska-Nosek K, Szymonski M, Lipiec E, Sofinska K. Molecular spectroscopic markers of abnormal protein aggregation. Molecules 2020;25.
[82] Pelton JT, McLean LR. Spectroscopic methods for analysis of protein secondary structure. Anal Biochem 2000;277:167–76.
[83] Milosevic J, Prodanovic R, Polovic N. On the protein fibrillation pathway: oligomer intermediates detection using ATR-FTIR spectroscopy. Molecules 2021;26.
[84] Sun Y, Kakinen A, Wan X, Moriarty N, Hunt CPJ, Li Y, Andrikopoulos N, Nandakumar A, Davis TP, Parish CL, Song Y, Ke PC, Ding F. Spontaneous formation of

beta-sheet nano-barrels during the early aggregation of Alzheimer's amyloid beta. Nano Today 2021;38.
[85] Vosough F, Barth A. Characterization of homogeneous and heterogeneous amyloid-beta42 oligomer preparations with biochemical methods and infrared spectroscopy reveals a correlation between infrared spectrum and oligomer size. ACS Chem Nerosci 2021;12:473–88.
[86] Celej MS, Sarroukh R, Goormaghtigh E, Fidelio GD, Ruysschaert JM, Raussens V. Toxic prefibrillar alpha-synuclein amyloid oligomers adopt a distinctive antiparallel beta-sheet structure. Biochem J 2012;443:719–26.
[87] Conway KA, Harper JD, Lansbury Jr PT. Fibrils formed in vitro from alpha-synuclein and two mutant forms linked to Parkinson's disease are typical amyloid. Biochemistry 2000;39:2552–63.
[88] Mishra R, Sellin D, Radovan D, Gohlke A, Winter R. Inhibiting islet amyloid polypeptide fibril formation by the red wine compound resveratrol. Chembiochem 2009;10:445–9.
[89] Mishra R, Bulic B, Sellin D, Jha S, Waldmann H, Winter R. Small-molecule inhibitors of islet amyloid polypeptide fibril formation. Angew Chem 2008;47:4679–82.
[90] Meersman F, Smeller L, Heremans K. Comparative Fourier transform infrared spectroscopy study of cold-, pressure-, and heat-induced unfolding and aggregation of myoglobin. Biophys J 2002;82:2635–44.
[91] Raskovic B, Popovic M, Ostojic S, Andelkovic B, Tesevic V, Polovic N. Fourier transform infrared spectroscopy provides an evidence of papain denaturation and aggregation during cold storage. Spectrochim Acta A Mol Biomol Spectrosc 2015;150:238–46.
[92] Ami D, Natalello A, Taylor G, Tonon G, Maria Doglia S. Structural analysis of protein inclusion bodies by Fourier transform infrared microspectroscopy. Biochim Biophys Acta 2006;1764:793–9.
[93] Doglia SM, Ami D, Natalello A, Gatti-Lafranconi P, Lotti M. Fourier transform infrared spectroscopy analysis of the conformational quality of recombinant proteins within inclusion bodies. Biotechnol J 2008;3:193–201.
[94] Lomont JP, Rich KL, Maj M, Ho JJ, Ostrander JS, Zanni MT. Spectroscopic signature for stable beta-amyloid fibrils versus beta-sheet-rich oligomers. J Phys Chem B 2018;122:144–53.
[95] Roeters SJ, Iyer A, Pletikapic G, Kogan V, Subramaniam V, Woutersen S. Evidence for intramolecular antiparallel beta-sheet structure in alpha-synuclein fibrils from a combination of two-dimensional infrared spectroscopy and atomic force microscopy. Sci Rep 2017;7:41051.
[96] Nabers A, Ollesch J, Schartner J, Kotting C, Genius J, Hafermann H, Klafki H, Gerwert K, Wiltfang J. Amyloid-beta-secondary structure distribution in cerebrospinal fluid and blood measured by an Immuno-infrared-sensor: a biomarker candidate for Alzheimer's disease. Anal Chem 2016;88:2755–62.
[97] Barth A. Infrared spectroscopy of proteins. Biochim Biophys Acta 2007; 1767:1073–101.
[98] Shivu B, Seshadri S, Li J, Oberg KA, Uversky VN, Fink AL. Distinct beta-sheet structure in protein aggregates determined by ATR-FTIR spectroscopy. Biochemistry 2013;52:5176–83.
[99] Amenabar I, Poly S, Nuansing W, Hubrich EH, Govyadinov AA, Huth F, Krutokhvostov R, Zhang L, Knez M, Heberle J, Bittner AM, Hillenbrand R. Structural analysis and mapping of individual protein complexes by infrared nanospectroscopy. Nat Commun 2013;4:2890.
[100] Stetefeld J, McKenna SA, Patel TR. Dynamic light scattering: a practical guide and applications in biomedical sciences. Biophys Rev 2016;8:409–27.
[101] Bhattacharjee S. DLS and zeta potential—what they are and what they are not? J Control Release 2016;235:337–51.

[102] Hill SE, Robinson J, Matthews G, Muschol M. Amyloid protofibrils of lysozyme nucleate and grow via oligomer fusion. Biophys J 2009;96:3781–90.
[103] Soto C. In vivo spreading of tau pathology. Neuron 2012;73:621–3.
[104] Sgourakis NG, Lange OF, DiMaio F, Andre I, Fitzkee NC, Rossi P, Montelione GT, Bax A, Baker D. Determination of the structures of symmetric protein oligomers from NMR chemical shifts and residual dipolar couplings. J Am Chem Soc 2011;133:6288–98.
[105] Zhou C, Qi W, Lewis EN, Carpenter JF. Characterization of sizes of aggregates of insulin analogs and the conformations of the constituent protein molecules: a concomitant dynamic light scattering and Raman spectroscopy study. J Pharm Sci 2016;105:551–8.
[106] Butler HJ, Ashton L, Bird B, Cinque G, Curtis K, Dorney J, Esmonde-White K, Fullwood NJ, Gardner B, Martin-Hirsch PL, Walsh MJ, McAinsh MR, Stone N, Martin FL. Using Raman spectroscopy to characterize biological materials. Nat Protoc 2016;11:664–87.
[107] Shashilov VA, Lednev IK. 2D correlation deep UV resonance raman spectroscopy of early events of lysozyme fibrillation: kinetic mechanism and potential interpretation pitfalls. J Am Chem Soc 2008;130:309–17.
[108] Qiang W, Yau WM, Lu JX, Collinge J, Tycko R. Structural variation in amyloid-beta fibrils from Alzheimer's disease clinical subtypes. Nature 2017;541:217–21.
[109] Kurouski D, Sorci M, Postiglione T, Belfort G, Lednev IK. Detection and structural characterization of insulin prefibrilar oligomers using surface enhanced Raman spectroscopy. Biotechnol Prog 2014;30:488–95.
[110] Bernado P, Akerud T, Garcia de la Torre J, Akke M, Pons M. Combined use of NMR relaxation measurements and hydrodynamic calculations to study protein association. Evidence for tetramers of low molecular weight protein tyrosine phosphatase in solution. J Am Chem Soc 2003;125:916–23.
[111] Fawzi NL, Ying J, Torchia DA, Clore GM. Kinetics of amyloid beta monomer-to-oligomer exchange by NMR relaxation. J Am Chem Soc 2010;132:9948–51.
[112] Fawzi NL, Ying J, Ghirlando R, Torchia DA, Clore GM. Atomic-resolution dynamics on the surface of amyloid-beta protofibrils probed by solution NMR. Nature 2011;480:268–72.
[113] Feng Y, Taraban MB, Yu YB. Water proton NMR—a sensitive probe for solute association. Chem Commun 2015;51:6804–7.
[114] Taraban MB, Truong HC, Feng Y, Jouravleva EV, Anisimov MA, Yu YB. Water proton NMR for in situ detection of insulin aggregates. J Pharm Sci 2015;104:4132–41.
[115] Taraban MB, DePaz RA, Lobo B, Yu YB. Water proton NMR: a tool for protein aggregation characterization. Anal Chem 2017;89:5494–502.
[116] Paravastu AK, Qahwash I, Leapman RD, Meredith SC, Tycko R. Seeded growth of beta-amyloid fibrils from Alzheimer's brain-derived fibrils produces a distinct fibril structure. Proc Natl Acad Sci U S A 2009;106:7443–8.
[117] Wickramasinghe A, Xiao Y, Kobayashi N, Wang S, Scherpelz KP, Yamazaki T, Meredith SC, Ishii Y. Sensitivity-enhanced solid-state NMR detection of structural differences and unique polymorphs in pico- to nanomolar amounts of brain-derived and synthetic 42-residue amyloid-beta fibrils. J Am Chem Soc 2021;143:11462–72.
[118] Walsh P, Simonetti K, Sharpe S. Core structure of amyloid fibrils formed by residues 106–126 of the human prion protein. Structure 2009;17:417–26.
[119] Savastano A, Jaipuria G, Andreas L, Mandelkow E, Zweckstetter M. Solid-state NMR investigation of the involvement of the P2 region in tau amyloid fibrils. Sci Rep 2020;10:21210.

[120] Wu X, Ma Y, Zhao K, Zhang J, Sun Y, Li Y, Dong X, Hu H, Liu J, Wang J, Zhang X, Li B, Wang H, Li D, Sun B, Lu J, Liu C. The structure of a minimum amyloid fibril core formed by necroptosis-mediating RHIM of human RIPK3. Proc Natl Acad Sci U S A 2021;118.

[121] Suzuki Y, Brender JR, Hartman K, Ramamoorthy A, Marsh EN. Alternative pathways of human islet amyloid polypeptide aggregation distinguished by (19)F nuclear magnetic resonance-detected kinetics of monomer consumption. Biochemistry 2012;51:8154–62.

[122] Suzuki Y, Brender JR, Soper MT, Krishnamoorthy J, Zhou Y, Ruotolo BT, Kotov NA, Ramamoorthy A, Marsh EN. Resolution of oligomeric species during the aggregation of Abeta1–40 using (19)F NMR. Biochemistry 2013;52:1903–12.

[123] Larda ST, Simonetti K, Al-Abdul-Wahid MS, Sharpe S, Prosser RS. Dynamic equilibria between monomeric and oligomeric misfolded states of the mammalian prion protein measured by $^{19}$F NMR. J Am Chem Soc 2013;135:10533–41.

[124] Pandya A, Howard MJ, Zloh M, Dalby PA. An evaluation of the potential of NMR spectroscopy and computational modelling methods to inform biopharmaceutical formulations. Pharmaceutics 2018;10.

[125] Pedersen JT, Heegaard NH. Analysis of protein aggregation in neurodegenerative disease. Anal Chem 2013;85:4215–27.

# CHAPTER 15

# Multimodal spectroscopic methods for the analysis of carbohydrates

**Nidhi Sharma[a], Himanshu Pandey[b], Amit Kumar Sonkar[c], Manjul Gondwal[d], and Seema Singh[a]**

[a]Department of Chemistry, School of Applied and Life Sciences, Uttaranchal University, Dehradun, India
[b]Department of Chemistry, Ben-Gurion University of the Negev, Beer-Sheva, Israel
[c]Department of Biochemistry, All India Institute of Medical Sciences, Guwahati, India
[d]Department of Chemistry, Laxman Singh Mahar Government Post Graduate College (Soban Singh Jeena University, Almora), Pithoragarh, India

## 1. Introduction

Carbohydrates are biomolecules that are virtually found everywhere in biological systems. Chemically, these organic biomolecules are made of carbon, hydrogen, and oxygen in the form of $(CH_2O)_n$ (hydrate of carbon) in the ratio of $C_n$:$H_{2n}$:$O_n$. These are a very diverse class of macromolecules that can be broadly classified as simple carbohydrates consisting of monosaccharides (as a single unit, e.g., glucose, fructose, galactose), disaccharides composed of two units (e.g., sucrose, lactose, and maltose are the dimers of glucose-fructose, glucose-galactose, and glucose-glucose units, respectively), and complex carbohydrates generally known as polysaccharides, made up of hundred to several thousand monosaccharide units (e.g., glycogen, starch, and fibers). Further classification of polysaccharides depends on their molecular arrangement, bonding, solubility, surface interaction, and nutrition effect (e.g., digestible or nondigestible).

Because of their chemical and structural diversity, carbohydrates play vital roles in numerous biological processes. Carbohydrates account for around 40%–80% of the total energy needs of the human body and are the primary source of energy for the brain [1]. They are the key backbone of DNA and RNA; play a central role in cell signaling; control blood glucose, metabolism, and serum cholesterol; and influence colonic microflora, laxation, molecular recognition, immunity, and inflammation. They also act as the building subunits of bioscaffolding frameworks in plants and insects. Biologically, carbohydrates are also found conjugated to other biomolecules

*Advanced Spectroscopic Methods to Study Biomolecular Structure and Dynamics*
https://doi.org/10.1016/B978-0-323-99127-8.00019-2

such as proteins, lipids, and phenols as glycoproteins, glycolipids, and glycophenols, respectively, and have a significant influence on their bioactivity. For example, glycosylation and deglycosylation influence the biological activity of proteins that can be utilized to provide new technologically attractive properties to proteins and have been projected as strategic targets for controlling and regulating the immunoreactive action of key food allergens [2–4]. Polysaccharides are condensation polymers of monosaccharides made of either one repeating monosaccharide unit (homopolysaccharides) or a mixture of a variety of monosaccharides (heteropolysaccharides). They can be linear (unbranched) or branched. α-1,4 linkages are present in the linear unbranched chain, whereas branched polysaccharides are linearly connected by α-1,4 linkage, and branching occurs at α-1,6 glycosidic bonds.

Different approaches have been carried out to analyze the composition and structural fundamentals of carbohydrates. These include chromatography, mass spectrometry (MS), nuclear magnetic resonance (NMR) spectroscopy, vibrational spectroscopy, X-ray diffraction (XRD), and so on. The following topics of this chapter focus on various aspects of the spectroscopic analysis of carbohydrates, the working principles applied to characterize simple to complex carbohydrates, and the structural information they provide about carbohydrates when combined with proper separation/fractionation or other spectroscopic methods.

## 2. Sample preparation for the spectroscopic analysis of carbohydrates

A crucial step while analyzing carbohydrates is their separation from conjugated lipids and proteins and fractionation to obtain a particular carbohydrate fraction from complex carbohydrate samples. The following section briefly describes the approaches commonly employed to extract and fractionate carbohydrate samples for analysis.

### 2.1 Extraction and purification

Simple filtration, ultrafiltration, or membrane-based filtration and extraction procedures, such as liquid-liquid extraction (LLE) and solid-phase extraction (SPE), are routinely utilized as cleanup and to concentrate carbohydrates for analysis [5–7]. The solvent and the column are selected based on the properties of the samples to be analyzed, with C18 columns being the most common ones. However, C8, porous graphitic carbon, and cation-exchange columns are frequently employed [8–11]. Supercritical fluid extraction

(SFE) has also been reported to fractionate different carbohydrates efficiently [12,13].

### 2.1.1 Pressurized liquid extraction

Because of advantages such as a faster extraction process and higher yields, pressurized liquid extraction (PLE) is becoming a favored method for extracting carbohydrate sample components from natural products [14–16].

### 2.1.2 Field flow fractionation

To fractionate large polysaccharides such as cellulose, starch, pullulan, and so on, the field flow fractionation (FFF) technique, which is based on electric, thermal, magnetic, or gravitational field applied perpendicular to a laminar flow, is often employed.

### 2.1.3 Chromatographic procedures

To fractionate, separate, and determine the molecular mass of oligo- and polysaccharides, size exclusion chromatography (SEC) is the most preferred technique [17,18]. High-performance liquid chromatography (HPLC), which provides both qualitative and quantitative information and, importantly, does not require derivatization of carbohydrates, is the official technique for routine sugar analysis recommended by the Association of Official Analytical Chemists International (AOAC) [19]. Polysaccharides dissolved in specific solvent systems in combination with appropriate chemical modifications can also be analyzed for structural homogeneousness, branching patterns, molecular mass variations, and aggregation by high-performance SEC (HP-SEC) equipped with multiangle laser-light scattering (MALLS) or refractive index (RI) detectors [20]. Ion-exchange chromatography (IEC) is another chromatography-based technique broadly applied to fractionate and analyze oligo- and monosaccharides.

### 2.1.4 Other important fractionation techniques

The carbon fractionation method that consists of carbohydrate adsorption in an activated charcoal column is also an important technique for carbohydrate fractionation based on the degree of polymerization [21]. This technique could be combined with other fractionation methods (e.g., PLE) for the appropriate characterization of carbohydrates in a subsequent analytical phase [14,22]. Other remarkable methods capable of differentiating different saccharides or even the epimers employ molecularly imprinted polymers (MIPs).

## 2.2 Purity estimation

Traditional total and/or reducing sugar assays and assessing the presence of other contaminating products are typically employed for this purpose. In a pure polysaccharide sample, estimation of reducing sugars presents an idea about the quantity of reducing end groups. Colorimetry-based iodine test [23], anthrone or phenol-sulfuric acid assay [24], *p*-hydroxybenzoic acid hydrazide [25], 2,2′-bicinchoninate [26], and 3,5-dinitrosalicylic acid [27] are the commonly used methods for these purposes. In addition, chromatography-based GlcN assay after total hydrolysis [28] and acetic acid assay after deacetylation [29,30] are commonly used to estimate the purity of chitin-based compounds. Moreover, to estimate impurities such as protein and phenols, Kjeldahl [31,32], Lowry [32,33], Bradford, and Folin-Ciocalteu assays are commonly employed.

## 2.3 Chemical modifications for analysis

Once pure samples are prepared, carbohydrate samples can be analyzed as they are or may require further treatments such as hydrolysis, derivatization, and so on to assist ensuing analytical procedures. Prior to a chromatographic analysis and linkage analysis of complex carbohydrates, chemolytic methods based on chemical or enzymatic cleavage of glycosidic bonds joining different monosaccharides are frequently followed. Strong acidic conditions are generally used for the total hydrolysis employing trifluoroacetic acid, formic acid, hydrochloric acid, or sulfuric acid under heating [34]. Different glycosidic bonds linking monosaccharide units in oligosaccharides have differential susceptibility to hydrolysis, resulting in partial degradations under controlled conditions producing monosaccharides and oligosaccharides whose analysis generates information that could lead to the elucidation of polysaccharide structures. The resulting mono-, di-, or oligosaccharides can be chemically modified for further analysis. Because most carbohydrates are neutral or hydrophilic, derivatization also produces polar and charge variations in compounds that help chromatographic separation and analysis. The monosaccharides and oligosaccharides can be reduced, followed by acetylation or sialylation that generates volatile derivatives such as alditol acetates and trimethylsilyl or trifluoro acetyl ether mixtures, respectively, that can be analyzed through gas chromatography (GC) equipped with a flame ionization detector (FID) or mass spectrometry (MS) detector. The procedure for *O*-methyl glucitol derivatives of different glucopyranose units of β-D-glucans that can be analyzed by GC is discussed in detail elsewhere [35].

For other analytic techniques, such as capillary electrophoresis (CE) or HPLC, it is essential to increase the sensitivity of detection, which is usually performed by precolumn derivatization using UV-absorbing or fluorescent labels [36,37]. These procedures work on the generation of imine (Schiff base) by the treatment of the carbonyl group present on the carbohydrate molecule with primary amines. This Schiff's base, on subsequent reduction, produces N-substituted glycosylamine [36–38].

# 3. Advanced analysis of carbohydrates

## 3.1 High-performance liquid chromatography

In recent times, HPLC equipped with advanced spectroscopic detectors has become a method of choice for the isolation and separation of carbohydrates and their derivatives. Samples containing a mixture of polysaccharides can be separated by using the interaction of the analytes with the stationary and mobile phases. Here, the relative affinity of the mobile and stationary phases decides the speed of separation of the analyte [39]. Depending on stationary and mobile phases used for the analysis, various mechanisms for the separation of polysaccharide mixtures with HPLC techniques have been used, such as reversed-phase HPLC (RP-HPLC), hydrophilic interaction liquid chromatography (HILIC), anion-exchange chromatography (AEC), and SEC.

In normal-phase HPLC, the hydrophilic stationary phase and nonpolar mobile phase are used for separation of analytes, but in the case of RP-HPLC, a polar and hydrophilic mobile phase and a hydrophobic and nonpolar stationary phase are used, and the rate of elution depends on the interaction of the eluent with the mobile phase and with hydrophobic groups of stationary phases. Mainly C8 or C18 columns are used for RP-HPLC separation of monosaccharides [40,41]. The separation of free carbohydrates takes a longer time because of the polarity. This problem can be overcome by the derivatization of carbohydrates, which increases the hydrophobic interaction with the stationary phase and reduces the elution time. Moreover, the polar stationary phase, similar to those used in the normal-phase HPLC, and the mobile phase similar to those used in RP-HPLC, are used in HILIC. The separation in HILIC is based on the configuration and the number of polar groups present [42].

Acidic or anionic saccharides are separated through IEC. The separation depends upon the interaction of negatively charged analyte ions with the cationic stationary phase. Water-based solvent systems with bases such as sodium hydroxide (NaOH) and sodium acetate ($CH_3COONa$) are used with the mobile phase. Positively charged quaternary ammonium-modified crosslinked polystyrene is used as a stationary phase for high-pH AEC. Under mild and neutral conditions, glycoproteins, uronic acids, sulfated saccharides, and sialic acids become negatively charged and behave as acidic saccharides. These acidic saccharides can be efficiently analyzed and separated using AEC [43].

SEC, also known as gel filtration chromatography (GFC), is another common chromatographic technique used to separate polysaccharides. Separation through SEC is based on the size of the molecule. SEC has become an essential method to determine the molecular mass and molecular mass distribution of carbohydrates and their derivatives, such as alginate, starch, chitosan, fructan, hyaluronic acid, proteoglycans, and glycoproteins [44,45]. Technological advances have allowed the manufacturing of SEC analytical columns that have permitted its usage in HPLC-based procedures.

### 3.1.1 HPLC detectors

Traditionally, photometric detections based on UV-visible or fluorescence spectroscopy are the commonly used detection methods integrated into the HPLC systems, but because carbohydrates have low detection sensitivity in these regions of the electromagnetic spectrum, precolumn derivatization is routinely performed for the efficient detection of carbohydrates. Generally, derivatizing reagents such as 1-phenyl-3-methyl-5-pyrazolone and *p*-nitrobenzoyl chloride, which react with saccharides under mild conditions and provide sensitive UV detection at ~245 nm, are employed to confer UV absorbance [46]. To avoid laborious derivatization of carbohydrates, detectors based on RI and evaporative light scattering detectors (ELSDs) have been integrated into HPLC systems. ELSD, in combination with high-performance anion-exchange chromatography (HPAEC), has an advantage over RI detectors as it allows gradient elution but can be used only for non-volatile compounds [1,47]. HPLC, in combination with isotope ratio mass spectrometry (IRMS), can be utilized for the analysis of carbohydrates, which measures the abundance of $^{13}C$ in different compounds [48].

## 3.2 Mass spectrometry

High sensitivity, resolution, and easy connection to detectors make GC coupled with mass spectrometers a dominant analytic procedure for the analysis of complex mixtures of polysaccharides. GC-MS separates the derivatives of monosaccharides and small oligosaccharides, whereas HPLC-MS efficiently analyses the molecular conjugates and small and large oligosaccharides.

The HPLC-MS technique is used for the faster identification and analysis of unknown carbohydrates within a shorter time and less laborious fractionation. GC-MS conventionally analyzes monosaccharides and small oligosaccharides but requires derivatization, whereas HPLC-MS helps analyze molecular conjugates and small and large oligosaccharides. The rapid identification of unknown carbohydrates is feasible with the HPLC-MS technique without consuming much time and laborious fractionation, derivatization, and purification processes. Owing to the low volatility and high polarity of carbohydrates, electrospray ionization (ESI) is generally chosen over atmospheric pressure chemical ionization (APCI) and atmospheric pressure photoionization (APPI) methods [49]. In coupling techniques, the liquid chromatography-MS (LC-MS) technique produces real-time results; however, several obstacles need to be overcome to permit the unique online coupling of HPLC with MS, including the changes in working pressures of two systems and the mismatch of MS with the distinctive LC mobile phases. Recently, HPLC coupled with ESI-tandem mass spectrometry (ESI-MS/MS) has evolved as a more powerful tool for analyzing carbohydrates than APCI because it is more suitable for the analysis of nonpolar and thermally stable small molecules rather than carbohydrates [49]. During the past few decades, capillary LC-MS analysis has advanced as a popular analysis method as it requires a very small volume of samples for injection [50].

Matrix-assisted laser desorption/ionization (MALDI), which characteristically generates the singly charged species, integrated with MS (MALDI-MS), is also a useful technique and has been employed to analyze large carbohydrates with simpler sample preparation [51]. The online LC-MALDI-MS approach comprises continuous flow, in which the matrix solution and the LC eluent are mixed and focused on the flow probe [51]. For accurate mass analysis, quadrupole-orbitrap (Q-Orbitrap) MS and high-resolution time-of-flight MS (TOF-MS) are also used [52].

## 3.3 Infrared spectroscopy

The infrared (IR) spectroscopy technique plays an important role in the structure elucidation of several classes of compounds. Similar to organic molecules, carbohydrates also contain functional groups that have characteristic absorption at specific wavelengths in the IR region of the electromagnetic spectrum [53–55]. This rapid, nondestructive, and easily accessible technique requires a change in the dipole moment for the structural analysis of the carbohydrates/compounds. The IR region of 12,500 to 10 $cm^{-1}$ (wavelengths from 800 nm to 1 mm) can be divided into three subregions in the electromagnetic spectrum: far-IR (FIR) (400 to 10 $cm^{-1}$, i.e., wavelengths from 2.5 to 25 μm), mid-IR (MIR) (4000 to 400 $cm^{-1}$, i.e., wavelengths from 2.5 to 25 μm), and near-IR (NIR) (12,500 to 4000 $cm^{-1}$, i.e., wavelengths from 800 to 2500 nm). Among these subregions, the MIR region has applications in the structural analysis of the molecules as the vibrational frequencies in this region fall in the same range as the vibrational frequencies of bonds present in the organic molecules. Particularly, between 1200 and 800 $cm^{-1}$, the wavenumber range in MIR is considered the identification region, also known as the "fingerprint region" for carbohydrates [56]. This region provides information about the major functional groups (—OH, —CHO, —CO—, etc.) or chemical groups present in the carbohydrates. An IR spectrophotometer is used to obtain an IR spectrum of any sample as a function of radiation frequencies by recording the absorption of IR radiation by the sample. The absorption of a particular wavelength is specific to a particular bond. The typical IR spectra for carbohydrates are listed in Table 1 [57].

Because of the complicated IR spectra of carbohydrates, the interpretation of their spectral peaks becomes difficult. To simplify the interpretations of peaks, Fourier transformation IR (FTIR) spectroscopy is used, which has the additional advantage of having a spectrum with a well-divided absorption region of distinct functional groups [58,59]. This technique is useful for the identification of sugar rings (furan and pyran) and anomeric configurations of glycosidic bond linkages present even in the high-molecular weight fractions of carbohydrates isolated from biological systems. There are mainly two spectral regions used for the structural identification of the carbohydrates, that

**Table 1** The IR frequencies of common functional groups of polysaccharides.

| Common functional groups | Wavenumber ($cm^{-1}$) | Intensity |
|---|---|---|
| O—H stretching | 3000–3500 | Broad, strong |
| C—H asymmetric and symmetric stretching | 2800–3000 | Sharp, occasionally double overlapping with O—H |
| C—H bending | 1380 | Weak |
| —COO— asymmetric and symmetric stretching | 1600–1630, 1400 | Strong, weak |
| Fingerprint region of carbohydrates | 900–1280 | Strong |

is, the "sugar region" (950–1200 $cm^{-1}$) and the "anomeric region" (950–750 $cm^{-1}$). The sugar region contains intense bands because of the overlapping of the stretching vibrations of C=O and C—C groups in the glycosidic bonds and pyranoid rings. Pyranose and furanose forms of sugars show three absorption peaks and two peaks within 1000–1200 $cm^{-1}$. The anomeric region shows weak and less intense vibrational bands. The band at about 890 $cm^{-1}$ indicates the existence of the β-configuration, whereas the α-configuration shows weak bands around 890–920 $cm^{-1}$. The presence of both α- and β-configurations in the sugar molecules is confirmed by the presence of multiple bands in a region between 920 and 840 $cm^{-1}$ [35] (Fig. 1).

IR spectroscopy has become an important tool for the identification of different polymorphs of large carbohydrates such as cellulose. Adaptation of different forms of cellulose chains and intra- and intermolecular hydrogen bonds induces various crystalline forms of cellulose. There is a difference in the absorptions of the —OH stretching band in IR spectra of cellulose crystalline polymorphs. Cellulose $I_{\alpha}$ shows absorptions at 3240 and 750 $cm^{-1}$, and cellulose $I_{\beta}$ shows absorptions at 3270 and 710 $cm^{-1}$ frequencies. The shifting of the —C—H stretching band to higher wavenumbers (from 2900 $cm^{-1}$) with low intensity indicates the loss of crystallinity or

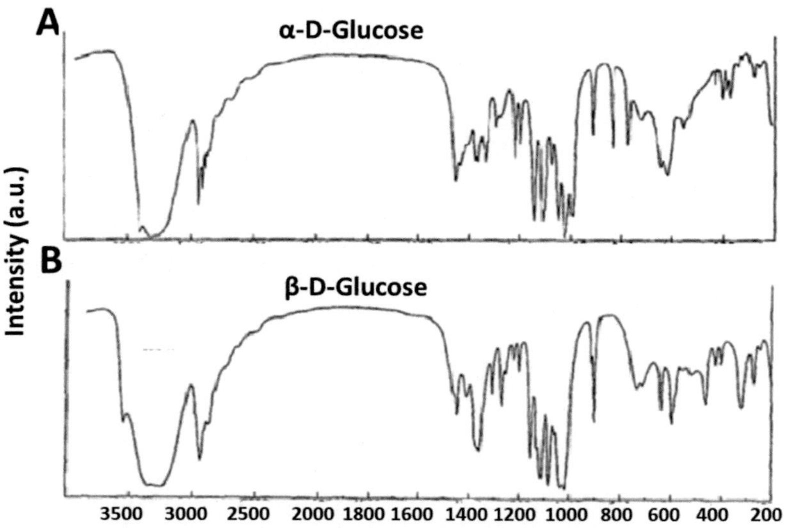


**Fig. 1** (A and B) IR spectra of α- and β-D-glucose. *(Reproduced with permission from Dujardin N, Dudognon E, Willart JF, Hédoux A, Guinet Y, Paccou L, et al. Solid state mutarotation of glucose. J Phys Chem B 2011;115. https://doi.org/10.1021/jp109382j.)*

shifting toward the amorphous cellulose. Even some bands become less intense or disappear at 900–1500 $cm^{-1}$ [60–62]. The decrement in the intensity of the absorption band, also known as the marker of the crystallinity band at 1430 $cm^{-1}$ for —$CH_2$ scissoring vibrations, specifies the reduction in the crystallinity degree. On the other hand, the increment in the intensity of the band at 898 $cm^{-1}$ for —C—O—C— stretching vibrations of glycosidic bonds, also known as the amorphous state marker, confirms the amorphous cellulose [63].

High sensitivity and minimal sample requirement make IR analysis an effective technique for identifying the different substituted groups. Progress of the substitution reaction can also be monitored in the case of the synthetic substituted polysaccharides as a substitution to the polysaccharide backbone by any group leads to the generation of the corresponding vibrational bands in the IR spectra [64,65]. Commonly used substituents are sulfonyl (—$SO_2$—), selenyl (—$SeO_2$—), acetyl (—CO—), phosphoryl (—$PO_3^{2-}$—), methyl (—$CH_3$), and carboxymethyl (—$CH_2COOH$) groups, and their characteristic wavenumbers are summarized in Table 2.

Although absorption of IR radiations of a particular wavelength is widely implemented to identify types of anomeric rings, functional and substituted groups, hydrogen bonding, and crystal allomorphs, the overlapping of absorption peaks makes it challenging to analyze the sample. These lead to incomplete or sometimes misleading information to identify the structure of the compound. Hence, the data for structural and molecular identification of polysaccharides only by IR are unreliable. To overcome the complexity and improve the accuracy of the data, other spectrochemometric tools such as Raman, NMR, MS, and so on should be employed in conjunction with IR spectroscopy.

## 3.4 Raman spectroscopy

Raman and IR spectroscopic techniques complement each other with respect to the structural elucidation of any molecule (Fig. 2). Similar to IR, Raman spectroscopy also provides data in less time with easy sample preparation and is nondestructive. The incident light of a wavelength of 750–850 nm scatters inelastically. These scattered radiations are collected and dispersed and generate spectra that have the information necessary for the structural analysis of the sample. Change in the polarizability because of molecular vibrations is a necessary condition for a molecule to be Raman-active, rather than the change in dipole moment for a molecule

**Table 2** Common substitution groups in polysaccharides and their corresponding characteristic frequencies in wavenumbers.

| Common substitution groups | Wavenumbers | References |
|---|---|---|
| Sulfonyl (—$SO_2$—) | 1200–1270 cm$^{-1}$ and 1010–1060 cm$^{-1}$ (S=O stretching) and 900–800 cm$^{-1}$ (—C—O—S— stretching) | [65,66] |
| Selenyl (—$SeO_2$—) | 1080 and 850 cm$^{-1}$ (Se=O stretching) and 610 cm$^{-1}$ (—C—O—Se— stretching) | [67] |
| Phosphoryl (—$PO_3^{2-}$—) | 1250 cm$^{-1}$ (P=O stretching), 915 cm$^{-1}$ (—C—O—P— bending) | [68,69] |
| Acetyl (—CO—) | *O*-linked: 1735 cm$^{-1}$ (C=O stretching), 1365–1380 cm$^{-1}$ (symmetric $CH_3$ bending), 1247 cm$^{-1}$ (C—O vibration) | [70] |
| | *N*-linked: 1640 cm$^{-1}$ (C=O stretching), 1365–1380 cm$^{-1}$ (symmetric $CH_3$ bending), 1550 cm$^{-1}$ (C—N stretching) | [71] |
| Carboxymethyl (—$CH_2COOH$) | 1600 and 1420 cm$^{-1}$ (—COO stretching), 1250 and 1070 cm$^{-1}$ (O—C—O stretching) | [72] |
| Methyl (—$CH_3$) | Disappearance of the band at 3400 cm$^{-1}$ (—OH stretching) and intensity increases of the band at 2930 cm$^{-1}$ (—$CH_2$— and —$CH_3$) | [73] |

to become IR-active. This is the reason why the C—C and C—H vibrations are more easily observed in the Raman spectrum. Because the incident radiations are weakly scattered in water, it adds to the advantage of its effective use for the biological sample analysis as water does not create any hindrance in Raman scattering from solutes in an aqueous solution. Also, the Raman spectrum has a better resolution than an IR spectrum [74]. The Raman spectra in Fig. 2 show one and three peaks for α-D-glucose and β-D-glucose, respectively.

Raman absorption generates an intense band at 470–485 cm$^{-1}$ of amylase and amylopectin (α-D-glucans) at the C-4 position, whereas absorption at 540–545 cm$^{-1}$ has been observed in the case of dextran. These bands are absent in pullulan, which is a polysaccharide polymer made of maltotriose units. Raman spectroscopy and IR are also applied to understand retrogradation in starch gel, quantify the amount of amylase present in starch, and

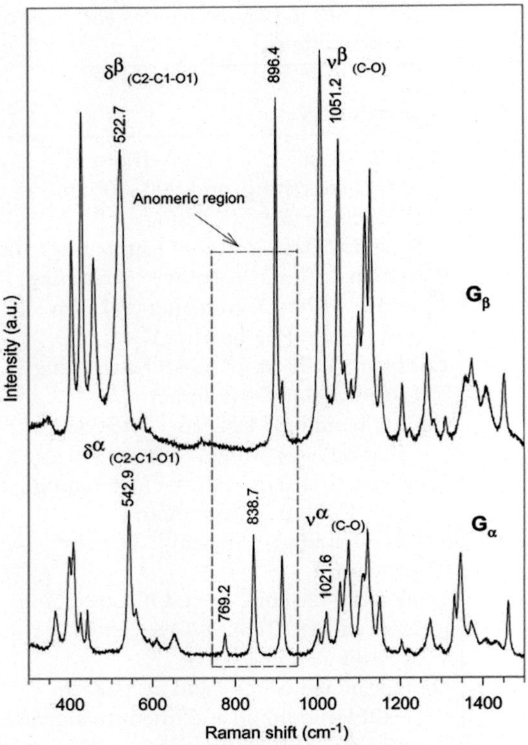


**Fig. 2** Raman spectra of α- and β-D-glucose (Gα and Gβ). *(Reproduced with permission from Wells HA, Atalla RH. An investigation of the vibrational spectra of glucose, galactose and mannose. J Mol Struct 1990;224. https://doi.org/10.1016/0022-2860(90)87031-R.)*

identify structural differences between celluloses from plants. The crystal structures and allotropes are also identified through this technique [75–80].

## 3.5 Nuclear magnetic resonance spectroscopy

NMR spectroscopy is a powerful technique for the analysis of the structure and purity of carbohydrates. This spectroscopic method has proven its applicability for the analysis of simple as well as complex carbohydrates [81]. In addition, it helps in the investigation of carbohydrates attached to proteins and other physiologically relevant components. Various NMR approaches have been employed to study carbohydrates, such as one-dimensional (1D), two-dimensional (2D), $^{13}$C, and $^{15}$N solid-state NMR [82–85] and $^{1}$H, $^{31}$P, and $^{13}C$ liquid-state NMR [86–91]. Commonly used solvents in carbohydrate liquid-state NMR spectroscopy are $D_2O/DCl$, $D_2O/CD_3COOD$, and $D_2O/DCOOD$. Ionic liquids such as 1-*n*-butyl-3-methylimidazolium chloride ([C4mim]Cl) have also been successfully used as solvents for solution NMR studies, as carbohydrates from biological sources such as fruit pulp can be easily dissolved in it [92]. NMR is sensitive enough to determine the anomeric configuration of glucopyranose units and identify the connecting glycosidic bonds [93] (Fig. 3). For example, the NMR resonance signal for anomeric proton and carbon peaks at 4.9–5.1 and 98–100 ppm, respectively, indicating α-D-glucans, whereas the signal for the respective β-D-glucans peaks at 4.3–4.6 and 103–104 ppm. The carbon signal downfield shift by ~5–10 ppm with respect to the 1-methylglucoside anomer designates the involvement of the particular carbon in the glycosidic bond. Furthermore, advanced 2D correlation NMR methods (such as COSY/DQF-COSY, NOESY, and HSQC) have been developed to solve complicated overlapping resonance signals, which deliver a greater amount of information about the molecule than 1D NMR and are helpful in studying complex molecules (Table 3). The following experiments are usually used sequentially to interpret such complicated resonance signals [35]

1. **COSY and TOCSY:** for assigning $CH_2/CH$ proton resonance.
2. **HETCOR, HMQC, and HSQC:** for assignment of the corresponding carbon signals.
3. **NOESY, ROESY, and HMBC:** for assignment of intra- and interunit interactions (Fig. 3).

Based on these assignments, structural models are prepared for the corresponding molecules. For example, different NMR experiments have identified polysaccharides from hazelnut cell walls, tomatoes, and mushrooms [94]. Using correlation NMR spectroscopy and deuterium exchange together with an insensitive nuclei enhancement by polarization transfer (INEPT) experiment, resonance signals of glucan fragments and branching glucan units of amylopectin were assigned [95,96].

Several β-D-(1 → 3, 1 → 6)-linked glucans were analyzed by 2D correlation NMR spectroscopy [97]. However, it is noteworthy that for molecules that have solubility issues such as chitins (with a high degree of acetylation), only solid-state NMR can be employed, which are mostly recorded with magic-angle spinning (MAS) or cross-polarization (CP). MAS relies on averaging dipolar interactions and chemical shift anisotropy and provides highly resolved resonance spectra, whereas CP improves sensitivity by reducing the relaxation delay by magnetization transfer from the proton

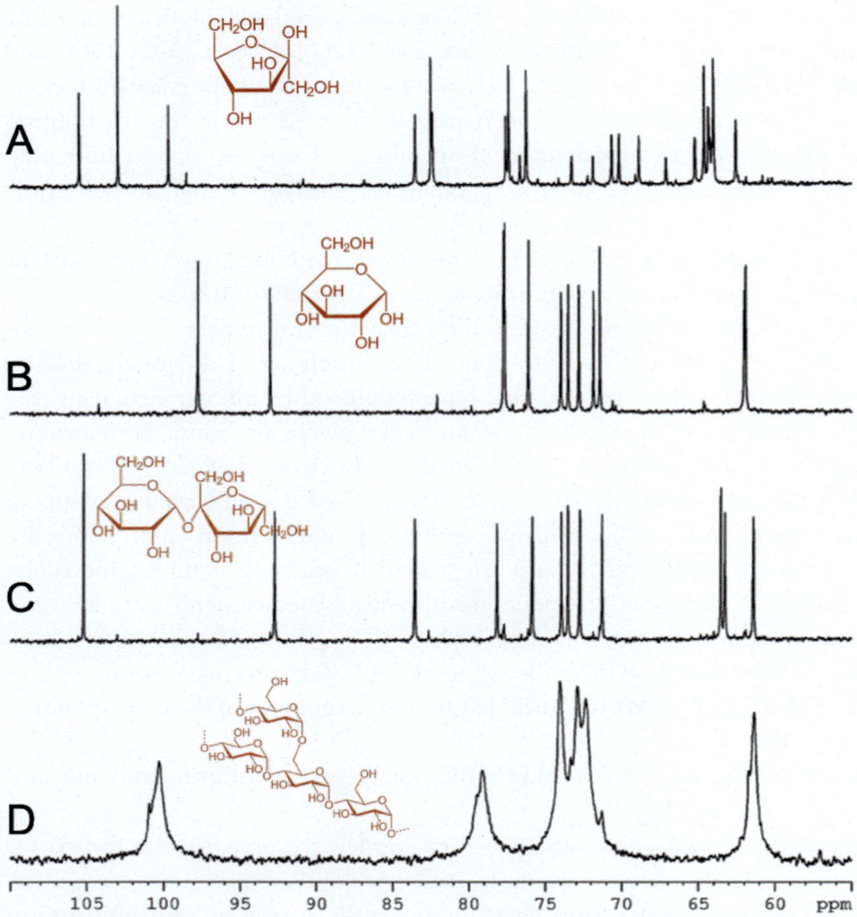


**Fig. 3** $^{13}$C NMR spectra of (A) fructose, (B) glucose, (C) sucrose, and (D) amylopectin recorded in d 1-*n*-butyl-3-methylimidazolium chloride ([C4mim]Cl/DMSO-$d_6$) solution. *(Reproduced with permission from Fort DA, Swatloski RP, Moyna P, Rogers RD, Moyna G. Use of ionic liquids in the study of fruit ripening by high-resolution 13C NMR spectroscopy: "Green" solvents meet green bananas. Chem Commun 2006. doi:https://doi.org/10.1039/b515177p.)*

to the $^{13}$C spins [98]. The carbohydrates whose samples cannot be prepared for the solution NMR analysis can be chemically modified or partially hydrolyzed to obtain the soluble components that can be analyzed by NMR spectroscopy. For example, with the Smith degradation of glycans, a linear polysaccharide is obtained that could be analyzed by $^{13}$C NMR in combination with other analytical methods. A polysaccharide from the

**Table 3** Commonly used two-dimensional NMR experiments and their applications.

| NMR experiment | Application in solving structure |
|---|---|
| HSQC (heteronuclear single quantum coherence) | Determination of specific carbon-linked protons |
| COSY (correlation spectroscopy) and DQF-COSY (double quantum filtered correlation spectroscopy) | Determination of spin-spin coupled protons (two to three bonds apart) |
| NOESY (nuclear Overhauser effect spectroscopy) | Presents correlations of all protons that are close enough for dipolar interaction by coupling through space |

ascomycetous fungus was characterized as linear (1→3)-β-D-glucan with Smith degradation and a subsequent $^{13}C$ NMR analysis [99]. Owing to the 100% isotopic abundance and the high magnetogyric ratio, $^{31}P$ NMR spectroscopy is another strong NMR technique. However, because carbohydrates usually do not contain phosphorus atoms, phosphorus-containing derivatives of carbohydrates can be characterized using liquid-state or solid-state $^{31}P$ NMR spectroscopy [100].

NMR techniques are also employed to analyze chemical and physiochemical processes by examining intermediate or final products [100]. Kim et al. reported NMR as a tool to study pinitol, an antidiabetic agent, from economically viable natural sources [101]. The molecular ordering of cellulose from the leaves of *Arabidopsis thaliana* has also been characterized using solid-state $^{13}C$ NMR [102]. Sophisticated analytical methods such as NMR are very helpful in extracting information about molecular structures along with the degree of extraction, concentration, and purity estimations. Moreover, NMR-based analysis can also be used for in situ characterization of naturally occurring carbohydrates.

## 3.6 X-ray diffraction analysis

XRD is a vital tool for identifying the structure, lattice parameters, and arrangement of atoms in crystalline molecules. This technique is also effectively used for the structural analysis of polysaccharides, particularly starch. XRD analysis of the crystal structure of starch classified it into A, B, and C types. Monoclinic unit cells, hexagonal unit cells, and a mixture of monoclinic and hexagonal unit cells are present in A-type, B-type, and C-type starch, respectively [103–105]. A few studies reported that the structural information obtained from XRD analysis and IR analysis of starch does not match each other. For example, wheat starch (A-type) and potato starch (B-type) have the same IR spectral features but have different XRD data. Because of controversy in the information provided by XRD and IR, researchers have concluded that XRD is a more suitable technique for the polymorph and crystal structure analysis of starch than the IR method [106,107].

## 3.7 Multidimensional techniques for carbohydrate analysis

There are a lot more spectroscopic approaches that can be applied in combination with other techniques for the characterization of carbohydrates. Scanning electron microscopy (SEM), an important technique for the visual analysis of surface morphology, is usually employed with energy-dispersive spectroscopy (EDAX), transmission electron microscopy (SEM-TEM system), or inductively coupled plasma spectroscopy (ICP). For example, SEM has been employed for characterizing the surface of crab chitin and chitosan [108–110]. Similarly, EDAX has shown its importance in the determination of the metal uptake mechanism of polysaccharides [111]. Circular dichroism spectroscopy, which measures structural asymmetry, has been used in the investigation of the degree of *N*-acetylation of complex carbohydrates [112]. NMR spectroscopy, along with nuclear Overhauser effect-based molecular dynamics (MD) simulations and molecular modeling, is another tool for studying interactions between carbohydrates and proteins [113]. 2D GC, in combination with time-of-flight MS (GC-GC-TOFMS), has been used to analyze carbohydrates from plant samples [114]. A combination of RP-HPLC and HILIC with online integrated pulsed electrochemical detection-mass spectrometry (IPED-MS) has been used for the complete analysis of carbohydrates [115]. Microarray technologies have also shown great potential in complementing traditional glycobiology analysis. A quantum dot and gold nanoparticle-fabricated chip technology has been successfully applied to analyze the degree of glycosylation in proteins [116]. Integration of such technologies with biosensors and MS has generated tremendous enthusiasm among researchers and is anticipated to evolve further and provide efficient analytical tools for glycobiology.

# 4. Conclusions

Carbohydrates are biomolecules with unique and diverse structures and interesting properties. Analysis of carbohydrates employs several approaches for the estimation of their purity, molecular mass, and configuration.

Spectroscopic methods play a dominant role in the structural analysis of monosaccharides, disaccharides, oligosaccharides, and their derivatives. Moreover, several nonspectroscopic analytical techniques such as thermogravimetric analysis (TGA), differential scanning calorimetry (DSC), isothermal titration calorimetry (ITC), HPLC, and SEC are also routinely used. Both spectroscopic and nonspectroscopic methods have their own set of advantages and limitations; hence, a combination of spectroscopic, chemical, and separation techniques is better to obtain a more comprehensive characterization of carbohydrates.

## References

[1] Muir JG, Rose R, Rosella O, Liels K, Barrett JS, Shepherd SJ, et al. Measurement of short-chain carbohydrates in common Australian vegetables and fruits by high-performance liquid chromatography (HPLC). J Agric Food Chem 2009;57. https://doi.org/10.1021/jf802700e.

[2] Corzo-Martínez M, Moreno FJ, Villamiel M, Harte FM. Characterization and improvement of rheological properties of sodium caseinate glycated with galactose, lactose and dextran. Food Hydrocoll 2010;24. https://doi.org/10.1016/j.foodhyd.2009.08.008.

[3] Amigo-Benavent M, Athanasopoulos VI, Ferranti P, Villamiel M, del Castillo MD. Carbohydrate moieties on the in vitro immunoreactivity of soy β-conglycinin. Food Res Int 2009;42. https://doi.org/10.1016/j.foodres.2009.03.003.

[4] van de Lagemaat J, Manuel Silván J, Javier Moreno F, Olano A, Dolores del Castillo M. In vitro glycation and antigenicity of soy proteins. Food Res Int 2007;40. https://doi.org/10.1016/j.foodres.2006.09.006.

[5] Kamada T, Nakajima M, Nabetani H, Iwamoto S. Pilot-scale study of the purification and concentration of non-digestible saccharides from yacon rootstock using membrane technology. Food Sci Technol Res 2002;8. https://doi.org/10.3136/fstr.8.172.

[6] Mullin WJ, Emmons DB. Determination of organic acids and sugars in cheese, milk and whey by high performance liquid chromatography. Food Res Int 1997;30. https://doi.org/10.1016/S0963-9969(97)00026-4.

[7] la Torre GL, la Pera L, Rando R, lo Turco V, di Bella G, Saitta M, et al. Classification of Marsala wines according to their polyphenol, carbohydrate and heavy metal levels using canonical discriminant analysis. Food Chem 2008;110. https://doi.org/10.1016/j.foodchem.2008.02.071.

[8] Castellari M, Versari A, Spinabelli U, Galassi S, Amati A. An improved HPLC method for the analysis of organic acids, carbohydrates, and alcohols in grape musts and wines. J Liq Chromatogr Relat Technol 2000;23. https://doi.org/10.1081/JLC-100100472.

[9] Lehtonen P, Hurme R. Liquid chromatographic determination of sugars in beer by evaporative light scattering detection. J Inst Brewing 1994;100. https://doi.org/10.1002/j.2050-0416.1994.tb00834.x.

[10] Megherbi M, Herbreteau B, Faure R, Salvador A. Polysaccharides as a marker for detection of corn sugar syrup addition in honey. J Agric Food Chem 2009;57. https://doi.org/10.1021/jf803384q.

[11] Kitahara K, Copeland L. A simple method for fractionating debranched starch using a solid reversed-phase cartridge. J Cereal Sci 2004;39. https://doi.org/10.1016/S0733-5210(03)00069-9.

[12] Montañés F, Fornari T, Martín-Álvarez PJ, Montilla A, Corzo N, Olano A, et al. Selective fractionation of disaccharide mixtures by supercritical $CO_2$ with ethanol as co-solvent. J Supercrit Fluids 2007;41. https://doi.org/10.1016/j.supflu.2006.08.010.
[13] Montañés F, Fornari T, Martín-Álvarez PJ, Corzo N, Olano A, Ibáñez E. Selective recovery of tagatose from mixtures with galactose by direct extraction with supercritical $CO_2$ and different cosolvents. J Agric Food Chem 2006;54. https://doi.org/10.1021/jf0618123.
[14] Ruiz-Matute AI, Ramos L, Martínez-Castro I, Sanz ML. Fractionation of honey carbohydrates using pressurized liquid extraction with activated charcoal. J Agric Food Chem 2008;56. https://doi.org/10.1021/jf8014552.
[15] Ruiz-Matute AI, Sanz ML, Corzo N, Martín-Álvarez PJ, Ibáñez E, Martínez-Castro I, et al. Purification of lactulose from mixtures with lactose using pressurized liquid extraction with ethanol-water at different temperatures. J Agric Food Chem 2007;55. https://doi.org/10.1021/jf070018u.
[16] Mendiola JA, Herrero M, Cifuentes A, Ibañez E. Use of compressed fluids for sample preparation: food applications. J Chromatogr A 2007;1152. https://doi.org/10.1016/j.chroma.2007.02.046.
[17] Ward RM, Gao Q, de Bruyn H, Gilbert RG, Fitzgerald MA. Improved methods for the structural analysis of the amylose-rich fraction from rice flour. Biomacromolecules 2006;7. https://doi.org/10.1021/bm050617e.
[18] Hernández O, Ruiz-Matute AI, Olano A, Moreno FJ, Sanz ML. Comparison of fractionation techniques to obtain prebiotic galactooligosaccharides. Int Dairy J 2009;19. https://doi.org/10.1016/j.idairyj.2009.03.002.
[19] Ulberth F, Wrolstad RE, Acree TE, Decker EA, Penner MH, Reid DS, et al. Handbook of food analytical chemistry—water, proteins, enzymes, lipids, and carbohydrates. In: Wrolstad RE, Acree TE, Decker EA, Penner MH, Reid DS, Schwartz SJ, Shoemaker CF, Smith DM, Sporns P, editors. Analytical and bioanalytical chemistry, vol. 387. Wiley; 2007. https://doi.org/10.1007/s00216-007-1138-x.
[20] You SG, Lim ST. Molecular characterization of corn starch using an aqueous HPSEC-MALLS-RI system under various dissolution and analytical conditions. Cereal Chem 2000;77. https://doi.org/10.1094/CCHEM.2000.77.3.303.
[21] Morales V, Sanz ML, Olano A, Corzo N. Rapid separation on activated charcoal of high oligosaccharides in honey. Chromatographia 2006;64. https://doi.org/10.1365/s10337-006-0842-6.
[22] Wei YA, Hendrix DL, Nieman R. Diglucomelezitose, a novel pentasaccharide in silverleaf whitefly honeydew. J Agric Food Chem 1997;45. https://doi.org/10.1021/jf970228e.
[23] Martinez C. Determination of amylose in flour by a colorimetric assay: collaborative study. Starch/Staerke 1996;48. https://doi.org/10.1002/star.19960480303.
[24] Dubois M, Gilles KA, Hamilton JK, Rebers PA, Smith F. Colorimetric method for determination of sugars and related substances. Anal Chem 1956;28. https://doi.org/10.1021/ac60111a017.
[25] Lever M. A new reaction for colorimetric determination of carbohydrates. Anal Biochem 1972;47. https://doi.org/10.1016/0003-2697(72)90301-6.
[26] Doner LW, Irwin PL. Assay of reducing end-groups in oligosaccharide homologues with 2,2′-bicinchoninate. Anal Biochem 1992;202. https://doi.org/10.1016/0003-2697(92)90204-K.
[27] Miller GL. Use of dinitrosalicylic acid reagent for determination of reducing sugar. Anal Chem 1959;31. https://doi.org/10.1021/ac60147a030.

[28] Chen GC, Johnson BR. Improved colorimetric determination of cell wall chitin in wood decay fungi. Appl Environ Microbiol 1983;46. https://doi.org/10.1128/aem.46.1.13-16.1983.

[29] Ekblad A, Nasholm T. Determination of chitin in fungi and mycorrhizal roots by an improved HPLC analysis of glucosamine. Plant Soil 1996;178. https://doi.org/10.1007/BF00011160.

[30] Holan Z, Votruba J, Vlasáková V. New method of chitin determination based on deacetylation and gas-liquid chromatographic assay of liberated acetic acid. J Chromatogr A 1980;190. https://doi.org/10.1016/S0021-9673(00)85512-0.

[31] Lynch JM, Barbano DM. Kjeldahl nitrogen analysis as a reference method for protein determination in dairy products. J AOAC Int 1999;82. https://doi.org/10.1093/jaoac/82.6.1389.

[32] Jernejc K, Cimerma A, Perdih A. Comparison of different methods for protein determination in Aspergillus niger mycelium. Appl Microbiol Biotechnol 1986;23. https://doi.org/10.1007/BF02346058.

[33] Lowry OH, Rosebrough NJ, Farr AL, Randall RJ. Protein measurement with the Folin phenol reagent. J Biol Chem 1951;193. https://doi.org/10.1016/s0021-9258(19)52451-6.

[34] Herrero M, Cifuentes A, Ibáñez E, del Castillo MD. Advanced analysis of carbohydrates in foods. In: Otles S, editor. Methods of analysis of food components and additives. 2nd ed. CRC Press; 2011.

[35] Synytsya A, Novak M. Structural analysis of glucans. Ann Transl Med 2014;2. https://doi.org/10.3978/j.issn.2305-5839.2014.02.07.

[36] Molnár-Perl I. Role of chromatography in the analysis of sugars, carboxylic acids and amino acids in food. J Chromatogr A 2000;891. https://doi.org/10.1016/S0021-9673(00)00598-7.

[37] Lamari FN, Kuhn R, Karamanos NK. Derivatization of carbohydrates for chromatographic, electrophoretic and mass spectrometric structure analysis. J Chromatogr B Analyt Technol Biomed Life Sci 2003;793. https://doi.org/10.1016/S1570-0232(03)00362-3.

[38] Ruhaak LR, Zauner G, Huhn C, Bruggink C, Deelder AM, Wuhrer M. Glycan labeling strategies and their use in identification and quantification. Anal Bioanal Chem 2010;397. https://doi.org/10.1007/s00216-010-3532-z.

[39] Lee TD. Introduction to modern liquid chromatography, third edition. J Am Soc Mass Spectrom 2011;22. https://doi.org/10.1007/s13361-010-0021-8.

[40] Bean SR, Ioerger BP, Blackwell DL. Separation of kafirins on surface porous reversed-phase high-performance liquid chromatography columns. J Agric Food Chem 2011;59. https://doi.org/10.1021/jf1036195.

[41] Kelebek H, Selli S, Canbas A, Cabaroglu T. HPLC determination of organic acids, sugars, phenolic compositions and antioxidant capacity of orange juice and orange wine made from a Turkish cv. Kozan. Microchem J 2009;91. https://doi.org/10.1016/j.microc.2008.10.008.

[42] Jandera P. Stationary and mobile phases in hydrophilic interaction chromatography: a review. Anal Chim Acta 2011;692. https://doi.org/10.1016/j.aca.2011.02.047.

[43] Bruggink C, Maurer R, Herrmann H, Cavalli S, Hoefler F. Analysis of carbohydrates by anion exchange chromatography and mass spectrometry. J Chromatogr A 2005;1085. https://doi.org/10.1016/j.chroma.2005.03.108.

[44] Churms SC. Chapter 8 Modern size-exclusion chromatography of carbohydrates and glycoconjugates. J Chromatogr Libr 2002;66. https://doi.org/10.1016/S0301-4770(02)80033-1.

[45] Churms SC. High performance hydrophilic interaction chromatography of carbohydrates with polar sorbents. J Chromatogr Libr 1995;58. https://doi.org/10.1016/S0301-4770(08)60508-4.
[46] Zhang L, Xu J, Zhang L, Zhang W, Zhang Y. Determination of 1-phenyl-3-methyl-5-pyrazolone-labeled carbohydrates by liquid chromatography and micellar electrokinetic chromatography. J Chromatogr B Analyt Technol Biomed Life Sci 2003;793. https://doi.org/10.1016/S1570-0232(03)00373-8.
[47] Davis F, Terry LA, Chope GA, Faul CFJ. Effect of extraction procedure on measured sugar concentrations in onion (Allium cepa L.) bulbs. J Agric Food Chem 2007;55. https://doi.org/10.1021/jf063170p.
[48] Abramson FP, Black GE, Lecchi P. Application of high-performance liquid chromatography with isotope-ratio mass spectrometry for measuring low levels of enrichment of underivatized materials. J Chromatogr A 2001;913. https://doi.org/10.1016/S0021-9673(00)01032-3.
[49] Domon B, Costello CE. A systematic nomenclature for carbohydrate fragmentations in FAB-MS/MS spectra of glycoconjugates. Glycoconj J 1988;5. https://doi.org/10.1007/BF01049915.
[50] Zhang S, Li C, Zhou G, Che G, You J, Suo Y. Determination of the carbohydrates from Notopterygium forbesii Boiss by HPLC with fluorescence detection. Carbohydr Polym 2013;97. https://doi.org/10.1016/j.carbpol.2013.05.041.
[51] Lin X, Xiao C, Ling L, Guo L, Guo X. A dual-mode reactive matrix for sensitive and quantitative analysis of carbohydrates by MALDI-TOF MS. Talanta 2021;235. https://doi.org/10.1016/j.talanta.2021.122792.
[52] Horatz K, Giampà M, Karpov Y, Sahre K, Bednarz H, Kiriy A, et al. Conjugated polymers as a new class of dual-mode matrices for MALDI mass spectrometry and imaging. J Am Chem Soc 2018;140. https://doi.org/10.1021/jacs.8b06637.
[53] Kacuráková M, Capek P, Sasinková V, Wellner N, Ebringerová A. FT-IR study of plant cell wall model compounds: pectic polysaccharides and hemicelluloses. Carbohydr Polym 2000;43. https://doi.org/10.1016/S0144-8617(00)00151-X.
[54] Prado BM, Kim S, Özen BF, Mauer LJ. Differentiation of carbohydrate gums and mixtures using Fourier transform infrared spectroscopy and chemometrics. J Agric Food Chem 2005;53. https://doi.org/10.1021/jf0485537.
[55] Hong T, Yin J-Y, Nie S-P, Xie M-Y. Applications of infrared spectroscopy in polysaccharide structural analysis: progress, challenge and perspective. Food Chem X 2021;12. https://doi.org/10.1016/j.fochx.2021.100168.
[56] Ozaki Y. Infrared spectroscopy—mid-infrared, near-infrared, and far-infrared/terahertz spectroscopy. Anal Sci 2021;37. https://doi.org/10.2116/analsci.20R008.
[57] Guo Q, Cui SW, Kang J, Ding H, Wang Q, Wang C. Non-starch polysaccharides from American ginseng: physicochemical investigation and structural characterization. Food Hydrocoll 2015;44. https://doi.org/10.1016/j.foodhyd.2014.09.031.
[58] Reich P. Infrared and Raman spectroscopy. Methods and applications. Z Phys Chem 1998;205. https://doi.org/10.1524/zpch.1998.205.part_1.127a.
[59] Agarwala UC, Nigam HL, Agrawal S. Infrared spectroscopy of molecules. World Scientific; 2014.
[60] Renois-Predelus G, Schindler B, Compagnon I. Analysis of sulfate patterns in glycosaminoglycan oligosaccharides by MS n coupled to infrared ion spectroscopy: the case of GalNAc4S and GalNAc6S. J Am Soc Mass Spectrom 2018;29. https://doi.org/10.1007/s13361-018-1955-5.
[61] Makarem M, Lee CM, Kafle K, Huang S, Chae I, Yang H, et al. Probing cellulose structures with vibrational spectroscopy. Cellulose 2019;26. https://doi.org/10.1007/s10570-018-2199-z.

[62] Rongpipi S, Ye D, Gomez ED, Gomez EW. Progress and opportunities in the characterization of cellulose—an important regulator of cell wall growth and mechanics. Front Plant Sci 2019;9. https://doi.org/10.3389/fpls.2018.01894.

[63] Ciolacu D, Ciolacu F, Popa VI. Amorphous cellulose—structure and characterization. Cellul Chem Technol 2011;45.

[64] Luo M, Zhang X, Wu J, Zhao J. Modifications of polysaccharide-based biomaterials under structure-property relationship for biomedical applications. Carbohydr Polym 2021;266. https://doi.org/10.1016/j.carbpol.2021.118097.

[65] Caputo HE, Straub JE, Grinstaff MW. Design, synthesis, and biomedical applications of synthetic sulphated polysaccharides. Chem Soc Rev 2019;48. https://doi.org/10.1039/c7cs00593h.

[66] Korva HE, Kärkkäinen J, Lappalainen K, Lajunen M. Spectroscopic study of natural and synthetic polysaccharide sulfate structures. Starch 2016;68(9–10):854–63.

[67] Fiorito S, Epifano F, Preziuso F, Taddeo VA, Genovese S. Selenylated plant polysaccharides: a survey of their chemical and pharmacological properties. Phytochemistry 2018;153:1–10.

[68] Chen L, Huang G. The antiviral activity of polysaccharides and their derivatives. Int J Biol Macromol 2018;115:77–82.

[69] Wang J, Wang Y, Xu L, Wu Q, Wang Q, Kong W, et al. Synthesis and structural features of phosphorylated *Artemisia sphaerocephala* polysaccharide. Carbohydr Polym 2018;181:19–26.

[70] Lucas AJDS, Oreste EQ, Costa HLG, López HM, Saad CDM, Prentice C. Extraction, physicochemical characterization, and morphological properties of chitin and chitosan from cuticles of edible insects. Food Chem 2021;343:128550.

[71] Shi XD, Nie S, Yin J, Que Z, Zhang L, Huang X. Polysaccharide from leaf skin of *Aloe barbadensis* Miller: Part I. Extraction, fractionation, physicochemical properties and structural characterization. Food Hydrocoll 2017;73:176–83.

[72] Theis TV, Queiroz Santos VA, Appelt P, Barbosa-Dekker AM, Vetvicka V, Dekker RFH, et al. Fungal exocellular (1-6)-β-d-glucan: carboxymethylation, characterization, and antioxidant activity. Int J Mol Sci 2019;20(9):2337.

[73] Liu M, Fu L, Jia X, Wang J, Yang X, Xia B, et al. Dataset of the infrared spectrometry, gas chromatography-mass spectrometry analysis and nuclear magnetic resonance spectroscopy of the polysaccharides from *C. militaris*. Data Brief 2019; 25:104126.

[74] Arboleda PH, Loppnow GR. Raman spectroscopy as a discovery tool in carbohydrate chemistry. Anal Chem 2000;72. https://doi.org/10.1021/ac991389f.

[75] Dupuy N, Laureyns J. Recognition of starches by Raman spectroscopy. Carbohydr Polym 2002;49. https://doi.org/10.1016/S0144-8617(01)00304-6.

[76] Bulkin BJ, Kwak Y, Dea ICM. Retrogradation kinetics of waxy-corn and potato starches; a rapid, Raman-spectroscopic study. Carbohydr Res 1987;160. https://doi.org/10.1016/0008-6215(87)80305-1.

[77] Almeida MR, Alves RS, Nascimbem LBLR, Stephani R, Poppi RJ, de Oliveira LFC. Determination of amylose content in starch using Raman spectroscopy and multivariate calibration analysis. Anal Bioanal Chem 2010;397. https://doi.org/10.1007/s00216-010-3566-2.

[78] Szymańska-Chargot M, Cybulska J, Zdunek A. Sensing the structural differences in cellulose from apple and bacterial cell wall materials by Raman and FT-IR spectroscopy. Sensors 2011;11. https://doi.org/10.3390/s110605543.

[79] Zhbankov RG, Firsov SP, Korolik EV, Petrov PT, Lapkovski MP, Tsarenkov VM, et al. Vibrational spectra and the structure of medical biopolymers. J Mol Struct 2000;555. https://doi.org/10.1016/S0022-2860(00)00590-1.

[80] Fechner PM, Wartewig S, Kleinebudde P, Neubert RHH. Studies of the retrogradation process for various starch gels using Raman spectroscopy. Carbohydr Res 2005;340. https://doi.org/10.1016/j.carres.2005.08.018.
[81] Hounsome N, Hounsome B, Tomos D, Edwards-Jones G. Plant metabolites and nutritional quality of vegetables. J Food Sci 2008;73. https://doi.org/10.1111/j.1750-3841.2008.00716.x.
[82] Heux L, Brugnerotto J, Desbrières J, Versali MF, Rinaudo M. Solid state NMR for determination of degree of acetylation of chitin and chitosan. Biomacromolecules 2000;1. https://doi.org/10.1021/bm000070y.
[83] Yu Ge, Morin FG, Nobes GAR, Marchessault RH. Degree of acetylation of chitin and extent of grafting PHB on chitosan determined by solid state 15N NMR. Macromolecules 1999;32. https://doi.org/10.1021/ma9813338.
[84] Raymond L, Morin FG, Marchessault RH. Degree of deacetylation of chitosan using conductometric titration and solid-state NMR. Carbohydr Res 1993;246. https://doi.org/10.1016/0008-6215(93)84044-7.
[85] Pelletier A, Lemire I, Sygusch J, Chornet E, Overend RP. Chitin/chitosan transformation by thermo-mechano-chemical treatment including characterization by enzymatic depolymerization. Biotechnol Bioeng 1990;36. https://doi.org/10.1002/bit.260360313.
[86] Yun Yang B, Montgomery R. Degree of acetylation of heteropolysaccharides. Carbohydr Res 1999;323. https://doi.org/10.1016/S0008-6215(99)00242-6.
[87] Jaiswal N, Raikwal N, Pandey H, Agarwal N, Arora A, Poluri KM, et al. NMR elucidation of monomer–dimer transition and conformational heterogeneity in histone-like DNA binding protein of Helicobacter pylori. Magn· Reson Chem 2018;56. https://doi.org/10.1002/mrc.4701.
[88] Weinhold MX, Sauvageau JCM, Kumirska J, Thöming J. Studies on acetylation patterns of different chitosan preparations. Carbohydr Polym 2009;78. https://doi.org/10.1016/j.carbpol.2009.06.001.
[89] Lebouc F, Dez I, Madec PJ. NMR study of the phosphonomethylation reaction on chitosan. Polymer (Guildf) 2005;46. https://doi.org/10.1016/j.polymer.2004.11.017.
[90] Vårum KM, Antohonsen MW, Grasdalen H, Smidsrød O. Determination of the degree of N-acetylation and the distribution of N-acetyl groups in partially N-deacetylated chitins (chitosans) by high-field n.m.r. spectroscopy. Carbohydr Res 1991;211. https://doi.org/10.1016/0008-6215(91)84142-2.
[91] Rinaudo M, Ledung P, Gey C, Milas M. Substituent distribution on O, N-carboxymethylchitosans by $^{1}$H and $^{13}$C NMR. Int J Biol Macromol 1992;14(3):122–8.
[92] Fort DA, Swatloski RP, Moyna P, Rogers RD, Moyna G. Use of ionic liquids in the study of fruit ripening by high-resolution 13C NMR spectroscopy: "Green" solvents meet green bananas. Chem Commun 2006. https://doi.org/10.1039/b515177p.
[93] Mulloy B. High-field NMR as a technique for the determination of polysaccharide structures. Appl Biochem Biotechnol Part B Mol Biotechnol 1996;6. https://doi.org/10.1007/BF02761706.
[94] Dourado F, Vasco P, Barros A, Mota M, Coimbra MA, Gama FM. Characterisation of Chilean hazelnut (Gevuina avellana) tissues: light microscopy and cell wall polysaccharides. J Sci Food Agric 2003;83. https://doi.org/10.1002/jsfa.1287.
[95] Falk H, Stanek M. Two-dimensional 1H and 13C NMR spectroscopy and the structural aspects of amylose and amylopectin. Monatsh Chem 1997;128. https://doi.org/10.1007/BF00807088.
[96] Isogai A. NMR analysis of cellulose dissolved in aqueous NaOH solutions. Cellulose 1997;4. https://doi.org/10.1023/A:1018471419692.

[97] Kim YT, Kim EH, Cheong C, Williams DL, Kim CW, Lim ST. Structural characterization of β-D-(1 → 3, 1 → 6)-linked glucans using NMR spectroscopy. Carbohydr Res 2000;328. https://doi.org/10.1016/S0008-6215(00)00105-1.

[98] Duarte ML, Ferreira MC, Marvão MR, Rocha J. Determination of the degree of acetylation of chitin materials by 13C CP/MAS NMR spectroscopy. Int J Biol Macromol 2001;28. https://doi.org/10.1016/S0141-8130(01)00134-9.

[99] Barbosa AM, Steluti RM, Dekker RFH, Cardoso MS, Corradi Da Silva ML. Structural characterization of Botryosphaeran: a (1 → 3;1 → 6)-β-D-glucan produced by the ascomyceteous fungus, Botryosphaeria sp. Carbohydr Res 2003;338. https://doi.org/10.1016/S0008-6215(03)00240-4.

[100] Kumirska J, Czerwicka M, Kaczyński Z, Bychowska A, Brzozowski K, Thöming J, et al. Application of spectroscopic methods for structural analysis of chitin and chitosan. Mar Drugs 2010;8. https://doi.org/10.3390/md8051567.

[101] Kim JI, Kim JC, Kang MJ, Lee MS, Kim JJ, Cha IJ. Effects of pinitol isolated from soybeans on glycaemic control and cardiovascular risk factors in Korean patients with type II diabetes mellitus: a randomized controlled study. Eur J Clin Nutr 2005;59. https://doi.org/10.1038/sj.ejcn.1602081.

[102] Newman RH, Davies LM, Harris PJ. Solid-state 13C nuclear magnetic resonance characterization of cellulose in the cell walls of Arabidopsis thaliana leaves. Plant Physiol 1996;111. https://doi.org/10.1104/pp.111.2.475.

[103] Wu AC, Witt T, Gilbert RG. Characterization methods for starch-based materials: state of the art and perspectives. Aust J Chem 2013;66. https://doi.org/10.1071/CH13397.

[104] Han Z, Shi R, Sun DW. Effects of novel physical processing techniques on the multistructures of starch. Trends Food Sci Technol 2020;97. https://doi.org/10.1016/j.tifs.2020.01.006.

[105] Wang S, Xu H, Luan H. Multiscale structures of starch granules. In: Wang S, editor. Starch structure, functionality and application in foods. Springer; 2020.

[106] Sevenou O, Hill SE, Farhat IA, Mitchell JR. Organisation of the external region of the starch granule as determined by infrared spectroscopy. Int J Biol Macromol 2002;31. https://doi.org/10.1016/S0141-8130(02)00067-3.

[107] Warren FJ, Gidley MJ, Flanagan BM. Infrared spectroscopy as a tool to characterise starch ordered structure—a joint FTIR-ATR, NMR, XRD and DSC study. Carbohydr Polym 2016;139. https://doi.org/10.1016/j.carbpol.2015.11.066.

[108] Trimukhe KD, Varma AJ. A morphological study of heavy metal complexes of chitosan and crosslinked chitosans by SEM and WAXRD. Carbohydr Polym 2008;71. https://doi.org/10.1016/j.carbpol.2007.07.010.

[109] Yen MT, Yang JH, Mau JL. Physicochemical characterization of chitin and chitosan from crab shells. Carbohydr Polym 2009;75. https://doi.org/10.1016/j.carbpol.2008.06.006.

[110] Yen MT, Mau JL. Physico-chemical characterization of fungal chitosan from shiitake stipes. LWT Food Sci Technol 2007;40. https://doi.org/10.1016/j.lwt.2006.01.002.

[111] Kamst E, van der Drift KMGM, Thomas-Oates JE, Lugtenberg BJJ, Spaink HP. Mass spectrometric analysis of chitin oligosaccharides produced by Rhizobium NodC protein in Escherichia coli. J Bacteriol 1995;177. https://doi.org/10.1128/jb.177.21.6282-6285.1995.

[112] Kittur FS, Vishu Kumar AB, Tharanathan RN. Low molecular weight chitosans—preparation by depolymerization with Aspergillus niger pectinase, and characterization. Carbohydr Res 2003;338. https://doi.org/10.1016/S0008-6215(03)00175-7.

[113] Colombo G, Meli M, Cañada J, Asensio JL, Jiménez-Barbero J. Toward the understanding of the structure and dynamics of protein-carbohydrate interactions: molecular dynamics studies of the complexes between hevein and oligosaccharidic ligands. Carbohydr Res 2004;339. https://doi.org/10.1016/j.carres.2003.10.030.

[114] Pierce KM, Hope JL, Hoggard JC, Synovec RE. A principal component analysis based method to discover chemical differences in comprehensive two-dimensional gas chromatography with time-of-flight mass spectrometry (GC × GC-TOFMS) separations of metabolites in plant samples. Talanta 2006;70. https://doi.org/10.1016/j.talanta.2006.01.038.
[115] Louw S, Pereira AS, Lynen F, Hanna-Brown M, Sandra P. Serial coupling of reversed-phase and hydrophilic interaction liquid chromatography to broaden the elution window for the analysis of pharmaceutical compounds. J Chromatogr A 2008;1208. https://doi.org/10.1016/j.chroma.2008.08.058.
[116] Kim YP, Park S, Oh E, Oh YH, Kim HS. On-chip detection of protein glycosylation based on energy transfer between nanoparticles. Biosens Bioelectron 2009;24. https://doi.org/10.1016/j.bios.2008.07.012.

# CHAPTER 16

# Integration of spectroscopic and computational data to analyze protein structure, function, folding, and dynamics

**Kavya Prince[a], Santanu Sasidharan[a], Niharika Nag[b], Timir Tripathi[b,c], and Prakash Saudagar[a]**
[a]Department of Biotechnology, National Institute of Technology Warangal, Warangal, India
[b]Molecular and Structural Biophysics Laboratory, Department of Biochemistry, North-Eastern Hill University, Shillong, India
[c]Regional Director's Office, Indira Gandhi National Open University (IGNOU), Regional Center Kohima, Kohima, India

## 1. Protein structures: A race with time

The most diverse and versatile group of biological molecules are proteins, and they are unique in that they are composed of only 20 standard amino acids. Ranging from the physiological responses to the structure stabilization in cells, proteins play vital functional roles that help maintain homeostasis. For the proper functioning of a cell, proteins must be folded into a specific conformational structure [1]. In the 1960s, Anfinsen discovered that a denatured protein possesses the capacity to spontaneously and rapidly refold to its native conformation [2]. This puzzled the scientists about what causes or drives the folding of a protein to its active or native conformation. To make the scenario more complicated, it was shown that even a small protein of 100 amino acids in length could fold into a plethora of conformations, considering the rotational degrees of freedom around each bond. This puzzling fact led scientists to believe that protein folding is not a random process but follows a predetermined folding pathway(s) [3]. In simple words, the linear chain of amino acids not only codes for its native structure but also defines the mechanism of folding that the protein should follow to attain the said native structure. For the past 50 years, the folding mechanism of proteins has been considered from multiple perspectives, including amino acid codes and thermodynamic, spectroscopic, and microscopic views. Today, the field

*Advanced Spectroscopic Methods to Study Biomolecular Structure and Dynamics*
https://doi.org/10.1016/B978-0-323-99127-8.00018-0

of protein folding and dynamics has progressed by leaps and bounds, especially with the advent of advanced and sophisticated technologies.

Although the field has developed well, exactly how the protein folding proceeds is still unknown, that is, the transition of the linear amino acid chain to a complete tertiary structure remains elusive to researchers as the process is very fast. Therefore, solving the protein folding problem requires a multidisciplinary endeavor with expert contributions from biologists, chemists, physicists, and even mathematicians. Contemporarily, the experimental methods to determine the structure of proteins such as nuclear magnetic resonance (NMR) spectroscopy, cryo-electron microscopy (cryo-EM), and X-ray crystallography have been central to the field of structural biology. Apart from these, there are experimental methods that can derive structural information rather than the structure itself. These data are not meaningful or have a limited meaning on their own, but when combined, they provide a lot of information regarding their behavior and dynamics. The experimental methods, such as X-ray crystallography, Fourier transform infrared (FTIR) spectroscopy, circular dichroism (CD) spectroscopy, fluorescence spectroscopy, small-angle X-ray scattering (SAXS), and mass spectroscopy, provide information including size, shape, solvent accessibility, protein composition, distances and contacts, orientation, flexibility, local environment, spatial density, motions, and so on [4]. Although these experiments provide certain information, they are ambiguous without the three-dimensional structural data.

On the other hand, the computational prediction of protein structure has evolved as an alternative method over the past 20–30 years. Several online tools have been developed for the structural modeling of proteins. The overall process of computational prediction follows three steps: protein modeling, protein docking, and molecular dynamics (MD). The protein modeling step is the most predictive, considering the myriads of backbone conformations possible [3]. Therefore, the step in computational prediction is completely stochastic and is based on Monte Carlo methods, where the backbone atoms of a protein are modeled based on fragments already available in the protein data bank (PDB). The models are then scored based on steric hindrances and clashes. Then, the proteins are docked against each other, and their interactions are analyzed. These docked complexes are further subjected to MD ranging from nanoseconds (ns) to microseconds (μs) to simulate the in vitro conditions.

This chapter focuses on how the different spectroscopic data can be used along with computational data to understand protein structure and dynamics. The different spectroscopic and computational techniques are discussed initially, followed by how the integration of the data can advance the knowledge. We also discuss the concepts with the help of case studies used in laboratory situations. The chapter will allow researchers to design and integrate spectroscopic and computational experiments to better picture the protein structure and dynamics.

## 2. Spectroscopic tools to study protein structure and dynamics

### 2.1 CD spectroscopy

CD spectroscopy utilizes circularly polarized light to assess the chirality of molecules and has versatile applications in protein structure analysis [5–7]. Because chiral centers of proteins and peptide systems have differential absorption of right and left circularly polarized light, this method is used to determine their secondary and tertiary structures. One of the components of the circularly polarized light is absorbed by the protein molecule more than the other rendering the transmitted electric field oscillating in an elliptical pattern. CD spectroscopy quantifies the ellipticity angle of this oscillation as a function of the wavelength of the light. The far-UV region of the electromagnetic spectrum is absorbed by the amide groups of proteins [8]. This absorption leads to three different electronic transitions at 220, 190, and 140 nm. The transitions at 190 nm ($NV_1$) couple among identical peptide groups in the protein, and thus, the N-fold degenerate excited state is split into N-levels where N signifies the number of amide bonds. Then, the symmetry of the protein determines whether the transitions are allowed or forbidden. Thus, the secondary structure affects the coupling interactions and gives rise to CD signatures, which are specific to the structural motif. A positive peak at 190 nm and two negative peaks at 208 and 222 nm are obtained for α-helices, whereas β-sheets present with a positive peak at around 198 nm and a negative peak at around 216 nm. Random coils show a negative peak around 200 nm. CD spectroscopy can also be used to study the tertiary structure of proteins as interactions between nearby aromatic residues as exciton couples between 215 and 230 nm [9]. Furthermore, changes under the environmental conditions and the consequent changes in the protein structure can be analyzed with CD spectroscopy, thereby allowing temperature or pH denaturation for the determination of the thermodynamics of the folding process [10–12]. For a detailed discussion on the use of CD spectroscopy to study biomolecular structure and dynamics, please refer to Chapter 3.

## 2.2 Fluorescence spectroscopy

A fluorescence experiment for protein and peptide analysis is dependent on three factors: the dynamic nature of the signal, its localization, and its redundancy. The method relies on the ability of the fluorophore to display fluorescence. The fluorophore tryptophan shows an emission maximum of different wavelengths depending on its location in the protein. When it is in the core of a protein, tryptophan emits maximally at around 330–335, 340–345 nm when near the surface of proteins, and around 350–355 when exposed to free water molecules. The tryptophan fluorescence is quenched by other amino acids [13–18], water [13,14], and disulfide bonds; this complicates the analysis of protein fluorescence but provides valuable structural information. The use of spectrally enhanced protein mutants is an emerging field in protein fluorescence, which takes advantage of tryptophan analogues that are incorporated into the native sequence as intrinsic fluorophores by biosynthesis or chemical synthesis [19–21]. For tryptophan-free proteins, tyrosine fluorescence is used. This method is used to study protein folding and determine the denaturation or renaturation by changing the temperature, pH, or solvents [22–24]. Further, it is also used to analyze protein-ligand interactions, membrane proteins, and peptides [25–27]. For a detailed discussion on the use of fluorescence spectroscopy to study biomolecular structure and dynamics, please refer to Chapter 2.

In Förster resonance energy transfer (FRET) spectroscopic methods, FRET tag pairs are attached to the termini of different subunits within a protein to determine their interactions. The tag pair consists of a donor and an acceptor fluorophore. The donor fluorophore, when excited, emits fluorescence which excites the acceptor fluorophore. The acceptor then emits its own characteristic fluorescence. The most commonly used FRET tags are cyan fluorescent protein (CFP) and yellow fluorescent protein (YFP), which emit at 480 and 520 nm, respectively [28]. The results are measured as FRET efficient ($E_{FRET}$), which is dependent on the distance between donor and acceptor tags and the number of acceptors, which can be used for modeling. Single-molecule FRET (smFRET) is used to model molecules as it determines the distance between the fluorophores. These data are used to provide restraints for protein-protein docking [29]. The benefit of using FRET for protein modeling is that it can be performed in vitro as in smFRET and in vivo, which is based inside the cell and provides information on protein–protein interactions [30,31]. The drawbacks of this method are that the fluorophores must be covalently attached to the protein, and it merely provides the distance information between the center of the fluorophores. Further, the derived distances can have high uncertainty; hence, the data are too sparse to be used without additional methods and have an issue of noise because of the need for attachment of the fluorescent tags.

## 2.3 NMR spectroscopy

One of the ways to determine the structure of a protein is by NMR spectroscopy. The structures of small proteins can be obtained from a series of different NMR experiments. For this, the primary requisite is labeling the proteins by expressing them in isotopically labeled media. This can be achieved using NMR-active $^{13}C$ and/or $^{15}N$ isotopes. The optimization and production of these proteins in large volumes are extremely challenging and overly expensive considering the utilization of the isotopically labeled medium. Several specific experiments are performed to determine the structure, but they may vary [32,33]. Once the experiment is completed, the obtained peaks are assigned using 2D HSQC (heteronuclear single-quantum coherence) to determine the sequence positions of the observed chemical shifts of amides. Following this, 2D NOESY (nuclear Overhauser effect spectroscopy) is used to determine the atoms that are close together in space. Other experiments can also provide additional restraints to the observed data [34]. Multiple experiments such as 3D HNCACB and 3D CBCA are required to assign the backbone peaks of the HQSC spectra. The process is cumbersome and takes several days to weeks to obtain the end results [34]. For a detailed discussion on the use of NMR spectroscopy to study biomolecular structure and dynamics, please refer to Chapters 4 and 5.

## 2.4 FTIR spectroscopy

FTIR spectroscopy is a well-known technique that probes molecular vibrations. The functional groups present in a compound can be associated with a unique infrared absorption band, which corresponds to the fundamental vibrations [35,36]. A nonlinear molecule with N atoms has 3N-6 vibrational motions (or fundamental motions or normal modes). The normal mode of vibration is deemed to be infrared-active if there is a change in the dipole moment during the vibration course. On the other hand, if a molecule has a center of symmetry, then the vibrations will also be symmetrical, rendering it infrared-inactive. However, asymmetric vibrations are detected. This selectivity property allows detecting all chemical groups in a sample, including amino acids and water molecules that are not observable in other spectroscopic techniques. The detection of these vibrational frequencies is expected in regions that are specific to a fundamental type of atoms or type of chemical bond. With these vibrational frequency data, the chemical groups are established with respect to the structural properties of a given residue. The sensitivity of the method is high, and changes as small as 0.2 Å in bond length can be detected. Therefore, a detailed analysis of the IR data is necessary to obtain meaningful results [37]. For a detailed discussion on the use of FTIR spectroscopy to study biomolecular structure and dynamics, please refer to Chapter 6.

## 2.5 Raman spectroscopy

Raman spectroscopy is a method of nondestructive spectroscopy based on the scattering technique. It is based on the Raman effect, where the frequency of the fraction of scattered radiation is different from the frequency of the monochromatic incident radiation. This is caused by the inelastic scattering of incident radiation, where interaction is observed with the vibrating molecules [38,39]. With these interactions and the scattered ratio frequency, one can probe the molecular vibrations. In principle, the sample is excited with a monochromatic laser beam, and the molecules in the sample interact with the light. The scattered light distinguishes from the incident light because of the inelastic scattering, which constitutes the Raman spectrum. Much of the scattered radiation has frequencies equal to that of the incident light, which is known as Rayleigh scattering. However, there is a small fraction of unique scattered radiation known as Raman scattering. When the frequency of the scattered light is lower than that of the incident light, a Stokes line appears in the Raman spectrum, and if it is higher than that of the incident light, an anti-Stokes line appears. This scattered light is measured at right angles to the direction of incident radiation [40–42]. Raman spectrophotometers can be both dispersive and nondispersive. Although the dispersive Raman spectrophotometer uses a prism or grating, the nondispersive spectrophotometer uses an interferometer. The spectrum is given as intensity-versus-wavelength shift and can record from 4000 to 10 $cm^{-1}$. The spectrum obtained is used to identify the vibrational modes of the molecule, which provide structural information about the protein. The Raman spectrum is simpler than the infrared spectrum because of the normal Raman overtones, where combinations and difference bands are rarely observed [38]. For a detailed discussion on the use of Raman spectroscopy to study biomolecular structure and dynamics, please refer to Chapter 7.

## 2.6 X-ray crystallography

Although recently, cryo-EM has gained popularity for determining protein structures, the most commonly used technique for determining the atomic structures of protein has been X-ray crystallography. The first step of X-ray crystallography involves the crystallization of proteins by adding precipitants such as salts and organic solvents. Ammonium sulfate and polyethylene glycols are the commonly used precipitants under precise conditions of pH, temperature, and protein concentration. This process is complicated, cumbersome, and time-consuming [43]. The success of protein crystallization is influenced by many factors such as precipitant concentrations, ionic strength, vibration, protein flexibility, protein purity, temperature, and so on. Because of this, some protein structures that have been published were proved partially or completely incorrect [44,45]. In X-ray crystallography, the structure determination is performed by observing the diffraction pattern results from the scattering of X-rays by a protein, ordered in a crystal lattice. Most proteins diffract to resolutions where heterogeneity is difficult to analyze and therefore approximated to a single average conformation, whereas in their native form, proteins are dynamic with many anisotropic motions and conformational states. Disregarding this structural heterogeneity introduces degeneracy into protein modeling, making the relative positions of each subunit prone to inaccuracy [46]. For a detailed discussion on the use of X-ray crystallography to study biomolecular structure and dynamics, please refer to Chapter 11.

## 2.7 Small-angle X-ray scattering

The overall shape and state of a protein in its native solution can be determined using SAXS. In this technique, X-rays are incident on a target protein, and the scattered rays are detected at small angles. The SAXS profile of the protein is given by the scattering profile of the sample subtracted by the profile of the buffer. The SAXS profile is used to generate a pairwise distribution function using the Fourier transform method [47]. The advantage of using SAXS over other methods is its ability to describe the structure of the protein in its native solution state. Unfortunately, this significantly reduces the resolution of the scattering profile and thus needs to be used in conjunction with other protein structure determination methods. Because of the topological information that can be derived from SAXS profiles, the data are significantly used in protein–protein docking [34]. Furthermore, SAXS profiles are used for modeling multidomain proteins, which are arduous to obtain using X-ray crystallography because of the presence of flexible linker regions, where only the structures of stable regions of the protein are obtained. These structures can be used along with SAXS data to model the flexible linker regions of the protein of interest [48]. For a detailed discussion on the use of SAXS to study biomolecules, please refer to Chapter 10.

## 3. Computational tools to study protein structure and dynamics

Nowadays, computational tools to study the structure and folding of proteins are common among researchers. With the introduction of high-performance computers, the amount of data that can be processed has grown exponentially. The data can range from sequence data to structural data. In computational protein structure prediction, different methods are available to predict the structure of a protein. It is essential to understand these methods to help researchers choose the best approach to derive the correct and required information. The section discusses the computational techniques employed right from the structure prediction until the structural MD simulation of proteins.

### 3.1 Modeling the protein structure

The first and foremost step in modeling the protein structure is choosing the appropriate method for predicting the structure. There are two general approaches, that is, template-free modeling and template-based modeling. Template-based modeling utilizes the data from solved structures available in databases as a template and predicts the structure of a query protein. On the other hand, template-free modeling does not rely on the data from already available structures, and the modeling is performed de novo. The two methods vary in theory, where the template-based modeling is based on the alignment of the template and the query sequence to model the query protein, whereas template-free modeling employs energy functions and conformational mapping for modeling purposes. In terms of accuracy, the template-based model has the upper hand because of its reliance on solved structure data, but this advantage is being diluted with the advent of machine learning, artificial intelligence, and fragment-based methods in protein structure prediction.

### 3.1.1 Template-based modeling

Template-based modeling is a simple method because the prediction is based on available structural data resources. Initially, the sequence of interest is aligned with the template sequence (sequence of solved structures from protein structure databases such as PDB) [22,24,49]. The alignment results serve as the primary information for the template-based modeling of the backbone atoms of the query protein. This method can also be used to model mutated residues as well as optimize the side chains of amino acids present in the protein. In some cases, the template sequence might exhibit less than 30% identity to the query protein or may be distantly related. The modeling in such cases requires advanced approaches such as multitemplate modeling and conformational sampling. Multitemplate modeling involves employing parts of two or more proteins as a template to model the query protein. This could provide a starting conformation from which several other conformers are generated by MD simulations and worked upon. As of today, roughly two-thirds of the protein families can be subjected to template-based modeling methods for predicting their structures.

### 3.1.2 Template-free modeling

The template-free modeling approach is chosen only when the query sequence does not have a template sequence available in the databases or if no global similarity is recorded. The modeling approach considers conformational sampling and several other factors such as target sequence, multiple sequence alignment, backbone torsional angles, distances between inter-residues, nonlocal features, residue–residue contacts, and secondary structure information. The method also considers the propensity of amino acids to form a secondary structure to predict the model structure. The template-free modeling uses a plethora of information and factors to predict the structure of the query protein. The fragment-based assembly is the most followed method for template-free modeling. In addition, software such as ROSETTA and I-TASSER also employ this method for better accuracy and prediction [50,51].

## 3.2 Molecular dynamics simulation of protein structure

Once the protein structure is modeled or extracted from the databases, the next step would be to check if the structure is at its energy minimum. The energy minimum means that in the structure of interest, the atomic forces are balanced, and hence, the structure is at its most stable conformation.

The goal of the energy minimization step is to attain a global minimum, but this is only theoretically possible. There are several local minima where the energy of a protein is close to zero, but the true minimum is hard to be defined in terms of structural conformation. The minimized structure provides information such as conformational changes, inter- and intraresidual contacts, energy contributions, binding energies, conformational changes because of mutations, and so on.

Once the energy-minimized structure is obtained either through different algorithms or through production simulations, the structure is then placed inside a box. The box (usually a dodecahedron) should be big enough to accommodate the protein and allow conformational changes. It is then solvated with water molecules, and the charges on the protein are neutralized using ions. The solvated and neutralized protein in the box is then subjected to temperature and pressure equilibration. The equilibration allows the protein to be relaxed and stabilizes the protein to a particular temperature and pressure. This is followed by the production simulation, where the protein is simulated for a period of time. Simulating for a larger time will allow identifying conformational changes appearing in the protein. In the case of binding studies, the larger the simulation, the better the chance of understanding the binding energies and interacting residues.

Trajectory analysis postsimulation is the key for such a computational study. This step is performed only when the production simulation is complete. Several analyses can be performed to analyze the trajectories; some of them include root-mean-square deviation (RMSD), root-mean-square fluctuations (RMSFs), radius of gyration (Rg), solvent accessible surface area (SASA), principal component analysis (PCA), hydrogen bond (both inter and intra), and so on. The choice of analysis depends on the type of study being conducted. In short, the simulation of a protein is based on identifying and extracting the energy-minimized structure and analyzing the trajectories of the simulated protein. One can refer to these references for further details on the use, applications, and analysis of MD simulations [52–54].

## 3.3 Secondary structure prediction using the DSSP tool

The directory of secondary structure of proteins (DSSP) is a tool used to measure the evolution of the secondary structure content of a protein with time. The tool is based on the DSSP library designed by Kabsch and Sander [55,56]. DSSP is a database for secondary structure assignments but does not predict the secondary structure of the proteins. In the analysis, the trajectory is broken down into individual residues, and secondary structures are assigned to them. This allows the user to visualize or quantify the secondary structure data of the protein.

## 4. Integrating spectroscopic data with computational data

The integration of spectroscopic and computational data was not a necessary step for researchers earlier, but it is now gaining increasing interest. The spectroscopic data can indeed provide information about the changes in the dynamics of the proteins and other macromolecules, but the changes at the atomic level can only be characterized using computational tools. The dynamics and data deciphered from the computational tools serve as corroborative information to the spectroscopic data. Together, the spectroscopic and computational tools can allow researchers to understand the macro and atomic level structural perturbations of the macromolecules when events such as ligand binding, protein–protein interactions, and so on occur. A schematic representation of the integration of spectroscopic and computational data is given in Fig. 1. Below are a few examples of how spectroscopic data can be scaffolded with computational data.

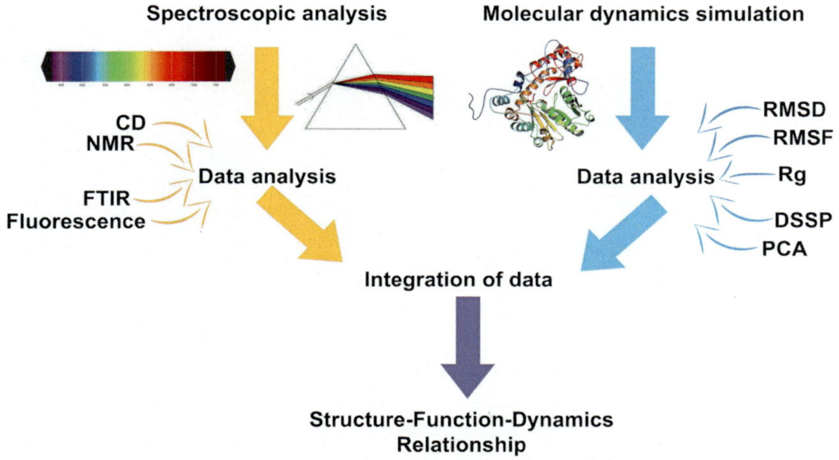


**Fig. 1** Schematic representation of the integration of spectroscopic and computational data.

# 5. Case studies

## 5.1 Integrating data from CD spectroscopy and computational analysis

CD is a commonly used spectroscopic technique to characterize the secondary structure of proteins. However, the secondary structure can also be determined using the GROMACS DSSP tool. In a study by Soufian et al., the structure of the protein aurein was analyzed using CD and DSSP tools [10]. The characteristics of interest were the hydrophobicity/hydrophilicity and solvent accessibility in the native and chirally inverted forms of the peptide. The results from CD spectra indicated that the native peptide had a similar hydropathy profile, amino acid composition, and periodicity of helices as the chirally inverted amino acid. These results were consistent with the MD results, which displayed a similar pattern in characteristics [10]. A similar study was conducted by Borocci et al. on the structural analysis of chionodracine-derived (Cnd) peptides [11]. CD was used to investigate the secondary structure of these peptides. The experiments indicated that in the absence of lipid media, the results were characteristic of a random coil structure. However, in the presence of POPC (1-palmitoyl-2-oleoyl-sn-glycero-3-phosphocholine) vesicles, mimicking the lipid bilayer of a eukaryotic cell, the CD spectrum shows that Cnd folds to a canonical α-helix structure. The CD spectrum of the Cnd peptide showed around 62% α-helix content in the peptide in the presence of POPC/POPG (1-palmitoyl-2-oleoyl-*sn*-glycero-3-phospho-(1′-rac-glycerol)) lipid (Fig. 2A).

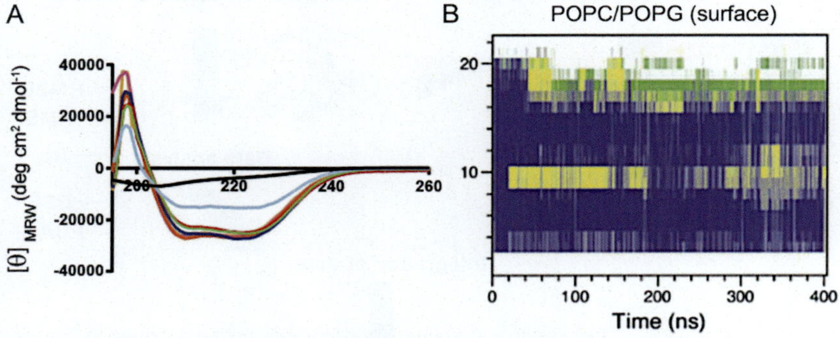

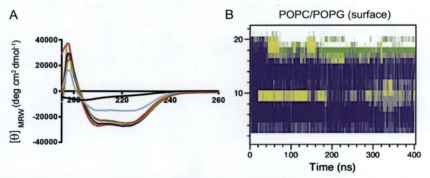


**Fig. 2** CD spectra and MD simulation. (A) CD spectra of Cnd peptide in the presence of POPC/POPG at varying concentrations. (B) Secondary structure of Cnd peptide in the POPC/POPG bilayer from 400 ns MD simulation. *(Figure reproduced with permission from Borocci S, Della Pelle G, Ceccacci F, Olivieri C, Buonocore F, Porcelli F. Structural analysis and design of chionodracine-derived peptides using circular dichroism and molecular dynamics simulations. Int J Mol Sci 2020;21(4):1401.)*

Parallelly, MD analyses showed comparable results in the presence of the bilayer and moderate conservation of helicity. The percentage of helicity was found to be 50% ± 5% in the presence of the POPC/POPG bilayer surface [11] (Fig. 2B). The results showed that both CD spectra and the MD simulation results are significantly close to each other and supplemented the understanding of the fold of the Cnd peptide at the molecular level. Studies on mechanistic insights into the urea-induced denaturation of human sphingosine kinase I (SphKI) are comparable to the work mentioned above [57]. They cross-examined the folding nature of the protein using CD and MD simulations. The folding pathway of SphKI showed a biphasic unfolding transition from native to denatured with an intermediate state. The corresponding MD simulations showed a similar transition ($N \rightleftharpoons I \rightleftharpoons D$) with increasing concentrations of urea.

## 5.2 Urea denaturation analysis: Integrating fluorescence and MD simulation data

Studying the effect of urea on protein structures using combined fluorescence spectroscopy and computer simulation can provide beneficial information. The effect of urea on the *Leishmania donovani* tyrosine aminotransferase (LdTAT) was studied by Sasidharan et al. using fluorescence spectroscopy and MD simulations [22,24]. Fluorescence spectroscopy showed that the intensity of fluorescence increased with the urea concentration and a three-state folding mechanism in the TAT enzyme. Further, analyzing the relationship between the fluorescence intensity and the concentration of urea at 346 nm revealed a sigmoid curve indicating that LdTAT had a three-state unfolding mechanism (Fig. 3A and B). In silico characterization of LdTAT was performed using RMSD analysis and Rg analysis, which showed that TAT possessed three phases during unfolding. Both the analyses also indicated the presence of a three-state folding mechanism (Fig. 3C and D) [22]. The analysis is a typical example of the integration of spectroscopic and computational data to understand the folding mechanism. In a similar study, the enzyme aminotransferase was truncated at both terminals, and fluorescence studies proved that the intermediate state present was a result of the C-terminal folding [24]. This was hypothesized earlier from the MD simulation data. Chen et al. studied the effects of lysozyme in increasing urea concentrations [58]. Here, the wavelength and intensity of fluorescence displayed a gradual decrease [58]. This was attributed to the tryptophan quenching property of water molecules in the solution, further illustrating the stability of proteins in low urea concentrations.

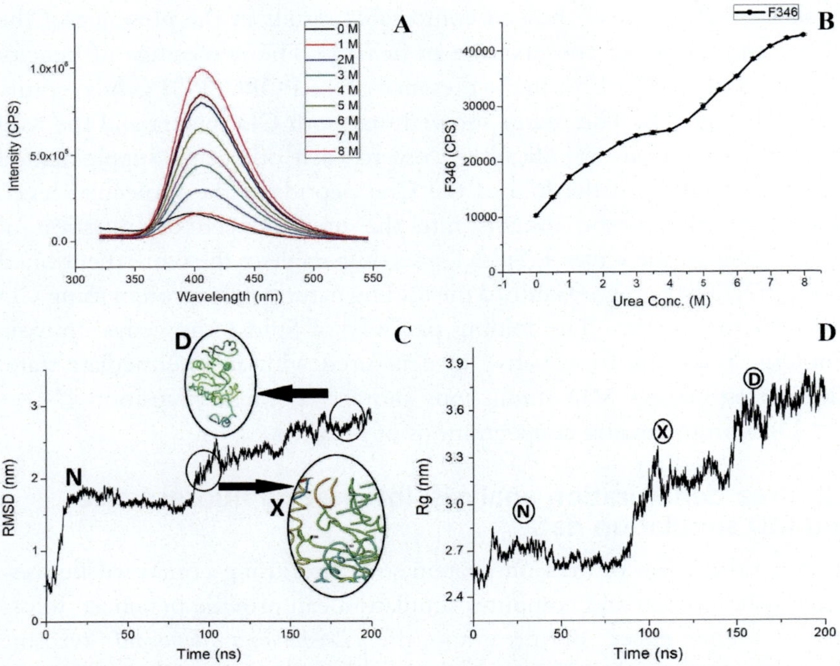


**Fig. 3** Fluorescence spectra and MD simulation: (A) Fluorescence spectra of tyrosine aminotransferase at increasing urea concentration. (B) Plot representing the three phases of unfolding in the presence of increasing concentrations of urea. (C) RMSD analysis of tyrosine aminotransferase in urea exhibiting the biphasic folding mechanism during 200 ns simulation. (D) Radius of gyration analysis of tyrosine aminotransferase depicting the unfolding of the protein in urea concentration. *(Figure reproduced with permission from Sasidharan S, Saudagar P. Biochemical and structural characterization of tyrosine aminotransferase suggests broad substrate specificity and a two-state folding mechanism in Leishmania donovani. FEBS Open Bio 2019;9(10):1769–83.)*

These results have been consistent with other studies. For instance, the results of bovine serum albumin (BSA) in urea solution were reported by Kumaran and Ramamurthy [59]. Here, an increase in the concentration of urea up to 6.0 M caused fluorescence quenching, along with a shift in the emission maximum toward the blue region. They concluded that the results were because of the presence of urea derivatives containing free N—H moieties, which quenched the fluorescence of BSA. In addition, the N—H moieties, as opposed to the carbonyl oxygen of urea derivatives, were responsible for the unfolding of tryptophan in BSA [59].

## 5.3 Integrating CD, fluorescence, and MD simulation data to understand the roles of specific mutations in protein structure, function, and dynamics

Shukla et al. demonstrated the unified use of the CD, fluorescence, and MD simulation techniques to understand the role of distant amino acids in regulating the structure and function of *Mycobacterium tuberculosis* isocitrate lyase (MtbICL) [60–62]. MtbICL was used to study the roles of specific residues in the catalytic function of the enzyme and evaluate its potential as a drug target. The function of a protein is dependent on conformational dynamics and structural stability. The amino acid sequence provides insights into the structural features of mutants, which is helpful in comprehending the regulation of enzymatic activities of proteins. The authors performed three point mutations far away from the active site and studied their effect on the conformation and activity of the MtbICL enzyme [60–62]. The three mutations, that is, F345A, L418A, and H46A, were present at a distance of 44, 30, and 12.9 Å from the catalytic site, respectively (Fig. 4).

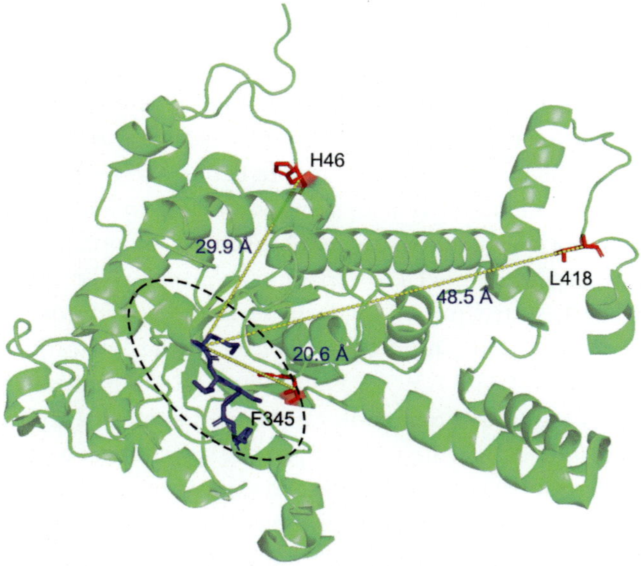


**Fig. 4** Structural location of the mutation sites in the structure of MtbICl. The structure of MtbICL is represented in a cartoon form and in green (PDB ID: 1F8I). The mutated residues (H43, F345, and L418) are shown in red, catalytic residues are shown in blue, and the active site is circled in black. The distance between the active site and the mutated residues is depicted in the figure.

The first part of the study was the mutation of F345A in MtbICL. The substitution of F345 with alanine led to the complete loss of activity [62]; however, the far UV-CD spectra showed that both the wild type and F345A mutant possessed a similar α/β structure, and minor perturbations were observed upon mutation. The tryptophan-based intrinsic fluorescence also showed an emission maximum at 336 nm for both the wild type and F345A variant, suggesting an insignificant structural change upon F345A mutation. Urea- and GdHCl-induced unfolding showed similar sigmoidal curves with a two-state process for both the wild-type and mutant structures. However, the $C_m$ value was recorded to be lower for F345A mutation, which indicated destabilization of the protein structure. The data from MD simulations corroborated the spectroscopic data by confirming the reduction in the structural plasticity of the enzyme upon F345A mutation.

The L418 residue is present far away from the active site $^{189}$KKCGH$^{193}$, but mutating it to L418A caused complete activity loss. The mutation, however, did not show any structural change, which was deduced from the GdnHCl and urea denaturation studies using CD spectroscopy and intrinsic tryptophan fluorescence. Both the wild type and the L418A variant exhibited a two-state unfolding process with a $C_m$ value of 2 M and 4 M for GdnHCl and urea-induced denaturation, respectively, indicating a similar unfolding profile. The MD simulations were performed to offer an insight into the atomistic details. Upon L418A mutation, there was a drastic increase in the conformational flexibility of the protein, which was attributed to the loss of function. The secondary structure prediction also indicated a decrease in the β-bridge, coil, and α-helices along with an increase in β-sheets and bends in the L418A variant. Far-UV CD spectroscopy predicted the structure of the wild type and the L418 variant to be principally composed of the α/β type secondary structure. The decrease in the activity of the enzyme upon mutation was understood to be a result of the loss of hydrophobic interaction between L418-A399/E394 residues [60]. This study is a classic example of how computational tools can be used hand-in-hand with spectroscopic data to decipher the mutation in the structure of a protein.

In another study, the authors performed a mutational analysis of MtbICL at H46A, which is present at the N-terminal. It should be kept in mind that all three mutants, including H46A, are nonactive-site mutants. The H46A mutant also did not show any catalytic activity. To further understand the role of H46A mutation, an all-atom MD simulation was conducted that showed that the structural plasticity of the enzyme was reduced as observed before in the L418A variant [61]. However, in contrast to the L418 variant, the secondary structures were similar in both the wild type and the mutant H46A. Yet again, the results suggested that the mutation of a critical residue in MtbICL causes loss of function because of reduced plasticity, but the overall secondary structure of the protein was conserved even after mutation. Overall, all three mutations in MtbICL cause loss of activity, whereas the structure remained stable, as shown by the CD, fluorescence, and MD simulation analyses.

## 6. Conclusion and future perspectives

We discussed various spectroscopic techniques that provide specific information for integrating with computational data. The challenges in unifying the data are the noise and experimental errors that cannot be normalized in the spectroscopic approaches. On the other hand, the modeling of macromolecules, especially de novo, might not be accurate, and therefore, the obtained results are speculative and unexpected at times. Although there are challenges in integration, there are ways to overcome them. To obtain more accurate and significant data, one can run repeated experiments and perform a multispectroscopic approach. In the computational simulations, the modeling can be conducted using better servers such as AlphaFold [63] to obtain accurate models and energy minimized independently with long production runs. However, there is always a scope for better spectroscopic experiments/techniques and computational methods that may lead to better integration methods. Like in other fields, it is not long before these techniques will be pipelined, and the data obtained from several approaches have to be integrated automatically for a better understanding of protein structure and dynamics. The current and future spectroscopy-computational technique integration holds immense potential to pave the way for better knowledge-driven structure and dynamics.

## References

[1] Singh DB, Tripathi T. Frontiers in protein structure, function, and dynamics. 1st. Singapore: Springer; 2020. p. 1–458.
[2] Anfinsen CB, Haber E, Sela M, White Jr F. The kinetics of formation of native ribonuclease during oxidation of the reduced polypeptide chain. Proc Natl Acad Sci U S A 1961;47(9):1309.
[3] Levinthal C. Are there pathways for protein folding? J Chim Phys 1968;65:44–5.
[4] Tripathi T, Dubey VK. Advances in protein molecular and structural biology methods. 1st ed. Cambridge, MA, USA: Academic Press; 2022. p. 1–716.

[5] Chen Y-H, Yang JT, Chau KH. Determination of the helix and β form of proteins in aqueous solution by circular dichroism. Biochemistry 1974;13(16):3350–9.
[6] Johnson Jr WC. Protein secondary structure and circular dichroism: a practical guide. Proteins Struct Funct Bioinf 1990;7(3):205–14.
[7] Kelly SM, Jess TJ, Price NC. How to study proteins by circular dichroism. Biochim Biophys Acta Proteins Proteom 2005;1751(2):119–39.
[8] Berova N, Nakanishi K, Woody RW. Circular dichroism: principles and applications. John Wiley & Sons; 2000.
[9] Grishina IB, Woody RW. Contributions of tryptophan side chains to the circular dichroism of globular proteins: exciton couplets and coupled oscillators. Faraday Discuss 1994;99:245–62.
[10] Soufian S, Naderi-Manesh H, Alizadeh A, Sarbolouki M. Molecular dynamics and circular dichroism studies on Aurein 1.2 and retro analog. World Acad Sci Eng Technol 2009;56:858–64.
[11] Borocci S, Della Pelle G, Ceccacci F, Olivieri C, Buonocore F, Porcelli F. Structural analysis and design of chionodracine-derived peptides using circular dichroism and molecular dynamics simulations. Int J Mol Sci 2020;21(4):1401.
[12] Tripathi T. Calculation of thermodynamic parameters of protein unfolding using far-ultraviolet circular dichroism. J Proteins Proteomics 2013;4(2):85–91.
[13] Burstein E. Intrinsic luminescence of proteins (origin and applications). Ser Biophys 1977;7.
[14] Bushueva T, Busel E, Burstein E. Relationship of thermal quenching of protein fluorescence to intramolecular structural mobility. Biochim Biophys Acta Protein Struct 1978;534(1):141–52.
[15] Colucci WJ, Tilstra L, Sattler MC, Fronczek FR, Barkley MD. Conformational studies of a constrained tryptophan derivative: implications for the fluorescence quenching mechanism. J Am Chem Soc 1990;112(25):9182–90.
[16] Gross M, Furter-Graves EM, Wallimann T, Eppenberger HM, Furter R. The tryptophan residues of mitochondrial creatine kinase: roles of Trp-223, Trp-206, and Trp-264 in active-site and quaternary structure formation. Protein Sci 1994;3(7):1058–68.
[17] Van Gilst M, Hudson BS. Histidine-tryptophan interactions in T4 lysozyme:'anomalous' pH dependence of fluorescence. Biophys Chem 1996;63(1):17–25.
[18] Chen Y, Barkley MD. Toward understanding tryptophan fluorescence in proteins. Biochemistry 1998;37(28):9976–82.
[19] Ross JA, Szabo AG, Hogue CW. Enhancement of protein spectra with tryptophan analogs: fluorescence spectroscopy of protein-protein and protein-nucleic acid interactions. Methods Enzymol 1997;278:151–90.
[20] Hogue CW, Rasquinha I, Szabo AG, MacManus JP. A new intrinsic fluorescent probe for proteins biosynthetic incorporation of 5-hydroxytryptophan into oncomodulin. FEBS Lett 1992;310(3):269–72.
[21] Ross J, Senear DF, Waxman E, Kombo BB, Rusinova E, Huang YT, et al. Spectral enhancement of proteins: biological incorporation and fluorescence characterization of 5-hydroxytryptophan in bacteriophage lambda cI repressor. Proc Natl Acad Sci 1992;89(24):12023–7.
[22] Sasidharan S, Saudagar P. Biochemical and structural characterization of tyrosine aminotransferase suggests broad substrate specificity and a two-state folding mechanism in *Leishmania donovani*. FEBS Open Bio 2019;9(10):1769–83.
[23] Sasidharan S, Saudagar P. Concerted motion of structure and active site charge is required for tyrosine aminotransferase activity in Leishmania parasite. Spectrochim Acta A Mol Biomol Spectrosc 2020;232;118133.
[24] Sasidharan S, Saudagar P. Mapping N-and C-terminals of *Leishmania donovani* tyrosine aminotransferase by gene truncation strategy: a functional study using in vitro and in silico approaches. Sci Rep 2020;10(1):1–15.

[25] Mocz G, Ross JA. Fluorescence techniques in analysis of protein–ligand interactions. In: Methods in molecular biology. Springer; 2013. p. 169–210.
[26] Groemping Y, Hellmann N. Spectroscopic methods for the determination of protein interactions. Curr Protoc Protein Sci 2005;39(1):20.8. 1–8.7.
[27] Raghuraman H, Chatterjee S, Das A. Site-directed fluorescence approaches for dynamic structural biology of membrane peptides and proteins. Front Mol Biosci 2019;96.
[28] Miller III D, Desai N, Hardin D, Piston D, Patterson G, Fleenor J, et al. Two-color GFP expression system for *C. elegans*. Biotechniques 1999;26(5):914–21.
[29] Brunger AT, Strop P, Vrljic M, Chu S, Weninger KR. Three-dimensional molecular modeling with single molecule FRET. J Struct Biol 2011;173(3):497–505.
[30] Okamoto K, Hibino K, Sako Y. In-cell single-molecule FRET measurements reveal three conformational state changes in RAF protein. Biochim Biophys Acta Gen Subj 2020;1864(2);129358.
[31] Qiao Y, Luo Y, Long N, Xing Y, Tu J. Single-molecular Förster resonance energy transfer measurement on structures and interactions of biomolecules. Micromachines 2021;12(5):492.
[32] Cavalli A, Salvatella X, Dobson CM, Vendruscolo M. Protein structure determination from NMR chemical shifts. Proc Natl Acad Sci 2007;104(23):9615–20.
[33] Purslow JA, Khatiwada B, Bayro MJ, Venditti V. NMR methods for structural characterization of protein-protein complexes. Front Mol Biosci 2020;7:9.
[34] Seffernick JT, Lindert S. Hybrid methods for combined experimental and computational determination of protein structure. J Chem Phys 2020;153(24);240901.
[35] Zeng XM, Martin GP, Marriott C. Particulate interactions in dry powder formulation for inhalation. CRC Press; 2000.
[36] Griffiths PR, De Haseth JA. Fourier transform infrared spectrometry. John Wiley & Sons; 2007.
[37] Berthomieu C, Hienerwadel R. Fourier transform infrared (FTIR) spectroscopy. Photosynth Res 2009;101(2):157–70.
[38] Settle F. Handbook of instrumental techniques for analytical chemistry. Arlington: National Science Foundation; 1997.
[39] Chalmers JM, Edwards HG, Hargreaves MD. Infrared and Raman spectroscopy in forensic science. John Wiley & Sons; 2012.
[40] Skoog DA, Holler FJ, Crouch SR. Principles of instrumental analysis. Cengage learning; 2017.
[41] Willard HH, Merritt Jr LL, Dean JA, Settle Jr FA. Instrumental methods of analysis. Wadsworth Publishing Co Inc; 1988.
[42] Smith E, Dent G. Modern Raman spectroscopy: a practical approach. John Wiley & Sons; 2019.
[43] Ferré-D'Amaré AR. Crystallization of biological macromolecules, by Alexander McPherson. 1999. Cold Spring Harbor, New York: Cold Spring Harbor Laboratory Press. Hardcover, 586 pp. $97. RNA 1999;5(7):847–8.
[44] Cross T, Opella S. Protein structure by solid state nuclear magnetic resonance: residues 40 to 45 of bacteriophage fd coat protein. J Mol Biol 1985;182(3):367–81.
[45] Nabuurs SB, Spronk CAM, Vuister GW, Vriend G. Traditional biomolecular structure determination by NMR spectroscopy allows for major errors. PLoS Comp Biol 2006;2(2);e9.
[46] DePristo MA, de Bakker PI, Blundell TL. Heterogeneity and inaccuracy in protein structures solved by X-ray crystallography. Structure 2004;12(5):831–8.
[47] Köfinger J, Hummer G. Atomic-resolution structural information from scattering experiments on macromolecules in solution. Phys Rev E 2013;87(5);052712.

[48] Hou J, Adhikari B, Tanner JJ, Cheng J. SAXSDom: modeling multidomain protein structures using small-angle X-ray scattering data. Proteins Struct Funct Bioinf 2020;88(6):775–87.
[49] Yang J-M, Tung C-H. Protein structure database search and evolutionary classification. Nucleic Acids Res 2006;34(13):3646–59.
[50] Song Y, DiMaio F, Wang RY-R, Kim D, Miles C, Brunette T, et al. High-resolution comparative modeling with RosettaCM. Structure 2013;21(10):1735–42.
[51] Yang J, Yan R, Roy A, Xu D, Poisson J, Zhang Y. The I-TASSER suite: protein structure and function prediction. Nat Methods 2015;12(1):7–8.
[52] Shukla R, Tripathi T. Molecular dynamics simulation of protein and protein-ligand complexes. In: Computer-aided drug design. Springer; 2020. p. 133–61.
[53] Shukla R, Tripathi T. Molecular dynamics simulation in drug discovery: opportunities and challenges. In: Singh SK, editor. Innovations and implementations of drug discovery strategies in rational drug design. Singapore: Springer Nature; 2021. p. 295–316.
[54] Padhi AK, Rath SL, Tripathi T. Accelerating COVID-19 research using molecular dynamics simulation. J Phys Chem B 2021;125(32):9078–91.
[55] Kabsch W, Sander C. Dictionary of protein secondary structure: pattern recognition of hydrogen-bonded and geometrical features. Biopolymers 1983;22(12):2577–637.
[56] Touw WG, Baakman C, Black J, Te Beek TA, Krieger E, Joosten RP, et al. A series of PDB-related databanks for everyday needs. Nucleic Acids Res 2015;43(D1):D364–8.
[57] Khan FI, Gupta P, Roy S, Azum N, Alamry KA, Asiri AM, et al. Mechanistic insights into the urea-induced denaturation of human sphingosine kinase 1. Int J Biol Macromol 2020;161:1496–505.
[58] Chen B, Zhang H, Xi W, Zhao L, Liang L, Chen Y. Unfolding mechanism of lysozyme in various urea solutions: insights from fluorescence spectroscopy. J Mol Struct 2014;1076:524–8.
[59] Kumaran R, Ramamurthy P. Denaturation mechanism of BSA by urea derivatives: evidence for hydrogen-bonding mode from fluorescence tools. J Fluoresc 2011;21(4):1499–508.
[60] Shukla H, Shukla R, Sonkar A, Tripathi T. Alterations in conformational topology and interaction dynamics caused by L418A mutation leads to activity loss of *Mycobacterium tuberculosis* isocitrate lyase. Biochem Biophys Res Commun 2017;490(2):276–82.
[61] Shukla R, Shukla H, Tripathi T. Activity loss by H46A mutation in *Mycobacterium tuberculosis* isocitrate lyase is due to decrease in structural plasticity and collective motions of the active site. Tuberculosis 2018;108:143–50.
[62] Shukla H, Shukla R, Sonkar A, Pandey T, Tripathi T. Distant Phe345 mutation compromises the stability and activity of *Mycobacterium tuberculosis* isocitrate lyase by modulating its structural flexibility. Sci Rep 2017;7(1):1–11.
[63] Jumper J, Evans R, Pritzel A, Green T, Figurnov M, Ronneberger O, et al. Highly accurate protein structure prediction with AlphaFold. Nature 2021;596(7873):583–9.

CHAPTER 17

# Advance data handling tools for easy, fast, and accurate interpretation of spectroscopic data

**Anand Salvi, Shreya Sarkar, Manish Shandilya, and Seema R. Pathak**
Department of Chemistry, Biochemistry and Forensic Science, Amity School of Applied Sciences, Amity University Haryana, Gurugram, India

## 1. Introduction

Spectroscopy is the scientific study of matter using electromagnetic radiation. The spectroscopic methods use the emission and absorption spectra of colored solutions and plant pigments, and Beer-Lambert's law is used for qualitative and quantitative analysis and kinetics investigations [1,2]. Spectroscopic techniques are based on molecule absorbing energy and then measuring how it affects the molecule. The spectrum generated by spectroscopic methods often yields specific information that may be collected and utilized to establish the molecule's structure.

However, modern scientific activities require online and interactive visualization of spectroscopic data to evaluate and understand the scientific data. Web-based solutions are advantageous because of their platform independence and low system requirements. In chemical research, information from $H^1$ and $C^{13}$, nuclear magnetic resonance (NMR), infrared (IR), electron paramagnetic resonance (EPR), and mass spectroscopic investigations is critical because these techniques are required for the structural identification of molecules. Commercial software can be used to analyze such data [3]. The web-based tools such as JSpecview, NMRView, MetaboAnalyst, MetaboHunter, COLMAR, jsNMR, and SpeckTackle are established tools for visualizing the NMR and MS spectra [4]. ChemSpectra is a web-based tool to visualize and analyze spectroscopic data, which analyzes data associated with a sample or reaction in the Chemotion electronic lab notebook (ELN) [3]. ChemSpectra is used in IR, mass, and NMR spectroscopy (one-dimensional $^1H$ and $^{13}C$) as analytical tools in synthetic organic

*Advanced Spectroscopic Methods to Study Biomolecular Structure and Dynamics*
https://doi.org/10.1016/B978-0-323-99127-8.00009-X

chemistry. The tool supports open file formats such as JCAMP-DX (jdx) and mz(*X*)ML [5]. Apart from the tools that help in processing the spectroscopic data, other tools help in the downstream processing of the data result, making the whole process more efficient. Tools such as MASCOT are used to sequence the peptide sequences [6]. MBROLE 2.0 is used for the proteomics data that rely on enrichment analysis [7]. Raman data are processed using the RAMANMETRIX [8], and the magnetic resonance spectroscopy (MRS) data can be easily interpreted using the INSPECTOR tool [9]. In this chapter, we discuss the various tools that are useful in analyzing different spectroscopic results.

## 2. Spectroscopic data handling tools

We now discuss various tools and software used to analyze different spectroscopic data efficiently.

### 2.1 Origin

Origin (https://www.originlab.com) is a computer tool that allows users to interactively graph and analyze scientific data. It is developed by the Origin Lab Corporation, USA, and works on Microsoft Windows. Beginners appreciate Origin's user-friendly interface [10]. Origin graphs and analysis results can be updated automatically in response to changes in data or parameters, allowing to create templates for repetitive tasks or run batch operations from the user interface without the need to program. Surface fitting, statistics, peak fitting, and signal processing are a few leading analytical tools and applications available in the Origin software [10].

#### *2.1.1 To install origin*

1. Use https://www.originlab.com to access the Origin website.
2. Click "Try Origin for Free" and fill out the relevant information to begin downloading.
3. After completing the form, you will receive an email from the server asking you to confirm your email address.
4. Run the installer, select "install the product," and when required, input the serial number and finish the installation.
5. After the installation is complete, launch Origin. Copy/paste the product key from your registered email into the product textbox in the License Activation window.
6. Then, to activate your Origin, click the "activate" button.

### 2.1.2 To use origin

1. Connect to an online file, a local PC or network file, a cloud file, or a database.
2. Comma-separated values (CSV), Excel, ASCII/Binary (using import Wizard filters), HTML, JSON, MATLAB, Origin Projects, and several other file types are supported for analysis.
3. Select data to be imported, such as an XML node, an HTML table, or an Origin project sheet.
4. To examine and alter the subset of data to be imported, use the "Data Navigator" panel in the worksheet.
5. The information about the connection and data selection is recorded in the worksheet/workbook.
6. Add the same type of connector to several worksheets in a workbook as well as various types of connectors to multiple workbooks. You have the option to leave the imported data related to all connectors or specific connectors out when saving the Origin project. This aids in the reduction of project file size. There will be no clearing of calculated results or graphs from results.
7. At any moment, you can reimport data from the source file as well as you can alter the source or data choices.

## 2.2 SigmaPlot

SigmaPlot (https://sigmaplot.com) is a scientific data analysis and graphing software program featuring a user-friendly interface and wizard technology to help users with their analysis and graphing needs [11]. The graphing capabilities of SigmaPlot allow us to effortlessly customize every graph detail and create graphs that are suitable for publication. Advanced curve-fitting capabilities and step-by-step instruction in executing over 50 commonly used statistical tests are among SigmaPlot's analytical features [11].

Systat Software Inc., through in-house development and numerous commercial and technological relationships with leaders in the Bioinformatics, Cheminformatics, and Lab Informatics industries, provides enterprise-level data management solutions. SYSTAT's SigmaPlot graphing software goes beyond conventional spreadsheets and helps show the work clearly and precisely [12]. Exceptionally good graphs can be created with SigmaPlot without spending too much time on the computer. SigmaPlot possesses strong interaction with Microsoft Office, allowing quick access to the data from Microsoft Excel spreadsheets [11].

### 2.2.1 To install SigmaPlot

1. Download the software and save the file to your computer.
2. Open the installer file and unzip it.
3. Remove SigmaPlot 12.0. Release Version from your computer.
4. If you make any changes to your files in the user>SigmaPlotSPW12 folder, make a backup.
5. Delete that <user>\SigmaPlot\SPW12 folder.
6. Install the application by double-clicking the EXE file.

### 2.2.2 To use SigmaPlot

The Award-winning boundary of SigmaPlot is where graph creation starts. From the Graph toolbar's easy-to-read icons, choose the graph type you want to build. The Graph Wizard is an interactive tool that guides you through the process of creating a graph. In no time, you can make captivating, publication-quality charts and graphs. More choices for charting, modeling, and graphing your technical data are available in SigmaPlot than in any other graphics software package. You may create many graphs per page, several pages per worksheet to compare, and multiple axes per graph and analyze trends in your data. With SigmaPlot's WYSIWYG page layout and zoom tools, you can accurately arrange various graphs on a page using built-in templates or your own page layouts.

SigmaPlot offers the exact technical graph type one requires for his research, ranging from simple two-dimensional (2-D) scatter plots to captivating contour plots. SigmaPlot also produces numerous intersecting three-dimensional (3-D) meshes with concealed line removal to help you see interactions in your 3-D data. One can discover the ideal visual representation of the data with multiple graphs and charts format to select from. SigmaPlot gives the freedom to personalize every aspect of the graph. The colors, typefaces, line thicknesses, add axis breaks, standard or asymmetric error bars, and symbols can be modified. You can double-click on any graph element to open the graph properties dialogue box. One can paste an equation, a symbol, a map, a picture, an illustration, or other images into the presentation to further customize the graph, chart, or diagram.

In the Graph Style Gallery, one can save all the attributes of his/her favorite graph style. By instantly recalling an existing graph type and applying its style to the dataset, you can improve the speed and efficiency of your analysis. It creates spectacular presentations and can further personalize graphs in drawing software and reports. With SigmaPlot's extensive graphic export options, you may save graphs for publishing in a journal or article.

Putting your findings on display and publishing them have become very easy with SigmaPlot. With SigmaPlot's Report Editor, you can create customized reports or embed your graphs in any Object Linking and Embedding (OLE) container such as Microsoft PowerPoint, graphics programs, or Word processors. To edit the graph directly within your paper, simply double-click it. You can also send your high-resolution graphs online easily and quickly share them with others. Instead of basic JPEG or GIF files, you can export charts as high-resolution, dynamic Websites. Viewers may use a Web browser to explore the data used to make graphs as well as zoom, pan, and print full-resolution photos. You can create live Web objects from your graphs mechanically or embed them in other Web pages. However, to limit data access to authorized persons, create an optional password before exporting your graph.

## 2.3 JCAMP-DX

The JCAMP-DX (Joint Committee on Atomic and Molecular Spectroscopy-Data Exchange) (https://iupac.org/what-we-do/digital-standards/jcamp-dx/) is a standard file format to exchange spectra and related chemical and physical information between data obtained from instruments of different manufacturers [13–20]. In spectroscopy, the data generated are in an instrument-dependent format, which is converted to an instrument-independent format by JCAMP-DX. JSpecView (http://jspecview.svn.sourceforge.net/viewvc/jspecview/dev2/) is an open-source project created to provide a teaching and research tool for displaying JCAMP-DX spectra. The adoption of open standards by the IUPAC (International Union of Pure and Applied Chemistry) and ASTM International (formerly known as the American Society for Testing and Materials) has dramatically improved the ability to transfer data between spectroscopic instruments of various origins, regardless of their application software and operating systems [13–20]. The JCAMP-DX protocols have made it possible to transfer and archive datasets from IR and Raman EPR, NMR, mass spectroscopy, ion-mobility spectroscopy, and so on. Robert Lancashire created this free standalone application to view JCAMP DX spectra. It can open files, zoom in and out, select peaks, and print them. This application has an advantage over the CHIME plug-in in that it supports printing. This program is incredibly compact and straightforward to use [13–20].

### *2.3.1 To install JCAMP-DX*

The website has two utilities available for download at the following URL (http://www.ansci.de/downloads/jcamp/). The source code for both the

JCAMP to ASCII executable and the JCAMP DLL is in the public domain, and you can easily set up the tools.

1. Download and unzip the campDLL.zip file.
2. Copy the "jcamp.dll" Dynamic Link Library (DLL) to the Origin installation directory (for example, C: Program FilesOriginLabOrigin7G) or a directory in the search path.
3. If you are unsure about your installation, look for *.OTW in the ORIGIN directory tree and copy the JCAMP.DLL to the top-level directory where these files are located.
4. It is necessary to copy the Origin Worksheet Template "JCAMP.OTW" to the Origin installation directory or a data directory.

### 2.3.2 To use JCAMP-DX

1. Open the "JCAMP.OTW" Origin Worksheet Template.
2. Press the "JCAMP-DX Import" button.
3. You can now travel throughout your environment or network in the same way until you find the JCAMP-DX file you want to import in a new window.
4. Select the file and choose "open." The first line of the JCAMP-DX file (##TITLE=...) will appear in the comment box if the "Show Info" check box is selected.
5. The abscissa values are entered in the first column, whereas the ordinates are entered in the second. The time is kept as a dataset label above the Y-column in either the JCAMP-DX LDR ##TIME= or ##LONG DATE=. The worksheet name is set to the name of the JCAMP-DX file, and the worksheet label is set to the title (##TITLE=).
6. After the data have been loaded, charts of the spectra can be generated. If IR spectra require reversed wavenumber presentation, do so by clicking on the *x*-axis in the plot, right-clicking to access the *x*-axis options box, and selecting settings, where you may change all the axis attributes, including the first and last values. The *x*- and *y*-axis labels are currently not read from the JCAMP-DX file; thus, they must be inserted manually by clicking on the labels in the plot and modifying them.
7. Multiple spectra can be handled that are stored in a composite JCAMP-DX file. The *x*-axis for all spectra in the dataset must be the same as Origin will utilize the *x*-axis from the first file for the rest of the data in the array. Only the *y*-values from the remaining spectra are then imported by the tool. The tool works flawlessly with up to 500 spectra and 2000 data points per spectrum (1 million *y*-values) [13].

### *2.3.3 JDXview*

JDXview is a freeware (open-source) Windows tool for viewing spectra in the JCAMP-DX format (https://homepage.univie.ac.at/norbert.haider/cheminf/jdxview.html). JDXview is a JCAMP-DX viewer that displays a variety of spectra. It provides graphics output in vector graphics format and allows zooming and measuring distances on a spectrum [5,21]. JDXview was created as an external "helper program" that can be used in conjunction with web browsers if they do not have a compatible plug-in installed [21,22]. Dr. Robert Lancashire's JCAMP-DX data viewer inspired the generation of JDX-view, and the new version is called the MDL-Chime browser plug-in. JDXview is a program that displays JCAMP-DX spectra (NMR, IR, MS, etc.). The specifications support uncompressed (AFFN) and compressed (ASDF, e.g., DIFDUP) formats. Some of the more recent JCAMP-DX features are also implemented. Furthermore, JDXVIEW provides minimal processing options, such as clipping a spectrum's range and saving the result in JCAMP-DX format. Two cursors that are positioned by clicking the left or right mouse button, followed by a click on the ">button", can be used to zoom in and measure distances [13,21].

#### To install JDXview

There is no need for any additional setup. Simply copy the jdxview.exe application file to a suitable place. If desired, you can add a desktop icon and/or a start menu entry. Associating the JCAMP-DX file type (extensions: jdx,.dx) with JDXview is handy. Add or update the settings for MIME type "chemical/x-jcamp-dx" and jdxview.exe (with its entire directory path if necessary) as the software to handle this file type for usage as a web browser helper application.

## 2.4 jsNMR

jsNMR (https://inano.au.dk/about/research-centers-and-projects/nmr/software/jsnmr/) is a cross-platform JavaScript-based spectrum viewer for one-dimensional (1-D) NMR spectra [23,24]. It contains both spectrum data and a viewer in a single HTML file. Apart from visualizing the spectrum, jsNMR can also convert the data into multiple file formats, such as SIMPSON, CSV, and PDF files. It also stores the raw data so that the spectrum may be examined in other programs at any time [23]. The following are common features of jsNMR [24].

### 2.4.1 Ways to load the spectrum

1. Drag your data into a zip file and drop it into the jsNMR window. It will be loaded if it contains files in well-known formats.
2. You have to sign up for the jsNMR mail. You can send spectra as attachments to spec@jsnmr.net and receive a link to the spectra's jsNMR production files after registration.

### 2.4.2 Zooming and panning

Zooming horizontally and vertically is accomplished by clicking the appropriate buttons and using the computer mouse or zoom and pinch gestures on the mobile. You can always use the full button to return to the full spectrum view.

### 2.4.3 Analysis and processing

Hold down the left mouse button on a computer and drag up or down to phase the spectrum with a zero-order phase shift. The zero-order phase shift position is saved as the pivot point if a first-order phase shift is required. Then move the pointer to another peak, hit the Alt key, and slide the mouse up or down to perform a first-order phase shift. One finger performs a zero-order phase shift on a mobile device, followed by two fingers performing a first-order phase shift. The pivot point is where your first finger is pressed.

Peak picking is performed by pressing the peak-picking button and selecting a peak with the mouse. Measure the peak by hovering your mouse over it. After this process, jsNMR will choose the nearby peak. The peaks are integrated using the mouse. By pushing and dragging the mouse on one side of the peak, you can move it to the other side. The chosen region will not be blended, and the outcome will be displayed. The weakest peak in the spectrum will be integrated. When working with a large dataset in JavaScript, for example, long reaction times may occur when the mouse is pressed. Use the compress dataset button to make a copy of the dataset with half the number of data points. jsNMR provides a dataset with a slight line broadening to avoid truncation wiggles, which implies that the intensities of the individual peaks may not be identical. Ensure that the new spectrum is of good quality before removing the old one. With datasets of 8k or fewer data points, jsNMR performs well; on mobile devices, there are even rarer.

### 2.4.4 Spectral management

- One can obtain information about a spectrum's spectral parameters. The spectral width (SW), the reference point (ref, the center of the spectrum), the number of points (np), and the title are all included.

- The data in a jsNMR file is a self-transforming dataset that can be saved as a CSV file, a new jsNMR file, or the raw data of the original spectrum. The latter option allows you to process the raw data with other software programs. Peak lists can be exported using CSV files.
- By hitting this button, you can delete the current spectrum. An asterisk (*) in front of the current spectrum's title indicates that it is active. Simply click the title of another spectrum to make it the current one.
- The current spectrum should be printed. This process may take some time.

#### *2.4.5 File types that are commonly used*

jsNMR currently supports the following file formats for 1-D NMR spectra: Bruker, Agilent, jsNMR, and SIMPSON. The data are kept in a separate directory for Bruker and Agilent. Before uploading, this directory must be compressed.

### 2.5 Unscrambler

Unscrambler (www.camo.com/unscrambler/) is a popular and widely used piece of software for scientists, researchers, and engineers who analyze vast and complicated datasets employing multivariate analysis and design of experiments as well as spectroscopy and chemometrics [25]. The Unscrambler's primary goal is to offer tools to assist in examining multivariate data. Finding variations, covariations, and other internal linkages in data matrices can be performed. It can determine the number of pure components in a mixture and estimate their concentration profiles and spectra [25].

#### *2.5.1 To install Unscrambler*

1. Save The Unscrambler X Standalone installer to a temporary folder from www.camo.com/unscrambler/.
2. Extract the files from the temporary folder, setup.exe, TheUnscramblerX.msi, and the installation documentation can all be found in the temporary folder. To begin the Unscrambler X installation, double-click setup.exe.
3. Before opening the file, the Windows installer displays a security warning. If you are a nonprivileged user, right-click setup.exe and choose Run as administrator from the context menu.
4. The Unscrambler X is installed via the Windows installer procedure. To continue with the installation, select Yes. The following window will pop up.

5. To view the End-User License Agreement, click Next (EULA).
6. To continue, please read the EULA carefully and accept the License Agreement terms. If you want a hard copy of the EULA, click the Print option. Once you accept the terms, the Next button will become active.
7. To see the compliance mode settings window, click Next.
8. If you wish certain restrictions to be imposed on the software, select Install in the compliance mode.
9. If you choose to install in the compliance mode, the Hide Login option will be enabled. Check the box if you want to use the Windows login details in The Unscrambler X without re-entering the password.
10. Next should be selected. It will be suggested that you use the default installation folder. By choosing Change, you can select a different installation folder.
11. Again, click Next and then install. Depending on your operating system and security settings, a Windows security popup may appear after clicking Install. To complete the installation, you must give the installer permission to make changes to your computer. A new window will appear once the setup is complete.
12. Finish by clicking the Finish button. The installation is completed.

### 2.5.2 To run Unscrambler

To use The Unscrambler X for the first time, you have to first activate it.

1. Press the Start button. Point the mouse to All Programs on the Start Menu.
2. Select the Unscrambler X 10.3 from the displayed submenu, and then select The Unscrambler X 10.3 from the presented submenu.
3. Depending on security settings, the following popup may appear in current Windows versions. To activate the device, select Yes.
4. Fill in the required fields; then click Activate. A congratulations message box will appear.
5. Click the OK button. The login screen for New Project-Unscrambler X will appear.

### 2.5.3 To use Unscrambler

1. Tools for importing and accessing data, tools for visualizing and altering data, and tools for analyzing data can all be found in the top menu. The project navigator is located on the left and allows you to simply move across your project.

2. The first step is to import your information into a scrambler. To do so, go to File, Import data, and choose a data format from the drop-down menu.
3. After that, you can plot to visualize your data before analyzing it. To create alternative plots, go to the plot menu on the main toolbar. Line graphs are handy when dealing with spectral data. As a result, each sample's columns are displayed as a spectrum. To get a quick summary of the data, descriptive statistics are often beneficial. For this, go to the main toolbar and select tasks, analyze, and descriptive statistics from the drop-down menus. You can look at a box plot, for example, to learn more about the variables.
4. Preprocessing or transformation is frequently required to remove extraneous information, such as light scattering effects. Transformation can be accessed from the toolbar by selecting jobs; for example, multiplicative scatter correction can be used to eliminate unwanted light scattering. The transform spectra can be shown and compared to the raw data visualization.
5. The next step is to undertake multivariate analysis using the analyze menu, which offers a variety of multivariate methods to choose from. If you want to perform a principal component analysis, which is a standard exploratory method, first select the data you want to analyze in the model in the model input tab, then select the number of selections you want to make for your analysis, and click finish to begin the study. Automatically, plots with an overview will display. The project navigator allows you to see and access the entire analysis output.

## 2.6 MBROLE 2.0

MBROLE 2.0 (http://csbg.cnb.csic.es/mbrole2/analysis.php) is a web-based server that analyzes and interprets a list of chemical compounds derived from a high-throughput metabolomics experiment in biological terms. This tool is responsible for performing functional enrichment analysis [7,26]. The last step in the transcriptomics and proteomics experiment is to interpret the results; this can be achieved by the functional enrichment that gives a functional interpretation of the results. The functional enrichment analysis involves the analysis of the assigned functional annotations to the result of the experiment (gene or proteins) and comparison of the frequency of the assigned annotation to the annotation of the background set, which helps in obtaining the annotations enriched in the result of the "omic" experiment (gene or protein being studied). Then this annotation is interpreted in biological terms such as which biological pathways or subpathways were affected, physiological locations, chemical classifications, taxonomies, interaction with enzymes, proteins, applications, and so on [7]. The MBROLE 2.0 is an advanced version of MBROLE that is provided with functional annotations to various new metabolites, a new supported compound identifier, automatic conversion of the identifier, emphasis on metabolite interactions such as metabolite-protein or drug-protein, and so on. It has a better user interface and intuitive presentation [7].

### 2.6.1 To use MBROLE 2.0

1. Go to the homepage of MBROLE 2.0 (http://csbg.cnb.csic.es/mbrole2/analysis.php).
2. Click on "Go to the Analysis."
3. Enter the compound set (list of compound IDs) in the box. Copy-paste or upload files of compound IDs, start with a new line for each ID, or use the tab.
4. For enrichment analysis, add vocabulary for every chemical or biological annotation. One can search for annotations in the search box. The annotations can be removed from the list by clicking on "x" beside them.
5. Now provide a background set (reference) to analyze the significance of an annotation to be enriched in the compound being studied.
6. The background set contains all the compounds that could be identified in the experiment. It can be a full database or organism-specific. One can also add an individual list of IDs as a background set.
7. The result is displayed in a tab that includes Annotation, Category, Set, Inset, *P*-value, and FDR correction.

## 2.7 MASCOT

MASCOT (https://www.matrixscience.com/search_form_select.html) is a widely used piece of software to determine the sequence of the peptides after mass spectrometric measurements [6]. In MS-based proteomics, one of the approaches for downstream bioinformatic analysis is the bottom-up approach, where the protein sample is digested with the proteolytic enzyme and then analyzed using a mass spectrometer. After MS analysis, the next step of sequencing the protein fragments is achieved by searching against the fragmentation spectra database [27]. MASCOT is a piece of software that uses a database matching approach, where all expressed or hypothetical protein sequences are digested in silico to generate a target database. It then calculates the peptide spectrum match (PSM) of all fragmentation spectra and theoretical fragmentation spectra information from the target database. The peptide with the highest PSM is used as a candidate for the experimental peptide (query peptide) [27]. MASCOT uses probability-based scoring, where the total score is based on the probability of an observed match being a random event. It produces probability-based scores by combining mass measurement, protein sequence information, proteolytically digested peptide molecular weight, tandem mass spectrometry data, and so on [27].

### 2.7.1 To use MASCOT

1. Open a web browser and search for https://www.matrixscience.com/search_form_select.html.
2. In the Sequence query section, click on "perform search."
3. On the new page, enter the search parameters—name, email ID, search title (a title that will appear at the top of the result), select database(s) (this may include the FASTA sequence of amino acids or nucleic acid or spectral library files or against an EST database.
4. To specify a particular species or a group of species, the "Taxonomy parameter" is used.
5. In "Fixed modification," the modification is applied whenever the specified residue or terminus arrives. Thus, it is applied universally.
6. In "Variable modification," all possible arrangements of the variable modifications are applied for the best match.
7. The "Peptide tool" is an error window on experimental peptide mass values, whereas the MS/MS tool is an error window on experimental MS/MS fragment ion mass values.
8. In "Peptide charge," mention the precursor peptide charge state in a sequence query. The query contains the list of peptide mass values, followed by the other qualifiers of the peptide.
9. Choose the instrument type and check whether the experimental mass value is monoisotopic or average.
10. Click on "Start search."

## 2.8 MaxQuant

The MaxQuant (https://www.maxquant.org/) tool provides a platform for the quantitative analysis of the MS-based proteomics data [27,28]. The quantitative analysis involves checking for the level of abundance of the proteins across the proteome in a given sample. It can analyze large-scale MS datasets using both label and label-free procedures. It has a set of algorithms that detects the mass and intensity of the peptide peaks in an MS spectrum to identify the protein. It searches for the peptide or fragment mass in organism-specific sequence data, followed by the assignment of the peptide score based on a probability-based approach. It also provides a viewer module for visualizing 3-D graphical models [28].

### *2.8.1 To use MaxQuant*

1. Download the MaxQuant software from https://www.maxquant.org/ and install it.
2. To enter data, select the raw files tab and click on "Load" to open a browser where the files to be loaded can be selected.
3. Click on the "Group-specific parameter" tab for the specification of the labels and enzyme with which protein was digested.
4. Click on the "Global parameters" to find the FASTA files and add them.
5. Click on the "start" button for the progression of the analysis.
6. The results can be viewed by clicking on the "Viewer" tab.

## 2.9 RAMANMETRIX

Raman spectroscopy is a chemometric technique that involves the usage of vibrational spectroscopy [8,29]. It is a label-free, nondestructive technique with huge applications in biomedical and clinical fields. The processing of Raman data is very complex, and thus, it is not much used in the areas of clinical research [8,29]. The RAMANMETRIX (https://ramanmetrix.eu/info/) is software that allows chemometric analysis in a user-friendly manner. The software provides a graphical user interface (GUI). RAMANMETRIX reduces the complexity of the RAMAN spectroscopy data. The tool is available in two versions of web application and software. The tool has a backend and a frontend that communicate by the application programming interface (API) [8]. The RAMANMETRIX software supports the data in different formats, but before importing, the files should be compressed into ZIP files. The software involves the step of pretreatment, where the variations in the device calibration and the cosmic ray noise are suppressed. It also provides the calibration of the wavenumber, dark background, and intensity. It also offers automatic despiking of the spectra and a robust baseline correction [8].

### *2.9.1 To use RAMANMETRIX*

1. The tool has various expertise levels, based on which the number of buttons appears in the main interface.

2. The import button facilitates the import of the spectral data.
3. In pretreatment step 1, despiking is performed, which involves the removal of the cosmic ray noise. Preprocessing step 1 involves the correction of the baseline.
4. In pretreatment step 2, the calibration of the wavenumber, dark background, and intensity is performed. Preprocessing step 2 involves the normalization of the data.
5. With the next step, quality filters can be applied.
6. The next step involves the construction of a model based on the data. The test data function is available for the expertise level below 1, and above 1, data analysis is possible.
7. From the Report icon, the summary of the result can be availed.

## 2.10 INSPECTOR

INSPECTOR (http://juchem.bme.columbia.edu/inspector-spectroscopy-software-launched) is free software that processes the magnetic resonance spectroscopy (MRS) data [9]. MRS is a powerful and noninvasive tool for studying the small molecules in tissues, tissue metabolism, and functions. The INSPECTOR tool has advanced data processing, analysis, and simulation capacities [9]. It can handle extensive data with consistency while managing its quality. It provides a user-friendly platform with visualization and automation potential. This tool entirely works through a GUI. It offers a set of pages for the manipulation of a specific aspect of the MRS signal [9]. The data page provides for the loading and averaging of the raw data. The processing page uses standard signal processing methods. The synthesis or Magnetic Resonance Spectrum Simulator (MARSS) page provides the simulation of the basis sets for metabolite quantification. It has a linear combination modeling (LCM) page to analyze the spectra. The tool also facilitates the reproducibility of the data processing by providing Protocol files that save the tool's workflow and save all the selected parameters [9].

### *2.10.1 To use INSPECTOR*

1. Download the software from the Columbia Technology Venture website. Also, the MATLAB Runtime environment is required to provide suitable software libraries.
2. The first page of the tool is the "Data Page," where the MRS data are loaded and preprocessed. Here, it ensures quality management, data correction, and data visualization.

3. The "Quality Assessment Tool" visualizes, quantifies, and manipulates the spectral data quality after combining the signals from individual receivers to form one free indication decay (FID) per experimental trace and repetition time.
4. On clicking the "Details" button, the "parameter page" is opened, where the parameters for the alignment tool can be set. The alignment tool helps in the evaluation of the frequency, amplitude, and phase before the combination of the individual traces. By clicking the "Align button," the parameters are aligned.
5. The average of the traces can be estimated by clicking the "Mean" button.
6. The alignment protocol can be saved by choosing the desired path in the protocol field and pressing the "save" button.
7. The "processing page" comes next, where a single spectrum or two spectra are processed. For this, the data are selected from the pull-down menu. The data from other files can be directly loaded onto the processing page using the "proc" option. The nonflat baseline of MRS can be corrected using the Baseline handling tool.
8. The "synthesis page" allows for the simulation and the processing of the singlet arrays.
9. The "Magnetic Resonance Spectrum Stimulator" page provides the different simulation methods for MRS metabolites.
10. The "Linear Combination Modeling" page contains methods for processing and loading the averaged spectra, basis set creation and management, LCM analysis adjustment, and export of the analysis results.

## 3. Conclusions

In this chapter, we discussed software tools that can be used to analyze various spectroscopic results. Every spectroscopic approach produces different bits of information that may be collected and used to determine the structure of the molecule. Modern scientific research necessitates the online and interactive visualization of scientific data for analysis. Besides those that aid in the processing of spectroscopic data, other tools undoubtedly aid in the downstream processing of the data output, making the entire process more efficient.

## References

[1] Lambert JH. Photometry, or, on the measure and gradations of light, colors, and shade : translation from the Latin of Photometria, sive, De mensura et gradibus luminis, colorum et umbrae. Illuminating Engineering; 2001.

[2] Beer A, Beer P. Determination of the absorption of red light in colored liquids. Ann Phys Chem 1852;86(5):78–88.
[3] Huang YC, Tremouilhac P, Nguyen A, Jung N, Bräse S. ChemSpectra: a web-based spectra editor for analytical data. J Cheminform 2021;13(1).
[4] Mohamed A, Nguyen CH, Mamitsuka H. NMRPro: an integrated web component for interactive processing and visualization of NMR spectra. Bioinformatics 2016;32 (13):2067–8.
[5] Bramer SV. Tools for processing and interpreting spectral data. In: 2004 Fall ConfChem: teaching computing in chemistry courses; 2004.
[6] Perkins D, Pappin D, Creasy D, Cottrell J. Probability-based protein identification by searching sequence databases using mass spectrometry data. Electrophoresis 1999;20 (18):3551–67.
[7] López-Ibáñez J, Pazos F, Chagoyen M. MBROLE 2.0-functional enrichment of chemical compounds. Nucleic Acids Res 2016;44(W1):W201–4.
[8] Storozhuk D, Ryabchykov O, Popp J, Bocklitz T. RAMANMETRIX: a delightful way to analyze Raman spectra. Available from: arXiv:2201.07586; 2022.
[9] Gajdošík M, Landheer K, Swanberg KM, Juchem C. INSPECTOR: free software for magnetic resonance spectroscopy data inspection, processing, simulation and analysis. Sci Rep 2021;11(1):1–16.
[10] Origin(Pro). Version 2022. Northampton, MA: OriginLab Corporation; 2022.
[11] SigmaPlot. Version 12.3 (2013). San Jose, CA: Systat Software, Inc.; 2013.
[12] Stein PG, Matey JR, Pitts K. A review of statistical software for the Apple Macintosh. Am Stat 1997;51(1):67–82.
[13] Lancashire RJ. The JSpecView project: an open source java viewer and converter for JCAMP-DX, and XML spectral data files. Chem Cent J 2007;1(1):1–11.
[14] Davies AN, Lambert J, Lancashire RJ, Lampen P, Conover W, Frey M, et al. Guidelines for the representation of pulse sequences for solution-state nuclear magnetic resonance spectrometry (IUPAC Recommendations 2001). Pure Appl Chem 2001;73 (11):1749–64.
[15] Baumbach JI, Davies AN, Lampen P, Schmidt H. JCAMP-DX. A standard format for the exchange of ion mobility spectrometry data (IUPAC recommendations 2001). Pure Appl Chem 2001;73(11):1765–82.
[16] Lampen P, Lambert J, Lancashire RJ, McDonald RS, McIntyre PS, Rutledge DN, et al. An extension to the JCAMP-DX standard file format, JCAMP-DX V.5.01. Pure Appl Chem 1999;71(8):1549–56.
[17] Lampen P, Hillig H, Davies AN, Linscheid M. JCAMP-DX for mass spectrometry. Appl Spectrosc 1994;48(12):1545–52.
[18] Davies AN, Lampen P. JCAMP-DX for NMR. Appl Spectrosc 1993;47(8):1093–9.
[19] Gasteiger J, Hendriks BMP, Hoever P, Jochum C, Somberg H. JCAMP-CS: a standard exchange format for chemical structure information in computer-readable form. Appl Spectrosc 1991;45(1):4–11.
[20] McDonald RS, Wilks PA. JCAMP-DX: a standard form for exchange of infrared spectra in computer readable form. Appl Spectrosc 1988;42(1):151–62.
[21] JDXview Homepage. Available from: https://homepage.univie.ac.at/norbert.haider/cheminf/jdxview.html.
[22] Veltri P. Algorithms and tools for analysis and management of mass spectrometry data. Brief Bioinform 2008;9(2):144–55.
[23] Pellet N. jsGraph and jsNMR—advanced scientific charting. Challenges 2014;5 (2):294–5.
[24] Vosegaard T. jsNMR: an embedded platform-independent NMR spectrum viewer. Magn Reson Chem 2015;53(4):285–90.
[25] Helland K. UNSCRAMBLER 11, version 3.10: a program for multivariate analysis with PLS and PCA/PCR. J Chemometr 1991;5(4):413–5.

[26] Marco-Ramell A, Palau-Rodriguez M, Alay A, Tulipani S, Urpi-Sarda M, Sanchez-Pla A, et al. Evaluation and comparison of bioinformatic tools for the enrichment analysis of metabolomics data. BMC Bioinf 2018;19(1):1–11.
[27] Chen C, Hou J, Tanner JJ, Cheng J. Bioinformatics methods for mass spectrometry-based proteomics data analysis. Int J Mol Sci 2020;21(8).
[28] Tyanova S, Temu T, Cox J. The MaxQuant computational platform for mass spectrometry-based shotgun proteomics. Nat Protoc 2016;11(12):2301–19.
[29] Raman CV, Krishnan KS. The negative absorption of radiation. Nature 1928;122 (3062):12–3.

# Index

Note: Page numbers followed by *f* indicate figures and *t* indicate tables.

## D

## J



## L

## Q



## R

## S

Printed in the United States
by Baker & Taylor Publisher Services